# THE DOWNY MILDEWS

From the
. . . . Rosen

Property of
RIVERSIDE COUNTY
AGRICULTURE COMMISSIONER

# THE
# DOWNY MILDEWS

*Edited by*

## D. M. SPENCER

*Glasshouse Crops Research Institute, Littlehampton, Sussex*

1981

## ACADEMIC PRESS

London   New York   San Francisco

*A Subsidiary of Harcourt Brace Jovanovich, Publishers*

Property of
RIVERSIDE COUNTY
AGRICULTURE COMMISSIONER

ACADEMIC PRESS INC. (LONDON) LTD.
24–28 Oval Road
London, NW1


*U.S. Edition published by*
ACADEMIC PRESS INC.
111 Fifth Avenue,
New York, New York 10003

Copyright © 1981 by Academic Press Inc. (London) Ltd.

All Rights Reserved
No part of this book may be reproduced in any form by photostat, microfilm, or any other
means, without written permission from the publishers

The Downy Mildews.
1. Mildew
I. Spencer, D. M.
632′.4′52    SB741.M65
ISBN 0–12–656860–X

*Library of Congress Catalog Card Number:* 81–66686

Text set in 10/12pt Monophoto Times Roman by
Northumberland Press Ltd, Gateshead, Tyne and Wear,
and printed by Fletcher and Son Ltd, Norwich

# Contributors

BULIT, J.  *INRA Bordeaux, Station de Pathologie Végétale, 33140 Pont de la Maye, France*

BYFORD, W. J.  *Broom's Barn Experimental Station, Higham, Bury St Edmunds, Suffolk, UK*

CHANNON, A. G.  *The West of Scotland Agricultural College, Auchincruive, Ayr, UK.*

COHEN, Y.  *Department of Life Sciences, Bar-Ilan University, Ramat-Gan, Israel*

CRUTE, I. R.  *National Vegetable Research Station, Wellesbourne, Warwick, UK*

DIXON, G. R.  *School of Agriculture, Aberdeen, Scotland, UK*

DUNLEAVY, J.  *Department of Botany and Plant Pathology, Iowa State University, Ames, Iowa, USA*

INGRAM, D.  *The Botany School, Downing Street, Cambridge, UK*

KENNETH, R. G.  *Department of Plant Pathology and Microbiology, The Hebrew University of Jerusalem, Rehovot, Israel*

KREMHELLER, H. Th.  *Bayerische Landesansalt fur Bodenkultur und Pflanzenbau, Munchen-Freising, Federal Republic of Germany*

LAFON, R.  *INRA Bordeaux, Station de Pathologie Végétale, 33140 Pont de la Maye, France*

MATTHEWS, P.  *John Innes Institute, Norwich, Norfolk, England*

MICHELMORE, R. W.  *The Botany School, Downing Street, Cambridge, UK*

PALTI, J.  *Agricultural Research Organisation, Bet Dagan, Israel*

POPULER, C.  *Station de Phytopathologie, Centre de Recherches Agronomiquesde l'Etat, Gembloux, Belgium*

RENFRO, B.  *The Rockefeller Foundation, P.O. Box 2453, Bangkok, Thailand*

ROTEM, K.  *Agricultural Research Organisation, Bet Dagan, Israel*

ROYLE, D. J.  *Department of Hop Research Wye College (University of London) Near Ashford, Kent, UK*

SACKSTON, W. E.  *Department of Plant Sciences, Macdonald Campus, McGill University, St Anne de Bellevue, Quebec, Canada*

SARGENT, J.  *ARC Weed Research Organisation, Yarnton, Oxford, UK*

SCHILTZ, P.  *Tobacco Experimental Institute, Bergerac, France*

SCHWINN, F. J.  *Agricultural Division, Ciba-Geigy Ltd, Basel, Switzerland*

SHANKARA BHAT, S.  *The University of Mysore, Manasagangotri, Mysore 570006, India*

SHAW, C. G.  *Department of Plant Pathology, Washington State University, Pullman, Washington, USA*

STUTEVILLE D. L.  *Department of Plant Pathology, Kansas State University, Manhattan, Kansas 66506, USA*

TOMMERUP, I. C.  *Department of Soil Science and Plant Nutrition, Institute of Agriculture, University of Western Australia, Nedlands, Australia 6009*

VIENNOT-BOURGIN, G.  *8, Square Vauban 78220, Viroflay, France*

VIRANYI, R.  *Research Institute for Plant Protection, Budapest, Hungary*

WELTZIEN, H.C.  *The International Centre for Agricultural Research in the Dry Areas, P.O. Box 5466, Aleppo, Syria*

WHEELER, B. E. J.  *Imperial College Field Station, Silwood Park, Berkshire, England*

# Preface

When Academic Press invited me to look into the possibility of preparing "a new treatment of the mildews" it was an easy matter to decide that the powdery mildews and the downy mildews should be dealt with separately. It was equally easy, since I had more experience with powdery mildews to decide that "The Powdery Mildews" should come first, but before this second book "The Downy Mildews" could be started it had first to be determined what exactly constituted the downy mildews.

Ainsworth and Bisby's Dictionary of the Fungi defines a downy (or false) mildew as a disease caused by one of the Peronosporaceae. This volume could therefore have had as its title "The Peronosporaceae", and this might have been helpful since some diseases caused by members of the family are not called downy mildews; blue mould of tobacco is a case in point. However, based on the dictionary definition, this volume is devoted to the Peronosporaceae whether or not the diseases they cause are commonly called downy mildews. This having been said, there is a particularly notable omission—the genus *Phytophthora*, were it to be included, would overshadow the rest of the genera. Indeed *P. infestans* or *P. cinnamomi* could each provide enough material to fill a book and therefore in letters inviting contributions it was made clear to authors that work on *Phtophthora* species was not to be included except for purposes of comparison.

Some downy mildews have been important for as long as their host crops have been cultivated, others have become important only relatively recently but then, for various reasons, have exploded, devastatingly. Tobacco blue mould and vine downy mildew are two diseases of particular importance in these respects and they, accordingly occupy an important part of the opening chapters.

As with "The Powdery Mildews" the aim of this treatize is to present an integrated and thought-provoking discussion of our present knowledge of the downy mildews. Presented in the one volume is up-to-date information on the disease of a wide variety of plants accompanying an examination of the progress made in recent years in fundamental studies on some of the fungi involved. This desire for an overall comprehensive cover of the subject made it obvious that only a multi-author approach could be adopted and in order to provide the optimum treatment of the various subjects the contributions have been obtained from an international selection of authors. In this context, some authors, in pleading for more space than they had been allocated, pointed out that this was their first opportunity to produce a definitive review article on their particular topic.

The book is in two parts; the first chapters provide general reviews of history, taxonomy, distribution, epidemiology, host range, genetics, breeding, host–parasite relations and control. These are all applicable to downy mildews in general but some of the topics dealt with here make particular reference to individual diseases. The second group of chapters deals with the downy mildews of individual crops or commodity groups.

It has emerged that in some ways the downy mildews are more difficult to deal with than are the powdery mildews. These latter display similar symptoms on a wide range of plant species and thus lend themselves to a generalized treatment whereas the downy mildews tend towards diversity in form and function. Some authors have therefore found difficulty in preparing their material to fit the editor's chosen arrangement. To their credit all of the authors tailored their material to meet the requirements of the book and diligent readers will note similar material being used in different chapters to illustrate different points. The danger of overlap between chapters seemed to be more likely here than with the powdery mildews but though there is a certain amount of duplication it is hoped that this serves to strengthen rather than weaken the linkage between the various chapters.

The criticism could well be made that some progressing fields of research are either not represented here or have been given scant attention. As seems to be inevitable in a venture such as this, the withdrawal of authors at a late stage is an occupational hazard resulting in unavoidable gaps, but almost equally alarming for an editor is the incredible tardiness of some writers. Though many chapters arrived precisely on schedule others arrived up to one year later than the agreed deadline. Nevertheless, the authors have borne the greatest share of the work in the preparation of this book and it is to them that credit will be due if the book proves acceptable to the readers.

I have done the minimum of editing and thus, to some extent, the style and coverage of each chapter varies according to the interests and approach of its author. It might be thought that I should have standardized the terms used in various aspects of structure and taxonomy. This, has been purposely avoided, not only because some of the subjects are under current or continual discussion but also because the individual authors are authorities in their own fields and therefore, in my view, they know best. Sackston, for example, points out that the causal organism of downy mildew of sunflower is now widely known in Europe as *Plasmopara helianthi* but continues as *P. halstedii* in the Americas. Likewise, many authors refer to the asexual spore as a conidium, presumably because it germinates by forming a germ tube, whereas Michelmore uses the term sporangium for the asexual spore of all downy mildew-causing fungi regardless of the mode of germination. The reader is referred to the authoritative treatment of this particular problem and of various taxonomic matters by Shaw (Chapter 2).

It has been my wish to draw together both pure and applied information on this diverse group of diseases, from a wide range of authorities. In doing this we have then, together, tried to present a review of the available literature, to give an insight into the current thinking of leading research workers, to show the extent and limitations of our present knowledge and, where appropriate, to erect signposts for future research. Thus, the more general student is given an overall picture of the factors that are important in downy mildew diseases, but it is hoped that the more serious student will be provided with enough points to embark on a deeper specialist study of his chosen subject.

The Director of the Glasshouse Crops Research Institute, Dr D. Rudd-Jones was instrumental in getting this venture started and for this, as well as his continuing support, I am most grateful. My thanks go again to my wife Eileen for her patience, encouragement and involvement in the preparation of the book and finally I wish to record the help given in various ways by David, Jane and Michael.

JULY 1981                                    D. M. SPENCER
                                 *Glasshouse Crops Research Institute*
                                 *Rustington, Littlehampton, Sussex*

# Contents

## 5. Epidemiology of Downy Mildews
### C. POPULER

## 6. Role of Wild Hosts in Downy Mildew Diseases
### B. L. RENFRO and S. SHANKARA BHAT

## 7. Cytology and Genetics of Downy Mildews
### I. C. TOMMERUP

## 8. The Biochemistry of Host–Parasite Interaction

D. S. INGRAM

## 9. Sexual and Asexual Sporulation in the Downy Mildews

R. W. MICHELMORE

## 10. The Fine Structure of the Downy Mildews

J. A. SARGENT

## 11. The Host Specificity of Peronosporaceous Fungi and the Genetics of the Relationship between Host and Parasite

I. R. CRUTE

## 12. Breeding for Resistance to Downy Mildews

PETER MATTHEWS

# 13. Control of Downy Mildews by Cultural Practices

J. PALTI and J. ROTEM

# 14. Chemical Control of Downy Mildews

F. J. SCHWINN

## 15.  Downy Mildew of Brassicas

A. G. CHANNON

## 16.  Downy Mildew of Cucurbits

YIGAL COHEN

## 17.  Downy Mildew of Forage Legumes

D. L. STUTEVILLE

## 18. Downy Mildews of Graminaceous Crops
### R. G. KENNETH

## 19. Downy Mildew of the Hop
### D. J. ROYLE and H. TH. KREMHELLER

## 20. Downy Mildew Diseases Caused by the Genus *Bremia* Regel
### I. R. CRUTE and G. R. DIXON

## 21. Downy Mildew of Onion
F. VIRÁNYI

## 22. Downy Mildews of Ornamentals
B. E. J. WHEELER

## 23. Downy Mildews of Peas and Beans

G. R. DIXON

## 24. Downy Mildew of Soybean

J. M. DUNLEAVY

# 25. Downy Mildews of Beet and Spinach

W. J. BYFORD

# 26. Downy Mildew of Sunflower

W. E. SACKSTON

## 27. Downy Mildew of Tobacco

PIERRE SCHILTZ

## 28. Downy Mildew of the Vine

R. LAFON and J. BULIT

## Subject Index

Chapter 1

# History and Importance of Downy Mildews

## G. VIENNOT-BOURGIN

*8, Square Vauban 78220 Viroflay, France*

## I. Introduction

The term "mildew" was employed initially in the United States to denote a group of parasitic fungi with little in common except their appearance as a delicate outgrowth (either white, or highly coloured in the case of moulds) caused by the proliferation and fructification of mycelium on the surface of necrotic tissue. It was Riley (1886) who first proposed a distinction between the "downy mildews" produced by certain of the Peronosporaceae, and the "powdery mildews" caused by the Erysiphaceae, and use of the term "potato blight" to describe *Phytophthora infestans* (Mont.) de Bary. The term mould was reserved for outgrowths of different aspect such as the blue mould produced by *Penicillium* and grey mould caused by *Botrytis cinerea* Pers. The generalization of these names is not absolute however. Thus spinach leaf mould is the common name for *Peronospora spinaciae* Laub. and tobacco blue mould for *Peronospora tabacina* Adam, whilst the compact crusts produced by the Capnodialeae and the Meliolaleae are referred to as dark mildews.

The term "mildew", followed by "downy mildew" was very quickly adopted in Europe at the time of the introduction of *Plasmopara viticola* (Berk. & Curt.) Berl. & de Toni from North America. At about the same time Planchon (1879) suggested calling the disease "false oidium or mildew". This term is still employed in French-speaking countries in reference to potato mildew (Pythiaceae) and most of the Peronosporaceae, whilst in German the similar terms Falscher-Mehltau (Peronosporaceae) and Echter-Mehltau (Erysiphaceae) are employed.

The downy mildews constitute a distinct fungal group as a result of both their morphology and epidemic character. Certain of them are very dangerous parasites, being detrimental to crops of considerable economic and social importance, e.g. vine, tobacco, beet, sunflower, and soybean.


## II. Epidemic and Pandemic


The presence of mildew in a crop is initially confined to sporadically distributed centres of infection the development of which depends both on the existence of an active inoculum and on environmental conditions particularly favourable to infection over short distances. The "infection centre" may only be a single plant initially, resulting from the germination of a conidium or oospore, or from the growth and development of mycelium contained in the seed or conserved in a latent state in some perennial organ.

In the second phase the epidemic nature of the disease becomes apparent. Variations in humidity cause the conidia to be detached and turbulence phenomena are involved in their distribution in considerable numbers. Waggoner and Taylor (1958) estimated that a leaf-spot of *Plasmopara halstedii* (Farl.) Berl. & de Toni covering a 200 cm² area on sunflower produced 200 000 spores daily. The resulting successive infections can lead to the entire destruction of a seeded or transplanted crop. In the case of *P. halstedii* it was estimated in Canada that systemic infection via the roots can destroy 95% of affected plants.

In the complete absence of human intervention the mildew story would be quite straightforward. In reality, however, as is also the case with many other diseases, a normally "indigenous" disease can very often become pandemic as a result of the inadvertent transference of inoculum to regions far from the original source. Marked changes in land use, the introduction of new crops and the need to increase yields by resorting to imported seed and plants have resulted in the development of pandemics. The explanation for this transformation is most often an ignorance of the capacity of the parasite to remain viable and to be transported over great distances in perennial plant parts, either as mycelium or as oospores. This frequently results in a considerable extension of the area already covered by the parasite over both continuous and fragmented geographical zones.

Although *Peronospora schachtii* Fck. normally overwinters in the leaves of beet tops conserved in silos it was originally imported into Australia in oospore-containing seed (Magee, 1935). According to Campbell (1922), *Peronospora aestivalis* Syd. was introduced into Italy in infected seeds of lucerne originating from Argentina. Ciccarone (1952) reported that *Peronospora pisi* Syd. also made its appearance in Italy in plants grown from Dutch seed. Although *Plasmopara halstedii* only rarely persists as mycelium in the embryonic tissue, both hyphae and oospores have been observed in the pericarp of sunflower achenes (Delanoe, 1972; Cohen and Sackston, 1973). It was demonstrated by Leppik (1962a, b) that a race of this particular *Plasmopara* was introduced into Iowa in imported seed from Turkey and eastern Europe. These examples explain the increasing attention now paid to seed testing. It is known, for example, that *Peronospora manshurica* (Naum.) Syd. of soybean, which has occasionally been reported in Europe, produces oospores embedded in the seed teguments and that these spores have a longevity of at least 8 years (Neergaard, 1977).

The brutal extension in geographical distribution of a number of different mildew species during the second half of the nineteenth century together with the extensive crop destruction that very often resulted were sufficient to motivate all sorts of investigations. Some of these illustrate one of the most brilliant periods in the systematic mycology of parasitic micromycetes. Others, which constitute the basis of modern phytopathology, resulted in the discovery of a number of chemical treatments which could be applied as preventive measures to cultivated plants. The studies of *Plasmopara viticola* (Berk. & Curt.) Berl. & de Toni and of *Peronospora tabacina* Adam serve as demonstrations in this respect.

## III. Vine Mildew

### A. Coming from America

It is generally accepted that grape mildew *Plasmopara viticola* has always been an endemic parasite in North America like powdery mildew (*Uncinula necator* (Schw. Burr.), black rot *Guignardia bidwelli* (Ell.) Viala & Ravaz, and the phylloxera aphid *Viteus viticola* (Fitch).

Numerous species and varieties of *Vitis* have been identified in North America. These include *V. berlandieri* Planch., the little mountain grape found in the Texas hills and northern Mexico, *V. aestivalis* Michx., or summer grape, *V. cordifolia* Michx., *V. riparia* Michx., the riverside grape, *V. labrusca* L. the fox grape, etc. All these species, together with their varieties and hybrids, are able to harbour mildew although its virulence is generally low. It was mainly *Vitis labrusca* that the American viticulturists employed initially to constitute the first vineyards for wine and table grape production.

One consequence of the extended cultivation of vines in north America was the increased possibility of mildew attack. At the same time the contagiousness of the disease and its epidemic capacity were demonstrated although its exact nature and origin remained unknown. Knowledge of plant diseases still relied, in fact, on a totally empirical approach at this time, any observed organic or physiological changes usually being explained as resulting from the adverse effects of certain natural factors. This was certainly so for mildew. It no doubt explains the name of "sun scald" used to describe its effects in the Missouri region. Often quite marked differences were observed which could be related to the variety or time of year. These are illustrated in the names of brown rot, soft rot, grey rot, and common rot used in particular to describe the destructive effect of the mildew on grapes. During an important scientific mission to North America in 1887, Viala (1893) made a study of these various aspects and concluded that the parasite was well established throughout southern Canada and most of the United States, with the exception of the dry regions of California, Oregon, Arizona and New Mexico.

The effects of mildew in American vineyards resulted in considerable yield losses and, to a large extent, the parasite was responsible for limiting the further extension of vine cultivation. It certainly prevented the use of certain cultivars in those northern regions with extensive stretches of water where the damp climate was highly favourable to mildew development. Finally it made introduction of the European vine (*Vitis vinifera* L.) into the eastern USA a very risky operation due to the great susceptibility of this species to the parasite.

Although it cannot be said that the means of limiting mildew development were understood at this time it had been observed that a relatively high night temperature followed by abundant moisture from the morning dew favoured the appearance of disease symptoms. Certain American viticulturists tried to avoid the condensation of moisture on the grapes by devising means of sheltering or covering the vine plants.

It was in 1837 that Schweinitz discovered the true cause of grape mildew, attributing it to the development of a microscopic fungus. He classified it as a mildew from the appearance of the efflorescence that developed on the underside of leaves, and identified it as *Botrytis cana* Lk., in the tribe Cymosae, initially reported on decaying leaves (E. Fries, *Systema mycolo-*

*gicum*, 1832). A little later on (1855) the great mycologist Berkeley, after studying samples of diseased vine brought from America by Curtis identified the fungus as a separate species in the genus *Botrytis*, and called it *B. viticola* Berk. & Curt. The identity of the parasite described by Schweinitz and *B. viticola* was established by Farlow (1876) who at the same time produced a very detailed description and reported that the fungus was common throughout the Atlantic and Central States of the USA, not only on *Vitis labrusca, aestivalis, cordifolia, vulpina = riparia*, but also on cultivated varieties.

In 1863 de Bary published his remarkable work on the development of fungal parasites. He excluded a large number of species from the *Botrytis* genus. Thus *B. viticola* was transferred to the genus *Peronospora* defined by Corda in 1837, under the name of *Peronospora viticola* de Bary, in the sub-family Peronosporae of the Peronosporaceae. In 1879 Planchon in his study of American vines mentioned "within the various cryptogams causing the stunting of certain varieties, one which could be termed false oidium, known as mildew to the Americans, and *Peronospora viticola* to the Botanists."

This name persisted until 1888. Recognition of the genus *Plasmopara*, established by Schroeter, led to the final naming of grape mildew as *Plasmopara viticola* (Berk. & Curt.) Berlese & de Toni (1888). The subdivision of the *Plasmopara* genus into *Rhysotheca* (presence of zoospores) and *Plasmopara sensu stricto* (absence of zoospores), proposed by Wilson (1907) has not been adopted.

The unfortunate introduction of vine mildew into Europe was of cardinal importance in the history of this parasite. It took place only a short time after European agriculture had suffered two consecutive economic and social catastrophes. The first occurred between 1830 and 1845, depending on the region, and followed the destruction of potato crops, initially in western Europe, by the leaf blight caused by *Phytophthora infestans* (Mont.) de Bary. Harvest destruction, farmers' ruin, famine and the resulting exodus and emigration necessary to survival, have remained deeply engraved in the memories of the peasant classes. For a great many years the use of potato varieties recognized, through natural selection, as being resistant or tolerant to mildew, were the only means available for reconstituting this staple crop.

Almost as alarming a situation arose when a few years later, from 1851 onwards, the vine powdery mildew became established throughout the grape-growing regions of France and the Mediterranean basin. This "pernicious mould", declared Berkeley in 1847 in his description of *Oidium tuckeri* Berk., originated in the United States and was inadvertently introduced in 1845 into the Margate vinery near to London. It caused extensive damage during 1852 and 1853, but most especially during 1854, creating despair throughout the peasant world. Luckily the curative action of sulphur was known and from 1850 onwards recognition of the effectiveness of this element was demonstrated in its widespread utilization over extensive areas planted with vines.

It was the import of American plant material (the intention being to discover or breed vines that were resistant to powdery mildew) that led to the introduction of other parasites of this crop, in particular of phylloxera and downy mildew. Between 1863 and 1881 a root-attacking strain of phylloxera totally destroyed the vineyards of Europe and Algeria. Since then its existence has been reported throughout the world. Successful reconstitution of these vineyards following the disastrous "phylloxera crisis" resulted from the considerable efforts of numerous vine researchers and technicians. It was achieved by grafting the European vine on to resistant American rootstocks and later by using directly productive material obtained from hybridizations between American and European vines.

The suddenness and gravity of the mildew invasion of Europe and Asia Minor can be explained essentially by the fact that at this time vine cultivation in these regions depended almost entirely on the vinaceous qualities of *Vitis vinifera* L. This climbing plant, originating south of the Caspian Sea, had been known since prehistoric times for the quality of its table grapes and currant grapes, but above all for its wine. Progressively mastered by pruning and training techniques it had not suffered from any serious physiological disorder or parasitic disease until the end of the last century. *Vitis vinifera* suffers from phylloxera attack to a much greater extent than the American vines, however, and is also highly susceptible to attack by mildew.

The massive import of grape seeds, rooted plants and vine cuttings, resistant to phylloxera but likely to be contaminated with mildew and thus to cause its introduction into European vineyards, was denounced as a dangerous operation. As early as 1873 Cornu drew the attention of botanists and viticulturists to a "fungus particularly injurious to American vines". Following his publication of the Peronosporaceae of France, Cornu (1978) repeated his fears that "the introduction of mildew from American vines could take place at any time".

The urgent need to replant the vineyards prevailed nonetheless and in consequence the devastating invasion of vine mildew, firstly in France and then throughout the wine-producing areas of Europe, took place. It was in August 1878 that Deluze (*in* Galet, 1977) discovered the first centre of mildew infection on Jacquez, an American vine growing in a vineyard in the Coutras region near to Libourne in the Gironde. Specimens of infected leaves were sent to Planchon, Director of the School of Pharmacy, Montpellier, who identified it as mildew. Other infection centres were soon reported from Nerac in the Lot-et-Garonne. On 16 September mildew was discovered at Saintes (Charente maritime) while Millardet (1882a), once a pupil of Montagne and de Bary, observed it on seedlings of *Vitis aesti-valis* in the "Gironde Farming Association" nursery at Floirac.

With disconcerting rapidity various French vineyards were progressively infected. During October 1879 Pirotta reported the parasite in the Pavie province of Italy. In 1880 at the same time as it was wreaking havoc in the Gironde region and certain vineyards in the Gard (near to Aigues Mortes), mildew was continuing to spread in France and in Spain where according to Viala (1893) it was found by Panchon (1880) near to Barcelona, in Switzerland, in Germany on the banks of the Rhine, in the Austrian Tyrol, from north to southern Italy and in Greece. According to Trabut it was also in 1880 that *Plasmopara viticola* crossed the Mediterranean and arrived on the outskirts of Algiers. In 1885 it was responsible for serious losses in Tuscany and Lombardy. Its effects were recognized as disastrous in a great many French vineyards, especially in the Gironde, during 1886. In 1887 it reached the Caucasus while the British Isles were infected a little later. Mildew was reported for the first time in Britain by Cooke (1894) in two distinct localities. From then on it was no longer observed until 1926 when a severe attack was reported (Harrison and Ware, 1926) in a garden in Wye, Kent. According to Moore (1949, 1959) it reappeared during 1932 on the ornamental species *Vitis coignetiae* Pull. in Surrey and again in 1936. In Norway, where vines are grown under glass, mildew has been known to exist since 1893 (Wille, 1893).

Extension of the geographical distribution of *Plasmopara viticola* was also evident in other vine-growing regions of the world. According to Brunneman (*in* Sorauer, 1908, p. 157) it has existed in Brazil since 1893. It has been reported in S. Africa (1907), Australia (1916–1917), and finally more recently in New Zealand (1926).

## B. Bordeaux mixture

The dramatic situation facing viticulture as a result of the development of mildew in Europe inspired numerous attempts to overcome the disease before a rational control method was finally introduced. This was based on the toxicity of copper salts to fungal spores, a property which seems to have been demonstrated for the first time in 1761 by the German Schulthess. In 1801 Prevost recommended its use to destroy the germination power of bunt spores in wheat. We owe it to the Frenchman Proust (1800) to have been the first to propose reducing the phytotoxicity of copper sulphate by combining it with lime.

The utilization of copper salts in vineyards was a very ancient practice associated with two totally different aspects of vine cultivation. The first was employed from 1831 onwards by viticulturists in the Beaune region of the

Bourgogne. It consisted of soaking the poles and stakes (made of white-wood or pine in preference to oak), which were used to support the vine, and protect it from rot, in a solution of copper sulphate for 4 days before putting them in place. The same treatment was used for the ties of rye straw, osier or hemp fibres used to attach the vine shoots, these being left to soak in a 10% solution of copper sulphate. From 1883 onwards the Bourgogne viticulturists noted the efficiency of sulphated vine stakes in diminishing the significance of mildew. This method of control was tried in a number of vineyards although certain practical difficulties were encountered.

The second mode of using copper salts depended on the well-known reputation of "vitriol" (or solution of copper sulphate in water) as a violent poison to man. This particular property had always been exploited in the Medoc area to dissuade thieves. At the approach of the grape harvest those vines nearest to the edge of the vineyard were sprinkled by means of a twig broom with a solution of copper sulphate thickened with milk of lime. As early as 1882 Millardet (1882a, b) Professor at the Faculty of Science at Bordeaux, noticed while surveying the Bordeaux vineyards that "vitriola-tion", the term given to the application of this Medoc mixture, offered protection to mildew. Aided with numerous field observations by David, responsible at that time for a large Bordeaux vineyard, and helped in the laboratory by the chemist Gayon, Millardet considered that a practical treat-ment against mildew should aim not to destroy the parasite in infected leaves (which seemed impossible without killing the leaves as well), but instead should anticipate its possible development by covering the leaf surface with various substances capable of inducing a loss of spore vitality or at least of impeding spore germination. Numerous publications demonstrating the efficiency of treatments, which to be effective had to be preventive, are associated with this period. At the same time the innocuity of cupric salts with regard to wine quality was finally established. It was 1 April 1885 when Millardet published his initial formulation of "Bordeaux Mixture" which, in the same year, was applied to 250 000 vines in an extensive vine-yard in the Medoc area. In 1887 Masson at Beaune formulated the "Bour-gogne Mixture" in which the copper sulphate was neutralized with sodium or potassium carbonate.

Although the nature of the disease had been scientifically established, numerous agricultural technicians would not accept that mildew had been introduced from the American continent. Certain of them alleged that the effects observed on leaves and fruits had always existed in an intermittent manner. Others tried to make out that these were not caused by the develop-ment of a parasite but were rather the adverse effects of particularly un-favourable climatic conditions. Thus in Algeria destruction of the foliage by mildew was initially attributed to the combined effects of the Sirocco wind coming from the Sahara and the sun (Trabut, 1880). These claims were refuted by Cornu (1882) who, basing his affirmations on continuous observations made between 1872 and 1878, insisted that mildew had not existed in France before that time. He had a thorough knowledge of the parasite having examined samples in the herbaria of the National Museum of Natural History in Paris and received others from Farlow and Cooke.

Diversification of the compounds and techniques followed and recommendations to viticulturists were increased. In this respect the need to inform vine-growers of the seasonal development of grape mildew, and of other pests, led in 1898 to the creation of the first Agricultural Advisory Station at Cadillac in the Gironde.

At the same time numerous research projects were initiated to determine the biology of *Plasmopara viticola*. Amongst the more recent, mention should be made of the work permitting the production of pure cultures of the fungus on vine callus inoculated with conidia. Morel (1948) was the first to keep the parasite active under such conditions for a month. Boubals (1957), using a similar technique, showed that the most rapid infection occurred in tissue cultures of *Vitis vinifera*.

## C. Losses

Viticulturists are now in possession of efficient control methods and, in fact, the vine mildew story could stop there. However the anxiety of viticulturists and the efforts made to control mildew are justified when one remembers that, since its introduction into Europe, this parasite has frequently caused tremendous damage and that each year it could, given favourable climatic conditions, more or less annihilate a certain proportion of the grape harvest.

The pernicious character of vine mildew makes it difficult to put the losses caused by this parasite into figures. In those regions where climatic conditions are favourable to its development and where susceptible vine varieties are grown, damage in the short term is expressed as a decrease in the volume of grapes harvested which in turn leads to a reduction in wine yield and relatively substantial changes in its composition and quality. Also by provoking early shedding of the foliage, mildew attacks lead to a loss in vine vigour and a reduction in the amount of vegetation produced during the following year.

Certain numerical data help to give some idea of the damage caused by wine mildew in North America. Thus in 1869 losses of the order of 70–80% of the harvest were recorded in North Carolina. The rapid improvement of viticulture in the United States has greatly reduced the economic significance of mildew. According to statistics published in 1965 by the Department of Agriculture (USDA), for the period 1951–1960, the annual grape harvest was 3 165 000 tons representing 96% of the total production of North and Central America. California alone produced 2 862 000 tons. The percentage losses due to virus diseases were around 20% whilst 0·4% of the yield reduction could be attributed to mildew.

1882 was a disastrous year for French viticulture as the effects of phylloxera attack were added to those of mildew. In 1886 the cupric treatments were only in their early stages of development so that losses were considerable due to the widespread occurrence of mildew. In 1891 and 1892 Linhart and Mezey (1895) reported serious losses in Hungary and in 1895 the wine yields in Italy (according to Caruso), as in France, were particularly poor. France suffered numerous "mildew years" when prolonged periods of rain encouraged development of the parasite. Such was the case in 1910 when, according to Galet (1977), the harvest was only 28 million Hℓ and in 1915 when it was a mere 20·5 million Hℓ instead of the normal 40–50 million. During 1912 and 1913 serious losses were reported in Baden. In 1930 and 1932 severe attacks of mildew swept through certain vineyards in the south of France reducing the harvest by a third. In 1957 the French vines again suffered disastrous attacks of mildew, and in 1977, despite the efficacy of the anti-cryptogamic compounds employed, 12–15 treatments were required to reduce attacks in the south of France.

## IV. Tobacco Blue Mould

### A. Coming from Australia

The story of tobacco blue mould begins at the centres of origin of various edible and ornamental species of the genus *Nicotiana*, i.e. principally in South America but also in the Pacific Islands and in Australia.

In 1891 Spegazzini described a *Peronospora nicotianae* Speg. that he found in the vicinity of Buenos Aires in Argentina on *Nicotiana longiflora* Cav. According to the original description this parasite causes the appearance of small round spots on the leaves. These steadily get larger and become covered with a greyish efflorescence consisting of a mass of conidiophores. Gäumann (1923) also assigned a *Peronospora* harvested by Farlow (1886) on *Nicotiana glauca* in southern California to this species as well as a blue mould found on young plants of *N. bigelovii* Wats. in Nevada. But as Stevens and Ayres (1940) pointed out, this classification is questionable due to the absence of the characteristic oospores. Later on, certain authors refuted the validity of *P. nicotianae* as a separate species.

The tobacco disease, known as blue mould, was first reported in 1892 in Queensland (north-eastern Australia) by Cooke who attributed it to *Perono-*

*spora hyoscyami*. In 1914 its existence was apparent in the State of Victoria, the yield losses varying in significance with the nature of the pathotype present. In certain years they represented as much as 20–30% of the harvest, resulting from the accumulation of damage to young nursery plants prior to transplanting and to attacks in the open field where destruction of the foliage was particularly marked.

In 1933 Adam demonstrated some slight differences in morphology between the fungus causing blue mould and *Peronospora hyoscyami*. These differences were considered sufficient to justify the creation of *Peronospora tabacina* Adam. as a separate species. This opinion was contested a few years later. Only the nature of the plant hosts permitted such a distinction and Skalicky (1964), Hill (1966) and Shepherd (1970) attributed this to the existence of four "special forms" within *Peronospora hyoscyami*.

| | Sporulation on: | | |
|---|---|---|---|
| f. sp. | *Hyoscyamus niger* | *Nicotiana langsdorfii* | *Nicotiana* hybrid |
| *hyoscyami* | + | O | O |
| *tabacina* | O | + | O |
| *hybrida* | O | + | + |
| *velutina* | O | O | + |

As far as the Australian continent was concerned, the endemic character of blue mould disappeared. As a consequence, the parasite began to appear in regions far distant from one another. This new type of behaviour was linked firstly to the possibility of wind transport of the conidia and secondly to conservation of the fungus in the form of latent mycelium in the seed (Pont and Hughes, 1961, Borovskaya, 1965). It is to these two modes of dissemination that the abrupt appearance of blue mould in North and South America can be attributed.

In these two sub-continents *Peronospora tabacina* behaves essentially as a parasite of young tobacco seedlings. The disease was first discovered in Florida and Georgia during the spring of 1921. It occurred as sporadic "infection centres" which were energetically destroyed with the result that during the subsequent 10-year period no further damage caused by mildew was reported from these regions. In contrast, from 1931 onwards the epidemic character of tobacco mildew started to become apparent. Between 1934 and 1949 it was progressively reported from Ontario and the Eastern States of the USA right through to Texas. During 1935 it was found in Brazil in the Grande do Sul State where tobacco cultivation is very important. In 1938 it was observed in Canada, in 1939 in north-west Argentina, in 1953 in the Santiago region of Chile and finally, since 1957, it has been known to exist in Cuba.

The introduction of blue mould into Europe could have occurred several years before its eventual appearance in Great Britain during 1957, had the

Portuguese Quarantine department not intercepted and destroyed tobacco seed to be used in the production of new hybrids on recognizing the seed to be contaminated with the latent mycelium of mildew (Oliveira, 1963, *in* Neergaard, 1977).

The initial infection centres of tobacco mildew were observed under glass and resulted from the unfortunate import from Australia to Great Britain of a particularly virulent pathotype to be used in a company's fungicide trials (Klinkowsky and Schmiedeknecht, 1960; Klinkowski, 1961, 1970). In the initial stages of attack the tobacco mildew developed on species and varieties of ornamental tobacco and on *Nicotiana glutinosa* kept under glass to maintain numerous virus strains. The English virologists had to destroy the *Nicotiana* plants several times in succession as a result of the sudden and severe attacks by *Peronospora*. During July 1959 tobacco mildew made its appearance in Holland in the Utrecht region following an exchange of plant material destined for virological studies. It was reported under similar conditions a little later in Baden-Wurtemberg while it continued to spread from its initial centres of infection in the Dutch, Belgian and German crops.

After overwintering in a vegetative state in the residues of contaminated crops, or in the form of oospores in fragments of dry leaves, the mildew developed again during 1960 when its widespread occurrence, of catastrophic dimension, was initiated from infection centres in western Flanders, Baden-Wurtemberg, Schleswig-Holstein and then in Lower Saxony (Kröber, 1961). In July of the same year tobacco mildew appeared in Central France (Loire region) in an isolated infection centre, and a little later was reported in the area around Strasbourg (Vuittenez, 1961). Alsace suffered widespread attack and losses were estimated at around 4 billion (old) French francs. Towards the end of 1960 there was an invasion which affected the whole of Poland, northern Hungary and Italy, where blue mould was reported around Salerno in the November. It also destroyed tobacco cultivated under glass in Montpellier.

From these now well-established infection centres tobacco mildew continued, during 1961, to extend over larger and larger areas. In March it was observed in Corsica and Algeria, and later in Spain, the Balearics, Sicily, Turkey, Greece, Bulgaria and Romania. The Soviet Republics of Moldavia, Ukraine, Bielorussia, and Lithuania were infected, it was reported in northern Sweden while new infection centres developed in Great Britain (Corbaz, 1960, 1964; Populer, 1964, 1965).

By 1962, that is 5 years after its initial appearance in Europe, the parasite had covered the entire European tobacco-growing area. It had also extended its former geographic distribution, having invaded North Africa, central Spain, Crete, Israel, the Crimea and the borders of the Caspian Sea in northern Iran (Niemann and Zalpoor, 1963).

Comparative studies of the *Peronospora tabacina* pathotype that had invaded the American continent and the one introduced into Europe revealed marked differences in behaviour. Whereas the former was mainly a parasite of young plants growing in nurseries, the European pathotype attacked both young plants and those in the height of their development, being especially injurious to these latter.

## B. Losses

During the 1957–1962 period the losses due to *Peronospora tabacina* were extremely high everywhere. They were evident in the number of contaminated leaves totally unsuitable for tobacco manufacture and also in the mediocre quality of the tobacco obtained. Many of the crops in the Federal Republic of Germany, as in East Germany, which had been invaded in their entirety during 1960, were destroyed. Numerous planters gave up this crop, the yield per hectare of which had diminished by 42% at this time. The quality of the tobacco obtained was such that the average selling price fell by 20% so that the total harvested product from the Baden and Palatinat regions was about 58% lower than normal. Whereas 64000 ha had been planted with tobacco in the Federal Republic of Germany during 1960, no more than 4000 ha remained by 1961. Comparable figures were recorded in other regions of Europe. Thus in Flanders and northern France, during 1960, harvest losses of the order of 90% were noted and 80% in Alsace. Considering the countries affected by mildew as a whole, the overall losses for 1961 were more than 24% of the normal production.

From 1962 onwards, thanks to measures taken in Europe by the International Union of Tobacco Planters (UNITAB) and the Centre of Cooperation for Scientific Research on Tobacco (CORESTA), a net diminution in the damage caused by *Peronospora tabacina* has been observed. In the United States the 1965 report published by the Department of Agriculture (USDA) states that annual losses due to blue mould are estimated at around 5·5% of the harvest.

## V. Conclusions

In the same way that we have traced the histories of *Plasmopara viticola* and *Peronospora tabacina* so we could have traced that of hop mildew, *Pseudoperonospora humuli* (Miyb. & Tak.) G. W. Wilson. Two explanations are offered for the introduction of this fungus into Europe. Some pathologists believe that the parasite came from the Caucasus and central Europe, and from 1895 onwards was responsible for the red rust disease of hop cones in Alsace (Blattny, 1927; Hampp, 1931). Other pathologists consider that this mildew was introduced from Japan (where it was found on wild and cultivated hops and described by Miyabe and Takahashi (1905)), firstly into Great Britain during 1920 in infected cuttings (Salmon and Wormald, 1923), then rapidly throughout Europe.

It is almost always man's intervention, whether inevitable or inadvertent, that initiates and facilitates the transport and dissemination over long distances of agents causing plant diseases.

Only by respecting rigorous measures of practical protection and understanding the dangers that uncontrolled importation can pose will new disasters be avoided or retarded.

## References

Adam, D. B. (1933). *J. Dept. Agric. Victoria* **31**, 112–146.
de Bary, A. (1863). *Ann. Sci. nat. Bot.* **IV**, Sér. XX, 5–148.
Berkeley, M. J. (1847). *Gardeners' Chronicle* **48**, 779.
Berkeley, M. J. and Curtis, M. A. (1855). *In* Ravenel Fungi caroliniana, **5**, 90
Berlese, A. N. and de Toni, J. B. (1888). *In* "Sylloge Fungorum" (P. A. Saccardo), Vol. VII. Pars. I, 239.
Blattny, C. (1927). *Trav. Inst. Rech. Agron. Répubi. Tchécoslovaque*. Prague.
Borovskaya, P. (1965). *Zashch. Rast. Vredit. Bolez* **10** (3), 44–45.
Boubals, D. (1957). *C. r. Acad. Sci., Paris* **214**, 1816–1818.
Campbell, L. (1922). *Ann. Bot.* **15**, 283–284.
Caruso, G. (1895). *Boll. Entomol. Agrar. Pathol. Veget. Padova* **II**, 68.
Ciccarone, A. (1952). *Ann. Sper. Agrar.*, N.S.6, 1065–1067.
Cohen, Y. and Sackston, W. E. (1973). *2nd Int. Congr. Pl. Pathol.* Abstr. Papers, no. 0446.
Cooke, M. C. (1892). "Handbook of Australian fungi". London.
Cooke, M. C. (1894). *Gardeners' Chronicle* **15**, ser. 3, 689.
Corbaz, R. (1960) *Rev. Rom. Agric.* **16**, 101–104.
Corbaz, R. (1964). *Phytopathol. Z.* **51**, 190–191.
Corda, A. C. J. (1837). *Icones Fungorum*. I, Prag.
Cornu, M. (1873). *Recueil des Savants Etrangers* **22**, 35–36.
Cornu, M. (1878). *C.r. Acad. Sci., Paris* **87**, 18 Nov.
Cornu, M. (1882). "Observations sur le Phylloxéra et sur les Parasites de la Vigne." Paris.
Delanoe, D. (1972). *Inform. Techniques* CETIOM, **22**, 1–49.
Farlow, W. G. (1876). *Bull. Bussey Inst.* **I**, 426.
Farlow, W. G. (1886). *Proc. Am. Soc. Adv. Sci.* **34**, 300–303.
Galet, P. (1977). "Les Maladies et Parasites de la Vigne". Montpellier.
Gäumann, E. (1923). Beiträge zu einer Monographie der Gattung *Peronospora* Corda. Beitr. *Kryptogamenflora Schweiz*. Zurich.
Harrison, R. M. and Ware, W. M. (1926). *Gardeners' Chronicle* **80**, 448–449.
Hill, A. V. (1966). *Coresta Bull.* **I**, 7–15.
Hampp, H. (1931). *Allg. Brau. Hopfenztg.* **17**, 819–821.
Klinkowski, M. (1961). *Deutsch. Landwirt.* **12**, 229–232, 237–239.
Klinkowski, M. (1970). *Ann. Rev. Phytopathol.* **8**, 37–60.
Klinkowski, M. and Schmiedeknecht, M. (1960). *NachrBl.d.PflSchutzdienst, Berlin* **14**, 61–74.
Kröber, H. (1961). *NachrBl. dt. PflSchutzdienst, Berlin* **13**, 69–70.

Leppik, E. E. (1962a). *Bull. phytosanit. F.A.O.* **10**, 126–129.
Leppik, E. E. (1962b). *Rep. N. Central Region. Intr. Sta.* Ames, Iowa, 11.
Linhart, G. and Mezey, G. (1895). "Krankheiten der Weinrebe". Magyar-Ovar.
Magee, C. J. (1935) *Agric. Gaz. New South Wales* **46**, 571–572.
Masson, G. (1887). *Progrès Agric. Vitic.* 35–320.
Millardet, A. (1882a). *Congr. Int. Phylloxérique de 1882.*
Millardet, A. (1882b). *J. Agric. Pratique* **34**, 267.
Millardet, A. (1885). *J. Agric. Pratique* **1**, 553.
Millardet, A. and Gayon, U. (1887). *J. Agric. Pratique* **51**, 698–704, 728–732, 765–769.
Miyabe, K. and Takahashi, Y. (1905–1906). *Trans. Sapporo Nat. Hist. Soc.* **1**, Part 2.
Moore, W. C. (1949). *Trans. Br. Myc. Soc.* **34**, 95–99.
Moore, W. C. (1959). "British parasitic fungi." Univ. Press, Cambridge.
Morel, G. (1948). *Ann. Epiphyties* **14**, 127–229.
Neergaard, P. (1977). Seed Pathology, Mac Millan Press, London.
Niemann, E. and Zalpoor, N. (1963). *Entomol. Phytopath. Appl.* Téhéran **21**, 5–12.
Pirotta, R. (1879). *C.r. Acad. Sci. Paris* **89**, 697.
Planchon, J. (1879). *C.r. Acad. Sci. Paris* **89**, 600–608.
Planchon, J. (1880). *La Vigne Américaine* 11–14.
Pont, W. and Hughes, I. K., (1961). *Queensl. J. Agric. Sci.* **18**, 1–31.
Populer, C. (1964). *Bull. Inst. Agr. Stat. Rech. Gembloux* **32**, 339–365.
Populer, C. (1965). *Parasitica* **21**, 37–39.
Prevost, B. (1807). "Mémoire sur la cause immédiate de la carie ou charbon des blés et de plusieurs autres maladies des plantes et sur les préservatifs de la carie". Paris.
Proust, E. (1800). *In* "Les fongicides". Lhoste, J. and Lambert, J. (1970). Institut de Phytopharmacie, Marseille.
Riley, C. V. (1886). *Rural New Yorker* **45**, 72–87.
Salmon, E. S. and Wormald, H. (1923). *J. Minist. Agric.* **30**, 430–435.
Schulthess, H. H. (1761). *Naturforsch. Gesellschaft in Zurich* Bd. **1**, 497–506.
Schweinitz, L. D. (1837). "Synopsis fungorum Amér. boréale", 2663, no. 25.
Shepherd, C. J. (1970). *Trans. Br. Myc. Soc.* **55**, 233–256.
Skalicky, V. (1964). *Acta Univ. Carol. Suppl.* 25–90.
Sorauer, P. (1908). Handbuch der Pflanzenkrankheiten. II Bd., Berlin.
Spegazzini, C. (1891). *Revista Argentina Hist. Natural* **I** 28–38.
Stevens, N. E. and Ayres, J. C. (1940). *Phytopath.* **30**, 648–688.
Trabut, L. (1880). *Bull. Ass. Sci. algérienne* **3**, 242–244.
Viala, P. (1893). "Les Maladies de la Vigne". Montpellier.
Vuittenez, A. (1961). *B.T.I.*, no. 158, 1–7.
Waggoner, P. E. and Taylor G. S. (1958). *Phytopath.* **48**, 46–51.
Wille, N. (1893). *Botan. Notizar* 1–11.
Wilson, G. W. (1907). *Bull. Torr. Bot. Club* **34**, 387–416.

Chapter 2

# Taxonomy and Evolution

C. G. SHAW

*Department of Plant Pathology, Washington State University, Pullman, Washington, USA*

Within the Peronosporaceae there exist obvious evolutionary sequences from zoosporangia to true conidia. The series on broad-leafed hosts, from *Bremiella* and *Plasmopara* which produce sporangia containing zoospores, through *Pseudoperonospora*, to *Peronospora* spp. which produce conidia wh.ch germinate by germ tubes, has been recognized for many years, but the full significance of the morphological and physiological differences has not been appreciated. A parallel series exists within the downy mildews on grasses.

Whether the asexual propagules of downy mildews are sporangia or conidia can be determined not only by germinating them, but more readily by morphological examination. Sporangia have a distinct apical modification of the wall; the sporangia are actually operculate; the operculum is lens-shaped. The operculum swells during the final stages of maturation; typically it is completely detached by the increasing osmotic pressure inside the sporangium immediately prior to initial swarming and escape of the zoospores. After release of zoospores, the empty sporangium is broadly poroid.

The structure of the apical region of the sporangia of *Phytophthora* spp. was clearly illustrated by Blackwell and Waterhouse (1931); the structure of sporangia in the Peronosporaceae is the same, although rarely do either drawings or photographs of sporangium-producing species indicate the structure of the apical region. Dogma (1975) has questioned the application of the term "operculum" to this modified apical region of the sporangium. He continues to refer to it as a "discharge papilla, dissolution of which effects zoospore discharge". Actually, rupture of the apical region is a rapid process, in which by circumscissile dehiscence, the inner and outer walls of

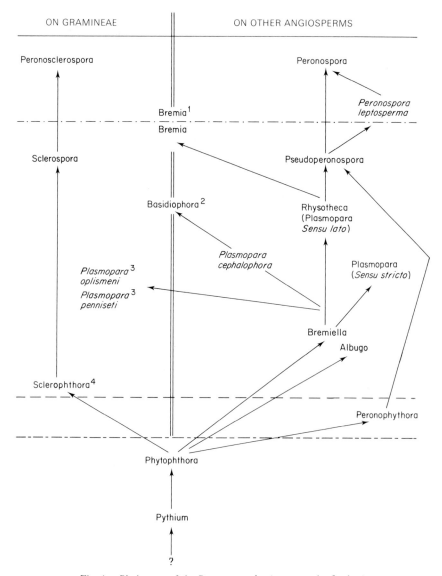

Fig. 1.   Phylogeny of the Peronosporales (see opposite for key).

the sporangium break below the operculum, and the central, lens-shaped portion of the operculum is lifted off and characteristically floats away. Zoospores then escape through the resultant pore. In rare instances dehiscence is incomplete, and the swollen but undissolved, lens-shaped operculum can be seen still adhering laterally to the pore formed by the opening. Typically, during final maturation, the operculum enlarges (giving the illusion of dissolving) and thus the apex of the sporangium appears papillate. The papillate nature of mature sporangia in the Peronosporaceae has been stressed repeatedly; the significant structural features which result in the papillate character have been ignored almost completely. Indeed, immature sporangia are structurally operculate but characteristically not papillate.

Sporangia in the Peronosporaceae are typically as described above in *Sclerophthora* and *Sclerospora* on Gramineae, and in *Bremiella*, *Plasmopara* (including *Rhysotheca*), *Basidophora*, *Pseudoperonospora* and *Bremia*, almost all species of which occur on Dicotyledoneae (see Fig. 1 for exceptions). In contrast, *Peronospora* and *Peronosclerospora* (Shaw, 1978) do not produce sporangia; there is no modification in the apical region of the asexual reproductive propagule. With loss of the operculum there has been a concomitant loss of the ability to produce zoospores. The propagule of the anamorph (asexual stage) has become a conidium which must germinate by a germ tube. Production of germ tubes is not limited to the apical region; the germ tube may emerge anywhere on the circumference of the conidium.

Wilson (1907) recognized the significance of the method of germination

---

| | Above = | Germination always direct (germ-tube; no operculum; conidia). |
| —.—.—.—.—.— | Below = | Germination usually indirect (zoospores; operculum present; sporangia). |
| | Above = | Obligate parasites. |
| | | |
| — — — — — | Below = | Facultative saprophytes. |
| | Above = | Determinate conidiophores or sporangiophores. |
| | | |
| — — — — — | Below = | Indeterminate sporangiophores. |

[1] = In the asexual state an operculum is present, and thus a sporangium is produced. Although liberation of zoospores has been reported (Milbrath) most workers report germ-tube germination. One species of *Bremia* is reported on Gramineae.

[2] = One species of *Basidiophora* is reported on Gramineae.

[3] = Two species of *Plasmopara* are reported on Gramineae.

[4] = In *Sclerophthora* sporangiophores develop basipetally; in all other genera development of sporangiophores and conidiophores is typically basifugal.

of asexual structures in distinguishing genera in the Peronosporaceae when he segregated those *Plasmopara* species which release zoospores from those that release a plasma from the sporangium. He retained in *Plasmopara sensu stricto* only the two species, *P. pygmaea* and *P. densa*, which release a plasma from the sporangium. This plasma is stated to germinate subsequently by a germ tube (Wilson, 1907). In contrast, the great majority of species referred to *Plasmopara* do not produce a plasma; rather, they produce zoospores which are released apically from operculate sporangia. Wilson transferred these species to *Rhysotheca*, and designated *R. viticola* as the type species. Subsequent authors have not accepted the distinctions made by Wilson, and the genus *Rhysotheca* is not recognized by other taxonomists of the Peronosporales. Interestingly enough, the structure of the sporangium in both species of *Plasmopara sensu stricto* is also operculate.

The liberation of a plasma through the operculum is reminiscent of the escape of the sporangial contents into a vesicle in *Pythium*. *Plasmopara sensu stricto*, however obviously differs from *Pythium* not only in the structure of the sporangiophores, but also in that the vesicle of *Pythium* shortly releases zoospores rather than a germ tube as from the plasma of *Plasmopara pygmaea* or *Plasmopara densa*.

The evolution of conidia from sporangia has resulted in both advantages and disadvantages. The most obvious disadvantage is a reduction in the number of propagules produced. Each sporangium is capable of forming 16–32 zoospores, each of which, after encystment, can produce a germ tube capable of penetrating a susceptible host. With conversion of sporangia into conidia, and with no significant change in the number of conidia produced per conidiophore versus the number of sporangia produced per sporangiophore, the total amount of inoculum has been reduced to one sixteenth, or less, of what it was before.

Reduction in the total amount of tissue devoted to reproduction is characteristic of all evolutionary sequences. The change from holocarpic to eucarpic organisms in both the Algae and the Fungi are examples of this; other series end in the Spermatophytes in the Plant Kingdom, and in the Mammalia in the Animal Kingdom.

Reduction in number of propagules is more than counterbalanced by the many advantages correlated with adaptation to a drier habitat. Water, indeed a continuing film of water, is essential for development of sporangiophores (5–8 h), maturation of sporangia (3–6 h), release and dispersal of sporangia (1–3 h), release and swarming of zoospores ($\frac{1}{2}$–1 h), encystment ($\frac{1}{2}$ h), and germ tube production and penetration (2 h). The total time required from the initiation of sporulation through to infection is 12 to more than 20 h; the time required is inversely correlated with temperatures over the range of 10–20°C (Blaeser and Weltzien, 1978; Cox and Large, 1960; Crosier, 1934). Only during the period when sporangia are being released and dispersed can there be a discontinuum in the presence of free water. Even the dispersal period for sporangia (1–3 h) requires either water (rain) or high relative humidity (r.h. of 90% or more). Longer periods of dry weather result in desiccation and death of sporangia. Thus, the distances (both temporal and spatial) over which sporangia can function are limited.

In contrast, the conidium accomplishes germination and infection much more quickly. Zoospores need not be differentiated, released, allowed to swarm and encyst. Some 2–4 h less is required for conidia to function compared to sporangia. Consequently, it is more likely that infection will be consummated by conidia than by sporangia between the time the former are produced and the moisture continuum is terminated. Of even greater importance is the fact that conidia of *Peronspora* and *Peronosclerospora* can withstand considerable desiccation, and remain viable for 2–3 days under field conditions. Thus, the total sequence from phore initiation through penetration of the germ tube need not be a single continuum in the presence of moisture or r.h. above 90% values. (Relative humidity of 90% or more in the ambient air normally implies free moisture on the vegetation composing the canopy below the point at which r.h. is being determined.)

*Bremia* presents an interesting situation. Milbrath (1923) reported "direct germination of conidia by means of a germ tube" and "production of zoospores" in California. "Sporangia produced during cool months, December to March inclusive, formed zoospores more readily than those collected during the warm season of the year." Darkness and 10°C were most favourable for zoospore formation. Neither previous nor subsequent workers have reported zoospore release. Although Milbrath's drawings do not indicate an operculum, the latter is discernible, though indistinct, on turgid sporangia of *B. lactucae*. Indeed, the operculum appears vestigial, giving increased credence to consistent reports of direct germination by other workers. Possibly, Milbrath mistook zoospores of some chytridiaceous fungus for those of *Bremia lactucae*. At 10°C germination of any chytridiaceous restings spores contaminating the leaves from which he obtained conidia of *B. lactucae* would be favoured. With an operculate mechanism still present, but probably nonfunctional, and germ tube germination occurring anywhere on the circumference, the propagule in question might be called either a sporangium or a conidium. On the basis of its mode of germination, the name conidium is preferred.

Waterhouse (1973) is the author of the most recent taxonomic treatment of the Peronosporales. Figure 1, together with its legend, presents my concepts of phylogeny in the order. Briefly, *Pythium* is considered the most primitive genus, with the sporangium typically releasing its contents into a vesicle in which zoospores are differentiated. The sporangiophores are hyphoid and hardly distinct from vegetative hyphae. Some *Phytophthora* species produce fairly well differentiated, indeterminate sporangiophores, and swellings (apophyses) occur on the sporangiophores of some species at the points of sporangial attachment. Species of both genera are facultative saprophytes, and thus are less physiologically specialized than genera in the Albuginaceae and Peronosporaceae, even though physiological races are commonly distinguished in *Phytophthora infestans* (Gallegly, 1968). Other genera assigned to the Pythiaceae are discussed by Waterhouse (1973). Except for *Sclerophthora* and *Peronophythora*, other genera assigned to this family do not seem to be directly involved in the evolutionary lines leading to the Peronosporaceae and consequently are not discussed further here.

*Sclerophthora* has been assigned to the Pythiaceae by Waterhouse (1973). In spite of careful consideration of the points raised by her, I prefer to place it low in the hierarchy of the Peronosporaceae (Shaw, 1976). *S. macrospora* has been cultured, especially from rice in Japan, and formation of sporangia in culture has been reported (Tokura, 1975). However, Tokura (1975) also reported that attempts to reinfect rice with these isolates were unsuccessful. No other species of *Sclerophthora* have yet been grown in axenic culture.

In articles published by Payak *et al.* (1970) and Shaw (1970), the former authors stated "*Sclerophthora*: sporangiophores determinate" (p. 184), while the latter author stated "Sporangiophores indeterminate ... *Sclerophthora*" (p. 366). Payak (personal communication) informed me that the basis for their statement was the cited abstract published by Thirumalachar (1969).

Thirumalachar and Shaw were two of the three authors who established the genus *Sclerophthora* (Thirumalachar *et al.*, 1953). Reference to that publication clarifies the apparent discrepancy. Apical growth of the sporangiophore is determinate and is terminated by a single large sporangium. However, subsequently additional sporangia are produced in basipetalous succession; the result is a cluster of three to five sporangia. Thus, development of the terminal sporangium does not end sporangiophore development. Consequently, overall development of the sporangiophore and its sporangia was considered indeterminate in the sense of both Snell and Dick (1957, p. 180) who include the phrase "without definite margin or edge", and Ainsworth (1971, p. 290) who includes "edge not well defined". These phrases more appropriately apply to margins of leaf spots as indicated by Ainsworth. I now recognize the confusion resulting from application of the term indeterminate to the sporangiphore of *Sclerophthora* and support its description as determinate.

The basic and significant point is that the sequence of sporangial development is entirely different in *Sclerophthora* and *Phytophthora*. In *Phytophthora*, sporangiophores develop indeterminately in the sense that this term is usually applied to sporangiophores and conidiophores. Furthermore, sequential sporangial formation and elongation of the sporangiophore is acropetalous, not basipetalous as in *Sclerophthora*.

These differences in sporangiophoric development, combined with the typically obligate parasitism of all species of *Sclerophthora* and the differences in oogenesis between species of *Sclerophthora* and *Phytophthora* not only justify the continued recognition of two distinct genera, but also placement of *Sclerophthora* in the Peronosporaceae, rather than the Pythiaceae (Waterhouse, 1973).

Recently Ko *et al.* (1978) have rediscovered the genus *Peronophythora* for which they erected the family Peronophythoraceae. The intermediate position of this genus is clearly indicated by its characters: saprophytic as in the Pythiaceae, sporangiophores determinate as in the Peronosporaceae and branched in a manner intermediate between typical species of *Plasmopara* and *Peronospora*. These characters have been considered in positioning this genus in Fig. 1.

*Albugo*, the only genus in the Albuginaceae, is obligately parasitic; the sporangia are catenulate on short, unbranched sporangiophores aggregated into a sorus. The asexual state is unique for the order, and must have segregated from the *Pythium–Phytophthora* line of development rather early. All other genera of obligate parasites within the Peronosporales are placed in the Peronosporaceae. Althought one may philosophize endlessly as to what constitutes obligate parasitism (Shaw and Yerkes, 1959), I do not consider growth on a very special synthetic medium of one or a few genotypes of a species to justify removing that species (or all the species of the same family or order) from the ranks of obligate parasites. We must look to the means by which nutritional requirements are satisfied in nature and on that basis, the Peronosporaceae are obligate parasites!

There are two parallel lines of development in the Peronosporaceae. One, consisting of *Schlerophthora*, *Sclerospora* and *Peronosclerospora*, is found exclusively on Gramineae; the other occurs primarily on other angiosperms, although a very few species (probably rather recently from an evolutionary standpoint) have become established on Gramineae. These species are: *Plasmopara oplismeni*, *P. penniseti*, *Basidiophora butleri*, and *Bremia graminicola* (including *B. graminicola* var. *indica*). One species, *Peronospora destructor*, occurs on *Allium* spp. (Liliaceae); otherwise the *Bremiella*, *Plasmopara*, *Pseudoperonospora*, *Peronospora* series has developed on Dicotyledoneae.

*Bremiella* (Wilson, 1914) is placed at the base of the developmental series on Dicotyledoneae because, although the sporangiophore is determinate, each ultimate branch is swollen at the apex to form an apophysis reminiscent of the swellings which subtend sporangia on the branches of the indeterminate sporangiophores of *Phytophthora*. *Plasmopara* species (including *Rhysotheca*) typically produce sporangiophores which branch sequentially (three to six times) at right angles; the ultimate branches occur in groups of two to five. In *Bremia* the branching has become dichotomous, and a cupulate swelling or apophysis ( = ganglion) subtends three to five ultimate branches. In *Plasmopara cephalophora*, branching is much reduced, while in *Basidiophora* branching has been all but eliminated, and the ultimate branches ( = sterigmata) are crowded on to the apex of the clavate sporangiophore. I have seen a single sporangiophore of *Basidiophora entospora* that was bifurcate, indicating its relationship to species such as *Plasmopara cephalophora*.

*Pseudoperonospora* produces dichotomously branched sporangiophores comparable to the conidiophores of *Peronospora* species; the asexual reproductive structures, however, are sporangia with apical opercula and these release the zoospores.

*Peronospora* represents the culmination of the series on Dicotyledoneae (*P. destructor* occurs on Liliaceae, as mentioned earlier). The asexual propagules have a uniform wall without a modified apex, and always germinate by germ tubes. Since the mechanism essential for zoospore release has been lost, these propagules and those of *Peronosclerospora* are true conidia.

Once the level of obligate parasitism has been reached, further evolution of obligate parasites typically is in conjunction with that of their hosts (Fischer and Shaw, 1953). Great leaps from one family of higher plants to another are not hypothesized. Consequently, the jump of one or two species of *Bremia*, *Basidiophora* and *Plasmopara* to the Gramineae and of *Peronospora* to Allium spp. may seem improbable. Yet it must be remembered that at this point in evolution the extant flora gives us no more than a single cross section through the evolutionary tree. In some instances downy mildews have persisted on many host species and genera within one host family (e.g. Cruciferae) and there are also downy mildews on one or more related host families (e.g. Fumariaceae and Resedaceae) (Gäumann, 1923). In other instances a single downy mildew (e.g. *Plasmopara viticola*) has persisted on one or two genera (e.g. *Psedera* and *Vitis*) within one host family (Vitaceae) no close relatives of which are now parasitized by downy mildews. This may explain the isolated occurrence of *Peronospora destructor*, for example, only on *Allium* spp. in the Liliaceae.

However, we must also remember the great plasticity of these pathogens. Maize (*Zea mays*) is native to the Western Hemisphere (Shaw, 1975); subsequent to the discovery of the New World, maize was transported and grown in many tropical and temperate countries of the world. Except for an occasional report of the incidental occurrence of *Sclerospora graminicola* on maize, no downy mildew was reported thereon in the Western Hemisphere until 1961 (Reyes *et al.*, 1964). Yet, prior to 1961 seven other downy mildews attacked corn in the Eastern Hemisphere; some of these are extremely virulent on maize and cause great losses. Since its introduction into Mexico, *Peronosclerospora sorghi* has spread as far north in the US as Indiana. The ability of different downy mildews to attack an introduced crop in several different localities around the world, and the ability of one of them, when introduced onto another continent, to spread from subtropical areas into temperate regions are indicative of adaptability. Such plasticity may also account for the occurrence of *Bremia, Basidiophora* and *Plasmopara* spp. on Gramineae.

Accepting the production of zoospores as primitive, the sequence of evolution indicated in Fig. 1 is proposed. The zoosporic genera are mostly temperate in their distribution, although some are reported from tropical and subtropical areas. *Sclerophthora* is represented by five species; the geographic distributions of three, as far as now known, are restricted to type localities in temperate regions. One species, especially its variety, *S. rayssiae*, var. *zeae*, has a somewhat wider distribution from Israel (temperate) mainly into the northern states of India (also temperate). *S. macrospora*, reported on 140 hosts (see later), is also primarily temperate although reported from some tropical and subtropical areas.

*Sclerospora graminicola* is also primarily temperate in its distribution, being circumpolar on species of *Setaria* in the Gramineae. However, it has become adapted to tropical habitats, particularly on bajra ( = pearl millet, *Pennisetum typhoides*; Safeeulla, 1976). Other genera of the Peronosporaceae, with the exception of *Peronosclerospora*, are also primarily temperate in their geographical distribution. Species of *Peronosclerospora* would seem to have originated relatively recently in tropical regions of the Far East, particularly the Malay Archipelago and the Philippines and to have spread from there. In spite of obvious adaptation to warmer and drier habitats, *Peronospora* must be judged to be temperate in its distribution on the basis of existing records. More complete cataloguing of the parasitic fungi of the tropics and subtropics might modify this concept.

Looking at evolution from another aspect, it is interesting to note that some species of *Plasmopara* occur on woody plants (e.g. *P. viticola* on *Vitis* and *Psedera* in the Vitaceae; *P. ribicola* on *Ribes* in the Saxifragaceae and *P. viburni* on *Viburnum* in the Caprifoliaceae); in contrast I am unaware of any report of a species of *Peronospora* on a woody host.

The distinction emphasized herein between sporangia and conidia is somewhat clouded by frequent reports in the literature of sporangia germinating by germ tubes. There is no question but that sporangia of *Phytophthora infestans* may germinate by a germ tube, particularly at higher temperatures (Cox and Large, 1960; Crosier, 1934). Significantly, illustrations of the process in *P. infestans* show the germ tube emerging from any point on the circumference of the sporangium. Except in *Bremia* spp., most reports of germ tube germination in sporangium-producing species are doubtful. Pupipat (1976) has provided an excellent illustration of what is undoubtedly the basis for most such reports. Zoospores of *Sclerophthora rayssiae* var. *zeae* have been observed escaping from the sporangium. Zoosporangia have been seen within which one zoospore has encysted, and the cyst has subsequently germinated. The germ tube took the path of least resistance and grew through the apical pore left by detachment of the operculum on the sporangium. The pore is larger than the diameter of the germ tube. Reports of direct germ tube germination are suspect in most sporangial Perono- sporaceae because the germ tube is consistently described or illustrated as emerging apically, not indiscriminately at any point on the circumference of the sporangium as in *Phytophthora infestans*, or of the conidium as in conidia of *Peronosclerospora philippinensis* (Weston, 1920).

Species concepts in the Peronosporaceae have been as diverse as in other groups of obligate parasites (Fischer and Shaw, 1953; Shaw, 1965). Gäumann's (1923) treatment of *Peronospora* (c 360 species) is the result of combining a narrow species concept with results of some cross inoculation studies and extensive biometric measurements of a very limited number of specimens (one or two) from each host species. The recognition of only one species occurring on the Cruciferae (*Peronospora parastica*) and one on the Chenopodiaceae (*P. farinosa*) results from a much more conservative species concept based on the examination of numerous collections on each host species (Shaw and Yerkes, 1959). They stated that,

> In the Peronosporaceae neither morphology nor physiologic specialization alone provides a satisfactory basis for species delimitation. Physiologic specialization "should" be utilized in conjunction with, but subordinate to, morphology. When downy mildews assignable to the same genus occur on hosts of one family and differ morphologically without intergradation, these entities constitute distinct species. If however, morphologic variation on a single host includes the variations found on other hosts of the same family, only one species is recognized. Or if morphologically the material from various hosts within one family forms a gliding, overlapping series, all of it is considered cospecific. This concept retains the species as the basic taxon, within which infraspecific taxa can be erected if necessary.

Structures, (conidia and conidiophores or sporangia and sporangiophores, as the case may be) produced externally on hosts are very variable structures, the dimensions of which may be greatly modified by environmental factors (Richards, 1939). The extent of these variations can only be determined by examining a large number of collections on each host, or by inoculating a given host under a variety of controlled environmental conditions. The mode of branching, shape of the ultimate branches, and the presence or absence of apophyses or ganglia on the phores show much less variation than does size of phores and sporangia or conidia. In contrast to conidia and coni- diophores, the dimensions of oogonia and oospores formed within host tissues vary much less. The exosporium of the oospores is irregularly thickened and ridged in many species; in some species the exosporium is distinctly sculptured, e.g. the reticulate exosporium of *Peronospora viciae* on *Pisum sativum*. Where present, such markings constitute the most diagnostic and immutable character found in the Peronosporaceae.

While the trend in species delimitation within the Peronosporaceae has been away from the narrow concepts of Gäumann, there is one obvious "composite" species in the family. Downy mildews collected on 140 different Gramineae are all referred to as *Sclerophthora macrospora* (Pupipat, 1975; Safeeulla, 1976). Critical examination of oospores from a number of these hosts reveals considerable morphologic variability—too much to be encompassed within a single species. As yet no one has critically compared sexual and asexual structures of the downy mildews on all these hosts; so they remain encompassed within a single binomial.

At the opposite extreme, only oospores have been reported for *Sclerospora iseilematis* and *S. northii*. Without the imperfect state, the affinities of these species cannot be ascertained. They may belong in *Sclerospora*, *Sclerophthora* or *Peronosclerospora* (or possibly even elsewhere), but, until the asexual states are discovered, there is no point in removing them from *Sclerospora*. Kenneth, by careful re-examination of the original collections of *S. farlowii* has found the anamorph thereof; that state permits the fungus to be referred to *Sclerophthora* (Kenneth, personal communication).

Thus far the anamorph has been emphasized. The sexual stage provides little elucidation of phylogenetic relations. In the Peronosporales, oospores germinate in three ways. In the Albuginaceae production of a sessile vesicle in which zoospores are differentiated is typical (Alexopoulos and Mims, 1979). In the Pythiaceae and Peronosporaceae each oospore may produce a large, operculate germ sporangium on a short unbranched sporangiophore, or a germ tube which infects directly. Germ tube germination is most frequently reported in *Pythium* and *Phytophthora* spp., although some species of *Phytophthora*, e.g. *P. infestans*, more typically produce a germ sporangium.

In *Sclerophthora macrospora* and *Plasmopara viticola* a germ sporangium is normally produced. In contrast, oospores of sporangial *Sclerospora graminicola* typically produce germ tubes upon germination (McDonough, 1937). Thus, there is no correlation between asexual reproduction by sporangia and the production of a germ sporangium from the oospore. Oospores of *Bremia* (Morgan, 1978), *Peronosclerospora* (Pratt, 1978), and *Peronospora* (McKay, 1937) germinate by germ tubes; there would seem to be a consistent correlation in these genera between conidial and oosporic germination by germ tubes.

Germination of oospores is confounded by the fact that other Oomycetes, Chytridiomycetes, Fungi Imperfecti, and bacteria may parasitize them (Kenneth *et al.*, 1975; Person *et al.*, 1955; Pratt, 1978; Sneh *et al.*, 1977). The report of the production of zoospores by oospores of *Peronospora tabacina* (Lucas and Person, 1954) was subsequently corrected when it was determined that a species of *Phlyctochytrium* had attacked the oospores

(Person *et al.*, 1955). The line drawings and cytological studies of McDonough (1937) who reported germ tube formation for the oospores of *Sclerospora graminicola* are more convincing than the photographs presented by Pande (1972). In the process of germination, the oospore wall becomes thinner (McDonough, 1937; Morgan, 1978). The endosporium completely disappears, and the wall of the germ tube fills the rupture of the remaining exosporium.

Since oospores may be attacked by hyperparasites, previous reports of oospore germination require critical re-evaluation. In that Fungi Imperfecti attacking oospores produce hyphae suggestive of germ tubes, reports of germ tube germination for oospores must also be evaluated critically.

Sculpturing of the exosporium was used by Schroeter (1886) to recognize subgenera ("Gruppen") within *Peronospora*, the exosporium being verrucose, tuberculate or reticulate in the Calothecae and smooth or irregularly thickened and angular in the Leiothecae. Fischer (1892) recognized sections ("Untergruppen") within each of these two subgenera. These infrageneric taxa are useful, and the characters upon which they are based may well be indicative of relations within *Peronospora*. More detailed comparisons between species are needed to test this hypothesis. Similarly, a critical comparison of oospore markings in *Sclerophthora macrospora* might well serve as a basis for distinguishing species within that complex.

Currently, taxonomy in the Peronosporaceae is fairly stable. The species now recognized, particularly the pathogenic species that have been extensively investigated, have stood the test of critical comparison, and the criteria of utility applied by plant pathologists. Certainly, we need to know more about all species in the family. As yet, a few species of *Sclerospora* have unknown imperfect states, and the oospores of a few species in other genera remain undescribed; the discovery of these will amplify our knowledge and taxonomic concepts. Axenic culture (Safeeulla, 1976) will assist not only plant pathologists, but taxonomists as well by making possible the comparison of morphologic structures produced under controlled environments, and by extending our knowledge of the physiology of species. The taxonomy of the group will continue to change as our total knowledge increases.

## References

Ainsworth, G. C. (1971). Dictionary of the Fungi. Ainsworth and Bisby 6th edn., p. 663. Comm. Mycol. Inst. Kew, Surrey.

Alexopoulos, C. J. and Mims, C. W. (1979). "Introductory Mycology". 3rd edn., pp. 632. John Wiley and Sons, New York.

Blackwell, E. M. and Waterhouse, G. M. (1931). *Trans. Br. Mycol. Soc.* **15**, 294–320.

Blaeser, M. and Weltzien, H. C. (1978). *Zeit. Pflanzenkrankheiten Pflanzenschutz*, **85**, 155–161.

Cox, A. E. and Large, E. C. (1960). U.S. Dept. Agric. Handbook, **174**, pp. 230. Washington, D.C.

Crosier, W. (1934). *Cornell. Univ. Agric. Exp. Mem.* **155**, pp. 40. Geneva, N.Y.

Dogma, I. J., Jr. (1975). *Trop. Agric. Res.* Tokyo, **8**, 47–55.

Fischer, A. (1892). Phycomyceten. *In* "KryptogamenFlora von Deutschland, Oesterrich und der Schweiz". L. Rabenhorst, 2nd edn., Band I, Abt. IV, pp. 505. Leipzig.

Fischer, G. W. and Shaw, C. G. (1953). *Phytopathology.* **43**, 181–188.

Gallegly, M. E. (1968). *Ann. Rev. Phytopathol.* **6**, 375–393.

Gäumann, E. (1923). *Beit. KryptogamenFlora Schweiz* **5(4)**, 1–360.

Kenneth, R., Cohen, E. and Shanor, G. (1975). *Phytoparasitica* **3**, 70. (Abstract).

Ko, W. H., Chang, H. S., Su, H. J. and Chen, C. C. (1978). *Mycologia* **70**, 380–384.

Lucas, G. B. and Person, L. H. (1954). *Pl. Dis. Reptr.* **38**, 243–244.

McDonough, E. S. (1937). *Mycologia* **29**, 151–173.

McKay, R. (1937). *Nature* **139**, 758–759.

Milbrath, D. G. (1923). *J. Agric. Res.* **23**, 989–994.

Morgan, W. M. (1978). *Trans. Br. Mycol. Soc.* **71**, 337–339.

Pande, A. (1972). *Mycologia* **64**, 426–430.

Payak, M. M., Renfro, B. L. and Sangam Lal (1970). *Indian Phytopathol.* **23**, 183–197.

Person, L. H., Lucas, G. B. and Koch, W. J. (1955). *Pl. Dis. Reptr.* **39**, 887–888.

Pratt, R. G. (1978). *Phytopathology* **68**, 1606–1613.

Pupipat, U. (1975). *Trop. Agric. Res.* Tokyo **8**, 63–80.

Pupipat, U. (1976). *Kasetsart J.* **10**, 106–110, Plate I, Fig. 9, 11.

Reyes, L. D., Rosenow, D. T., Berry, R. W. and Futrell, M. C. (1964). *Pl. Dis. Reptr.* **48**, 249–253.

Richards, M. C. (1939). *Bull. Cornell Univ. Agric. Exp. Stn.* **718**, pp. 29. Geneva, N.Y.

Safeeulla, K. M. (1976). "Biology and Control of the Downy Mildews of Pearl Millet, Sorghum and Finger Millet", p. 304. Wesley Press, Mysore, India.

Schroeter, J. (1886). Fam. Peronosporacei. *In* "KryptogamenFlora von Schlesien. Dritter Band". F. Cohn, Erste Halfte, 228–252. Breslau.

Shaw, C. G. (1965). *Phytopathology* **55**, 819–821.

Shaw, C. G. (1970). *Indian Phytopathol.* **23**, 364–370.

Shaw, C. G. (1975). *Trop. Agric. Res.* Tokyo **8**, 47–55.

Shaw, C. G. (1976). *Kasetsart J.* **10**, 85–88.

Shaw, C. G. (1978). Mycologia 70, 594–604.

Shaw, C. G. and Yerkes, W. D., Jr. (1959). *Phytopathology* **49**, 499–507.

Sneh, B., Humble, S. J. and Lockwood, J. L. (1977). *Phytopathology* **67**, 622–628.

Snell, W. A. and Dick, E. A. (1957). "A Glossary of Mycology", pp. 171. Harvard Univ. Press, Cambridge, Mass.

Thirumalachar, M. J. (1969). *Indian Phytopathol.* **22**, 155. (Abstract).

Thirumalachar, M. J. Shaw, C. G. and Narasimhan, J. J. (1953). *Bull. Torrey Bot. Club* **80**, 299–307.

Tokura, R. (1975). *Trop. Agric. Res.* Tokyo **8**, 57–60.

Waterhouse, G. M. (1973). *In* "The Fungi: an Advanced Treatise. Vol. IVB. A Taxonomic Review with Keys: Basidiomycetes and Lower Fungi" (Eds G. C. Ainsworth, F. K. Sparrow and A. S. Sussman), 165–183. Academic Press, New York and London.

Weston, W. H., Jr. (1920). *J. Agr. Res.* **19**, 97–122.

Wilson, G. W. (1907). *Bull. Torrey Bot. Club* **34**, 387–416.

Wilson, G. W. (1914). *Mycologia* **6**, 192–210.

Chapter 3

# Geographical Distribution of Downy Mildews

## H. C. WELTZIEN

*The International Center for Agricultural Research in the Dry Areas, P.O. Box 5466, Aleppo, Syria*

## I. Introduction

It is now generally understood that the geographical distribution patterns of plant diseases and their causal agents are not only determined by the geographical distribution of their host plants. Environmental factors are equally significant for the actual occurrence of plant diseases (Weltzien, 1972). As the pathogen is dependent on the host, but the host not dependent on the pathogen, the environmental range of the host distribution circumscribes the maximum potential area of distribution for the pathogen.

If the pathogen is endemic to the area, or has been transported with or to its host, it may be found sporadically wherever the host is grown, but its economic importance may or may not be significant. This significance is governed by environmental and host factors. Delineation of the following three zones of disease significance have been suggested as sufficient for most purposes (Weltzien, 1978a):

    (a) areas of main damage;
    (b) marginal damage;
    (c) sporadic attack.

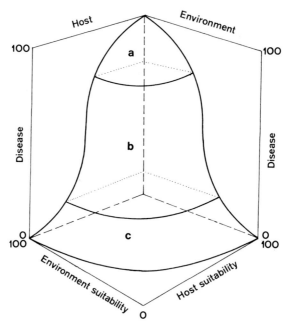

Fig. 1.   Model for disease occurrence and intensity as dependent on host and environment:
(a) area of main damage; (b) area of marginal damage; (c) area of sporadic attack.

To explain the geographical distribution patterns, we may therefore omit
discussion of the host plant's requirements, and consider the environmental
constraints for the pathogen only. The actual disease occurrence and intensity is then determined by host and environment as independent variables
with each factor governing epidemiological curves, if the other factor is considered constant. The situation can best be described by a three-dimensional
model as given in Fig. 1. In this model, each of the two factors are
actually integrations of a number of subfactors, such as plant density and
susceptibility for the host, or temperature, wetness and antagonists, for the
environment. The three different areas mentioned above can then be defined
as different proportions of the three-dimensional surface (Fig. 1).

If taxonomically related groups of pathogens do show similar reactions
to environmental factors, discussing their geographical distribution together
may be justified. This has been tried for the powdery mildews (Weltzien
1978 b), partly based on world-wide surveys and literature compilation by
Hirata (1966). A similar comprehensive study for the downy mildews does
not exist. The principle of geographic distribution for these fungi will therefore be explained on some selected host–parasite combinations representing
various patterns.

## II. Principles of Peronosporaceae Epidemics

The one factor that is common to all Peronosporaceae is that they all require wet plant surfaces for conidial infection (Royle, 1976). Dew, though adequate for infection (Bonde *et al.*, 1978), does not allow the spread of the inoculum and therefore only contributes greatly to the development of epidemics in exceptional cases. Thus rainfall, or in some cases sprinkler irrigation, can be considered to be the limiting factor for most downy mildew epidemics based on conidial infection (Palti and Rotem, 1971). If the vegetative period of the host falls largely into a dry season, the disease can usually not reach epidemic proportions.

Assuming adequate leaf wetness, the prevailing temperature during the wetness period will decide if there is infection or not. The average temperature (x) during the wetness time (y) results in a temperature sum x.y. As far as we know, the minimum temperature sum required for infection is constant (x.y = constant). This gives us a reliable parameter for disease prognosis and may also be helpful in understanding geographic distribution patterns (Blaeser and Weltzien, 1977).

Other supplementary factors often need consideration. Sporulation, for example, is often restricted to high relative humidity and darkness (Cohen *et al.*, 1971). Likewise infected transplants may result in what may be called a "pseudo-epidemic", even in the absence of infection conditions: whereas the survival of soil-borne oospores can be a major factor in determining the onset of an epidemic. The spread of downy mildew pathogens to hitherto uninvaded areas has given rise to some of the most spectacular plant disease cases documented in history.

## III. The World-wide Distribution of Downy Mildews on Maize

Though the downy mildews of dicotyledonous plants are most serious in temperate zones, the main damage area of the downy mildews on Gramineae is clearly in the tropics and sub-tropics. This can probably be explained by the significantly thermophilic reaction of oospore germination (Shurtleff, 1973). Of the eight species recently renamed by Shaw (1978), we will consider here as examples the four important maize-pathogens *Peronosclerospora sorghi, P. sacchari, P. philippinensis* and *P. maydis*. Frederiksen and Renfro (1977) surveyed the world literature and concluded that these species probably originated in Asia or Oceania. If this hypothesis is accepted, a map of distribution and spread may be drawn for these four species (Fig. 2). It

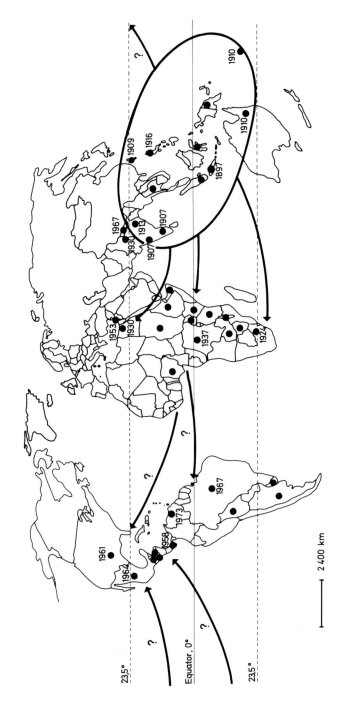

Fig. 2. Probable origin, spread and distribution of maize downy mildew.
Large circled area: Area of origin
→: Direction of spread
●: Countries with disease reported and year of first record.

shows clearly that all early records were from East Asia and India. The pathogens appear to have been endemic there on wild hosts and attacked maize after its introduction from the Americas. They then may have spread to the West and reached Africa. They arrived in the Americas between 1958 and 1961. Whether this occurred westward across the Atlantic or eastward across the Pacific is unknown.

The map also shows very clearly that these diseases are almost exclusively confined to the tropics, even though their hosts are much more widespread. The few recorded cases in sub-tropical, dry summer climates represent only scattered attack areas of limited economic significance. Why this is the case is not completely clear. Sporulation is reported to be most favoured by darkness, continuous wetness and temperatures between 10 and 25°C (Grower Sowell and Harris, 1969; Exconde *et al.*, 1968; Tarr, 1962; Shurtleff, 1973). As these various conditions can only coincide during the night, most temperate climates only occasionally provide good sporulation conditions. As conidial germination occurs over a range of temperatures, it may well be less significant than sporulation for disease occurrence. The other major sources of inoculum, soil-borne oospores, are dependent on relatively high soil temperatures for successful infection. Temperatures between 15 and 32°C are mentioned in the literature (Tarr, 1962; Shurtleff, 1973). Oospores may survive for up to 5 years but are also subject to attack by hyper-parasites (Kenneth and Shahor, 1975). Thus long-term optimum conditions for epidemics include soil temperatures up to 32°C, wet and warm nights, and a wetness period at temperature ranges up to 30°C during daytime; conditions which are common in the tropics.

## IV. Downy Mildew of Grapes, *Plasmopara viticola*

Grapevine cultivation, found in all continents, was reported to cover 10·4 million ha in 1976 (Anonymous, 1979).

The largest area was cultivated in Europe with 7·5 million ha, followed by Asia with 1·5, the Americas with 0·9, Africa with 0·5 million ha, and some smaller areas in Australia. *Plasmopara viticola* has threatened the large European vineyards ever since the pathogen was introduced into France from the USA in 1878. It has since been recorded in 83 countries (Commonwealth Mycological Institute, 1967) and from all grape growing areas.

As with downy mildew of maize, its economic significance varies widely, and a recent study on the biology and epidemiology of the pathogen has furnished much of the information needed for a better understanding of this phenomenon. (Blaeser, 1978; Blaeser and Weltzien, 1977, 1978). It was shown that *Plasmopara* epidemics are based on various environmental factors.

Sporulation occurs only in darkness, and between 17 and 27°C, if the relative humidity is at least 98%. This temperature is definitely not lower than that reported as suitable for most downy mildews of Gramineae in the tropics and does not help our understanding of the actual areas of disease occurrence. Sporangium survival, on the other hand, is favoured by low temperatures between 10 and 15°C. Temperatures above 20°C sharply limit the survival of these spores which quickly die at 30°C under all relative humidities. Hence spores, once produced, are much favoured by temperate climates.

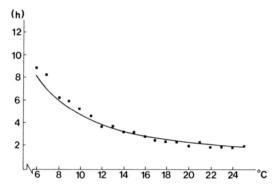

Fig. 3.    Requirements of temperature (°C) and leaf wetness duration (h) for *Plasmopara viticola* infections, if spores are applied as a suspension.

Infection conditions are based on the usual temperature–wetness relationship. Infection occurs between 6 and 25°C with wetness duration requirements between 9 and 2 h. The resulting temperature sum is statistically constant with 50 C x h. thus the hyperbole $y = 50 \cdot 1/x$ describes this relationship with (y) being the wetness duration required for the temperature (x) (Fig. 3). The relatively short wetness requirements at the temperatures frequently encountered in temperate climates, explain the dangers and the epidemic potential of this pathogen. However, sporulation is often suppressed by low temperature or lack of high relative humidity at night and this limits the pathogen's spread.

We may therefore postulate that typical *Plasmopara* localities have frequent summer rains at temperatures below 25°C and regular high humidity at night with temperatures above 15°C. It is difficult to correlate these data with the climatograms by Walter and Lieth (1966), as daily maximum and mimimum temperatures are not given. We must therefore select all those areas as main damage areas, where the rainfall curve is well above the temperature curve and the latter is not higher than about 20°C ( = monthly mean) (Fig. 4). This is definitely true for Germany and Australia,

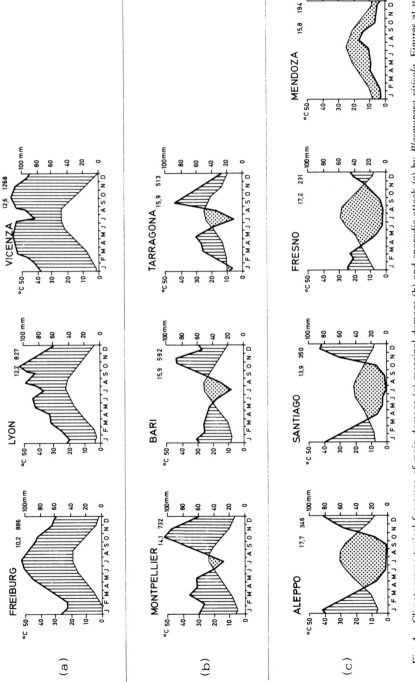

Fig. 4. Climatograms typical for areas of main damage (a) marginal damage (b) and sporadic attack (c) by *Plasmopara viticola*. Figures at upper right = several years average temperatures (°C) yearly total rainfall (mm).

with Freiburg as a typical example. In France, the Rhone Valley follows the same pattern (Lyon), and in northern Italy Vicenza is a good example (Fig. 4).

Less frequent epidemics may be expected, if summer dryness becomes more regular. Here years with epidemics may be followed by years without them. Typical diagrams for these marginal damage areas (Zone b) are presented for Montpellier in southern France and Bari in southern Italy. Spain (Tarragona) and many locations in Portugal and Yugoslavia follow the same pattern. In studying the diagrams, the time factor is also important. Early leaf development may coincide with rainy seasons, while later dryness and high daytime temperatures stop the spread of the epidemic. If much of the foliage develops in the dry season, we can expect areas of scattered attacks (Zone c). This is very typical for the southern and eastern Mediterranean, and some parts of Greece, especially Crete. Typical examples are Algeria, Syria, Central Lebanon, Israel, Cyprus and some parts of Turkey. Also, here the winter rains usually remain ineffective, though in some exceptional years limited epidemics may develop locally.

There are only a few grapevine growing areas in the world, such as Mendoza in Argentina, where relative dryness prevails throughout the year and *Plasmopara* danger is minimal (Fig. 4). Certain areas in Chile, as well as those in California and South Africa, follow the Mediterranean pattern. In Australia there is a main damage area (Maitland West) and a marginal damage area (Clare). An outline of the predictions of potential *Plasmopara* threat on a world map of grape cultivation, is given in Fig. 5. It seems to be a good illustration of the non-uniform geographical distribution of disease intensity.


## V. Blue Mould of Tobacco, *Peronospora tabacina*


The Peronosporaceae, being highly pathogenic, fast-multiplying obligate parasites, must be able to colonize suitable environments quickly after they are introduced. This is especially true if their conidia are distributed by wind as in the case of *Phytophthora infestans* and *Peronospora spp.* While the historically famous potato late-blight epidemic is only poorly documented, we are fortunate that the introduction and spread of *Peronospora tabacina* into Europe has been followed closely by competent scientific observation. Most of this information was collected and evaluated by Populer (1964).

The pathogen was first reported in south-east Australia in 1890 and slowly moved to Western Australia where it was found in 1950 (Klinkowski, 1961). In the USA, some areas of endemic infection have been described where native *Nicotiana* spp. served as hosts for *P. tabacina*. It is still unclear

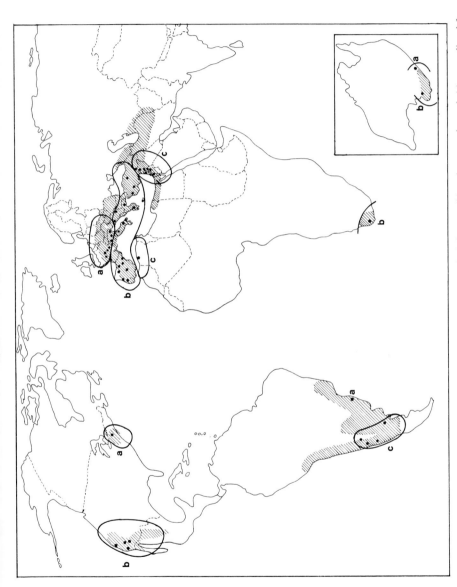

Fig. 5.   World distribution of grape cultivation and areas of main damage (a) marginal damage (b) and scattered attack (c) as predicted from climatograms for some of the areas. ●: Locations for which climatograms have been evaluated.

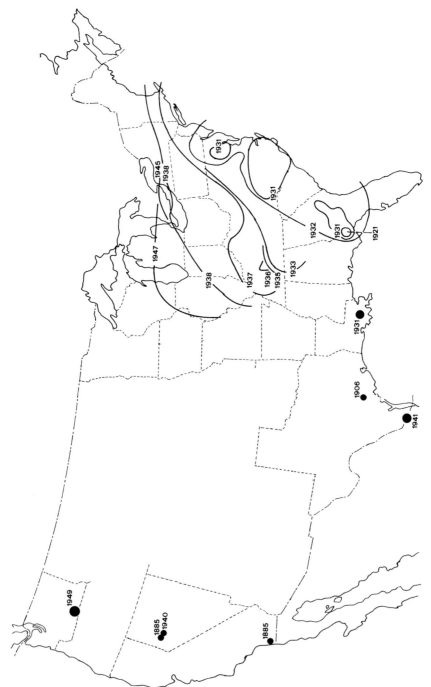

Fig. 6. *Peronospora tabacina* in North and South America. ● endemic on wild hosts with year of record ~ isolines of spread.

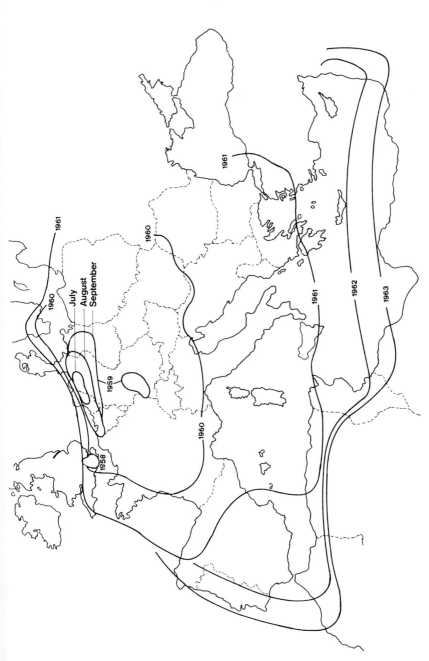

Fig. 7. *Peronospora tabacina* in Europe and the Near East ~ isolines of spread between 1958 and 1963.

whether these represent an ecotype which is different from the forms in
Australia. In any case, the pathogen did not reach the tobacco fields in the
south-eastern states before 1921, when it was first recorded on the
Florida–Georgia border (Fig. 6). It was then eradicated but reappeared in
1931, spread slowly northwards and by 1947 it had reached north-east USA.

As far as we know, *P. tabacina* entered Europe from Australia via England
in October 1958. From there it was obviously spread to the continent
through leaf material exchanged between virus research laboratories, and
became established during the summer of 1959. It only took 4 more years
to complete the spread of the epidemic between the Caspian Sea and
southern Scandinavia (Fig. 7), with distances of up to 400 km being covered
per month.

The question as to what may have finally limited the spread of the pathogen
has often been asked. Overwintering is obviously possible everywhere, either
on surviving host plants or through oospores in the soil. Humidity and
wetness can rarely limit infection in seed beds, which are either glass
covered or watered frequently. It is much more likely that high air temperature in summer is the limiting factor, as the optimum range for the
development of the pathogen is between 10 and 20°C. Thus, Populer (1964)
suggests that within an area where the average daily maximum temperature
exceeds 30°C, the disease cannot spread in the field and is confined to the
seed bed, from where it is taken to the fields by transplanting.

## VI. Conclusions

The downy mildews, though closely related to each other by morphological
and physiological characteristics, are not restricted in their distribution
pattern to any specific part of the globe. Only very dry areas can be ruled
out but otherwise they extend from the tropics to the northern and southern
temperate zones. However, each genus or species seems to have its own
typical ecological requirements. These have not yet been worked out in all
cases, though it is important that this should be done. As in the case of
*Plasmopara*, one may be able to modernize plant protection systems by
adequate prognosis systems, and the same principle may be valid for other
downy mildew diseases. Thus, in the long run, geographical distribution
research on downy mildews may prove to be a major tool in modern plant
disease management.

# References

Anonymous (1979) "Der Weinbau in der Welt 1976" Weinbaujahrbuch 209–210.

Blaeser, M. (1978) Untersuchungen zur Epidemiologie des falschen Mehltaus an Weinreben *Plasmopara viticola* (Berk & Curt. ex de Bary) Berl. et de Toni. Dissertation. Bonn, pp. 127.

Blaeser, M. and Weltzien, H. C. (1977) *Med. Fac. Landbouw Rijksuniv. Gent.* **42**, 967–976.

Blaeser, M. and Weltzien, H. C. (1978). *Z.Pfl. Krankh.* **85**, 102–107.

Bonde, M. R., Schmitt, C. G. and Dapper, R. W. (1978) *Phytopathology* **68**, 219–222.

Cohen, Y., Perl, M. and Rotem, J. (1971) *Phytopathology* **61**, 594–595.

Commonwealth Mycological Institute (1967) "Distribution Maps of Plant Diseases" Map 221, 2nd edn.

Exconde, O. R., Adversario, J. Q. and Advincula, B. A. (1968/69) *Philip. Agriculturist* **52**, 189–199.

Frederiksen, R. A. and Renfro, B. L. (1977) *Ann. Rev. Phytopathol.* **15**, 249–275.

Grower Sowell, Jr., and Harris, H. B. (1969) *Pl. Dis. Reptr.* **53**, 4.

Hirata, K. (1966) "Host Range and Geographical Distribution of Powdery Mildews." Niigata University, Japan.

Kenneth, R. G. and Shahor, G. (1975) *Trop. Agr. Res.* Series No. 8, 125–127.

Klinkowski, M. (1961) *Die Deutsche Landw.* **12**, 229–239.

Palti, J. and Rotem, J. (1971) *Z. Pfl. Krankh.* **78**, 495–501.

Populer, C. (1964) *Bull. Inst. Agron. Gembloux* **32**, 339–508.

Royle, D. J. (1976). *Ann. Appl. Biol.* **84**, 277–278.

Shaw, C. G. (1978). *Mycologia*, **70**, 594–604.

Shurtleff, M. G. (1973) *Am. Phytopath. Soc.* 64 pp.

Tarr, S. A. J. (1962) "Diseases of Sorghum, Sudangrass and Broom Corn." 380 pp.

Walter, H. and Lieth, H. (1966). "Klimadiagramm-Weltatlas" S. Fischer, Jena.

Weltzien, H. C. (1972). *Ann. Rev. Phytopath.* **10**, 277–298.

Weltzien, H. C. (1978a) *In* "Plant Disease—An Advanced Treatise". (Horsfall and Cowling, Eds) 339–358. Academic Press. New York and London.

Weltzien, H. C. (1978a) *In* "Plant Disease—An Advanced Treatise". (Horsfall and Cowling, Eds) 339–358. Academic Press. New York and London.

Chapter 4

# The Distribution of Downy Mildew genera over the families and genera of Higher Plants

J. PALTI[1] and R. KENNETH[2]

[1]*Agricultural Research Organization, Bet Dagen, Israel*
[2]*Faculty of Agriculture, Hebrew University of Jerusalem, Rehovot, Israel*

## I. Introduction

The Peronosporaceae constitute one of the largest groups of fungi parasitic on flowering plants. Many species of the family are highly destructive to important crops, which are mostly herbaceous, such as maize, tobacco, cucurbits, legumes, beets, roses, and many others, but also a few fruit crops, such as grapes.

In 1966, Hirata published a comprehensive list of powdery mildews and their hosts, and used this list as a basis for some generalizations concerning the distribution of the Erysiphaceae in their host families. No work on this line appears to have been attempted for the Peronosporaceae, and

what is being presented here is much more modest: an outline of the distribution of the genera of downy mildews on their hosts in the various classes,
orders, families, tribes and genera. This includes an attempt to trace relationships among the principal genera of downy mildews affecting dicotyledons (*Peronospora, Plasmopara, Bremia, Pseudoperonospora*) and the place
accorded to the various orders of their hosts in the evolutionary system
proposed by Cronquist (1957). This is followed by a survey of the distribution of downy mildew genera within those plant families in which five or
more host genera are affected.

The taxonomy of the flowering plants referred to in this survey follows
that found in the 1964 edition of Engler's "Syllabus der Pflanzenfamilien".
We are aware of the fact that more recent taxonomic treatises are available
for a number of major families, but think it best to follow a single, widely
available source for all the families dealt with here. The exception is the part
dealing with the evolutionary aspect of flowering plants where, for obvious
reasons, Cronquist's system has been followed.

## A. Sources

The sources from which records of Peronosporaceae and their hosts have
been gleaned are too numerous to be cited in their entirety in the list of
references. They consist in part of checklists of plant disease occurrence in
about 60 countries, and of lists of downy mildews published in various con-
texts, foremost among them those issued by Wilson (1907, 1908), Shaw (1951,
1955), Shaw and Yerkes (1951, 1952) and Yerkes and Shaw (1959) in the
United States, by Ciferri (1961) in Italy, by Savulescu and Rayss (1930, 1935)
in Romania, and by Rayss (1938, 1945) in Israel.

Important sources of relevant records have also been the collections at
the Commonwealth Mycological Institute, Kew, Great Britain, and at the
New York Botanical Gardens, USA.

A publication of great value to the study has been the momentous compilation by Stevenson (1926) of economic plant diseases not at that time
found or widely distributed in the United States.

## II. Overall Host Range of the Peronosporaceae

The overall host range of the Peronosporaceae in families and genera of the
Dicotyledoneae and Monocotyledoneae is indicated in Table 1. Out of the
total of 330 families recognized by Engler (1964), 56 families serve as hosts
for downy mildews, an overall rate of 17%. For the sake of comparison it

TABLE 1

Total number of families and genera in the classes of the Angiospermae, and number of families and genera in which hosts of Peronosporaceae have been recorded

| | Dicotyledoneae | | Monocoty-ledoneae | Total |
|---|---|---|---|---|
| | Archichla-mydeae | Sympetalae | | |
| *Families* | | | | |
| Total No. | 214 | 63 | 53 | 330 |
| No. with host records | 35 | 19 | 2 | 56 |
| % with host records | 16·4% | 30·2% | 3·8% | |
| *Genera* | | | | |
| Total No. | 5140 | 3690 | 2630 | 11460 |
| No. with host records | 328 | 165 | 68 | 561 |
| % with host records | 6·4% | 4·1% | 2·6% | |

may be noted that, according to Hirata (1966), Erysiphaceae have been recorded in about 150 families (45%).

The low number of families with downy mildew hosts in the monocotyledons is striking (3·8%), whereas powdery mildews have been found in 15% of monocotyledonous families.

The percentage of host genera recorded for downy mildews in the two subclasses of the Dicotyledoneae differs appreciably, and is even lower in the Monocotyledoneae.

## III. Distribution of Downy Mildew Hosts over the Orders and Families of Dicotyledoneae and Monocotyledoneae

Is the distribution of the genera of Peronosporaceae related to the evolutionary system of the orders and families of the Dicotyledoneae? We have examined this question by the use of the system proposed by Cronquist (1957). Figure 1 indicates the host families, if any, in each order in this system, as well as the downy mildew genera recorded in these families. This figure can be interpreted as follows, as far as the distribution of hosts of *Peronospora*, *Plasmopara* and *Pseudoperonospora* is concerned:

(1) Hosts of *Peronospora* are distributed over the entire system; at least one downy mildew host is included in every order in which downy mildews are known to occur, except the Celastrales (Fam. Vitaceae), and there are 8 orders which contain hosts of *Peronospora* only.

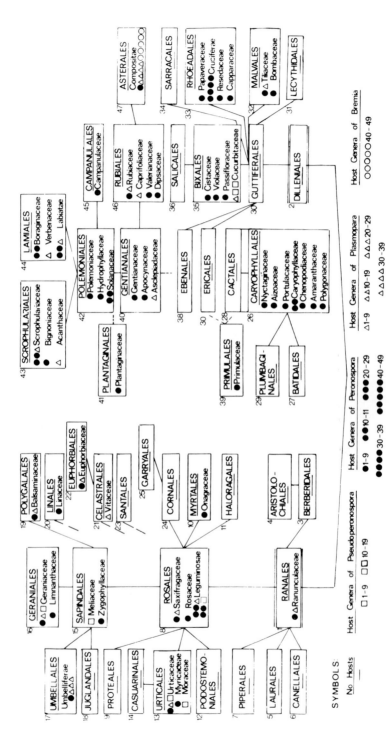

Fig. 1. The distribution of four genera of downy mildews over the orders and families of Dicotyledoneae, arranged by the evolutionary system proposed by Cronquist (1957).

*Peronospora* is markedly preponderant in the orders evolved from the Dilleniales (right half of Fig. 1), with the notable exception of the Cucurbitaceae and Compositae in their respective orders.

(2) Hosts of *Plasmopara* show two main groupings: in many of the orders related to the Rosales, and in those evolved from the Gentianales.

(3) Hosts of *Pseudoperonospora* are clearly grouped in orders derived directly from the Rosales, with the major exception of the many hosts of this genus in the Cucurbitaceae.

The distribution of hosts of the various downy mildew genera in the 56 families mentioned in Table 1, is further detailed in Tables 2 and 3, which also indicate the number of host genera affected:

(a) The preponderance of *Peronospora* is again in evidence in the Dicotyledoneae, with affected families and genera numbering 31 and 179, respectively in the Archichlamydeae, and 15 and 82, respectively, in the Sympetalae, for a total of 46 families and 253 genera. In addition, *Peronospora* affects 1 genus in 1 monocotyledonous family (Liliaceae).

(b) *Plasmopara* has a narrower distribution in 12 families and 46 genera of the Archichlamydeae, and 9 families and 45 genera of the Sympetalae, for a total of 21 families and 91 genera for all Dicotyledoneae. *Plasmopara* also affects 2 genera in the Gramineae.

(c) *Pseudoperonospora* has been recorded in dicotyledons only on hosts belonging to the Archichlamydeae, affecting 21 genera in 6 families.

(d) All other genera of Peronosporaceae are strictly limited in the number of families they attack, but may affect numerous host genera: thus *Bremia* has been recorded in 45 genera of Compositae and 2 of Gramineae, and in the Gramineae *Sclerophthora* affects 63, *Sclerospora* 8, and *Peronosclerospora* 13 host genera. *Basidiophora* is on record in hosts of the Compositae (5 genera) and Gramineae (1 genus). *Bremiella* affects only 1 genus in each of the Violaceae and the Umbelliferae.

## IV. Distribution of Downy Mildew Hosts Within Their Families

A study has been made of the distribution of hosts of various downy mildew genera within the host families in the order given by Engler (1964). The purpose was to determine:

(a) where a single genus of downy mildew attacks a given family; whether attack was or was not limited to certain taxa within the family.

(b) where two or more genera of downy mildew attack a single family, whether these genera differed in their distribution on the hosts within that family.

TABLE 2

Distribution of Peronosporaceae over families of Dicotyledoneae

| Host family | Number of host genera affected | | | | |
| --- | --- | --- | --- | --- | --- |
| | *Perono-spora* | *Plasmo-para* | *Pseudo-peronospora* | *Bremia* | *Other* |
| Acanthaceae (S)[a] | | 2 | | | |
| Aizoaceae (A) | 1 | | | | |
| Amaranthaceae (A) | 2 | | | | |
| Apocynaceae (S) | 1 | | | | |
| Asclepiadaceae (S) | | 1 | | | |
| Balsaminaceae (A) | 1 | 1 | | | |
| Bignoniaceae (S) | 1 | | | | |
| Bombaceae (A) | 1 | | | | |
| Boraginaceae (S) | 14 | | | | |
| Campanulaceae (S) | 2 | | | | |
| Capparaceae (A) | 2 | | | | |
| Caprifoliaceae (S) | | 1 | | | |
| Caryophyllaceae (A) | 16 | | | | |
| Chenopodiaceae (A) | 7 | | | | |
| Cistaceae (A) | 1 | | | | |
| Compositae (S) | 4 | 33 | | 39 | *Basidio-phora* 5 |
| Cruciferae (A) | 46 | | | | |
| Cucurbitaceae (A) | | 2 | 14 | | |
| Dipsaceae (S) | 5 | | | | |
| Euphorbiaceae (A) | 1 | 1 | | | |
| Gentianaceae (A) | 3 | | | | |
| Geraniaceae (A) | 1 | 1 | 1 | | |
| Hydrophyllaceae (S) | 3 | 1 | | | |
| Labiatae (S) | 11 | 1 | | | |
| Leguminosae (A) | 32 | 1 | 1 | | |
| Limnanthaceae (A) | 1 | | | | |
| Linaceae (A) | 1 | | | | |
| Meliaceae (A) | | | 1 | | |
| Moraceae (A) | | | 2 | | |
| Myricaceae (A) | 1 | | | | |
| Nyctaginaceae (A) | 4 | | | | |
| Onagraceae (A) | 4 | 1 | | | |
| Papaveraceae (A) | 10 | | | | |
| Passifloraceae (A) | 1 | | | | |
| Plantaginaceae (S) | 1 | | | | |
| Polemoniaceae (S) | 5 | | | | |
| Polygonaceae (A) | 5 | | | | |
| Portulaceae (A) | 3 | | | | |
| Primulaceae (S) | 3 | | | | |

TABLE 2 (Continued)

| Host family | Perono-spora | Plasmo-para | Pseudo-peronospora | Bremia | Other |
|---|---|---|---|---|---|
| Ranunculaceae (A) | 9 | 9 | | | |
| Resedaceae (A) | 1 | | | | |
| Rosaceae (A) | 9 | | | | |
| Rubiaceae (S) | 6 | 1 | | | |
| Saxifragaceae (A) | 2 | 1 | | | |
| Scrophulariaceae (S) | 13 | 4 | | | |
| Solanaceae (S) | 11 | | | | |
| Tiliaceae (A) | 1 | 2 | | | |
| Umbelliferae (A) | 1 | 21 | | | Bremiella 1 |
| Urticaceae (A) | 2 | 1 | 2 | | |
| Valerianaceae (S) | 2 | | | | |
| Verbenaceae (S) | | 1 | | | |
| Violaceae (A) | 1 | | | | Bremiella 1 |
| Vitaceae (A) | | 5 | | | |
| Zygophyllaceae (A) | 1 | | | | |

[a] A = Archichlamydeae, S = Sympetalae.

TABLE 3

Distribution of Peronosporaceae over families of Monocotyledoneae

| Family | Number of genera affected by | | | |
|---|---|---|---|---|
| | Sclerophthora | Sclerospora | Peronosclerospora | Others |
| Gramineae | 63 | 8 | 13 | Basidiophora 1 Bremia 1 Plasmopara 2 |
| Liliaceae | | | | Peronospora 1 |

## A. Host families attacked chiefly by downy mildews of a single genus

Polygonaceae (40 genera): the 5 genera affected by *Peronospora* are distributed over 2 of the 3 subfamilies, the Eriogonoideae and the Polygonoideae.

Chenopodiaceae (100 genera): the 7 genera affected by *Peronospora* are limited to 4 of the 8 tribes of this family, *viz.* the Chenopodieae, Beteae, Camphorosmeae, and Suaedeae.

Papaveraceae (47 genera): the 10 genera affected by *Peronospora* are distributed over all 3 subfamilies.

Saxifragaceae (80 genera): of the 17 subfamilies, 1 (Saxifragoideae) has 2 genera affected by *Peronospora*, another (Ribesionideae) has 1 genus affected by *Plasmopara*, while there are no hosts in the remaining 15 subfamilies.

Caryophyllaceae (80 genera); the 16 genera affected by *Peronospora* are distributed over all 3 subfamilies and over 7 of their 9 tribes.

Cruciferae (*ca.* 350 genera): the 46 genera affected by *Peronospora* are distributed over 8 of the 15 tribes (no pattern recognizable).

Rosaceae (*ca.* 100 genera): the 9 genera affected by *Peronospora* are concentrated in 5 of the 8 tribes of the subfamily Rosoideae, while the subfamilies, Spiraeoideae, Maloideae and Prunoideae are not affected.

Leguminosae (more than 600 genera): the 32 genera affected by *Peronospora* belong to one (the last) of the three subfamilies, the Faboideae. The only genera of Leguminosae in which downy mildews other than species of *Peronospora* have been recorded both belong to the Caesalpinioideae, a subfamily which comprises no hosts of *Peronospora*. In this subfamily *Cassia* has twice been recorded as the host of an unidentified species of *Pseudoperonospora* (Orieux and Felix, 1968; unpublished record in CMI Herbarium, Kew, UK) and *Plasmopara cercidis* C. G. Shaw, has been recorded on *Cercis canadensis* (Shaw, 1951). No downy mildew has been recorded in the subfamily Mimosoideae.

Cucurbitaceae (*ca.* 100 genera): the 14 genera attacked by *Pseudoperonospora* are distributed over 4 of the 5 tribes, with no hosts in the Fevilleae.

Polemoniaceae (18 genera): the 4 genera attacked by *Peronospora* belong to the last 2 of the 5 tribes, the Polemonieae and the Gilieae.

Boraginaceae (100 genera): the 14 genera attacked by *Peronospora* belong to the last (and largest) of the 4 subfamilies, *viz.* the Boraginoideae.

Labiatae (200 genera): the 11 genera affected by *Peronospora* belong to 3 of the 9 subfamilies. Eight of these genera are in the subfamily Stachyoideae.

Solanaceae (85 genera): the 11 genera affected by *Peronospora* are distributed over all 5 tribes.

Umbelliferae (300 genera): the 21 genera affected by *Plasmopara* are distributed over the largest (260 genera) subfamily, the Apioideae.

**B. Host families attacked by two or more genera of downy mildews**

Ranunculaceae (*ca.* 50 genera): the 9 genera which are attacked either by

*Peronospora* or *Plasmopara*, or both, are distributed over the same 2 out of 3 subfamilies, the Helleboroideae and Ranunculoideae and over all 4 of their tribes.

Scrophulariaceae (*ca.* 200 genera): the 13 genera attacked by *Peronospora* are distributed in the 2 largest of the 3 subfamilies, the Scrophularioideae and the Rhinanthoideae, in 9 out of 15 tribes (showing no pattern). The 4 genera attacked by *Plasmopara*, however, are all concentrated in a single tribe (Rhinantheae) of 1 subfamily, in which there is also 1 host genus of *Peronospora*.

Onagraceae (*ca.* 21 genera): of the 9 tribes, 7 have no genera affected, but 1 genus in the Epilobieae is affected by *Plasmopara*, and 3 genera in the Oenothereae are affected by *Peronospora*.

Compositae (*ca.* 920 genera): the distribution of 4 downy mildew genera attacking hosts in this family appears in Table 4. The data presented there support the following conclusions:

(1) *Plasmopara* is preponderant in the first 5 tribes of the Asteroideae, then diminishes in frequency and is found on only one genus of the Cichorioideae.

TABLE 4

Genera of downy mildews recorded on Compositae, arranged by subfamilies and tribes

| Subfamily and tribe | Number of genera affected by species of | | | |
| --- | --- | --- | --- | --- |
| | *Plasmopara* | *Bremia* | *Basidiophora* | *Peronospora* |
| Asteroideae | | | | |
| Vernonieae | 1 | | | |
| Eupatorieae | 2 | | | |
| Astereae | 4 | 2 | 3 | |
| Inuleae | 1 | 1 | | |
| Heliantheae | 15 | 5 | 2 | |
| Helenieae | 1 | 1 | | |
| Anthemideae | 1 | | | 4 |
| Senecioneae | 4 | 4 | | |
| Calenduleae | 1 | 1 | | |
| Arctoteae | 1 | 1 | | |
| Cardueae | 1 | 9 | | |
| Mutisieae | | | | |
| Cichorioideae | | | | |
| Cichorieae | 1 | 5 | | |
| | 33 | 39 | 5 | 4 |

TABLE 5

Number of genera of Gramineae attacked by various downy mildews (arrangement of assemblages and tribes according to Prat, 1960)

| Assemblages and tribes | Sclerophthora | Sclerospora | Perono-sclerospora | Others |
|---|---|---|---|---|
| Oryzoids | | | | |
| Oryzeae | 3 | | | |
| Festucoids | | | | |
| Festuceae | 7 | | | |
| Triticeae | 5 | | | |
| Aveneae | 5 | (1)? | 1[a] | |
| Phalarideae | 3 | | | |
| Chloridoids | | | | |
| Eragrosteae | 7 | | | Basidiophora 1 |
| Chlorideae | 4 | | | |
| Arundinoids–Danthonoids | | | | |
| Arundinelleae | 1 | | | |
| Arundineae | 1 | | | |
| Panicoids | | | | |
| Paniceae | 11 | 4 | (1)? | Plasmopara 2 |
| Andropogoneae | 15 | 2[a] | 9[b] | Bremia 1 |
| Maydeae | 1 | 2 | 3 | |
| Total | 63 | 9 | 13 + (1)? | |

[a] By inoculation.
[b] Two of these genera by inoculation.

(2) *Bremia*, on the other hand, has fewer host genera in the first tribes of the Asteroideae, is prominent in one of its last tribes (Cardueae), and is especially prevalent in the Cichorioideae.

(3) *Basidiophora* is limited to 2 tribes of the Asteroideae, and *Peronospora* to 1 of its tribes.

Gramineae (*ca.* 700 genera): the distribution of the genera of downy mildews affecting hosts in the Gramineae over the assemblages and tribes of this family is indicated in Table 5. This shows that the hosts of the highly destructive *Peronosclerospora* mildews are concentrated in the Andropogoneae and Maydeae. It is important to point out that the *Sclerospora* mentioned in the table is the single species *S. graminicola*, which causes particularly heavy losses in 3 genera of the Paniceae, (*Pennisetum*, *Setaria*, *Panicum*), but is rarely encountered in the Maydeae.

In contrast to the restricted host distribution of the other fungal genera, species of *Sclerophthora* have been found to affect all the 12 tribes that are hosts of downy mildews in the Gramineae.

The distribution of downy mildew genera shown in Table 5, does not include a small number of fungal species named by various authors on the basis of descriptions of oospores alone, which is not sufficiently reliable.

## V. Relevance of Distribution of Host/Fungus Combinations to the Taxonomic Study of Higher Plants and Fungi

One of the aims of this chapter has been to determine whether the host distribution of downy mildews can be evaluated for taxonomic purposes. Studies carried out in depth on other fungus/host patterns have yielded profitable information regarding taxonomies of both fungus and plant host. The classical work of Savile on rusts of the genus *Allium* (1962) and of the Saxifragaceae (1961), and that of Watson (1972) on smuts of Gramineae have pointed out relationships between parasite and host, sometimes bringing into question hitherto accepted taxonomic classifications of some plant genera and assemblages and allowing a degree of predictiveness as to whether a particular fungal group could or could not be found on a particular plant group. Watson (1972) stated that "certain smut records for grasses are taxonomically so peculiar as to suggest the need for critical reconsideration".

For the downy mildews, we can see that they are restricted to dicotyledons, with the exception of the Gramineae and the genus *Allium*; that the genus *Peronospora* is found scattered throughout the orders of higher plants, whereas *Pseudoperonospora* affects only hosts in the Archichlamydeae. It is also clear that many genera are hosts of two or more downy mildew genera. When set alongside taxonomic views of Compositae, Leguminosae and Gramineae, records for downy mildews reveal the following correlations: (1) Compositae: the concentrations of *Bremia* in hosts of the Cichorioideae (Liguliflorae) of Old World origin, makes it appear probable that any further downy mildews found in this subfamily of 65 genera will belong to *Bremia* rather than to *Plasmopara*. (2) Leguminosae: Any new host genera affected by *Peronospora* are more likely to belong to the subfamily Faboideae than to the Mimosoideae (in which no downy mildews have so far been found) or to Caesalpinioideae, in which, so far, isolated cases of *Plasmopara* and *Pseudoperonospora* have been recorded, but no *Peronospora*. (3) Gramineae: species of *Peronosclerospora* are essentially limited to the tribes Andropogoneae and Maydeae (Panicoids), whereas the genus *Sclerospora* is restricted to Paniceae and Maydeae, evidence of a tropical or sub-tropical origin of these fungi. *Sclerophthora*, however, is widely distributed over almost all grass assemblages. Downy mildew records, therefore, attest, as do smut records (Watson, 1972), to the taxonomic usefulness of Festucoid and Panicoid assemblages; the downy mildews on Chloridoid grasses, however, have been found not to belong to *Sclerospora* or *Peronosclerospora* (Kenneth, 1979); such grasses are not essentially Panicoid, which is in contrast to the findings relating to smut patterns (Watson, 1972).

## Acknowledgements

The authors gratefully acknowledge the permission given by Mr A. Johnston, Director of the Commonwealth Mycological Institute, Kew, England, and by Dr C. T. Rogerson, Senior Curator of Cryptogamic Botany, New York Botanical Gardens, to use the herbaria and libraries at their institutes.

Thanks are also due to Mrs E. Segal for technical assistance in the preparation of this review.

## References

Ciferri, R. (1961). *Riv. Patol Veg. Padova* Ser. III, **1**, 333–348.
Cronquist, A. (1957). *Bull. Jardin Bot. de l'Etat à Bruxelles*, **27**, 13–40.
Engler, A. (1964). *Syllabus der Pflanzenfamilien II.* Band. Gebr. Bornträger, Berlin-Nikolassee.
Hirata, K. (1966). "Host range and geographical distribution of powdery mildews." Niigata University, Niigata, Japan.
Kenneth, R. (1979). *Phytoparasitica* **7**, 50 (Abstr.).
Orieux, L and Felix, S. (1968) Phytopath. Papers 7 *Comm. Mycol. Inst.* Kew, UK. pp. 48.
Prat, H. (1960). *Bull. Soc. Bot. France* **107**, 32–79.
Rayss, T. (1938). *Palestine J. Bot.*, Jerusalem Ser., **1**, 143–160.
Rayss, T. (1945). *Palestine J. Bot.*, Jerusalem Ser., **3**, 151–166.
Savile, D. B. O. (1961). *In*: "Recent advances in Botany", Univ. of Toronto Press, Toronto, Canada.
Savile, D. B. O. (1962). *Nature*, London, **196**, 792.
Savulescu, T. and Rayss, T. (1930). *Ann. Mycol.* **28**, 297–320.
Savulescu, T. and Rayss, T. (1935). *Ann. Mycol.* **33**, 1–21.
Shaw, C. G. (1951). *Mycologia*, **43**, 445–455.
Shaw, C. G. (1955). *Northwest Science* **29**, 76–83.
Shaw, C. G. and Yerkes, W. D. (1951). *Northwest Science* **25**, 75–82.
Shaw, C. G. and Yerkes, W. D. (1952). *Northwest Science* **26**, 19–21.
Stevenson, J. A. (1926). "Foreign plant diseases." USDA, Washington, 198.
Watson, L. (1972). *The Quarterly Review of Biology* **47**, 46–62.
Wilson, G. W. (1907). *Bull. Torrey Bot. Club* **34**, 387–416.
Wilson, G. W. (1908). *Bull. Torrey Bot. Club* **35**, 543–554.
Yerkes, W. D. and Shaw, C. G. (1959). *Phytopathology* **49**, 499–507.

Chapter 5

# Epidemiology of Downy Mildews

## C. POPULER

*Station de Phytopathologie,*
*Centre de Recherches Agronomiques de l'Etat, Gembloux, Belgium*

## I. Introduction

In epidemics of downy mildews, the pathogen population classically starts from a low level—the initial inoculum—and increases through successive cycles during the growing season of the host. Thus, downy mildews fall into the group described by Van der Planck (1963) as compound interest diseases.

During the first sixty years of the present century, research on the epidemiology of downy mildews was mostly concerned with the downy mildew of vine, and additionally of hop, with the aim of establishing forecasting systems. In the last two decades, there has been a steady increase of papers on the epidemiology of other mildews, with less preoccupation for forecasting and with a background of more sophisticated equipment such as growth-rooms and spore traps.

Seasonal increase of the pathogen population has been investigated much more thoroughly than initial inoculum and it constitutes the main subject of this chapter. The asexual cycle on which this seasonal increase is based, is divided hereunder into a multiplication phase and a dissemination phase. The multiplication phase is the sequence of events in the life of the pathogen on its host: infection, colonization, sporulation. The phase begins with one spore and ends with many spores. The dissemination phase concerns the transfer of these spores to new multiplication sites.

## II. The Multiplication Phase

### A. Spore viability and infection

#### 1. Spore viability and in vitro germination

The asexual spores of the downy mildews germinate either by germ tubes or by zoospores. Germ tube germination is the rule in the genera *Bremia* and *Peronospora* and in all the *Sclerosporas* excepting one species; in these cases, the spores are truly conidia. Milbraith (1923) reported zoospore production in *Bremia lactucae* at low temperatures but this has not been observed by more recent authors (Sargent and Payne, 1974). *Sclerospora graminicola* produces zoospores and so do the genera *Plasmopara and Pseudoperonospora*. Direct germination of sporangia by germ tubes occurs very occasionally in *Sclerospora graminicola* (Weston, 1924) and *Plasmopara halstedii* (Goossen and Sackston, 1968).

(a) Spore viability

Viability is usually expressed as the percentage of spores which are able to germinate or to infect at a given time. Published data often depict the decay of spore viability with time (Fig. 1). In other cases, this decay is more simply expressed as longevity, i.e. the maximum time at which a last few spores are still viable.

Under the conditions prevailing in the crop during the growing season or in approximately similar laboratory conditions, the conidia and sporangia of downy mildews stay viable for only a short time. The most extreme fragility in this group is displayed by the thin-walled, hyaline conidia of the *Sclerosporas*, which lose their viability a few hours after being produced (Dogma, 1975). In other downy mildews, viability was reported to extend over a few days or weeks. In *Peronospora viciae*, Pegg and Mence (1970) found some conidia were still alive after 2 weeks on the surface of intact pea leaves in the glasshouse and after 3 weeks on sporulating plants in a growth-room. Sporangia of *Plasmopara viticola* could survive for over a week at moderate temperatures, and attached sporangia survived longer than detached ones (Arens, 1929a; Schad, 1936; Grünzel, 1963; Blaeser and Weltzien, 1978, 1979). At temperatures around 20°C, attached or detached conidia of *Peronospora tabacina* (Kröber, 1965; Hill, 1969a, b) and attached sporangia of *Plasmopara halstedii* (Goossen and Sackston, 1968) also survived for 1 or 2

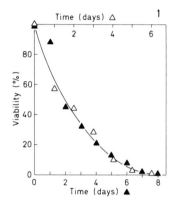

Fig. 1.   Viability of attached conidia of *Peronospora tabacina* on harvested tobacco leaves stored in an open shed, in Germany. (△) Summer; (▲) winter. Adapted from data by Kröber (1965).

weeks, while detached conidia of *Bremia lactucae* stayed viable for 1 or 2 weeks at 21°C and for more than 50 days at 2–10°C, a common temperature in winter and early spring lettuce crops in Holland (Verhoeff, 1960). The decrease in viability may be important during the first hours after sporulation or dispersal, as illustrated by Cohen and Rotem (1971a) for *Pseudoperonospora cubensis*, by Sonoda and Ogawa (1972) for *P. humuli* and by Fried and Stuteville (1977) for *Peronospora trifoliorum*.

The viability of conidia of *Bremia lactucae* (Verhoeff, 1960) and *Peronospora tabacina* (Hill, 1969a; Shepherd *et al.*, 1971) and of sporangia of *Plasmopara viticola* (Grünzel, 1963; Blaeser and Weltzien, 1978, 1979) was reported to decrease with increasing temperature, while Cohen and Rotem (1971a) found there was an optimum temperature for sporangium viability in *Pseudoperonospora cubensis*, which varied with air humidity. Shepherd (1962) and Shepherd *et al.* (1971) described a continuous decrease of conidium viability in *Peronospora tabacina* with increasing UV irradiation.

The role of humidity on spore viability is still a perplexing matter. Low humidities prolonged the viability of *Pseudoperonospora cubensis* sporangia trapped on plants (Cohen and Rotem, 1971a) and of detached conidia of *Peronospora tabacina* (Hill, 1962; Shepherd *et al.*, 1971), while high humidities were necessary for maintaining the viability of detached sporangia of *Pseudoperonospora humuli* (Sonoda and Ogawa, 1972). Even more confusing are conflicting reports that viability is increased by either high or low humidities for *Bremia lactucae* conidia (Verhoeff, 1960; versus Angel and Hill, 1931) and for *Plasmopara viticola* sporangia (Müller and Sleumer, 1934; Schad, 1936; Blaeser and Weltzien, 1978, 1979; versus Grünzel, 1963). Most of the data result from painstakingly detailed laboratory experiments and there is no apparent explanation for the discrepancies.

Conidia of *Peronospora tabacina* and sporangia of *Plasmopara halstedii* and *Pseudoperonospora cubensis* lose much of their capacity to germinate when sprayed on to leaves and allowed to dry immediately (Cohen and Rotem, 1971a; Cohen *et al.*, 1974).

Spores with a low viability not only germinate in smaller numbers but also more slowly, as demonstrated by Sonoda and Ogawa (1972) with *Pseudoperonospora humuli*. This implies that viability tests based on germination percentages should allow sufficient time for the least viable spores to germinate.

### (b) In vitro germination

Current practice with conidia of downy mildews is to germinate them on the surface of agar. Conidia have even been incorporated into melted agar at 40°C to be germinated at different temperatures after solidification of the medium, without apparent damage (Cruickshank, 1961a). Germination of conidia or sporangia in water is usually also successful. An exception is *Peronospora tabacina*, the conidia of which did not germinate and even lost their viability in a few hours when sown in water, the loss of viability being quickest at temperatures otherwise optimal for germination (Hill, 1969a). When sterile air was bubbled through a liquid medium, however, germination was as high as on agar (Shepherd, 1962).

A common procedure for collecting downy mildew spores for germination trials or for inoculation is to shake sporulating leaves in distilled water. In the case of *Peronospora tabacina*, conidial suspensions obtained by this method were found to contain substances washed from the leaf surface, which partly inhibited germination. Washing the conidia by alternate centrifuging and resuspending in water eliminated the inhibitor but also lowered the riboflavin content of the conidia so that they needed additional riboflavin in order to reach the high germination level of dry-collected conidia (Shepherd, 1962; Shepherd and Mandryk, 1962, 1963; Shepherd and Tosic, 1966). However, Kröber (1967) and Hill (1969a, b) obtained germination levels close to 100% with plain suspensions of unwashed conidia of the same fungus. Cruickshank (1961a) even reported that germination of *P. tabacina* conidia was stimulated by exudates from healthy tobacco leaves, though Shepherd (1962) found no such effect. Thus, the influence of substances from either healthy or sporulating tobacco leaves on the germinability of conidial suspensions of *P. tabacina* is still far from clear. The presence of an inhibitor in conidial suspensions was also observed in *Peronospora manshurica* (Pederson, 1961) and *Bremia lactucae* (Sargent and Payne, 1974) but not in *Peronospora viciae* (Pegg and Mence, 1970).

A gentle method for collecting the delicate conidia of the *Sclerosporas* is to spray sporulating leaves with a fine mist of water and collect the run-off (Schmitt and Freytag, 1974; Bonde *et al.*, 1978). An alternative technique is simply to brush them into water (Cohen and Sherman, 1977).

The time needed for downy mildew germination is variable. Conidia of *Sclerospora* species are discharged and germinate almost immediately after sporogenesis has been completed during the night, and penetrate the host tissues at the latest shortly after dawn (Dogma, 1975). They may even germinate while still attached to the conidiophore (Safeulla and Thirumalachar, 1955; Schmitt and Freytag, 1974). Leu and Tan (1970) reported that conidia of *S. sacchari* began to germinate 10 min after deposition on agar. In germination experiments with *S. sacchari* and *S. sorghi*, a proportion of the conidia were already germinated when the trials started and the terminal germination percentages were reached within 2 h in the range of optimal temperatures (Schmitt and Freytag, 1974; Bonde *et al.*, 1978; Bonde and Melching, 1979). In *Peronospora tabacina* and *P. viciae*, the same stage was reached in about twice as much time (Cruickshank, 1961a; Pegg and Mence, 1970). In *Sclerospora graminicola*, 95% germination was obtained within 30 min when sporangia were sown in water at 24–25°C (Safeulla and Shaw, 1963).

Light was reported to induce higher germination percentages in *Peronospora manshurica* and *P. trifoliorum* when applied before (Penderson, 1964) or during germination (Fried and Stuteville, 1977). With *Peronospora tabacina*, *P. viciae* and *Sclerospora sacchari*, there was no difference between germination percentages in light or darkness (Shepherd, 1962; Pegg and Mence, 1970; Leu and Tan, 1970).

Temperature requirements for germination *in vitro*, like infection, latent period and sporulation, reflect a general adaptation to the climatic area of the host plant. Pegg and Mence (1970), Verhoeff (1960) and Sargent and Payne (1974) found that *Peronospora viciae* and *Bremia lactucae*, the downy mildews of pea and lettuce, both crops of temperate climates, germinated within approximately the same range of 0–25°C, the optimum range for *P. viciae* being between 4 and 8°C. Isolates of Australian *Peronospora tabacina*, tobacco blue mould, had an optimum range of 8–27°C (Cruickshank, 1961a; Hill, 1969a). *Sclerospora sorghi*, a downy mildew of

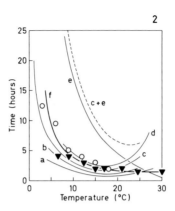

Fig. 2. Germination and infection in *Pseudoperonospora humuli*. (a–d) *In vitro* duration for c. 50% sporangium germination; (a) fresh sporangia, (b) 3 days old, (c) age unspecified, (d) 8–10 days old. (e) *In vitro* duration of zoospore motility. (f) Minimum duration of free moisture necessary to infect young hop leaves. Curves (a-e) from Zattler (1931). (▼) Data from Royle (1970) and (○) from Sonoda and Ogawa (1972) used for constructing curve (f).

sorghum and maize, and isolates from Taiwan of *S. sacchari* from sugarcane and maize, had an optimum range of 10–30°C (Weston and Uppal, 1932) and 12–32°C, respectively (Bonde and Melching, 1979). Moreover, temperature ranges for germination of a given downy mildew seem to vary with the geographical origin of isolates (Bonde *et al.*, 1978).

In such experiments, the percentage germination may be assessed at the end of the experiment only, or several times during the experiment. With the latter procedure, the optimum temperature range is usually narrow and high at first, while later in the experiment it broadens and drifts towards lower temperatures (Fig. 9a). In fact, the optimum range in the early stages of the experiment concerns the rate at which germination is initiated while the optimum range found at the end of the experiment is related to the terminal germination level (Fig. 9b). Most experimenters who assess germination on a single occasion have chosen consciously or instinctively a duration of time long enough to reach the terminal germination level at all temperatures. The difference between the two optimum ranges arises from the fact that the highest initial germination rate does not result in the highest terminal germination level.

In some germination experiments, germ tube lengths are measured concurrently with germination percentages. In the multi-stage experiment conducted by Cruickshank (1961a) with conidia of *Peronospora tabacina*, it is apparent that the optimum temperature range is initially the same for germination percentages and germ tube lengths, that it broadens more slowly for germ tube elongation but that the two optimum ranges are finally very similar. If the experiment had been conducted at an intermediate time only, the optimum temperature range would have seemed narrower and higher for germ tube lengths than for germination percentages. In view of this, it is doubtful whether there are any material differences in temperature requirements for germination and for germ tube elongation, as implied by some authors.

In the case of sporangium germination, the *in vitro* effect of temperature on the duration of zoospore motility is sometimes measured. This duration is shown in Fig. 2 to be about 5 h for zoospores of *Pseudoperonospora humuli* at 18°C. By contrast, Royle and Thomas (1973) observed that a large number of zoospores of the same fungus had already settled on stomata of hop leaves 4 min after their deposition and that settling was practically completed in the population of zoospores after 32 min at the same temperature of 18°C. Figure 2 shows that the duration of the whole infection process with *P. humuli* (curve f), which includes sporangium germination and zoospore encystment, germination and penetration, is about five times shorter than the total duration for *in vitro* germination and zoospore motility (curves c and e) and about equal to or even shorter than the sole duration of *in vitro* germination.

The underlying assumption in many *in vitro* germination studies is that the difference from germination on the host cannot be great. This seems to be approximately true as far as the temperature range is concerned, but it is far from certain for the time course of germination, which may be expected to be much accelerated by host stimulation. *In vitro* germination experiments are still less acceptable as a substitute for infection experiments, as infection incorporates not only *in vivo* germination, germ-tube growth and appressorium formation, but also penetration. Optimizing viability tests, characterizing fungal ecotypes and testing inhibitors are safer uses for *in vitro* germination studies than applying them to the epidemiological interpretation of field events.

Spore age influences the time course of *in vitro* germination (Fig. 2). It is, however, not always clear whether downy mildew spores used in short term viability tests or in germination experiments are a first crop produced by newly sporulating lesions or a blend of several successive spore crops on older lesions. According to Shepherd (1962), the germination of successive crops of *Peronospora tabacina* conidia would decrease markedly.

## 2. *Effect of environmental and host conditions on infection*

### (a) Light

Germ tubes of the Peronosporales, whether conidial or zoosporic, penetrate the host tissues between the epidermal cells, or directly into an epidermal cell, or through a stoma. Some downy mildews enter by only one method while other species are less specific.

Royle and Thomas (1971) demonstrated that the majority of zoospores liberated by sporangia of *Pseudoperonospora humuli* on the abaxial side of hop leaves in the light settled and encysted singly on stomata which were then penetrated by germ tubes. In darkness, zoospores settled more slowly and mostly between stomata and the number of stomata which were penetrated was much lower than in the light. This resulted in less infection in darkness. In *Plasmopara viticola* on vine leaves, one to eight zoospores settled on and penetrated each stoma. The difference between numbers of stomata penetrated in light and in darkness was less marked than in *Pseudoperonospora humuli* on hop leaves and the difference in infection was insignificant. Sonoda and Ogawa (1972), however, also obtained infections with *P. humuli* through the uninjured adaxial surface of hop leaves, where there are no stomata. In *Bremia lactucae*, a downy mildew of the conidial type, continuous light inhibited penetration of the fungus when applied during the first 24 h after inoculation but did not affect the development of the pathogen if applied thereafter (Raffray and Sequeira, 1971).

(b) Free moisture and temperature

Free moisture on the host surface is essential for the establishment of infection by the downy mildews. It is often designated as leaf-wetness, an appropriate expression when infection concerns leaves, and not stems, inflorescences or other plant parts.

(*i*) *Growth-room experiments*. The minimum duration of free moisture needed for infection in relation to temperature has been investigated in recent years for *Peronospora tabacina* (Hill, 1966; Kröber, 1967), *Pseudoperonospora humuli* (Royle, 1970; Sonoda and Ogawa, 1972), and *Plasmopara viticola* (Blaeser and Weltzien, 1979). The graphic relationship between minimum free moisture duration and temperature is a hyperbole-like curve for temperatures which do not exceed the optimum range (Figs 2 and 4); at higher temperatures, however, the curve is known to swerve sharply upwards. Royle (1970) complemented this relationship for *Pseudoperonospora humuli* with the free moisture duration required for severe infection with the same inoculum load (Fig. 4). Cohen (1977) and Bonde *et al.* (1978) went into still greater detail for *Pseudoperonospora cubensis* and *Sclerospora sorghi* by relating infection intensity to several leaf-wetness durations for each temperature. For *P. cubensis*, limited infection occurred after 2 h of wetness over a narrow optimum range of 20–25°C. After 6 h of wetness, infection rose suddenly to its ceiling level over the range of 15–20°C, while after 24 h of wetness the slow but steady infection increase at lower

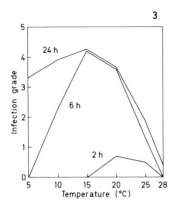
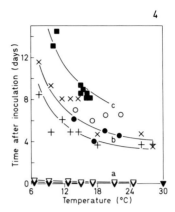

Fig. 3. Infection of leaves of young cucumber plants by *Pseudoperonospora cubensis* related to the duration of free moisture at different temperatures. Inoculum load: 1000 sporangia per cm² leaf area; infection grade evaluated visually as an approximation of the density of penetration points (Cohen, 1977).

Fig. 4. *Pseudoperonospora humuli* on young hop leaves. (a) Infection period: duration of free moisture for minimum (▼) and maximum (▽) infection. (b) Incubation period: time for appearance of first (+) and maximum (×) leaf spots. (c) Latent period: time for appearance of first sporulation; data from Zattler (1931) for potted cuttings of highly susceptible cultivar Hallertau (●) and less susceptible Spalt (○); first sporulation on unspecified variety as detected in the field by spore trapping (■), see Section III. B1 and Fig. 14. Data for (a) and (b) from Royle (1970) for potted cuttings of three highly susceptible cultivars; durations in (b) corrected to allow for a uniform 16 h infection period at 16–20 C; (▼) same data as in Fig. 2. No inoculum loads specified. The ordinates of the three hyperboles in (b) and (c) are in the ratios 1:1·3:2.

temperatures resulted in widening the optimum range to 5–20°C (Fig. 3). A similar drift and widening of the optimum temperature range has already been mentioned for *in vitro* germination. In *S. sorghi* (Bonde *et al.*, 1978), high levels of infection were reached after only 2 h of leaf-wetness over a large part of the temperature range. This is consistent with the field observation that the *Sclerosporas* infect their hosts in the short interval between their nocturnal dissemination and the evaporation of free moisture in the morning. With the other downy mildews cited above, some infection was obtained under optimal conditions with 2 h of free moisture or less, which is much shorter than *in vitro* germination experiments would have indicated.

Cohen and Sherman (1977) reported the still unexplained observation that the percentage of systemically infected plants obtained by spray-inoculating maize seedlings with *Sclerospora sorghi* increased with increasing free moisture durations up to 6–8 h but markedly decreased with longer durations. A similar observation had previously been made by Cohen and Sackston (1973) with sunflower seedlings dipped in an inoculum of *Plasmo-*

*para halstedii*; however, when sunflower seedlings were inoculated on the terminal bud, the percentage of systemically infected seedlings rose continuously over the whole inoculation period.

(*ii*) *Rain and dew in the field*. In non-arid areas, the available evidence points to rain rather than dew as the major cause of infection by downy mildews. In the French and German forecasting systems which were developed early in the present century against *Plasmopara viticola* on the basis of extensive field investigations, infection was attributed in practice only to rain followed by persistent surface wetness. This applied not only to primary infection where splash dispersal is needed to carry the zoospores released by oospores on the ground on to the foliage, but also to secondary infection by sporangia produced on the leaves (Müller and Sleumer, 1934; Müller, 1936; Schad, 1936). In recent investigations, Blaeser and Weltzien (1978, 1979) considered that secondary infection of vine by *P. viticola* was associated with rain, because dissemination in this pathogen occurred only during rain.

In England, Royle (1970) found that the major outbreaks of *Pseudoperonospora humuli* on hop during two successive growing seasons were traceable to rain and not to dew alone. He attributed this to the fact that rain not only wetted the leaves but also produced a peak of sporangium dispersal (see Section III.A). A complementary explanation (Royle and Thomas, 1971) came from the finding that fewer zoospores reached and penetrated a stoma in darkness (which is associated with dew formation) than in the light. However, this explanation does not hold for infection by *Plasmopara viticola*, which is relatively indifferent to light and darkness. In a subsequent 3-year experiment, in which potted hop-plants were exposed to natural infection in the field daily for several weeks, Royle (1973) confirmed that severe infection was always associated with prolonged rain-wetness. Moderate infection required rain-wetness except on a few occasions when dew extending over more than 8 h was also effective. Light infection was brought about equally by rain-wetness and by dew. Sonoda and Ogawa (1972) also reported that hop downy mildew spread rapidly with rain until early June in California and that dew alone during the following drier months allowed only limited infection although it lasted 5–7 h per night. The explanation was that few of the sporangia disseminated during daytime were still viable at the start of the nocturnal dew period.

While *Plasmopara viticola* and *Pseudoperonospora humuli* are zoospore formers, Hill *et al.* (1967) found that increases in the occurrence of *Peronospora tabacina*, a conidial downy mildew, were also related to rain in Australia.

By contrast, Palti and Chorin (1964) reported that the development of downy mildews in Israel during the rainless summer and autumn months depended on the duration of dew. This was 9–10 h in June–July when nights are short and morning evaporation is rapid, and 12–13 h in September–October with longer nights and slower evaporation. The spread of *Pseudoperonospora cubensis* under these conditions was explained by the finding that a proportion of the sporangia released during day-time could survive for more than 22 h—except at temperatures of 22°C and higher —and were consequently able to infect their host during the dew period on the following night (Cohen and Rotem, 1971a; Rotem *et al.*, 1971). In South Texas, dew periods of at least 5–6 h and of up to 12 h were also found to be causative in the epidemic development of *P. cubensis* during rainless periods (Thomas, 1977).

This suggests that differences in the appreciation of the role of dew might originate in the different durations of dew which the experimenters have had to deal with in their respective countries. The question is then why the dew period in the field should need to be so long to produce infection compared to the free moisture requirements of the pathogens as determined in the laboratory. One possibility is that spores which were exposed to dry air after dispersal need more time to germinate and infect. Arens (1929a) found that detached sporangia of *Plasmopara viticola* exposed to 60% relative humidity for 0, 2, 5, 15 min or 6 h germinated after 1 h, 2 h, 3 h 20 min, 5 h 30 min, and 5 h 45 min respectively. Another answer might lie in the possibility envisaged by Sonoda and Ogawa (1972) of a morning overlap of diurnal dissemination with the last hours of nocturnal dew, which might be sufficient to allow heavy infection. In Californian conditions, sporangial release in *Pseudoperonospora humuli*, occurred concurrently with dew for only 1 to 1 h 30 min, but in other climates the period of overlap might be longer. If this early morning infection were demonstrated to be real, it would extend to the whole group of the downy mildews the behaviour of the *Sclerosporas*, in which nocturnal or early morning infection immediately after spore release is known to occur commonly in the presence of dew (Dogma, 1975; Payak, 1975).

The sequence of field infection in the downy mildews may thus be summarized as follows: (1) spore release by rain and immediate infection in the ensuing surface wetness; high infection efficiency with evening rain or day-time permanent wet weather, which both ensure prolonged surface wetness; (2) day-time spore release and infection in dew deposited on the following night; low infection efficiency because of decreased inoculum viability; (3) overlap of the nocturnal dew period with nocturnal spore release in the *Sclerosporas* or with the early part of day-time spore release in the other downy mildews; efficiency depending on the persistence of dew in the morning.

(c) Inoculum load and host susceptibility

A given relationship between infection and the duration of free moisture at different temperatures is only valid for one set of conditions of inoculum load and host susceptibility.

Increasing inoculum loads result in an increase of the infection level (Hill, 1966; Kröber, 1969b; Cohen and Sackston, 1973; Cohen, 1977; Cohen and Sherman, 1977) and a reduction of the minimum free moisture duration necessary for infection (Kröber, 1967). The inoculum load is expressed as spores per drop in the case of drop inoculation and as spores per unit volume of suspension or unit leaf area when the inoculum is sprayed on to the host. Plotting the percentage of successful inoculum drops on a probit scale against the logarithm of the number of spores per drop was shown by Kröber (1969b) for two species of *Peronospora* to result in a straight line, the position and slope of which were characteristic of the pathogen and the host susceptibility. This log-probit linear relationship is a classic for representing the dose–response to fungicides. For sprayed inoculum, no comparable dose–response relationship seems to have been published for a downy mildew up to now.

It is debatable whether concentrations of hundreds or thousands of spores per drop or per $cm^3$ of inoculum as used in some inoculation experiments are realistic. In the field, spore deposits probably rarely approach comparable loads, except in seed-beds where sporulating and healthy leaves are crowded together. Hill (1969) commented on the large number of *Peronospora tabacina* conidia needed to obtain infection in his experiments compared to the extremely low spore concentrations in the air which are able to ensure infection in the field. He tentatively attributed the low efficiency of his inoculum to the washing of spores prior to their use. In later drop- and spray-inoculation experiments with unwashed spore suspensions, the same author found the minimum number of spores required to produce a lesion was 1·1–1·2 (Hill, 1969b).

Host susceptibility to infection changes with the age of the host plant or plant part and with environmental conditions. There are few specific studies on the changing susceptibility with host age where susceptibility to infection is dissociated from susceptibility to colonization. With *Peronospora tabacina*, Kröber (1967, 1969) demonstrated that the percentage of inoculation drops giving rise to a lesion decreased with the age of tobacco plants, while the minimum leaf-wetness duration increased. Hill (1969b) showed that infectibility decreased from the lowest leaf to the youngest top leaf of full-sized tobacco plants. The gradient was steeper on immature than mature plants, as susceptibility at all leaf levels decreased with plant age and especially on the lower leaf levels. A similar gradient was observed on young

3-leaved tobacco plants (Hill, 1966). Differences in susceptibility with leaf level were associated with the presence of a germination inhibitor on the leaf surface (Shepherd and Mandryk, 1963) and removal of this inhibitor by simulated rain increased susceptibility, especially in the younger top leaves (Hill, 1966, 1969b).

The infectibility of vine leaves by *Plasmopara viticola* is low at unfolding, increases to a maximum when leaf area reaches 20–25 cm$^2$ and then decreases again. The initial susceptibility increase is correlated with stomatal development (Müller and Sleumer, 1934). According to Schad (1936), vine leaves acquire their lowest susceptibility when their growth is terminated and become more susceptible again after shoot lignification. Similar patterns of infectibility changes, with age of host or host part, are known for diseases other than downy mildews (Populer, 1978).

Low soil moisture was reported by Royle and Thomas (1971) to close a proportion of stomata on hop leaves, which resulted in lower numbers of *Pseudoperonospora humuli* zoospores settling on stomata and lower infection percentages than with high soil moisture. The occurrence of rain a few days before infection was reported to increase the susceptibility of vine to *Plasmopara viticola* (Schad, 1936; Darpoux, 1943). No experiments were made to elucidate the relationship, but the preceding examples suggest either the elimination of an inhibitor from the leaves or an effect on stomatal opening.

## B. Colonization and symptom development

### 1. *Colonization, incubation period and latent period*

Colonization involves the vegetative growth of the pathogen inside the host. The incubation period is the time from the start of infection to the first appearance of symptoms, while the latent period is from the start of infection to the start of sporulation (Van der Planck, 1963). Local lesions caused by downy mildews are finite and do not enlarge after they have appeared, so that colonization coincides with the incubation period. In systemic infection, colonization continues after the first appearance of symptoms.

(a) Temperature and air humidity

With local lesions, the duration of the incubation period varies with temperature in much the same way as the duration of free moisture necessary for infection. The relationship is hyperbole-like for temperatures not exceeding the optimum range, as illustrated in Fig. 4 by Royle's

70                                    C. POPULER

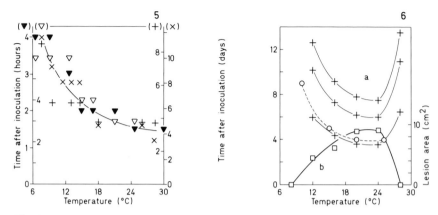

Fig. 5.   *Pseudoperonospora humuli.* Same data as in (a) and (b) in Fig. 4, on modified scales.

Fig. 6.   *Peronospora tabacina* lesions on young tobacco plants. (a) Incubation period ( + ); from
upper to lower curve, inoculum loads of 0·8, 80 and 8000 conidia per 100 cm² leaf area; data
recalculated from tables 7a and 7b in Hill (1966). Latent period ( ○ ), data from Jadot (1966b);
high inoculum load (personal communication). (b) Area per lesion, with low inoculum loads,
5–500 conidia per 100 cm² leaf area (Hill, 1966).

data (1970) on *Pseudoperonospora humuli*. At higher temperatures, the curve
slopes upwards, as shown by the data of Cohen (1977) for *Pseudoperonospora cubensis*, of Müller and Sleumer (1934) for *Plasmopara viticola*, of
Verhoeff (1960) for *Bremia lactucae* and of Hill (1966) for *Peronospora
tabacina* (Fig. 6). In each of these four local-lesion producing downy
mildews, the shortest incubation period in optimum conditions was 3 days.
Similar incubation curves for systemic colonization do not seem to have
been worked out.
    In experiments on the duration of incubation with temperature, infection
may either be established at the same temperature as that of the ensuing
incubation or at a selected temperature, known to give a desirably
uniform number of infections in all the incubation treatments. With the
latter method however, the observed incubation periods must be corrected
by allowing for the fraction of the incubation period covered by the
initial uniform infection period (Fig. 4).
    The duration of the incubation period was reported to be shorter at high
compared to low air humidities in *Plasmopara viticola* (Müller and Sleumer,
1934) and *Bremia lactucae* (Verhoeff, 1960), which correlates with a faster
mycelial growth in the leaf tissues. With *Plasmopara viticola*, the differences
were far from negligible, as the incubation period at 13·2–13·5°C was 6 days at
100% r.h. against 11–12 days at 80–90%, and at 28°C was 5–6 days at 100%
against 8 days at 70%. With *Pseudoperonospora humuli* however, Royle (1970)
found relative humidity had no marked influence on the incubation period.

Hill (1966) demonstrated that the area of individual lesions caused by *Peronospora tabacina* varied with the temperature applied during the incubation period (Fig. 6). This little-known relationship was confirmed by Rotem and Cohen (1970), who also showed that lesion appearance and sporulation were delayed and finally suppressed with increasing durations of exposure to 30–40°C intercalated in a 20°C incubation period.

With the downy mildews, infections which were initiated together do not become visible at the same time and symptoms are usually staggered over a few days. Royle (1970) measured the time both to first symptoms and generalized symptoms for *Pseudoperonospora humuli* on hop leaves (Fig. 4) and shoots. For shoots, the incubation was at least twice as long as that for leaves and the appearance of symptoms was also staggered over a longer period. Data on the delay between planting of maize seed infected by *Sclerospora maydis*, *S. philippinensis* and *S. sorghi* and the appearance of systemic infection symptoms, show a similar staggering (Advincula and Exconde, 1975–1976).

Hill (1966) noted that *Peronospora tabacina* on young tobacco plants sometimes sporulated on the day that symptoms were first observed, but more usually on the following day, and much later at low temperatures. A similar sequence was reported for *Plasmopara viticola* on vine leaves (Müller and Sleumer, 1934) and for *Sclerospora sorghi* in local lesions on diverse hosts (Kenneth, 1970). This sequence seems to be common with local lesions produced by downy mildews, though in some diseases, such as *Peronospora destructor* on onion, sporulation may occasionally be apparent before symptoms (Viranyi, 1975). Staggered lesion appearance logically implies staggered start of sporulation but no experimental data seem to be available on this.

(b) Inoculum load and host susceptibility

The incubation period was twice as long with low than with very high inoculum loads—applied either as a spray or as drops—and with short than with long leaf-wetness durations in leaf lesions obtained with *Peronospora tabacina* (Fig. 6), *P. viciae* (Mence and Pegg, 1971) and *Pseudoperonospora cubensis* (Cohen, 1977). Again it is questionable whether such high inoculum loads, and the shortened incubation periods which result, are realistic.

The incubation period of *Peronospora tabacina* lesions increased continuously with the age of tobacco plants and was about twice as long on mature tobacco plants as on 6-week seedlings (Populer, unpublished). There seems to be little information on this relationship in the downy mildew literature.

It is more clearly recognized that the area colonized by the downy mildews in their host generally decreases with increasing host tissue age. Seedling or shoot susceptibility to systemic colonization declines in a matter of days or weeks while local lesions are large on young tissues and tend to be smaller and finally reduced to flecks when established on older tissues.

In this respect, it must be noted that the expression "systemic infection", which is used in contrast to "local lesions", is somewhat misleading in epidemiology. Infection, in the sense of host penetration, is always local and the systemic character of a disease lies in the extent of host colonization. Thus, it would be more logical to speak of systemic disease or systemic colonization rather than systemic infection.

## 2. Symptom development

Once the local lesions of the downy mildews have become visible, the host tissues within the lesions become increasingly discoloured and finally necrotic. Rotem and Cohen (1970) and Cohen and Rotem (1971b) found that this symptom development in *Peronospora tabacina* on tobacco leaves and in *Pseudoperonospora cubensis* on cucumber leaves was most severe at temperatures higher than the optimum range for colonization. They concluded that colonization and symptom development were processes which responded to different environmental conditions. Previous investigations by Hill and Green (1965) on symptom development in and death of tobacco leaves infected by *Peronospora tabacina* in relation to temperature cannot be interpreted as either confirming or conflicting with this conclusion.

## C. Sporulation

## 1. Effect of environmental and host conditions

(a) Light

Weston (1923, 1924) and Yarwood (1937) were the first to point out that sporulation in the downy mildews occurred only under moist dark conditions following a period of light. Under continuous light or darkness and high humidity, there was little or no sporulation. This was demonstrated by Weston for several *Sclerosporas* and by Yarwood for *Bremia lactucae*, *Peronospora destructor*, *Plasmopara viticola* and *Pseudoperonospora humuli*. These relationships have since received more attention. The case of *Peronospora tabacina*, which was studied most closely, is treated here in some detail as it helps to piece together the sparse evidence available for other species.

(*i*) *Sporulation in* Peronospora tabacina. Uozumi and Kröber (1967) have quantified systematically a number of relationships linking sporulation in this fungus to light conditions at 20°C.

At this temperature, the intensity of sporulation induced in *P. tabacina* on tobacco leaves by a dark treatment increased with the duration of the previous photoperiod up to a photoperiod of 13 h; with longer photoperiods, sporulation intensity remained unchanged.

After a 13 h dry photoperiod, the minimum moist dark period for inducing sporulation was 2 h and maximum sporulation was induced after 5 h of darkness. After this dark induction period, sporulation proceeded equally well to its normal completion under moist conditions with either darkness or light. Lengthening the dark moist period beyond 5 h did not increase the amount of sporulation. Sporulation could also be induced by a dry dark period followed by a moist dark period but sporulation was maximal only if moist conditions set in within the first 5 h of darkness. If moist conditions began later in the dark period, the amount of sporulation which was induced declined progressively. A 4 h dry dark period followed by moist light conditions induced less sporulation than if followed by moist darkness. Cohen (1976) confirmed the latter finding and reported moreover that longer dry dark periods followed by moist light conditions resulted in decreasing sporulation intensity.

Uozumi and Kröber (1967) suggested that the slightly different time relationships reported by Cruickshank (1963) could have resulted from using other durations for the photoperiod preceding sporulation. A short period of light, as applied for examining or transferring samples during the dark induction was also reported to retard (Fried and Stuteville, 1977) or inhibit sporulation (Cohen *et al.*, 1978) and this may be a further cause of differences between authors.

When moisture was present from the start of the dark period, Uozumi and Kröber (1967) observed that the sporulation process was completed 24 h after the start of the previous photoperiod when the latter was shorter than 13 h, and 11 h after the start of the dark period when the previous photoperiod was equal to or longer than 13 h.

If this rule is applied to a field situation, for instance around 50° latitude and at the summer solstice, when night lasts approximately from 2000 h to 0400 h (local solar time; morning and evening twilight periods included in the night), then spores would be expected to be fully developed 11 h after sunset, i.e. at 0700 h. At the autumn equinox, with night lasting from 1800 to 0600 h, spores would have finished forming by 0600 h.

The relationship between sporulation intensity and the duration of the previous photoperiod explains why lesion-bearing leaves collected in the field or glasshouse in the afternoon or in the evening and subjected in the laboratory to moist darkness sporulate more profusely than when collected in the morning. If such leaves are incubated in saturated air and under a natural light cycle, sporulation follows the same periodicity as in the field, whatever the time of collection.

Under continuous light and in saturated air, *Peronospora tabacina* sporulates in decreasing amounts with increasing light levels (Cruickshank, 1963; Kröber, 1968). Conidiophores formed under these conditions are abnormal and the conidia are unpigmented, small and bottle- or eight-shaped (Jadot, 1966a). Blue and green light are more effective than red in effecting the inhibition of sporulation and the accompanying abnormalities (Cruickshank, 1963; Cohen, 1976).

All the above data on induction and inhibition apply to a temperature of 20°C. However, Cohen (1976) and Cohen *et al.* (1978) observed that blue light levels which inhibited sporulation completely at 20–24°C, did not inhibit and even slightly stimulated sporulation at 8–10°C; at 16°C, the response was intermediate. A similar response was found in *Pseudoperonospora cubensis* by Cohen and Eyal (1977).

To elucidate the temperature dependence of sporulation induction, a first step would be to verify the field realism of this non-inhibition or stimulation of sporulation by light at low temperatures. The time relationships linking sporulation to photoperiodicity and moisture would also need to be investigated in the laboratory at these low temperatures in the way Uozumi and Kröber (1967) explored them at 20°C. Even at 20°C, information is still lacking, which deprives the other data of much of their usefulness for interpreting field observations, namely the duration of saturation necessary for completing sporulation in relation to the starting time of saturation and with variants of the dark induction period.

*(ii) Other downy mildews.* Several relationships with light, closely similar to those described above for *Peronospora tabacina* have been demonstrated for *Pseudoperonospora cubensis* (Cohen and Rotem, 1970, 1971b; Cohen *et al.*, 1971; Cohen and Eyal, 1977). Fragmentary experimental evidence along the same lines is also available for *Bremia lactucae* (Verhoeff, 1960; Raffray and Sequeira, 1971), *Peronospora destructor* (Yarwood, 1943), *P. trifoliorum* (Fried and Stuteville, 1977), *P. viciae* (Pegg and Mence, 1970), *Plasmopara viticola* (Müller and Sleumer, 1934; Blaeser and Weltzien, 1978), *Sclerospora graminicola* (Safeulla and Thirumalachar, 1956), *S. sacchari* (Chang and Twu, 1972) and *S. sorghi* (Schmitt and Freytag, 1974). Some other observations on sporulation periodicity in the *Sclerosporas* were reviewed by Dogma (1975) and Payak (1975). Earlier experiments by Weston (1923, 1924) and Yarwood (1937) are mentioned above.

The *Sclerosporas* have a special position in that their spores are produced in a very few hours in the middle of the night and discharged immediately afterwards. After dawn, only scanty remnants of the previous night's crop are usually evident, as the conidiophores are killed by drying and disintegrate rapidly (Weston, 1923; Jones, 1971; Dogma, 1975; Payak, 1975). By comparison, conidia of *Peronospora destructor* and sporangia of *Pseudoperonospora humuli* were reported to be mature in the field only at about 0600 h (Yarwood 1937, 1943), much like *Peronospora tabacina*. Safeulla and Thirumalachar (1956), however, observed sporulation of *Sclerospora graminicola* and *S. sorghi* as late as midday on cloudy mornings following dry nights. The complete sporulation process, from the start of dark induction to spore maturation, may be completed in as little as 4–6 h in *S. sorghi* on maize and sorghum (Schmitt and Freytag, 1974; Payak, 1975) and 5–8 h in *S. sacchari* on maize (Payak, 1975; Bonde and Melching, 1979), against 8–12 h for *Peronospora tabacina* (Cruickshank, 1958; Uozumi and Kröber, 1967), *P. trifoliorum* (Fried and Stuteville, 1977) and *Pseudoperonospora cubensis* (Cohen *et al.*, 1971).

(b) Moisture and temperature

Decreasing sporulation with relative humidities under 100% was demonstrated for *Peronospora destructor* (Yarwood, 1943), *P. farinosa* (Kröber, 1968), *P. tabacina* (Cruickshank, 1958; Kröber, 1968), *P. trifoliorum* (Fried and Stuteville, 1977), *P. viciae* (Pegg and Mence, 1970) and *Plasmopara viticola* (Müller and Sleumer, 1934; Blaeser and Weltzien, 1978). The lower threshold under which there was no sporulation varied between 90 and 98% according to the species of downy mildew involved.

The effect of free water on the sporulation of downy mildews other than the *Sclerosporas*, is not well known. It was reported to inhibit sporulation in *Peronospora trifoliorum* (Fried and Stuteville, 1977) and in *P. tabacina* (Pinckard, 1942). Sonoda and Ogawa (1970) found that *Pseudoperonospora humuli* produced large hyaline sporangia on submerged lesions and small dark sporangia on dry lesions at 100% r.h.; both sporangium types were produced at the same speed and were infective. According to Verhoeff (1960), in *Bremia lactucae* sporulation occurred only when leaves were covered by a film of water, in contrast with Pegg and Mence (1970) who obtained sporulation in relative humidities between 91 and 100%.

For the *Sclerosporas*, it has been reported repeatedly since the early observations by Weston (1923, 1924) that sporulation in the field occurs only on host areas covered with heavy dew (Dogma, 1975; Payak, 1975). In the laboratory, *S. sorghi* did not sporulate on dry or submerged leaf areas, but only on areas covered with heavy dew (Safeulla and Thirumalachar, 1955). Dew chambers have been used for producing sporulation in *S. sacchari* (Bonde and Melching, 1979) and *S. sorghi* (Schmitt and Freytag, 1974).

Several authors have quantified final sporulation in relation to temperature during the moist dark sporulation period. The reported minimum, optimum and maximum temperatures may be low, as in *Peronospora destructor*, 7 C, 10–16 C and 22 C (Yarwood, 1943) and *P. viciae*, 4 C, 8–16 C and 20°C (Pegg and Mence, 1970), wide-ranging as in *P. tabacina*, 1–2°C, 15–23°C and 27°C (Cruickshank, 1961b) or high as in *Plasmopara viticola*, 15 C, 17–27°C and 29 C (Müller and Sleumer, 1934; Blaeser and Weltzien, 1978) and *Sclerospora sacchari*, 13°C, 22–25°C and 31 C (Chang and Wu, 1969).

The limited meaning of such temperature relationships, where the time variable is omitted, was discussed for germination and infection in Section II.A. In the present case, these relationships give a measure of the final sporulation intensity at each temperature provided that sporulation has been allowed a sufficient number of hours to reach its terminal level at all temperatures. In several of the experiments above, however, the time allowed for sporulation to develop is not clearly stated.

The time course of sporulation related to temperature, and to variants of the dark induction period, is a more meaningful relationship. It was quantified by Cruickshank (1958, 1963) and Uozumi and Kröber (1967) for *Peronospora tabacina*, by Fried and Stuteville (1977) for *P. trifoliorum* and by Cohen *et al.* (1971) for *Pseudoperonospora cubensis*, in each case at 20 C. The spore production curve of these three downy mildews showed a strikingly similar slope, with an interval of 2 h between the 10 and 90% production levels (Fig. 7). Cohen and Rotem (1969) also quantified more

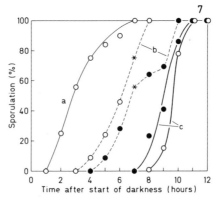

Fig. 7.    Time course of sporulation during the night in *Peronospora tabacina* on young tobacco plants at 20 C. (a) Number of conidiophores; (b) length of conidiophores; (c) number of conidia. (*) Conidiophores beginning to ramify. Previous light period of 13 h; the dark period was moist from the start (○) or dry for the first 4 h and then moist (●). Data from Uozumi and Kröber (1967).

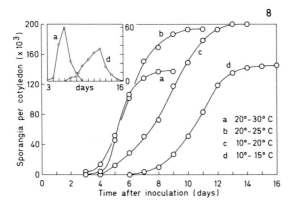

Fig. 8. Sporulation period of *Pseudoperonospora cubensis* lesions on young cucumber plants. Night (low) and day (high) temperatures during latent and sporulation periods, with 12 h photoperiod and high moisture on every night. Cumulative sporangium production calculated from daily production data (inset) in Cohen and Rotem (1971b).

completely the time course of sporulation in *Pseudoperonospora cubensis* at different temperatures ranging from 5 to 30°C. Further investigations might be useful to verify earlier reports that in some non-*Sclerospora* downy mildews, sporulation may not develop overnight at low temperatures or suboptimal humidities but may do so after several days (Arens, 1929).

Sporulation of *Peronospora tabacina* at a night temperature of 20°C was related by Cruickshank (1961b) to different preconditioning temperatures during the previous photoperiod. Compared to the relation with night temperatures during sporulation, there was a shift towards tolerance to higher temperatures, with a maximum at 35°C. However, in a subsequent paper, Rider *et al.* (1961) concluded from field evidence that day temperatures exceeding 30°C for more than 6 h inhibited sporulation on the following night. Yarwood (1943) held a similar opinion concerning *Peronospora destructor*.

Preconditioning experiments in which infected plants are subjected to different temperature regimes from inoculation onwards and are tested on the same day for sporulation have little meaning (Cruickshank, 1961b; Rotem and Cohen, 1970). The reason is that on the sporulation test day, lesions which have developed at different temperatures are at different stages of their sporulation potential (Fig. 8).

(c) Host condition

Cruickshank (1958) and Kröber (1968) found that sporulation of *Peronospora tabacina* on tobacco leaves and *P. farinosa* on beet leaves was maximal when the diffusion pressure deficit (DPD) of the host tissues was nil. Sporulation declined with increasing DPD and was completely inhibited around a 20 atm. DPD.

Decreasing sporulation intensity with increasing age of host plant or plant part was reported for *Peronospora tabacina* on tobacco (Kröber, 1968), *P. viciae* on pea (Mence and Pegg, 1971), *Pseudoperonospora cubensis* on cucumber (Cohen *et al.*, 1971) and *P. humuli* on hop (Royle and Thomas, 1971).

## 2. *The sporulation period*

In the experiments discussed in the preceding section, sporulation is a unique event observed at an arbitrary time after the end of the latent period. Downy mildews, however, are able to sporulate repeatedly for several successive nights on the same lesions. This was reported for *Peronospora destructor* (Yarwood, 1937), *P. farinosa* (Byford, 1967), *P. tabacina* (Cruickshank, 1961b), *P. trifoliorum* (Fried and Stuteville, 1977), *Plasmopara viticola* (Müller and Sleumer, 1934), *Pseudoperonospora cubensis* (Cohen and Rotem, 1971b), *Sclerospora philippinensis* and *S. graminicola* (Weston, 1923, 1924), and *Sclerospora sorghi* (Siradhana *et al.*, 1976).

Studies on the sporulation period are usually limited to establishing the duration of the latent period, i.e. the time from inoculation to the first sporulation (Figs. 4 and 6). There have been few endeavours to quantify the sporulation period any further. Cohen and Rotem (1971b) studied the sporulation on lesions of *Pseudoperonospora cubensis* on cucumber plants held at different temperature regimes and induced to sporulate each night. Sporulation was measured by subtracting counts of sporangia in evening samples from counts on samples taken the following morning. High temperature was associated with a short sporulation period and low temperature with a protracted sporulation period. The highest numbers of sporangia per sporulation period were produced at intermediate temperatures (Fig. 8). This method quantifies the sporulation potential of a generation of lesions without bias. Removing the sporophores on the lesions each day with a soft brush to observe the successive sporulation crops (Müller and Sleumer, 1934) is a relatively rough procedure. With field sampling (Byford, 1967), there is a risk of incorporating lesions from different generations. Inhibiting sporulation on growth-room plants by keeping them in dry air until tested for sporulation (Cruickshank, 1961b) may reflect the sporulation potential in an arid area with successions of dry nights rather than in the more usual conditions of alternating dry days and humid nights. A reasonable guess would seem to be that persistent obstruction of sporulation by dry nights prolongs the period during which lesions are able to sporulate.

Changes in the duration of the sporulation period of lesions and in the total spore output per lesion, with age of the host or host part, do not seem to have been investigated. For *Bremia lactucae* in the systemic state on young lettuce seedlings, Dickinson and Crute (1974) showed that sporulation was retarded and that spore output over the sporulation period decreased with increasing seedling age at inoculation time. Susceptibility to systemic colonization decreased concurrently. This is probably exemplary of downy mildews which produce systemic colonization only in very young host plants.

The development of sporulation in the case where systemic colonization proceeds during the whole host life cycle was illustrated in detail by Weston (1923) for *Sclerospora philippinensis*. On maize, sporulation occurs on discoloured leaf areas which reach their final whitish colour within 6 to 24 h after the appearance of the first difference in shade. These areas, which do not increase with time, are confined to the leaf base on lower leaves and run towards the tip on higher leaves. Sporulation occurs one day or more after the discolouration. It appears first on the lower leaves and develops progressively upwards on the younger leaves. On the first night, only scattered conidiophores are produced and successive sporulation crops become progressively more abundant on following nights. The daily spore output per plant increases as leaves unfold and grow, and decreases as successive leaves wither. The sporulation period is consequently shorter on short-cycled maize varieties and on plants infected at an early age, which succumb rapidly. On average, it lasts one or two months, whereas on sugarcane and sorghum, which die a few weeks after infection, the sporulation period is much shorter. In teosinte, on the contrary, sporulation may extend for several months, as new branches arise and new shoots are sent out from the base of the infected plant. In *Saccharum spontaneum*, a perennial wild grass, a similar process allows practically everlasting sporulation.

## D. Time relationships in the multiplication phase

### 1. Quantifying the multiplication phase in relation to time

(a) Diagrams and equations

A large part of the data discussed in the preceding pages of this chapter are presented in the original papers in the form of one of the three diagrams illustrated in Fig. 9, or alternatively as tabulated data which can be converted into one of these diagrams.

The variables in these diagrams are time, temperature and a parameter measuring one of the stages in the multiplication cycle. This parameter is designated by the letter "M" in the diagrams. Environmental factors other than temperature are not envisaged here.

Figures 9a and b are different arrangements of the same experimental data, which are represented by circles. Figure 9a presents the data as a series of temperature curves of M measured at different times while in Fig. 9b each curve illustrates the time course of M for a different temperature. Figure 9c is derived by simplification from Fig. 9b. The lower curve simply indicates the time when M begins to increase and the higher curve the time when M stops increasing. Black squares in Figs b and c illustrate this for one temperature.

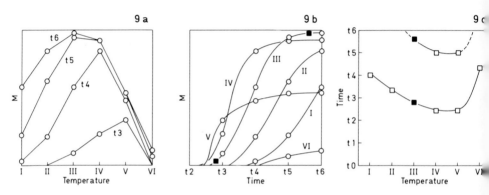

Fig. 9. Basic diagrams for representing the relationship between time, temperature and a parameter M measuring the development of a stage in the multiplication phase (see explanation in text).

Figure 9c is simpler to establish than Figs 9a or b because M is not quantified exactly. Figure 9a can be reduced to one temperature curve and Fig. 9b to the time course of M at one temperature only. These single-curve economy versions of Figs 9a and b, together with Fig. 9c, are commonly used by experimenters. Mathematicians who work on simulation models prefer the complete version of Fig. 9b, because they need to know how M changes with time at each temperature. A complete version of Fig. 9a is also utilizable for model making because it can easily be converted into a diagram of type 9b.

The curves in Fig. 9c are approximately hyperbolic between the minimum and optimum temperatures. Some authors have sought to relate the duration of free moisture necessary for infection by *Plasmopara viticola* (Blaeser and Weltzien, 1979) or the duration of the incubation period of the same fungus (Shatsky, 1935; Schad, 1943) to temperature by an hyperbolic function of the type $d \cdot t = C$ or $d(t-t_0) = C$, where $d$ is the duration in hours or days, $t$ is the temperature in degrees Centigrade, $t_0$ is the minimum temperature and $C$ is a constant expressed in degree-hours or degree-days. This expression is identical to the concept of the temperature sum. However, it does not apply to temperatures above the optimum. Another problem is that hyperbolic functions with different $t_0$ and $C$ values and also hyperbolic functions of other types may adjust equally well to the experimental data.

The variability of quantitative data on the multiplication phase, which is often large, is excellently illustrated by the data of Verhoeff (1960) on *Bremia lactucae.*

(b) Growth curves and cumulated frequency distributions

Parameter M in some experiments denotes the dimension of the individual lesion or the cumulative number of spores produced by a lesion. In these cases, the curves in Fig. 9b are truly fungal growth curves. An example of spore output quantified over the whole sporulation period is given in Fig. 8, in the form of a 9b type diagram. A 9a type diagram representing lesion dimension is included in Fig. 6. As the local lesions of downy mildews become visible only when colonization stops, there can be but one curve for lesion dimension on a 9a type diagram and there is no use for a 9b type diagram. This would not be the case for diseases where the growth of the colonization area is accompanied by the visible enlargement of a lesion, as for instance in potato late blight.

In other experiments, M denotes the number of infections per unit host tissue area, the percentage of successful inoculation drops, or the percentage of systemically infected plants. In these cases, M measures the frequency of infections, and the curves in Fig. 9b are not growth curves but cumulative frequency distributions of the Gaussian type. They illustrate the fact that among the spores which are deposited on the host surface, some are quick in achieving penetration and some are slower. The same is true in germination experiments, as the germination times of spores also follow a Gaussian-like distribution. Diagrams of type 9a for infection can be found in Cohen (1977) and Bonde *et al.* (1978), and for germination in Verhoeff (1960), Cruickshank (1961a), and Bonde and Melching (1979). Diagrams of type 9b are shown for germination by Verhoeff (1960), Shepherd (1962), Hill (1969a), Pegg and Mence (1970) and Schmitt and Freytag (1974). Diagrams of type 9c were used for infection by Royle (1970), Sonoda and Ogawa (1972), Blaeser and Weltzien (1979) and for germination by Zattler (1931), Müller and Sleumer (1934) and Schad (1936). Examples are shown in Figs 2, 3 and 4. In these diagrams, time is the duration of free moisture necessary for the establishment of a given level of infection. An appropriate expression for designating this duration is the infection period, by analogy with the incubation period, the latent period and the sporulation period, which are the succeeding stages in the multiplication phase.

Similarly, lesions which were initiated at the same inoculation time do not become visible all at the same time, nor do they start or stop sporulating simultaneously. In principle this staggering in time can also be represented by type 9a or 9b diagrams but it does not seem worthwhile to pursue this feature as lesion frequencies do not change after the end of the infection period. In practice, the end of the incubation period, or of the latent period is usually represented by a 9c type diagram. Examples of this are shown in Figs 4 and 6. A similar representation could be used for the end of the sporulation period.

Figure 7 was left out in the preceding discussion. Parameter M in curves "c" at the right of Fig. 7 represents the number of spores produced by a lesion during one night. The author expressed it in "per cent" but the initial counts made during the experiment were necessarily in numbers of spores. The terminal number of spores in Fig. 7 is the daily increment in a curve like those illustrated in Fig. 8 and so it is a fraction of a growth curve. Spore production is not continuous during the sporulation period, because it is regulated by the periodicity of light and humidity. When sporulation is induced at night, the sporophores do not emerge, grow and form spores synchronously, so that the curve in Fig. 7 is also a Gaussian distribution. But it has nothing to do with the previous Gaussian curves, where M was a number of lesions.

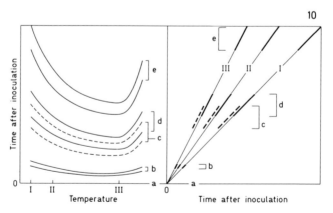

Fig. 10. Time relationships of successive stages in the multiplication phase. Diagrams established on the hypothesis that the transit times through successive stages at different temperatures are in the same ratios from beginning to end of the multiplication phase. (a) Inoculation time; (b) end of infection period, i.e. period during which free moisture is required; (c) end of incubation period; (d) end of latent period; (e) end of sporulation period. Staggering between lesions represented by the vertical distance between the two curves of each stage on the left diagram and by the length of the heavy lines on the right.

## 2. Differences in temperature requirements

The time relationships of the successive stages in the multiplication phase are grouped on Fig. 10 in a multi-stage 9c type diagram. In Fig. 4, we see that Royle (1970) produced experimental data for the two sets of curves concerning infection and incubation in *Pseudoperonospora humuli*. Figure 5 shows that these curves differ only by a scale coefficient. Put another way, if the durations of the infection period differ by a given factor between two temperatures, then the durations of the incubation period will differ by the same factor at the same temperatures. This is visualized in Fig. 10, where it is hypothesized that this proportionality also holds for the latent period and the sporulation period. A few data extracted from a paper by Cohen and Rotem (1971b) on *Pseudoperonospora cubensis* are shown in Fig. 11 to support this hypothesis. The implication is that temperature requirements do not change during the multiplication phase as far as speed of development through the successive stages of the multiplication phase is concerned. It does not imply that these temperature requirements are the same as for the proportion of spores which succeed in establishing infection, or for the dimension of the individual lesion or its spore output, which concern quantities of development and not speed of development. In Fig. 9b, the speed of development is highest at temperatures IV and V but the terminal level of M is highest at temperature III. This results in a shift of the optimum range with time in Fig. 9a, which has already been commented on in Section II.A, in connection with germination and infection.

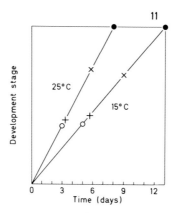

Fig. 11.   Time relationships of *Pseudoperonospora cubensis* lesions on young cucumber plants. Beginning (○) and end (●) of the sporulation period; see Fig. 8, curves (a) and (c). Grade 1 (+) and 3 (×) of symptom development, interpolated from data in Cohen and Rotem (1971b).

The proportionality hypothesis, as illustrated in Fig. 10, applies to lesions on a host substrate of uniform susceptibility. It does not apply as such to systemic disease, because systemic colonization and the associated sporulation proceed as long as the host plant is alive.

## III. THE DISSEMINATION PHASE

### A. Effect of weather factors on spore liberation

As the most detailed data available on the subject concern *Peronospora tabacina*, we shall examine this downy mildew first and discuss the other members of the group comparatively.

#### *1. Spore liberation in* Peronospora tabacina

The release of conidia in this species is bound for the main part to a regular diurnal periodicity but brief liberation peaks are also caused by rain and associated weather disturbances, as in many dry-spore fungi. When interpreting the results of spore trapping experiments in the field, the first logical step is to identify these peaks so as to separate them from the periodic pattern (Fig. 14). Spore dissemination diagrams on which this separation does not appear clearly can be very confusing.

(a) Liberation caused by rain

(*i*) *In the field.* The onset of rain, whether accompanied or not by wind gusts, sets off a short period of spore liberation (Hill, 1961). With intermittent rain, there is some liberation at the start of each shower. These successive releases are usually of decreasing size but they may increase when a light shower is followed by a much heavier one. Conidia can be liberated by exceedingly light rains and even by scattered rain-drops which are not detectable with a recording rain-gauge but only with a sufficiently sensitive leaf-wetness recorder of a type which discriminates rain-wetness from dew or fog. When rain falls steadily over several hours, the conidia of *P. tabacina* become almost absent from the air after the initial peak (Populer, 1966).

The number of conidia liberated by rain is usually higher during day-time than at night. It is also higher during days with high diurnal releases as the number of airborne spores in both periodic and rain releases is dependent on the amount of sporulation on the leaves. In both cases it is apparent that more spores are released by rain when the numbers ready to be liberated are high. Conidia liberated by rain represent a very low percentage—2 to

4%—of the total trapped during the epidemic season (Populer, 1966). They are, however, of prime importance for the establishment of infection (see Section II.A2b).

(ii) *Laboratory experiments.* Subjecting sporulating tobacco plants to falling weights (Cruickshank, 1958), falling sand or water drops (Hill, 1961) produces a sharp rise and fall of airborne conidia in a few minutes. This effect of falling water drops was also reported by Hirst and Stedman (1963) for *Pseudoperonospora humuli* on hop leaves and *Peronospora parasitica* on *Capsella bursa-pastoris.* With artificial wind, the number of airborne spores similarly rises and falls at each increase of air movement and remains low at constant wind speeds (Hill, 1961). This, together with the evidence from the field, indicates that rain and wind result in spore release in *P. tabacina* when the mechanical stimulus increases and not during periods of steady wind or rain.

(b) Diurnal periodicity

(i) *In the field.* Conidia of *P. tabacina* usually begin to be airborne in the morning and, in the earliest cases, soon after sunrise. This is associated with rising radiation and the concomitant increase in temperature and decrease in air humidity (Waggoner and Taylor, 1958; Hill, 1961; Populer, 1966; Valli, 1966). On rainy mornings, spore release may be postponed until the weather clears up, while very few spores are trapped on days with humidity persistently near saturation.

On days with continuous bright sun or a light haze or scattered clouds, the spore concentration in the air increases quickly and reaches a maximum in the second to fourth hour of release. This occurs several hours before maximum radiation and temperature and minimum humidity are reached. The decrease of spore concentration is slower than the increase and stretches over 4 to 8 h, after which a few spores may still be found in the air until late in the evening (Hill, 1961; Populer, 1966).

The typical form of the periodic spore release curve on a bright day is often obscured when hourly spore concentrations in the air are averaged over the trapping season. Averaging a choice of fair days without rain produces a curve which is closer to the individual daily curves, though somewhat smoothed out because the times of initial release and peak release vary from one day to another. Spore release by fair weather is best represented by a single-day curve or by an average curve over chosen days with peak release time taken as reference time (Fig. 12).

During days when bright periods alternate with long overcast periods, the spore release curve shows a succession of peaks corresponding to the sunny periods (Populer, 1966).

There are usually no conidia of *P. tabacina* to be trapped at night. A small number of spores may, however, be trapped throughout the night during a limited period in the middle of the epidemic, when a very large number of lesions are sporulating on the leaves (Hill, 1961; Populer, 1966).

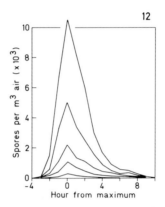

Fig. 12.   Diurnal spore liberation in *Peronospora tabacina*. Each curve is the average of several curves of similar magnitude on different days. The time of maximum spore release is taken as zero time (Populer, 1966).

(*ii*)  *Causes of periodicity.* Pinckard (1942) observed that conidiophores of *P. tabacina* and several other *Peronospora* species reacted instantaneously to dry air by twisting counter-clockwise when examined under a dissecting microscope. Spores were ejected by the spring-like action of entangled conidiophore branches as they disengaged, but forcible discharge from un-entangled sterigmata was also observed. On exposing the conidiophores to moist air, the twisting motion was reversed. These observations were con-firmed by Hill (1960) for *P. tabacina* and by Fried and Stuteville (1977) for *P. trifoliorum*. When sporulating tobacco leaf discs were transferred from saturation to different humidities between 90 and 50%, the number of spores of *P. tabacina* which were released increased as the humidity was lowered (Cruickshank, 1958). For *Pseudoperonospora humuli*, Sonoda and Ogawa (1972) made similar observations and Royle and Thomas (1972) reported that sporangium release was influenced only by changes in relative humidity and was unaffected by light in the range 0–4500 lx.

However, Hill (1960, 1961) reported that in glasshouse experiments, conidia of *Peronospora tabacina* were not trapped during a nocturnal decrease of relative humidity but only during the normal liberation period the follow-ing morning. As the overnight relative humidity was too low for sporulation, the day-time release was not the result of a renewed supply of spores. A similar situation was observed in the field, when the spore concentration of *P. tabacina* during a night with humidities as low as 55% did not exceed 150 spores m$^{-3}$ while it reached 1000 spores m$^{-3}$ the following day (Populer, 1966). Hill (1961) considered that loss of conidiophore turgidity and the associated spore release was not a direct reaction to ambient humidity but was due to the development of a diffusion pressure deficit in the leaf as a result of increasing radiation and temperature and of decreasing humidity. In Belgium, spore release of *P. tabacina* in the field occurred only when global radiation was 0·4–0·5 cal cm$^{-2}$ min$^{-1}$, or more, and the multiple-peak curves of spore release on days with changing weather fitted closely with the global radiation curves above this threshold (Populer, 1966). As the outgoing radiation is approximately 0·15–0·2 cal cm$^{-2}$ min$^{-1}$, the incoming global radiation, with a leaf-surface albedo of 0·2, must exceed 0·2–0·25 cal cm$^{-2}$ min$^{-1}$ before net radiation becomes positive and begins to exert an influence on the host plant or on sporulation structures.

The apparent conflict between the data on the effect of humidity versus radiation merit further consideration. In a temperate humid climate such as is experienced in Belgium, the effect of global radiation on evapotranspiration is about ten times larger than the combined effects of wind velocity and saturation deficit. In these conditions, the effect of global radiation on spore release in the field must be overwhelming if spore release is triggered by evapo-transpiration. In laboratory experiments, the radiation level is commonly much lower. An illumination level of 4500 lx, as cited above, represents approximately 2% of the global radiation received by a horizontal surface at noon during summer-time at 50° latitude. In such laboratory experiments, spore release would consequently be expected to show an exclusive relationship with air humidity.

## 2. Spore liberation in other downy mildews

A periodic spore release pattern similar to that of *Peronospora tabacina* was reported for *Peronospora manshurica* (Pederson, 1961), *P. viciae* (Pegg and Mence, 1972), *Bremia lactucae* (Fletcher, 1976), *Pseudoperonospora cubensis* (Schenck, 1968; Cohen and Rotem, 1971a; Thomas, 1977) and *P. humuli* (Royle, 1967; Mostade, 1971; Royle and Thomas, 1972; Sonoda and Ogawa, 1972). Spore liberation by rain has also been described for *Pseudoperonospora humuli* (Royle and Thomas, 1972) but not for the other downy mildews, though several were studied either in arid climates or in the glasshouse.

The time of earliest daily initial spore release varies appreciably between different species of downy mildews and among different authors for the same species. An unknown variable is the precision with which the time of initial spore release is established. When spores are counted at hourly intervals on slides exposed in an automatic volumetric spore trap, the count in any hour may include a proportion of spores captured one hour earlier or later and this proportion is related to the width of the traverse which is being scanned (Populer, 1965). Moreover, comparing initial times of spore release has little meaning unless the times of sunrise, which vary with latitude and season, are included in the comparison. In some papers, spore release time is expressed in summer time and this must be transformed into standard time. For places which are distant from the centre of their time-belt, standard time may differ from solar time by up to thirty minutes and this can also be allowed for. Figure 13, in which these transformations are incorporated, shows that initial

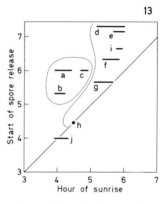

Fig. 13.   Starting time of diurnal spore liberation in some downy mildews related to hour of sunrise. (a–c) *Pseudoperonospora humuli* in England (Royle and Thomas, 1972), Belgium (Mostade, 1971), California (Sonoda and Ogawa, 1972). (d–h) *Peronospora tabacina* in Belgium (Populer, 1966), Australia, with distinct times for 3 trapping seasons (Hill, 1961), Connecticut (Waggoner and Taylor, 1958). (i) *Pseudoperonospora cubensis* in Israel (Cohen and Rotem, 1971a). (j) *Peronospora viciae* in England (Pegg and Mence, 1972). Hours of sunrise are given for the reported trapping periods. In each case, the indicated starting time of spore liberation is the earliest time recorded. All hours are in solar time (local meridian time).

spore release in *Peronospora tabacina* and *Pseudoperonospora cubensis* occurs shortly after sunrise, while for *Pseudoperonospora humuli* it is somewhat later. For *Peronospora viciae* the only available data place initial spore release also very close to sunrise. This species seems to be a quick as well as an early riser, since peak release is reported to occur 1–2 h after sunrise (Pegg and Mence, 1972). Thus, the speed of response of spore release to environmental changes is apparently specific to each of these downy mildews. Another conclusion from Fig. 13 is that averaging spore liberation times or curves over too long periods should be avoided, as it pools data related to quite different times of sunrise.

In the *Sclerosporas*, there is also a periodic pattern of spore release but it is mostly nocturnal, as spores are released immediately after being produced. In Israel, Kenneth (1970), trapped conidia of *S. sorghi* from midnight to 0500 h, with a maximum at 0200–0300 h. In Java, Semangoen (1970) observed spore release of *S. maydis* between 0200 and 0700 h, with the peak at 0300–4000 h. In Taiwan, spore release of *S. sacchari* began at 0100–0200 h and ended at 0300–0430 h (Chang and Wu, 1970). The spores of the *Sclerosporas* are liberated while humidity is near saturation and plants are covered with dew. They were reported to be forcibly discharged from the conidiophores (Weston, 1923; Kenneth, 1970; Jones, 1971). The spore release curve displayed by Semangoen (1970) for *S. maydis* in the laboratory displays symmetrical slopes for increase and decrease, which suggests that spore release follows a Gaussian-like distribution with time. If the hourly spore releases in this curve are accumulated, it is seen that the interval between the 10 and 90% levels is two hours, as in the sporulation curves obtained at 20°C for several non-*Sclerospora* downy mildews (Fig. 7). Thus, it seems that spore release in the *Sclerosporas* is not triggered by environmental factors but that it is the immediate consequence of spore maturation.

Weston (1923) observed that the tips of the conidiophores protruded slightly from the surface of the dew layer, so that sterigmata and conidia were free. The length of the conidiophores was adapted to the thickness of the moisture layer. Though the air on dewy nights is usually nearly still, *Sclerospora* conidia were observed to drift over at least 25 m. Dispersal over distances of $\frac{1}{2}$ mile or perhaps even more during nocturnal rain storms was also reported (Weston, 1923; Dogma, 1975).

For *Plasmopara viticola*, the number of sporangia trapped over a vine plot in Switzerland peaked at 1800–2000 h (Corbaz, 1972). In these experiments, actual spore counts were low and there was no attempt to correlate them with environmental changes. In Germany, Blaeser and Weltzien (1978) found no sporangia of *P. viticola* in the air but only in run-off water collected under the foliage. Spore release in this species as well as in some other downy mildews clearly needs further clarification.

## B. Changes in airborne spore numbers during the season

### 1. Relating spore concentrations in the air to previous events

The preceding section was concerned with changes in spore concentration with time of day. We shall now discuss daily spore releases as they vary during the epidemic season. Royle and Thomas (1972) showed that the daily peak concentrations of sporangia of *Pseudoperonospora humuli* in the middle section of a field epidemic were only predicted poorly by using multiple regression equations based on weather variables for the previous days. They concluded that biological variables were too important to allow for a purely meteorological prediction. These biological variables are the previous events in the epidemic development and the problem is to identify these events and relate them to the daily spore releases. This could be attempted in a "Grand Experiment" in the field, in which infection periods, latent periods and sporulation periods would be detected simultaneously with spore trapping. No data of this type seem to have been published yet. Another possible method is to identify these biological events on the spore trapping records and the accompanying weather records. Mostade (1971) published detailed spore trapping data on *Pseudoperonospora humuli* which are used in Fig. 14 to help discuss this method.

During the epidemic, the diseased leaf area displays a roughly logistic growth. The increments in the logistic curve, which represent new lesions, increase during the first half of the logistic curve and decrease thereafter. The general trend of spore release is consequently also to increase until the approximate middle of the epidemic and to decrease afterwards. Thus, logistic growth of the cumulated number of lesions is the first biological variable influencing spore concentration in the air. In Fig. 14, maximum spore release is observed during the first days of June.

A second biological variable concerns the observation that a lesion produces spores during a sporulation period of several days and that the daily amount of sporulation increases during the first days of the period and then decreases as shown by the inset in Fig. 8. If infection periods do not follow too closely on one another, distinct generations of new lesions can be observed to crop up at intervals of several days in the field. In these conditions, the sporulation periods corresponding to these generations must be identifiable in seasonal spore trapping records as successive waves of spore release. In Fig. 14, eight sporulation periods are identified.

A complication in the identification of sporulation periods arises from the fact that weather conditions are not equally favourable to sporulation on each night. The first actual sporulation in a potential sporulation period may be delayed by one or several dry nights after new lesions have emerged. On the other hand, Hill (1960) showed that lesions of *Peronospora tabacina* which had sporulated on one initial night released decreasing numbers of spores during several days afterwards, under conditions which inhibited any renewed sporulation. The consequence is that a dry night in the middle of a sporulation period results in a lowered spore release on the following day and not in a complete absence of airborne spores. In the epidemic represented in Fig. 14, we are luckily spared this problem, but the expression of sporulation periods 3, 4 and 5 may have been retarded by the previous dry nights. Thus, intensity of daily sporulation is the third variable influencing spore numbers in the air.

A fourth and last variable concerns the intensity of infection, which determines the number of lesions of a generation within the limits of logistic growth. To evaluate the weather conditions which prevailed at the time of

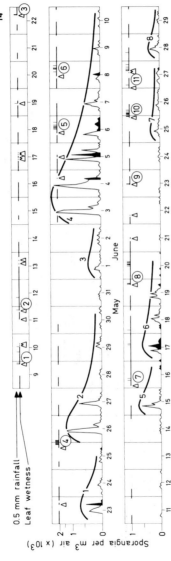

Fig. 14. Concentrations of *Pseudoperonospora humuli* sporangia in the air related to infection periods and sporulation periods. Numbers in circles = infection periods. Heavy lines = sporulation periods, identified by the same number as the infection from which they proceed. (▽) Onset of rain. Black areas under curve = sporangia released by rain. Spore trapping data from Mostade (1971).

infection, we must identify the infection periods. The knowledge we have of the duration of the incubation period or of the latent period (Fig. 4, curves b) should logically allow this identification but we must aware of a possible effect of differing host ages on the length of these periods. Additional help comes from the knowledge that major infections mostly occur during rain followed by prolonged wetness. Rains of this type are identified in Fig. 14. Sporulation periods 1 and 2 are too close to infection periods 3 and 4 to be related to them and there is therefore no other choice but to relate them to infection periods 1 and 2. Interconnecting the other infection and sporulation periods follows easily. The resulting latent periods are related to mean temperature in curve c of Fig. 4, which is proportional to curves b. Around 15°C, the latent periods (Fig. 4c) are approximately 3 days longer than Royle's incubation period for maximum leaf spots. We must allow 1 day between the appearance of lesions and first sporulation, which gives a difference of 2 days. The hop varieties are not cited in Mostade's paper (Fig. 14), but they could hardly be as susceptible as the highly susceptible varieties deliberately used by Royle for obtaining curves b in Fig. 4. They were also field plants while Royle's plants were potted cuttings. These are two reasons for a lengthened latent period and the third is that we averaged varying temperatures while Royle experimented with constant temperatures.

## 2. Spore concentration versus area dose

In spore trapping experiments with an automatic volumetric spore trap, the usual procedure is to transform spore counts into spores per $m^3$ of air and eventually to correct these concentrations for variations in trapping efficiency with wind-speed (Hirst, 1952).

Spores in the air are continuously driven away by wind, so that for maintaining a concentration of $\chi$ spores $m^{-3}$ around a spore trap in a wind of $u$ m s$^{-1}$, there must be a spore flow of $n = \chi.u$ spores $m^{-2}$ s$^{-1}$ through the plane perpendicular to wind direction. In given conditions of turbulence, this spore flow is proportional to the spore source output, expressed in spores s$^{-1}$. The spore flow through a unit area during a time of $t$ seconds is called the "Area Dose"; thus A.D. $= \chi.u.t$ (Gregory, 1961).

To calculate the area dose for 1 h, we multiply the mean hourly spore concentration in spores $m^{-3}$ by wind speed in m s$^{-1}$ × 3600. The total of hourly area doses over the sporulation period is proportional to the spore output of a generation of lesions, which in turn is proportional to the number of lesions of this generation. Thus, total area dose for successive sporulation periods is a rough estimate of the relative intensity of the corresponding preceding infections. If we want to relate numbers of airborne spores to previous events, area dose is evidently a more useful expression than spore concentration.

In principle, the absolute number of lesions per generation could also be calculated by transforming area dose into actual spore output of the plot or field with an appropriate formula (Gregory, 1961) and by dividing this plot or field output by the spore output per lesion. However, this would be difficult to apply with any precision because spore output per lesion varies with age of host or host part and sporulation conditions. Waggoner and Taylor (1958) inversely used area dose per day to calculate the number of spores per lesion released during a specific day by a known number of lesions of *Peronospora tabacina* in tobacco seed-beds.

### 3. Gradients of deposition and infection

In an early paper on spore trapping, Hirst (1953) stated that "area dose might well become a more convenient measure than spore number per cubic metre when more is known of the deposition of spores on crops". This remark alluded to the still more difficult problem of using spore trapping data of a given dissemination period to calculate the intensity of ensuing infection.

Knowledge about deposition processes and gradients was synthesized by Gregory (1961). Deposition gradients imply infection gradients. Infection gradients in the field were described for *Peronospora tabacina* by Waggoner and Taylor (1955) and Kosswig and Pawlik (1962), and for *P. manshurica* by Lim (1978). Relating the infection gradient to successive spore deposition gradients—and consequently to preceding spore releases—has not yet been achieved for a downy mildew. A major obstacle to this is the lack of evidence on how the sequence of dispersal, deposition, survival and infection develops after diurnal spore release on one hand and after release by rain and possibly accompanying wind gales on the other hand.

Hirst and Stedman (1963) considered that fungal spores liberated by rain would soon be washed from the air so that their dispersal would be important only for the local spread of disease. Van der Planck (1963) similarly speculated that dispersal of *Phytophthora infestans* during wet weather or on dewy nights probably increased foci in size and that sunny weather in the morning, while adverse for infection, would favour movement of sporangia. On the contrary, Hill (1966) held that diurnal dissemination of *Peronospora tabacina* ensured a slow, localized spread of disease while rainy, windy weather would disseminate conidia over a wide area. The contradiction between these authors may be partly due to the scale given to the word local.

## IV. Initial Inoculum

In warm climates with no season of plant dormancy, downy mildews may

maintain themselves throughout the year through successive conidial generations. Extreme cases are the downy mildew of maize in Java, *Sclerospora
maydis*, and its Philippine relative, *S. philippinensis*. These multiply through
the year in the conidial stage on maize which is continuously cropped in
both wet and dry seasons (Payak, 1975). Oospores are unknown in *S. maydis*
and their existence is uncertain for *S. philippinensis* (Dogma, 1975). In India,
*S. philippinensis* was found to infect maize at the start of the crop season
through conidial transfer from a weed, *Saccharum spontaneum* (Payak, 1975),
on which conidial sporulation is as persistly perennial as the grass itself
(Weston, 1923). Muskmelon, which is attacked by *Pseudoperonospora
cubensis*, cannot withstand winter in the Punjab, but another cucurbit, *Luffa
aegyptica*, is hardier and supports sporulating lesions of the same downy
mildew from autumn till spring. No oospores of the pathogen were found
in this area, though they are known in some other countries (Bains and
Jhooty, 1976). *Peronospora tabacina* also perpetuates during winter through
conidial multiplication on volunteer tobacco plants and on wild *Nicotiana*
species in Australia, in southern states of the USA and in Mediterranean
areas where these hosts are not exposed to freezing (Populer, 1964).

In most areas, however, the herbaceous hosts or host parts on which downy
mildews multiply do not withstand the adverse season and downy mildews
must either have a long-term multiplication system, allowing local perennation, or be brought in by long-distance dissemination, to provide inoculum
at the start of the growing season.

## A. Long-term multiplication

### 1. Survival in the asexual stage

A number of downy mildews have been demonstrated to survive as mycelium
in perennating plant parts or in seeds. This is the case with *Pseudoperonospora humuli* in hop rootstocks (Coley-Smith, 1965) and *Peronospora
destructor* in onion bulbs (Viranyi, 1975). Oospores are produced by these
two pathogens but are considered to be unimportant for disease transmission
compared to mycelial carry-over (Romanko, 1965; Golenia and Micinski,
1970; Nakov and Gyurov, 1972). *Peronospora farinosa* also overwinters as
mycelium in sugarbeet roots kept for seed in Byelorussia and Ukraine
(Gorovets and Marshakova, 1965; Tishchenko, 1966), while in Holland it
was presumed to be carried over by soil-borne oospores (Krijthe, 1965) and
in Australia by infected seed (Harrison, 1958).

*Plasmopara halstedii* may be present as mycelium or oospores in all parts
of dry sunflower seeds. Plants produced from such seeds rarely show
symptoms of systemic disease but they often carry "latent infection". In the

latter case, they may in turn produce infected seeds or infect the soil by allowing sporulation on their roots. Infected seeds can also contaminate healthy seedlings when they come into contact with the roots (Cohen and Sackston, 1974). Mycelium of *Sclerospora graminicola*, *S. maydis*, *S. philippinensis*, *S. sacchari* and *S. sorghi* is found in the pericarp or in the embryo of maize seeds from systemically colonized plants. Except apparently for *S. graminicola*, these infected seeds produce systemically diseased seedlings only when sown immediately after harvest and are not infective after normal storage (Sundaram *et al.*, 1973; Advincula and Exconde, 1975–1976). Whereas perpetuation in *S. maydis* and *S. philippinensis* is said to rely on conidial sporulation, oosporic perennation has been demonstrated for the other three downy mildews (Dogma, 1975; Payak, 1975).

The possibility of long-term conidial survival in the open has been demonstrated for several *Peronosporas*. In Australia, Hill (1962) found that less than 1% of *P. tabacina* conidia were still able to germinate after more than 100 days of storage at low humidities and constant temperatures in the laboratory. This was considered as a capacity for survival in the dry conditions of the regions in which *Nicotiana* species, the natural host plants, occur. In Germany, Kröber (1965, 1970) showed that conidia of *P. tabacina*, *P. parasitica* and *P. farinosa* stored in dry soil or dust in an open shed were still infective after several weeks in summer, and several months in winter. It was concluded that conidia of these fungi could possibly overwinter in barn soils. In moist soil, conidial longevity was shortened to a few days in summer and a few weeks in winter. Verhoeff (1960) also demonstrated that conidia of *Bremia lactucae* could survive up to 140 days at 2°C.

Thus, survival in the asexual stage in the downy mildews is not achieved by any particular morphological structure but by a simple stretching over time of the colonization subphase or of conidial longevity in special conditions.

It is apparent from the examples above that a downy mildew may have more than one mode of survival and that the mode considered as predominant may differ in separate geographic areas.

## 2. Oospores

Oospores are not known for all species of the downy mildews. In some species, they are reported to occur rarely or in small numbers but this apparent rarity may result from the search for oospores being made in an inadequate sample of the tissues of the host plant (Safeulla and Shaw, 1964). In *Bremia lactucae*, for example, oospores were repeatedly stated to be rare (Verhoeff, 1960; Ingram *et al.*, 1976) though as early as 1906 an investigator had noted that "they are produced in clusters and are plentiful in old and decayed stems" (Humphreys-Jones, 1971). In other species, they may be rare or non-existent in some areas or years and abundant in other places or times, as reported for *Plasmopara viticola* (Grünzel, 1961).

Oospores are usually disseminated in the soil with crop residues but they can also travel inside bulbs, as with *Peronospora destructor* (Blotnicka, 1974) or on seeds as with *Plasmopara halstedii* (Cohen and Sackston, 1974), *Peronospora manshurica* (Dunleavy, 1959; Pathak *et al.*, 1978), *P. tabacina* (Borovshaya, 1965) and *Sclerospora graminicola* (Kenneth, 1975).

The epidemiological importance of oospores is not necessarily related to their abundance and in some cases it may even prove difficult to demonstrate experimentally that they are able to germinate and infect. An illustration of this is furnished by the oospores of *Peronospora tabacina* which occur commonly but were found by Kröber (1969) to produce on the average one infection for $10^4$ to $10^6$ oospores in laboratory experiments in Germany. The number of primary foci detected in the whole of that country during the previous years was 1–5 per year. As tobacco does not overwinter in Germany, the fungus had no other means of survival than oospores—or possibly conidia—and the low number of foci was in accordance with the low infection efficiency of the oospores found in experiments (Kröber and Weinman, 1964). With oospores of *Plasmopara viticola*, on the contrary, germination and infection are obtained so easily that they were relied on for routine observations in the French forecasting system (Darpoux, 1949).

Knowledge of the effects of environmental and host conditions on oospore formation, longevity and infection and on the role of oospores in the development of epidemics of downy mildews is limited, except for *Plasmopara viticola*. The subject of oospores in this fungus was investigated more thoroughly and successfully than for any other downy mildew by German and French workers during the early part of this century. These early investigations unfortunately seem to be consistently ignored by present-day workers and therefore have been gradually sinking into oblivion.

(a) Oospore formation

There are indications that oospores are formed when conditions are unsuitable for asexual sporulation. In *Plasmopara viticola*, oospores are mostly formed in autumn in the mosaic-like flecks which result from late infections on ageing leaves; the density of oospores in these flecks can be as high as 250 per mm². They are found much less frequently in summer and then especially when low humidities inhibit asexual sporulation. In potted plants in the glasshouse, they are produced through the year on older leaves on which the lesions have remained small. They also occur more abundantly on *Vitis* species and vine cultivars which inhibit asexual sporulation than on hosts supporting a normal sporulation (Müller and Sleumer, 1934; Grünzel, 1961). Identical relationships with mosaic-flecks, leaf age, season and low humidity were described by Arens (1929b) for oospore formation in *Pseudoperonospora humuli*.

Oospore production was similarly reported to occur mostly at temperatures too high or humidities too low for conidial sporulation in *Peronospora viciae* (Pegg and Mence, 1970), on the lower, older leaves of field tobacco plants with *P. tabacina* (Peresypkin and Markhas'ova, 1966), on senescing cabbage leaves with *P. parasitica* (McMeekin, 1960) and at the end of the vegetative period of field plants of sunflower with *Plasmopara halstedii* (Tuboly, 1971). In *Sclerospora graminicola* on *Setaria viridis* and *S. sorghi* on sorghum and maize, oospore formation supersedes conidial sporulation as the host matures (Weston, 1924; Pratt and Janke, 1978).

These data suggest that the diagrams in Fig. 10 could be modified in the case of ageing host or host parts to accommodate an oospore production period.

(b) Oospore longevity

The longevity of oospores of *Plasmopara viticola* in the soil is normally limited to the spring and early summer following their late-season formation. Oospores which have been protected from drying during summer may exceptionally stay viable until spring of the second year (Müller and Sleumer, 1934; Zachos, 1959; Tsvetanov, 1976). In contrast, oospores of other downy mildews are generally reported to stay viable and infective in the soil for many years. This is the case with *Plasmopara halstedii* (Goossen and Sackston, 1968; Grinberg, 1972), *Peronospora manshurica* (Pathak *et al.*, 1978), *P. tabacina* (Borovskaya, 1968; Kröber, 1969a) and *P. destructor*, which apparently holds the record with an outdoor longevity of 25 years (McKay, 1957).

(c) End of dormancy, germination and infection

Oospores of *Plasmopara viticola*, produced in autumn, germinate under field conditions over a period of 2–3 months in late winter or spring. Times reported for this germination period vary from December–March in Greece to February–April in France and April–June in Germany. Experiments demonstrated that these oospores germinated earlier in the season, more abundantly and in a shorter time when they had previously been subjected to frequent rains and milder winter and spring temperatures. Oospores not exposed to rain during winter storage were not able to germinate. Cold was also reported to reduce the dormant period. The oospores of *P. viticola* germinate in wet weather, produce a short hypha terminated by a sporangium from which zoospores are released and splashed onto the vine leaves (Müller and Sleumer, 1934; Darpoux, 1943; Zachos, 1959). This germinating sequence

in the vineyard as well as the high percentage germination obtained by
Zachos (1959) with oospores stored in soil in the absence of plants, seem
to exclude the need for stimulation by contact with a plant part or plant
exudate.

Germination was induced in oospores of *Peronospora manshurica* by
washing for 1 week in running tap water, a treatment similar to rain exposure
in vine downy mildew. On agar, these oospores released an inhibitor of
conidial germination, which suggests that washing might break oospore
dormancy by eliminating an auto-inhibitor (Dunleavy and Snyder, 1962).

Extremely low numbers of infections were obtained by Kröber (1969a)
in Germany, with oospores of *Peronospora tabacina* which had been stored
in moderately moist soil in a glasshouse at 10–30°C for 8 to 50 months,
with a facultative 6 months exposure to field conditions in winter. In
Moldavia, however, Borovskaya (1968) reported that though the oospores
did not germinate after the first overwintering, they germinated well after
the second, and especially well after the third and fourth overwintering in
the soil. Furthermore high infection rates were obtained on tobacco. Initial
foci of tobacco downy mildew at the start of the growing season appear
consistently later in north-western Europe than in eastern Europe where
winter frosts are also much deeper and longer (Populer, 1964). Thus, there
is a suggestion that the earliness and frequency of initial foci in these areas
might be related to the response of oospores to climatic conditions.

With many downy mildews, infections by oospores are obtained easily
while attempts to germinate the oospores have had little or no success in
spite of various artificial treatments. This is reportedly the case with diverse
*Sclerosporas*, amongst which *S. sorghi* (Dogma, 1975). However, it was shown
recently that oospores of *S. sorghi* germinated consistently when maintained
between porous membranes adjacent to growing roots of several host and
non-host plants (Pratt, 1978). Germination of *Bremia lactucae* oospores was
also found to be stimulated by, though not dependent on, the presence of
germinating lettuce seed (Morgan, 1978). This approach clearly deserves to
be applied to other species of the downy mildews.


## B. Long-distance dissemination

In North America, the first outbreaks of *Peronospora tabacina* in the tobacco
crop occur each year in Florida and progress to Ontario between February
and June. In a series of papers, Valleau, followed by Stover and Koch (see
Populer, 1964) put forth the hypothesis that oospores and volunteer tobacco
plants are responsible for the early outbreaks in Florida and Georgia but
that in the other states of the Atlantic coast north of Georgia, initial foci
are regularly due both to oospores and also to airborne spores from the

south. In the more continental tobacco-growing states, extending from Tennessee to Ontario, it was considered that oospore survival was poor and that the disease consequently declines from year to year unless boosted by airborne spores from the south. Nusbaum (1975) similarly reported the overwintering of *Pseudoperonospora cubensis* in the southern regions of the USA and its northward progress during the growing season.

The uredospores of black rust, *Puccinia graminis*, have been demonstrated beyond doubt to travel up through the North American continent. Although the conidia of *Peronospora tabacina* may be presumed to be less robust than uredospores, tobacco blue mould was nevertheless observed to progress by up to 400 km per month in the direction of prevailing winds in Europe during its introduction phase in the early 1960s. Thus, there would seem to be a case for assuming long-distance conidial transport. A weak side in the hypothesis is that after its initial appearance in the Atlantic states from Florida to Maryland in 1931, *Peronospora tabacina* spread northwards very slowly and did not reach Ontario until 1974. Why the fungus should have needed 16 years to establish a route which it later covered in 4 months each spring is not evident but it strengthens the presumption that local perennation is the main cause of initial outbreaks. The annual south to north progression based on local overwintering is evidently the consequence of the gradual northward progress of isotherms in spring (Populer, 1964).

## V. Predictive and Descriptive Models

The very first forecasting systems to be set up in plant pathology concerned the downy mildew of vine, *Plasmopara viticola*. In Germany, Müller's incubation calendar method was put into practice as early as 1913 in the Baden area. The method was gradually perfected by a research team headed by Müller. Its final stage and the epidemiological basis of the method were described by Müller and Sleumer (1934) and summarized by Müller (1936). In France, several workers concurrently developed forecasting systems in different vine growing areas and the first official forecasting service was organized in 1912. A synthesis of the French systems was published by Darpoux (1943, 1949). The German and French systems much influenced one another and their final state is very similar. They illustrate perfectly the type of predictive model, integrating analytical knowledge on the relationships between infection, latent period and sporulation on the one hand and weather conditions on the other. As such, they are the true foregoers of modern simulation models, though the latter paradoxically seem not to have been applied yet to a downy mildew.

The essence of the French and German systems is to detect infection periods from the daily weather records. For infections considered as important, the

end of the latent period is forecast from a table—the "incubation calendar"—
in which the duration of the latent period is computed in relation to the
normal average temperatures from the beginning of May to the end of July.
A spray warning is then issued before the end of the predicted latent period,
so that the new sporulation wave emerges amid a protected foliage.
Susceptible stages of leaf and fruit and the ratio of unprotected on protected
leaf area are also taken into account for making the decisions.

In the French system, the maturation of oospores which will have been
stored in the open since late autumn is checked in the laboratory every
second or third day during the spring months. Weather conditions are
acknowledged as favourable for infection only when a sufficient percentage
of oospore germination, for instance 25%, has previously been reached in
the laboratory. The German system relies only on meteorological observations to determine the initial primary infection.

An interesting feature of these systems is that their principle implies that
successive infections give rise to separate, identifiable waves of sporulation,
as described in Section III. B1 in relation to spore dissemination. As a
corrective to this principle, Müller and Sleumer (1934) mentioned that
sporulation periods often overlap after the end of June but that after an
interruption caused by dry weather it is again easy to forecast the next
sporulation period.

After the initial organization of the French and German systems, a large
number of other forecasting methods for *Plasmopara viticola* were devised
in different vine-growing countries. During the last two decades, predictive
systems have also been described for *Plasmopara halstedii*, *Pseudoperonospora
humuli*, *Peronospora destructor* and *P. tabacina*. These forecasting systems
are too numerous to be listed here. They were reviewed for *Plasmopara viticola*
and *Peronospora tabacina* by Miller and O'Brien (1952, 1957). Later contributions can be traced easily in the *Review of Plant Pathology*.

Some of these systems were directly inspired by the French and German
models or were developed on a similar rationale without, however, improving
on or even reaching their level of elaborateness. Other systems are of the
empirical kind, being based on statistical correlations between disease
development and weather during the epidemic season or during the inoculum
survival season.

The empirical approach has been given new vigour in recent years by the
use of multiple regression analysis. This technique was applied successfully by
Royle (1973) to *Pseudoperonospora humuli* on hop. The best predictions were
given by an equation utilizing rain-wetness duration, rainfall amount and
airborne spore concentration while an equation based solely on relative
humidity and rainfall was only slightly inferior.

Descriptive mathematical models such as the logistic function do not seem
to have been utilized to characterize downy mildews epidemics, except by

Hill *et al.* (1967) in their work on the occurrence of *Peronospora tabacina* in a succession of years in Australia.


# VI. Outlook


The considerable volume of data which has accumulated on the epidemiology of the downy mildews is not distributed evenly over all the subjects. Research clearly tends to aggregate on some areas and to leave other areas under-investigated. As in many other scientific fields, research in epidemiology has developed by a kind of Brownian movement, implying duplication of effort and insufficient coherence. Duplication is not necessarily a fault because it is useful for verification, and Brownian movement has innovative virtues of its own. However, if we want epidemiological knowledge to become more firmly structured, more attention should be deliberately directed towards little investigated areas and more thought given to coherence.

There are important gaps in epidemiological knowledge on the downy mildews in the areas of changing host susceptibility with time and of oosporic multiplication. For many downy mildews, especially those with an early systemic stage, a good description of the successive disease facies during the seasonal cycle of the host would be an elementary requirement. New progress in the analysis of field epidemics would seem to be achievable by coordinated field observations and by exploiting the full potential of the volumetric spore trap instead of utilizing this instrument solely for the study of spore liberation. There are weak points where evidence is still inconclusive, in spite of repeated investigations; e.g the respective role of rain and dew in establishing infection.

There is a pressing need to integrate experimental data into a coherent whole. One condition of coherence is realism, which can be achieved by avoiding those experimental conditions which are not relevant to the field environment. Thus it is preferable to use *in vivo* rather than *in vitro* experiments, and to use dilute spore suspensions instead of highly concentrated inoculum. Perhaps the best aid to coherence is to devise experiments as if they were to provide parts of a simulation model. This requires primarily that time should always be included as one of the variables in the study of relationships between pathogen development and host or environmental conditions.


## References

Advincula, B. A. and Exconde, O. R. (1975–1976). *Philipp. Agric.* **59**, 244–255.
Angel, H. R. and Hill, A. V. (1931). *J. Counc. Sci. Ind. Res. Aust.* **4**, 178–181.

Arens, K. (1929a). *Jahrb. wiss. Bot.* **70**, 93–157.
Arens, K. (1929b). *Phytopathol. Z.* **1**, 169–193.
Bains, S. S. and Jhooty, J. S. (1976). *Indian Phytopathol.* **29**, 213–214.
Blaeser, M. and Weltzien, H. C. (1978). *Z. Pflanzenkr. Pflanzenschutz* **85**, 155–161.
Blaeser, M. and Weltzien, H. C. (1979). *Z. Pflanzenkr. Pflanzenschutz* **86**, 489–498.
Blotnicka, K. (1974). *Hodowla Rosl. Aklim. Nasienn.* **18**, 131–150.
Bonde, M. R. and Melching, J. S. (1979). *Phytopathology* **69**, 1084–1086.
Bonde, M. R., Schmitt, C. G. and Dapper, R. W. (1978). *Phytopathology* **68**, 219–222.
Borovskaya, M. (1965). *Zashch. Rast. Vred. Bolezn.* **10**, 44–45.
Borovskaya, M. (1968). *Mikol. Fitopatol.* **2**, 311–315.
Byford, W. J. (1967). *Ann. Appl. Biol.* **60**, 97–107.
Chang, S. C. and Twu, M. I. (1972). *Rep. Corn Res. Cent. Tainan Dais, Taiwan* **9**, 16–21.
Chang, S. C. and Wu, T. H. (1969). *Rep. Corn Res. Cent. Tainan Dais, Taiwan* **7**, 7–11.
Chang, S. C. and Wu, T. H. (1970). *Rep. Corn Res. Cent. Tainan Dais, Taiwan* **8**, 1–10.
Cohen, Y. (1976). *Aust. J. Biol. Sci.* **29**, 281–289.
Cohen, Y. (1977). *Can. J. Bot.* **55**, 1478–1487.
Cohen, Y. and Eyal, H. (1977). *Physiol. Pl. Pathol.* **10**, 93–103.
Cohen, Y. and Rotem, J. (1969). *Israel J. Bot.* **18**, 135–140.
Cohen, Y. and Rotem, J. (1970). *Phytopathology* **60**, 1600–1604.
Cohen, Y. and Rotem, J. (1971a). *Trans. Br. Mycol. Soc.* **57**, 67–74.
Cohen, Y. and Rotem, J. (1971b). *Phytopathology* **61**, 265–268.
Cohen, Y. and Sackston, W. E. (1973). *Can. J. Bot.* **51**, 15–22.
Cohen, Y. and Sackston, W. E. (1974). *Can. J. Bot.* **52**, 231–238.
Cohen, Y. and Sherman, Y. (1977). *Phytopathology* **67**, 515–521.
Cohen, Y., Perl, M. and Rotem, J. (1971). *Phytopathology* **61**, 594–595.
Cohen, Y., Perl, M., Rotem, J., Eyal, H. and Cohen, J. (1974). *Can. J. Bot.* **52**, 447–450.
Cohen, Y., Levi, Y. and Eyal, H. (1978). *Can. J. Bot.* **56**, 2538–2543.
Coley-Smith, J. R. (1965). *Ann. Appl. Biol.* **56**, 381–388.
Corbaz, R. (1972). *Phytopathol. Z.* **74**, 318–328.
Cruickshank, I. A. M. (1958). *Aust. J. Biol. Sci.* **11**, 162–170.
Cruickshank, I. A. M. (1961a). *Aust. J. Biol. Sci.* **14**, 58–65.
Cruickshank, I. A. M. (1961b). *Aust. J. Biol. Sci.* **14**, 198–207.
Cruickshank, I. A. M. (1963). *Aust. J. Biol. Sci.* **16**, 88–98.
Darpoux, H. (1943). *Ann. Epiphyt.* **9**, 177–205.
Darpoux, H. (1949). *Bull. Tech. Inf.* **41**, 403–411.
Dickinson, C. H. and Crute, I. R. (1974). *Ann. Appl. Biol.* **76**, 49–61.
Dogma, I. J. (1975). *Trop. Agric. Res. Ser.* No 8, 103–117.
Dunleavy, J. (1959). *Phytopathology* **49**, 537–538.
Dunleavy, J. and Snyder, G. (1962). *Phytopathology* **52**, 8.
Fletcher, J. T. (1976). *Ann. Appl. Biol.* **84**, 294–298.
Fried, P. M. and Stuteville, D. L. (1977). *Phytopathology* **67**, 890–894.
Golenia, A. and Micinski, B. (1970). *Rocz. Nauk. Roln.* Ser. E. **1**, 157–192.
Goossen, P. G. and Sackston, W. E. (1968). *Can. J. Bot.* **46**, 5–10.
Gorovets, V. K. and Marshakova, M. I. (1965). *Bot. Issled., Minsk* **7**, 53–59.
Gregory, P. H. (1961). "The Microbiology of the Atmosphere". Plant Science Monographs, Leonard Hill, London.
Grinberg, S. M. (1972). *Zashch. Rast., Mosk.* **17**, 19–20.
Grünzel, H. (1961). *Z. Pflanzenkr. Pflanzenschutz* **68**, 65–80.
Grünzel, H. (1963). *Nachrichtenbl. dtsch. Pflanzenschutzdienst, Berlin* **11**, 139–144.
Harrison, D. E. (1958). *J. Agric. Victoria* **56**, 675–677.
Hill, A. V. (1960). *Nature* **185**, 940.

Hill, A. V. (1961). *Aust. J. Biol. Sci.* **14**, 208–222.
Hill, A. V. (1962). *Nature* **195**, 827–828.
Hill, A. V. (1966). *Aust. J. Agric. Res.* **17**, 133–146.
Hill, A. V. (1969a). *Aust. J. Biol. Sci.* **22**, 393–398.
Hill, A. V. (1969b). *Aust. J. Biol. Sci.* **22**, 399–411.
Hill, A. V. and Green, S. (1965). *Aust. J. Agric. Res.* **16**, 597–607.
Hill, A. V., Paddick, R. G. and Green, S. (1967). *Aust. J. Agric. Res.* **18**, 575–587.
Hirst, J. M. (1952). *Am. Appl. Biol.* **39**, 257–265.
Hirst, J. M. (1953). *Trans. Br. Mycol. Soc.* **36**, 375–393.
Hirst, J. M. and Stedman, O. J. (1963). *J. Gen. Microbiol.* **33**, 335–344.
Humphreys-Jones, D. R. (1971). *Pl. Soil* **35**, 187–188.
Ingram, D. S., Tommerup, I. C. and Searle, L. M. (1976). *Ann. Appl. Biol.* **84**, 299–302.
Jadot, R. (1966a). *Parasitica* **22**, 55–63.
Jadot, R. (1966b). *Parasitica* **22**, 208–215.
Jones, B. L. (1971). *Phytopathology* **61**, 406–408.
Kenneth, R. (1970). *Indian Phytopath.* **23**, 371–377.
Kenneth, R. (1975). *CMI. Descr. Pathog. Fungi Bact.* No 452.
Kosswig, W. and Pawlik, A. (1962). *Z. Pflanzenkr. Pflanzenschutz* **69**, 462–465.
Krijthe, J. M. (1965). *A. Rep. 1964, Inst. Phytopathol. Res., Wageningen*, pp. 165.
Kröber, H. (1965). *Phytopathol. Z.* **54**, 328–334.
Kröber, H. (1967). *Phytopathol. Z.* **58**, 46–52.
Kröber, H. (1968). *Turk Cong. Microbiol.* (Istanbul, 12–14.9.1968) **13**, 177–200.
Kröber, H. (1969a). *Phytopathol. Z.* **64**, 1–6.
Kröber, H. (1969b). *Phytopathol. Z.* **66**, 180–187.
Kröber, H. (1970). *Phytopathol. Z.* **69**, 64–70.
Kröber, H. and Weinmann, W. (1964). *Phytopathol. Z.* **51**, 79–84.
Leu, L. S. and Tan, S. W. (1970). *Sugarcane Pathol. Newsl.* **5**, 16–21.
Lim, S. M. (1978). *Phytopathology* **68**, 1774–1778.
McKay, R. (1957). *Sci. Proc. R. Dublin Soc.* N.S. **27**, 295–307.
McMeekin, D. (1960). *Phytopathology* **50**, 93–97.
Mence, M. J. and Pegg, G. F. (1971). *Ann. Appl. Biol.* **67**, 297–308.
Milbraith, D. G. (1923). *J. Agric. Res.* **23**, 989–993.
Miller, P. R. and O'Brien, M. (1952). *Bot. Rev.* **18**, 547–601.
Miller, P. R. and O'Brien, M. (1957). *A. Rev. Microbiol.* **11**, 77–110.
Morgan, W. M. (1978). *Trans. B. Mycol. Soc.* **71**, 337–340.
Mostade, J. M. (1971). *Parasitica* **27**, 64–69.
Müller, K. (1936). *Z. Pflanzenkr. Pflanzenschutz* **46**, 104–108.
Müller, K. and Sleumer, H. (1934). *Landwirtsch. Jahrb.* **79**, 509–576.
Nakov, B. and Gyurov, L. (1972). *Rastit. Zashch.* **20**, 9–12.
Nusbaum, G. J. (1945). *Plant Dis. Rep.* **29**, 141–143.
Palti, J. and Chorin, M. (1964). *Phytopathol. Mediterr.* **3**, 50–56.
Pathak, V. K., Mathur, S. B. and Neergaard, P. (1978). *EPPO Bull.* **8**, 21–28.
Payak, M. M. (1975). *Trop. Agric. Res. Ser.* No 8, 81–91.
Pederson, V. D. (1961). *Diss. Abstr.* **22**, 703.
Pederson, V. D. (1964). *Phytopathology* **54**, 903.
Pegg, G. F. and Mence, M. J. (1970). *Ann. Appl. Biol.* **66**, 417–428.
Pegg, G. F. and Mence, M. J. (1972). *Ann. Appl. Biol.* **71**, 19–31.
Peresypkin, V. F. and Markhas'ova, V. A. (1966). *Zakhyst. Rosl.* 1966, 72–77.
Pinckard, J. A. (1942). *Phytopathology* **32**, 505–511.
Populer, C. (1964). *Bull. Inst. Agron. Stns Rech. Gembloux* **32**, 339–378, 435–508.
Populer, C. (1965). *Bull. Inst. Agron. Stns Rech. Gembloux* **33**, 388–404.

Populer, C. (1966). *Bull. Rech. Agron. Gembloux* **1**, 111–139.
Populer, C. (1978). *In*: "Plant Disease, An Advanced Treatise" (J. G. Horsfall and E. B. Cowling, Eds) Vol. II, 239–262. Academic Press, New York, San Francisco and London.
Pratt, R. G. (1978). *Phytopathology* **68**, 1606–1613.
Pratt, R. G. and Janke, G. D. (1978). *Phytopathology* **68**, 1600–1605.
Raffray, J. B. and Sequeira, L. (1971). *Can. J. Bot.* **49**, 237–239.
Rider, N. E., Cruickshank, I. A. M. and Bradley, E. F. (1961). *Aust. J. Biol. Sci.* **12**, 1119–1125.
Romanko, R. R. (1965). *Plant Dis. Rep.* **49**, 247–250.
Rotem, J. and Cohen, Y. (1970). *Phytopathology* **60**, 54–57.
Rotem, J., Palti, J. and Lomas, J. (1971). Final Rep. Res., Agric. Res. Organ., Volcani Center, Bet Dagan, Israel, 229 pp.
Royle, D. J. (1967). *Rep. Dep. Hop Res. Wye Coll. 1966*, 49–56.
Royle, D. J. (1970). *Ann. Appl. Biol.* **66**, 281–291.
Royle, D. J. (1973). *Ann. Appl. Biol.* **73**, 19–30.
Royle, D. J. and Thomas, G. G. (1971). *Physiol. Pl. Pathol.* **1**, 329–343.
Royle, D. J. and Thomas, G. G. (1972). *Trans. Brit. Mycol. Soc.* **58**, 79–89.
Royle, D. J. and Thomas, G. G. (1973). *Physiol. Pl. Pathol.* **3**, 405–417.
Safeulla, K. M. and Shaw, C. G. (1963). *Phytopathology* **53**, 1142.
Safeulla, K. M. and Shaw, C. G. (1964). *Phytopathology* **54**, 1436.
Safeulla, K. M. and Thirumalachar, M. J. (1955). *Phytopathology* **45**, 128–131.
Safeulla, K. M. and Thirumalachar, M. J. (1956). *Phytopathol. Z.* **26**, 41–48.
Sargent, J. A. and Payne, H. L. (1974). *Trans. Br. Mycol. Soc.* **63**, 509–518.
Schad, C. (1936). *Ann. Epiphyt.* **2**, 283–331.
Schad, C. (1943). *Ann. Epiphyt.* **9**, 19–25.
Schenck, N. C. (1968). *Phytopathology* **58**, 91–94.
Schmitt, C. G. and Freytag, R. E. (1974). *Pl. Dis. Rep.* **58**, 825–829.
Semangoen, H. (1970). *Indian Phytopathol.* **23**, 307–320.
Shatsky, A. L. (1935). *Pl. Prot., Leningrad* **6**, 75–85.
Shepherd, C. J. (1962). *Aust. J. Biol. Sci.* **15**, 483–508.
Shepherd, C. J. and Mandryk, M. (1962). *Trans. Br. Mycol. Soc.* **45**, 233–244.
Shepherd, C. J. and Mandryk, M. (1963). *Aust. J. Biol. Sci.* **16**, 77–87.
Shepherd, C. J. and Tosic, L. (1966). *Aust. J. Biol. Sci.* **19**, 335–337.
Shepherd, C. J., Simpson, P. and Smith, A. (1971). *Aust. J. Biol. Sci.* **24**, 219–229.
Siradhana, B. S., Dange, S. R. S., Rathore, R. S. and Jain, K. L. (1976). *Pl. Dis. Rep.* **60**, 603–605.
Sonoda, R. M. and Ogawa, J. M. (1970). *Mycologia* **62**, 1067–1069.
Sonoda, R. M. and Ogawa, J. M. (1972). *Hilgardia* **41**, 457–473.
Sundaram, N. V., Rama Sastry, D. V. and Nayar, S. K. (1973). *Indian J. Agric. Sci.* **43**, 215–217.
Thomas, C. E. (1977). *Phytopathology* **67**, 1368–1369.
Tishchenko, E. I. (1966). *Sakh. Svekla* **11**, 37.
Tsvetanov, D. (1976). *Gradinar. Lozar. Nauka* **13**, 137–143.
Tuboly, L. (1971). *Növényvéd. Kut. Intéz. Közl.* **5**, 51–62.
Uozumi, T. and Kröber, H. (1967). *Phytopathol. Z.* **59**, 372–384.
Valli, V. J. (1966). *Tobacco, N.Y.* **163**, 30–32.
Van der Planck, J. E. (1963). "Plant Diseases: Epidemics and Control". Academic Press, New York and London.
Verhoeff, K. (1960). *Tijdschr. Plantenziekten* **66**, 133–204.
Viranyi, F. (1975). *Növényvéd. Kut. Intéz. Évk.* **13**, 243–254.
Waggoner, P. E. and Taylor, G. S. (1955). *Pl. Dis. Rep.* **39**, 79–85.
Waggoner, P. E. and Taylor, G. S. (1958). *Phytopathology* **48**, 46–51.
Weston, W. H. (1923). *J. Agric. Res.* **23**, 239–277.

Weston, W. H. (1924). *J. Agric. Res.* **27**, 771–784.
Weston, W. H. and Uppal, B. N. (1932). *Phytopathology* **22**, 573–586.
Yarwood, C. E. (1937). *J. Agric. Res.* **54**, 365–373.
Yarwood, C. E. (1943). *Hilgardia* **14**, 595–691.
Zachos, D. G. (1959). *Ann. Inst. Phytopathol. Benaki* **2**, 193–355.
Zattler, F. (1931). *Phytopathol. Z.* **3**, 281–302.

Chapter 6

# Role of Wild Hosts in Downy Mildew Diseases

B. L. RENFRO[1] and S. SHANKARA BHAT[2]

[1]*The Rockefeller Foundation, GPO Box 2453, Bangkok, Thailand*
[2]*University of Mysore, Manasagangotri, Mysore 570 006, India*

## I. Introduction

The downy mildews are biotrophic organisms and hence are generally very specific to the type of hosts they attack, possibly because they have evolved in parallel with their hosts. Whereas certain downy mildew species have a very narrow host range, others have a wide host range which probably offers greater scope for widespread distribution and helps in their perpetuation.

Numerous papers have been published in the past few decades on the association of wild host plants with different downy mildew (DM) pathogens. The authors agree with Dinoor's statement (1974) that all wild plants are not technically weeds, nor are all weeds wild plants. Categorizing a particular plant species as a weed host is often complicated by the same species being considered non-economical in one place but economically important in another. Weston (1921a) suggested that "the wild grasses are the natural hosts of these oriental downy mildews from which they have passed and are passing to susceptible introduced crops such as maize". A similar view was held by Adam (1933) for tobacco DM (*Peronospora tabacina*).

Most of the DM pathogens develop systemically in plants, produce oospores and also very large numbers of asexual spores which are fragile, short-lived and believed to be disseminated only a few hundred metres by air currents in viable conditions. The principal means of seasonal carry-over is by internally borne mycelia and oospores. Both annual and perennial wild hosts, as well as cultivated plant species, are involved. Many of the DMs are most severe in the tropical and sub-tropical regions of the world where crop species are often grown under irrigation. This practice, together with volunteer plants, provides a means of the pathogens bridging dry and cool off-crop seasons.

Although many wild hosts have been reported for some DMs, in only a few cases has their role in disease dynamics been duly documented in the literature. The main purpose of this chapter is not simply to catalogue host ranges, but to present an overall survey of literature on wild hosts so as to assess the extent of their involvement in epidemiology.

## II. Diseases Caused by Common Genera

### A. *Basidiophora*

Each of the three species of *Basidiophora* has a narrow host range. *B. butleri* occurs on *Eragrostis* of the Gramineae and *B. kellermanii* on *Iva* of the Compositae. *B. entospora* had been reported to occur on eight genera of the Compositae in North America and to be important on China aster (*Callistephus hortensis* Cass. = *C. chinensis* Nees) in Texas, USA (Anon., 1937). The pathogen was reported to cause epidemics of DM on aster in Romania (Săvulescu and Săvulescu, 1952) and to occur on *Erigeron canadensis* Linn. though the role of this wild host was not described.

### B. *Bremia*

C. G. Shaw (personal communication, 1979) considered *Bremia* to be comprised of only one species although more than 20 species have been described by others. Regardless of species number, *B. lactucae* is the most economically important. It has a large host range among both cultivated and wild members of the Compositae (Ogilvie, 1943; Baudys, 1935; Crute and Davis, 1977) and it occurs on lettuce (*Lactuca sativa* L.) as well as on globe artichoke (*Cynara scolymus* L.), chicory (*Cichorium intybus* Linn.), endive (*C. endivia* L.) and on some ornamental plants of the family (Baudys, 1935). Many physiological races occur on lettuce and, furthermore, the races that attack one species apparently do not parasitize all other species of the same host genus (Crute and Davis, 1977; Ling and Tai, 1945; Ogilvie, 1946; Powlesland, 1954; Wild, 1947). Crute and Davis (1977) showed that isolates of *B. lactucae* from lettuce can cause DM only on species of *Lactuca* closely related to *L. sativa*. This information supports the earlier conclusion of Ogilvie (1943) that the presence of infected wild hosts near lettuce plantings constituted no danger. I. R. Crute (personal communication, 1979) stated that *B. lactucae* will attack other *Lactuca* species, but they are uncommon weeds and "it is highly unlikely that they play a role in the epidemiology of the (DM) disease in cultivated lettuce". The chief means of overwintering of the fungus is considered to be by oospores, though Weber and Foster (1928) reported DM on several wild plants allied to lettuce in Florida, USA and claimed they acted as sources of infection to produce severe losses in cultivated lettuce.

## C. *Peronospora*

Ainsworth (1971) stated that there are 75 widespread species of *Peronospora*. These species cause DM on a large number of economically important hosts and their wild relatives. However, wild hosts are of proven importance in agriculture only for blue mould (DM) of tobacco. Mycelia of the causal pathogen, *Peronospora tabacina*, perennate in plants of several wild *Nicotiana* species e.g. *N. glauca* R. Grah, *N. rustica* Linn., *N. repanda* Willd. ex Lehm. and *N. suaveolens* Lehm. and subsequently is the source of primary inoculum for plantings of cultivated tobacco (Pittman, 1930, 1932; Adam, 1933; Godfrey, 1941; Wolf, 1947; Pont and Hughes, 1961). Adam (1933) concluded that *P. tabacina* was indigenous on native wild tobacco in Australia and spread to tobacco cultivars introduced for commercial production. Pont and Hughes (1961) reported that blue mould was seed-borne in the wild host, *N. glauca*. The widely prevalent wild tobacco *N. repanda* in southern Texas, USA, was credited with the original widespread outbreaks of blue mould there, as well as to south-eastern USA (Wolf, 1947; Godfrey, 1941). In addition, they postulated that the fungus made a one-step migration to that area from southern California.

Other reports of wild hosts playing a significant role in the carry-over of *Peronospora* species are vague. The onion DM pathogen, *P. destructor*, was found to overwinter in Poland as viable mycelia in bulbs of a wild *Allium* species, which "appeared" to be the source of primary infection to cultivated onion (Rondomanski, 1973). Prior to 1938 *P. effusa* was only occasionally recorded on spinach in the Vancouver area of Canada, but was found to be prevalent on *Chenopodium album* L. that year and for the first time DM caused heavy damage to the canning crop of spinach (Eastham, 1939). McMeekin (1969) suggested that weeds might serve as hosts for strains of *P. parasitica* that attack commercially grown crucifers. *P. trifoliorum* was found to attack 10–60% of both wild and cultivated species of *Medicago* in Uzbekistan, USSR (Faizieva, 1968), but the role of these species in the epidemiology of cultivated lucerne was not ascertained.

Downy mildew of beet occurs world-wide and is a major disease. The pathogen *P. schachtii* (= *P. farinosa*) is capable of attacking all species and varieties of *Beta* and occurs naturally on wild beet in Europe, as well as, on sugar beet, garden beet and mangolds. While infection on wild beets has been implicated as a source of primary inoculum, Hughes (1959) and McKay (1952) discount their importance. McKay (1952) stated that while DM-diseased wild beets may occasionally act as a source of inoculum in maritime areas, it is a doubtful source of infection of any practical importance. W. J. Byford (personal communication, 1979) informed us that there is little evidence that wild hosts are of any importance for the spread of beet DM. Some weed beets originated from hybridization between cultivated and wild annual species in the Mediterranean region and are now a major weed problem in northern Europe. Byford pointed out the obvious danger of epidemics from this situation but that no associated outbreaks of DM had thus far been reported.

### D. *Peronosclerospora*

Like *Sclerospora*, *Peronosclerospora* is restricted to hosts of the Gramineae. Among the nine species reported, *P. maydis*, *P. sorghi*, *P. philippinensis* and *P. sacchari* incite DM diseases which are major constraints to production of commercial crops. Java DM (*P. maydis*) is largely restricted to Indonesia. It is reported to occur only on maize (*Zea mays* L.), teosinte (*Zea mexicana* (Schrad.) Reeves & Mangelsd.), a *Tripsacum* species and an interspecific *Pennisetum* cross (Semangoen, 1970; Pupipat, 1975). None of these wild species functions in the disease cycle on maize, nor does teosinte or *Tripsacum* occur naturally in Indonesia.

Sorghum DM (*P. sorghi*) is widespread and very destructive to maize and sorghum. The host range is limited to species of *Zea* and *Sorghum*, *Heteropogon contortus* Beauv.—Siradhana *et al.* (1980) have shown *Peronosclerospora heteropogoni* to be the maize DM fungus perpetuating on *Heteropogon contortus*, and *Panicum trypheron* Schult. The wild host, *Sorghum arundina-ceum* Roem. & Sch., was implicated as a collateral host as early as 1925 (Storey and McClean, 1930) in South Africa in furnishing inoculum to cultivated maize and sorghum. Later sorghum DM was stated to "spread from infected, poor perennial sorghum species" to kaffircorn and maize in South Africa by LeRoux (1961). Wild hosts are extremely important in the disease cycle of sorghum DM on corn and sorghum in Venezuela (Malaguti, 1976; Fernandez *et al.*, 1975; Frederiksen and Renfro, 1977). Malaguti (1976) found the wild perennial sorghums, *S. halepense* Pers., *S. verticilliflorum* Stapf and *S. arundinaceum*, were susceptible to DM, and were very common within and adjacent to cultivated fields. These perennial, rhizomatous weeds ensure fungal survival between seasons and provide constant inoculum, both as conidia and oospores.

Kenneth (1966) found that mycelia of *P. sorghi* perennate in underground parts of Johnson grass (*S. halepense*) which is the only known wild host in Israel but Kenneth and Klein (1969, 1970) claimed that "the great majority" of clones of Johnson grass are not infected and are considered resistant to sorghum DM. Williams and Herron (1974) concluded that shattercane or wild cane (*S. bicolor* Moench) was "a major means of oospore (of *P. sorghi*) overwintering" in Kentucky, USA, and a source of primary inoculum to maize grown along the Ohio River. Shattercane was also found to be infected and to harbour *P. sorghi* along the Wabash and Ohio Rivers in Indiana (Warren *et al.*, 1974) and Illinois (White *et al.*, 1978) and along the Republican River in Nebraska, USA (Partridge and Daupnik, 1979). The disease was expected to persist in these areas in shattercane and Johnson grass on which are formed both the conidia and oospores which readily attack maize and sorghum.

Sorghum DM is a major disease of maize, in Rajasthan, India, but there it neither attacks sorghum nor does it form oospores in maize tissue. *Hetero-pogon contortus*, a prevalent wild grass, is an important collateral host of *P. sorghi*. Oospores are produced abundantly in *H. contortus* and were described as the sole source of primary infection of DM on maize (Dange, 1976; Dange *et al.*, 1973; 1974).

*Dichanthium caricosum* L. Camus, a common wild perennial grass in Thailand, was found infected with what "resembled" *P. sorghi* (Wata-navanich *et al.*, 1976). Both conidia and oospores were produced and the former were infectious to maize and sorghum. However, the involvement of *D. caricosum* in the disease cycle on the two cultivated crops was not investigated.

*P. philippinensis* (Philippine DM) systemically infects wild sugarcane plants (*Saccharum spontaneum* Linn.) and persists in underground parts. Conidia formed on subsequent growth, constitute the primary inoculum for maize in northern India (Chona and Suryanarayana, 1955; Suryanarayana, 1961) and the Philippines (Exconde, 1970). In the Philippines, *Sorghum halepense* and *S. propinquum* (Kunth) Hitchc. serve as wild, perennial hosts for Philippine DM (Exconde, 1976). *Saccharum spontaneum* supports the seasonal carry over of *Peronosclerospora spontanea* also in the Philippines (Weston, 1921a) and Thailand (Titatarn and Syamananda, 1978) where it provides a source of primary inoculum to cultivated sugarcane and occasionally to maize.

Although sugarcane DM (*P. sacchari*) is a major disease of cultivated sugarcane and maize in several Asian and Oceanic countries, wild hosts are not implicated in the disease cycle. Teosinte is susceptible, but does not occur in the countries concerned but Johnson grass and Sudan grass were suspected to be collateral hosts by Leece (1941). Chang (1966) found that broom corn (*Sorghum bicolor* var. *nervosum*) and gama grass (*Tripsacum dactyloides* Linn.) were susceptible to sugarcane DM but were of no apparent importance in the disease cycle in Taiwan.


### E. *Plasmopara*


Wild hosts do not appear to serve a significant role in the epidemiology of DM diseases by *Plasmopara* spp. on cultivated crops. Among the 20 species reported (Ainsworth, 1971), *P. halstedii* on sunflower, *P. viticola* on grape vine and *P. nivea* on carrot, parsnip, parsley and other Umbelliferae are the species that cause important DM diseases on cropped plants.

Henry and Gilbert (1924) speculated that sunflower DM could become destructive in Minnesota, USA under favourable conditions because the fungus has many wild hosts. *Plasmopara halstedii* was reported on a large number of wild species in the Compositae, and on annual and perennial species of *Helianthus* by Orellana (1970). In Quebec, Canada the roots of dandelion (*Taraxacum officinale* Wigg.) and ragwort (*Senecio jacobaea* L.) from infested fields were found to harbour the pathogen (Anon., 1955). Leppik (1965) reported successful cross-inoculation in Iowa, USA between isolates from *Helianthus annuus* L. and *Dimorphotheca sinuata* DC., the latter a wild host which apparently does not occur in Iowa. Edward and Naim (1962) observed *P. halstedii* on *Vernonia cinerea* Less., a common weed in Uttar Pradesh, India. However, in none of the five cases cited has there been confirmation of the wild host being definitely involved in furnishing inoculum to cultivated sunflower. All perennial wild sunflowers tested in USSR by Krasnokutskaya *et al.* (1977) were found highly resistant to *P. halstedii*. W. E. Sackston (1979, personal communication) wrote that most of the results from cross inoculation indicate that DM from perennial species will not infect the annual sunflowers, and the pathogen from annual sunflower will not infect the perennial *Helianthus*.

*Plasmopara viticola* has a few wild hosts; e.g. *Ampelopsis veitchii* Hort., *Cissus* (*Vitis*) *caesia* Afzel., *Parthenocissus tricuspidata* Planch., *P. quinquefolia* Planch. and *Vitis coignetiae* Pulliat ex Planch. (Ducomet, 1925a; Deighton, 1938; Conners and Savile, 1944; Lustner, 1924; Simonet, 1925). The mycelium of *P. viticola* regularly overwinters in the cortical tissue and cane buds of the wild species, *Vitis californica* Benth., in California, USA, thereby perpetuating infection by downy mildew (Barrett, 1939). However, Barrett (1939) further reported that cultivated grapes, even in close proximity to severely infected wild grapes did not seem to contract downy mildew from this source.

Gooseberry DM (*P. ribicola*) was found to attack wild red currant (*Ribes rubrum* L.) heavily in France (Barthelet, 1930), but never to attack cultivated red currants in the same areas near Versailles and Meudon. All attempts to inoculate cultivated red currant with collections of *P. ribicola* from the wild species failed.

Kenneth and Kranz (1973) described *P. penniseti* on pearl millet (*Pennisetum typhoides* (Burm.) Stapf & Hubbard) from Ethiopia. It was found in the highlands in 1969 where pearl millet was introduced for experimental purposes in 1966. The DM disease was considered to have originated from one or more of the five wild species of *Pennisetum* native to the area.

### F. *Pseudoperonospora*

Among the five reported species of *Pseudoperonospora* (Ainsworth, 1971), *P. humuli* and *P. cubensis* which cause hop DM and cucurbit DM are the most economically important. Hop DM has been known to cause serious damage on wild hops (*Humulus* spp.) for several decades and is suspected to be endemic on them in England, Japan and USA (Salmon and Wormald, 1923). However, the wild hops in England are considered to be wild-growing "escape" plants of cultivated hops, *H. lupulus* L. (Salmon and Ware, 1924). Salmon (1926) found the wild-growing *H. lupulus* to be very prevalent in "hedges and waste places" in the hop growing areas and to be commonly infected. Inoculum from these plants was virulent on several cultivated hop varieties. The eradication of hops growing wild from hop growing vicinities was recommended by Salmon and Ware (1925a).

Nettles (*Urtica dioica* Linn. and *U. urens* Linn.) were found to be naturally infected by *P. humuli* in England. Cross-inoculation was successfully effected between isolates from wild nettles and cultivated hops by Salmon and Ware (1925b, 1926, 1928); however, they obtained no evidence that *P. humuli* could become permanently established on nettles. Hoerner (1940) successfully inoculated several species of *Cannabis*, *Celtis* and *Urtica* with *P. humuli* from cultivated hops in USA and considered *Pseudoperonospora cannabina*,

*P. celtidis* and *P. urticae*, their respective DM causal organisms, to be physiological races of a single species. Jones (1932) found the common nettle of British Columbia, Canada, *U. lyallii* S. Wats., susceptible to hop DM and cautioned that nettles may need to be eradicated from hop growing areas in the future. Conversely, Ducomet (1925b) in France, and Blattny (1927) in Czechoslovakia were unable to effect cross-inoculation between nettle and hops. Too much importance need not be attached to this apparently conflicting evidence since the situation concerning the importance of wild hosts to hop DM had been succinctly summarized by D. J. Royle (personal communication, 1979). He commented that *P. humuli* is "highly specialized to hop and no involvement of alternate hosts has ever been seriously proposed".

There is no evidence that any wild host is important in the disease cycle of *P. cubensis* even though it has been reported from some 40 wild and cultivated species of the Cucurbitaceae. Bains and Jhooty (1976) have recently concluded that in Punjab, India, *P. cubensis* "Perpetuates in the form of active mycelium on self-sown or cultivated sponge gourd vines", and they did not implicate any wild host species.

## G. *Sclerophthora*

Four species of *Sclerophthora* have been reported to attack members of the Gramineae. The crazy top organism, *S. macrospora*, has been reported on 141 species (Raghavendra, 1974) since Saccardo (1890) first reported *Sclerospora macrospora* on *Alopecurus* sp. in Australia. In nature, however, relatively few hosts have been found to play a cardinal role in the disease cycle on commercial crops. Two of the earlier reports were by Weston (1921b) and Peglion (1930); respectively, they considered the common weeds, "cheat" grass (*Bromus commutatus* Schrad.) involved in DM of wheat in areas subject to flooding in Tennessee and Kentucky, USA, and *Phragmites communis* Trin., in flood waters in central Italy, to furnish a constant source of oospores for infection of cultivated cereal crops. Several biennial and perennial wild grasses, especially *Agropyron tsukushiense* (Honda) Ohwi, *Alopecurus aequalis* Sobol., *Beckmannia syzigachne* Fernald, *Phalaris arundinacea* L. and *Poa annua* L., are important sources of primary DM infection of rice seedlings in nurseries in the Kyoto, Shiga, and Fukui prefectures of Japan (Katsura *et al.*, 1954). *Bromus inermis* Leyss., *Phalaris arundinacea* and *Poa pratensis* L. were considered the most important sources of inoculum in South Dakota, USA for infecting cereals and grasses (Semeniuk and Mankin, 1964). The pathogen was further reported to cross-infect with rice cultivars and with three prevalent and frequently infected wild annual grass species, *Eragrostis cilianensis* (All.) Link, *E. pectinacea* (Michx.) Nees and *Panicum capillare* L. (Semeniuk, 1976). Raghavendra (1974) observed six wild grasses around Ragi or finger millet (*Eleusine coracana* Gaertn.) fields in Mysore, India to harbour the pathogen. These annual and perennial wild grass hosts appear to serve an important role in perpetuating *S. macrospora* and as sources of primary inoculum, both sporangia and oospores.

*S. rayssiae* var. *rayssiae* causes moderate damage to cultivated barley (*Hordeum vulgare* L.) in northern Israel. It produces local lesions on leaves of two wild grasses, *H. spontaneum* C. Koch and *H. murinum* L. (Kenneth *et al.*, 1964), but the occurrence is considered unimportant. *S. rayssiae* var. *zeae* causes brown stripe downy mildew of maize which is a major disease in northern India. Singh *et al.* (1970) reported large crab grass (*Digitaria sanguinalis* (L.) Scop.) susceptible in nature and Bains *et al.* (1978) proved that this very prevalent wild grass species was a source of primary inoculum to maize. Chamswarng *et al.* (1976) reported brown stripe DM on white birdfoot grass (*D. bicornis* (Lamk.) Roem. & Sch. ex Loud.) in Thailand, but were unable to demonstrate its connection to maize.

## H. *Sclerospora*

Only one economically important species remains in the genus *Sclerospora* following transfer to *Peronosclerospora* of those species that produce conidia, i.e. asexual spores which germinate directly, (Shaw, 1978). Five other commercially unimportant species are retained in *Sclerospora* in which the asexual stage has not been discovered. *S. graminicola* causes "green ear" of the cultivated pearl millet (*Pennisetum typhoides*) and is commonly referred to as graminicola DM on the other host species of the Gramineae which it infects. The disease occurs on both wild and cultivated species of *Setaria, Panicum, Echinochloa, Pennisetum, Sorghum, Zea, Euchlaena, Saccharum, Agrostis* and *Chaetochloa*. However, the movement of inoculum from one host species to another in nature has not been well demonstrated. Melhus *et al.* (1928) found both *Setaria viridis* (Linn.) Beauv. and maize infected from overwintered oospores in Iowa, USA and successfully inoculated teosinte, maize and species of *Setaria, Saccharum* and *Sorghum*. In a reported case from India, no infection resulted from cross inoculation tests among isolates from pearl millet, *Setaria* and *Panicum* (Uppal and Desai, 1932).

## III. Management of Wild Hosts

Wild plants are older than agriculture itself where they are often present as weeds. As such they not only reduce yields, but they increase costs of production, increase labour and generally interfere with agricultural operations. In addition, some wild plants are collateral hosts of the DM pathogens and often provide a means of "bridging" unfavourable or un-cropped seasons. Obviously an important means of control for the DM diseases would be to disrupt this source of primary inoculum. Since the term "control" evokes a notion of finality which in practice is unlikely, the suppressive management of wild hosts is preferred in the present context even though eradication of wild hosts from the proximity of cultivated crops has been recommended for controlling certain DM diseases (Salmon and Ware, 1926; Jones, 1932; Pittman, 1932; LeRoux, 1961; Suryanarayana, 1961). Indeed, the eradication of particular wild species from certain localities is quite feasible provided appropriate manpower, funds and follow-up are allocated.

Wild host plants should be managed by conventional cultural, chemical and, if practical, biological means to interrupt pathogen survival between seasons. This would reduce the chance of an epidemic and significant yield loss from DM and help continue the world's abundant, increasing food supply. The objective should be to reduce the population of wild hosts to as low a level as is economically feasible and to prevent seed formation. Cultural means include crop rotation, deep ploughing, mechanical cultivation, hand hoeing, mowing, grazing and, occasionally, burning. A large number of selective, environmentally acceptable herbicides are available for management deployment.

## IV. Concluding Remarks

A perusal of the literature on the role of wild hosts reveals that some of them play a decisive role in the development of epidemics of certain DM diseases on commercially grown cropped plants. Wild plants have played a significant role in the origin and development of the DM pathogens and are important in their perpetuation. The epidemiological role that wild hosts fulfil has not been established for numerous important downy mildew diseases. This lack of information leaves a void, critical to understanding the dynamics involved and for disease control. Resolution of their importance to economic crop plants is therefore a challenge of the future.

### Acknowledgements

We express our sincere thanks to Dr M. A. Rau, Post-graduate Department of Botany, University of Mysore, Mysore, India for providing the botanical nomenclature of the host plant species, and to Dr C. G. Shaw for his critical review of the manuscript.

We are grateful to Dr T. Kajiwara, TARC, Japan for supply and translation of pertinent sections of articles published in Japanese and also to Drs W. J. Byford, A. G. Channon, I. R. Crute, J. M. Dunleavy, D. S. Ingram, R. G. Kenneth, J. Rotem, D. J. Royle, W. E. Sackston and W. C. Schnathorst for providing critical information and assessment of the role of wild hosts in their areas of crop specialization.

## References

Adam, D. B. (1933). *J. Dep. Agric. Victoria* **31**, 412–416.

Ainsworth, G. C. (1971). "Dictionary of the Fungi". Comm. Mycol. Inst., Kew, Surrey.

Anon. (1937). *Pl. Dis. Reptr.* **21**, 141 (Mimeo).

Anon. (1955). *Rep. Minist. Agric. Canada*, 166 pp.

Bains, S. S. and Jhooty, J. S. (1976). *Indian Phytopath.* **29**, 213–214.

Bains, S. S., Jhooty, J. S., Sokhi, S. S. and Rewal, H. S. (1978). *Pl. Dis. Reptr.* **62**, 143.

Barrett, J. T. (1939). *Phytopathology* **29**, 822–823 (Abstr.).

Barthelet, J. (1930). *Rev. Path. Veg. Ent. Agric.* **17**, 352–355.

Baudys, E. (1935). *Letak Fytopath. Sekce Semsk. Vyzk. Ust. Zemed. (Leafl. Phytopath. Sect. Reg. Agric. Exp. Sta.)* **93**, Brno. 2 pp.

Blattny, C. (1927). *Sb. Vyzk. Ust. Zemed.* RCS. 27a, 5–274; 297–299; 301–304.

Chamswarng, C., Pupipat, U., Sommartaya, T. and Renfro, B. L. (1976). *Kasetsart J.* **10**, 14–24.

Chang, S. C. (1966). *Rep. Corn Res. Center*, No. 4, 1.

Chona, B. L. and Suryanarayana, D. (1955). *Indian Phytopath.* **8**, 209–210.

Conners, I. L. and Savile, B.D.O. (1944). *Twenty-third A. Rep. Canadian Pl. Dis. Survey* **18**, 122 (Mimeo).

Crute, I. R. and Davis, A. S. (1977). *Trans. Br. mycol. Soc.* **69**, 405–410.

Dange, S. R. S. (1976). *Kasetsart J.* **10**, 121–127.

Dange, S. R. S., Jain, K. L., Siradhana, B. S. and Rathore, R. S. (1973). *Curr. Sci.* **42**, 834.

Dange, S. R. S., Jain, K. L., Siradhana, B. S. and Rathore, R. S. (1974). *Pl. Dis. Reptr* **58**, 285–286.

Deighton, F. C. (1938). *Rep. Dep. Agric. S. Leone*, 1937, 45–47.

Dinoor, A. (1974). *A. Rev. Phytopath.* **12**, 413–436.

Ducomet, V. (1925a). *Rev. Path. Veg. Ent. Agric.* **12**, 129–130.

Ducomet, V. (1925b). *Rev. Path. Veg. Ent. Agric.* **12**, 248–254.

Eastham, J. W. (1938–1940). *Rep. B.C. Dep. Agric.*, 1938, L42–L48, 1939; ibid., 1939, B57–B60, 1940.

Edward, J. C. and Naim, J. (1962). *Sci. Cult.* **28**, 190–191.

Exconde, O. R. (1970). *Indian Phytopath.* **23**, 275–284.

Exconde, O. R. (1976). *Kasetsart J.* **10**, 94–100.

Faizieva, F. (1968). *Uzbek. Biol. Zh.* **12**, 13–15.

Fernandez, A. B., Malaguti, G. and Nass, H. (1975). *Agronomia Trop.* **25**, 367–380.

Frederiksen, R. A. and Renfro, B. L. (1977). *A. Rev. Phytopath.* **15**, 249–275.

Godfrey, G. H. (1941). *Pl. Dis. Reptr.* **25**, 347–353 (Mimeo.).

Henry, A. W. and Gilbert, H. C. (1924). *Minn. Stud. Pla. Sci. Biol. Sci.* **5**, 295–305.

Hoerner, G. R. (1940). *J. Agric. Res.* **61**, 331–334.

Hughes, W. (1959). *Rep. Inst. Int. De Recherches Betteravieres, Brussels*.

Jones, W. (1932). *J. Inst. Brew.* **29**, 194–196.

Katsura, K., Tokura, R. and Furuya, T. (1954). *Sci. Rep. Fac. Agric. Saikyo Univ.* **6**, 49–66.

Kenneth, R. (1966). *Scr. Hierosol.* **18**, 143–170.
Kenneth, R. and Klein, Z. (1969). *Hassadeh* **49**, 17–20.
Kenneth, R. and Klein, Z. (1970). *Israel J. Agric. Res.* **20**, 183.
Kenneth, R. and Kranz, J. (1973). *Trans. Br. Mycol. Soc.* **60**, 590–593.
Kenneth, R., Koltin, Y. and Wahl. I. (1964). *Bull. Torrey Bot. Club* **91**, 185–193.
Krasnokutskaya, O. N., Ilatovskii, V. P., Slyusar, E. L. and Shkuropat, Z. Y. (1977). *Mikologiya Fitopatologiya* **11**, 149–155.
Leece, C. W. (1941). *Tech. Commun. Bur. Sug. Exp. Stns. Qd* **5**, 111–135.
Leppik, E. E. (1965). *Pl. Dis. Reptr* **49**, 940–942.
LeRoux, P. M. (1961). *Fmg S. Afr.* **37**, 85–88.
Ling, L. and Tai, M. C. (1945). *Trans. Br. Mycol. Soc.* **28**, 16–25.
Lüstner, G. (1924). *NachrBl. dt. Pflschutzdienst.* **4**, 74–75.
Malaguti, G. (1976). *Kasetsart J.* **10**, 160–163.
McKay, R. (1952). "Sugar Beet Diseases in Ireland". Irish Sugar Co., Dublin, 9–11.
McMeekin, D. (1969). *Phytopathology* **59**, 693–696.
Melhus, I. E., Van Haltern, F. H. and Bliss, D. E. (1928). Iowa Agric. Expt. Stn. Res. Bull. **111**, 297–338.
Ogilvie, L. (1943). *Rep. Agric. Hort. Res. Stn. Univ. Bristol*, 90–94.
Ogilvie, L. (1946). *Rep. Agric. Hort. Res. Stn. Univ. Bristol*, 147–150.
Orellana, R. G. (1970). *Bull. Torrey Bot. Club* **97**, 91–97.
Partridge, J. E. and Daupnik, B. L. (1979). *Pl. Dis. Rep* **63**, 154–155.
Peglion, V. (1930). *Boll. E. Staz. Pat. Veg.* N.S. **10**, 153–164.
Pittman, H. A. (1930). *J. Dep. Agric. West. Aust.* 2nd Ser. 7, 469–476.
Pittman, H. A. (1932). *J. Dep. Agric. West. Aust.* 2nd Ser. 9, 97–103.
Pont, W. and Hughes, I. K. (1961). *Qd. Agric. J.* **18**, 1–31.
Powlesland, R. (1954). *Trans. Br. Mycol. Soc.* **37**, 362–371.
Pupipat, U. (1975). *A. Rep. Thai. Nat. Corn and Sorghum Prog.* 287–292.
Raghavendra, S. (1974). Ph.D. Thesis, Univ. Mysore, India. 199 pp.
Rondomanski, W. (1973). *Rev. Pl. Path.* **52**, 293 (Abstr. 1369).
Saccardo, P. A. (1890). *Hedwigia* **29**, 154–156.
Salmon, E. S. (1926). *East Malling Res. Stn., Kent.* 31 pp.
Salmon, E. S. and Ware, W. M. (1924). *Gardeners Chronicle* **76**, 265.
Salmon, E. S. and Ware, W. M. (1925a). *J. Min. Agric.* **31**, 1141–1151; **32**, 30–36.
Salmon, E. S. and Ware, W. M. (1925b). *Ann. Appl. Biol.* **12**, 121–151.
Salmon, E. S. and Ware, W. M. (1926). *J. Min. Agric.* **33**, 149–161.
Salmon, E. S. and Ware, W. M. (1928). *Ann. Appl. Biol.* **15**, 352–370.
Salmon, E. S. and Wormald, H. (1923). *J. Min. Agric.* **30**, 430–435.
Săvulescu, T. and Săvulescu, O. (1952). *Acad. Romanian People's Repb.* pp. 327–457.
Semangoen, J. (1970). *Indian Phytopath.* **23**, 307–320.
Semeniuk, G. (1976). *Pl. Dis. Reptr* **60**, 745–748.
Semeniuk, G. and Mankin, C. J. (1964). *Phytopathology* **54**, 409–416.
Shaw, C. G. (1978). *Mycologia* **70**, 594–604.
Simonet. (1925). *J. Soc. Nat. Hort. France.* Ser. 4, **26**, 422–425.
Singh, J. P., Renfro, B. L. and Payak, M. M. (1970). *Indian Phytopath.* **23**, 194–208.
Siradhana, B. S., Dange, S. R. S., Rathore, R. S. and Singh, S. D. (1980). *Curr. Sci.*, **49**, 316–317.
Storey, H. H. and McClean, A. P. D. (1930). *Phytopathology* **20**, 107–108.
Suryanarayana, D. (1961). *Indian Fmg* **11**, 11.
Titatarn, S. and Syamananda, R. (1978). *Pl. Dis. Reptr* **62**, 29–31.
Uppal, B. N. and Desai, M. K. (1932). *Indian J. Agric. Sci.* **2**, 667–678.
Warren, H. L., Scott, D. H. and Nicholson, R. L. (1974). *Pl. Dis. Reptr* **58**, 430–432.
Watanavanich, P., Pupipat, U., Sommartaya, T., Choonhawongse, K. and Chamswarng, C. (1976). *Kasetsart J.* **10**, 25–31.

Weber, G. F. and Foster, A. C. (1928). *Fl. Agric. Stn. Bull.* **195**, 303–333.
Weston, W. H. Jr. (1921a). *J. Agric. Res.* **20**, 669–684.
Weston, W. H. Jr. (1921b). *U.S.D.A. Circular* **186**, 3–6.
White, D. G., Jacobsen, B. J. and Hooker, A. L. (1978). *Pl. Dis. Reptr* **62**, 720.
Wild, H. (1947). *Trans. Br. Mycol. Soc.* **31**, 112–125.
Williams, A. S. and Herron, J. W. (1974). *Pl. Dis. Reptr* **58**, 90–91.
Wolf, F. A. (1947). *Phytopathology* **37**, 721–729.

Chapter 7

# Cytology and Genetics of Downy Mildews

I. C. TOMMERUP

*Department of Soil Science and Plant Nutrition, Institute of Agriculture, University of Western Australia, Nedlands, Western Australia 6009*

## I. Introduction

Progression of downy mildew fungi through the various phases of their life cycles presents a complex developmental sequence. The changes are a result of their genetic competence reacting with the environment, which for these biotrophic fungi involves interactions with living plant cells. The developmental changes reflect the cytological differentiation that occurred earlier at the subcellular level and which must have a molecular basis. The ultimate aim is to understand them and the nature of the controlling mechanisms. Identification of the series of cytological events which comprise each developmental step, provides a basis for this.

## II. The Nucleus

A fundamental characteristic of downy mildew fungi is the mode of organization of their genetic material in the nucleus and the way in which it is distributed during division by mitosis and meiosis. The interaction of the nuclear and cell cycles during all the various developmental phases has genetic consequences.

### A. Mitosis

Quiescent conidia, the older portions of somatic hyphae, and haustoria of downy mildew fungi contain interphase nuclei which are subglobose, and have a uniformly particulate nucleoplasm with a large, diffuse electrondense nucleolus and a nucleus-associated organelle (NAO) (Chou, 1970; Coffee and Hickey, 1977; Ingram et al., 1976; Sargent and Payne, 1974). The NAO is a centriole (Sargent and Payne, 1974) like those of other Oomycetes (Heath, 1978). In germinating and germinated conidia (Sargent and Payne, 1974) and young hyphae (Ingram et al., 1976), the interphase nuclei have chromatin with variable and different staining intensities indicative of changes in structure. The altered structure may be different degrees of chromatin contraction and reflect the change from a quiescent to an active metabolic state in the spore. The degree of chromatin condensation is correlated with DNA template activity and seems to have a role in genome expression (Kwiatkowska and Maszewski, 1979; Nagl, 1978).

The onset of mitosis is recognized by the appearance of an intranuclear spindle and two NAOs (Ingram et al., 1976; Sargent et al., 1973, 1977; Tommerup et al., 1974). The nucleolus has a lateral position during mitosis and it constricts at telophase. Nucleolar fission (as opposed to dispersion) and reassembly result in continuous nucleolar activity during mitosis. At metaphase microtubules traverse the nucleus. There is an absence of a metaphase plate with the chromosomes appearing dispersed over the nuclear area. The length from the pole to the point of attachment of the microtubules to the chromosomes seems variable. During anaphase the microtubules are shorter and the chromosomes move to the poles. The structure and behaviour of the nuclei and NAOs during mitosis closely resembles that of other Oomycetes (Heath, 1978).

## B. Meiosis

Cytological evidence suggests that meiosis occurs in the oogonia and antheridia of *Bremia lactucae* (Sargent *et al.*, 1977) *Peronospora parasitica* (Sansome and Sansome, 1974) and *Sclerospora graminicola* (Sansome, 1966). Karyogamy occurs in the oosphere after fertilization, the only haploid nuclei being present in the gametangia. Microspectrophotometric analysis of nuclei in *B. lactucae* indicates that the DNA of mitotic telophase nuclei is 2X, the early meiotic prophase I nuclei 4X and the second division meiotic telophase nuclei 1X. This information is consistent with meiosis occurring in the gametangia. Genetical evidence supporting diploidy in the vegetative phase has been obtained from studies on the inheritance of virulence factors (Michelmore and Ingram, personal communication). All the findings are in agreement with the evidence and conclusions for other Oomycetes (Fincham *et al.*, 1979) and support the view that the vegetative phase is diploid with sexual fusion of haploid nuclei in the oosphere following gametangial meiosis.

Within individual oogonia and antheridia meiosis is a near synchronous event (Sansome and Sansome, 1974; Sargent *et al.*, 1977; Tommerup *et al.*, 1974). At early prophase I the chromatin is dispersed and there is a large lateral nucleolus. The nucleolus decreases in size during prophase and it has not been recognized in late metaphase to anaphase nuclei (Sargent *et al.*, 1977). At metaphase each polar NAO lies in a depression of the nuclear envelope which is entire. Median longitudinal sections indicate that the intranuclear microtubules of *B. lactucae* terminate in a structure resembling the kinetochore of *Saprolegnia* (Heath, 1978; Howard and Moore, 1970). The chromosomes of *B. lactucae* become organized in a metaphase plate. That arrangement appears to be a feature of meiotic metaphase in Oomycetes (Howard and Moore, 1970). At anaphase the nucleus of *B. lactucae* elongates and the chromosomes move to the poles.

The second meiotic division follows almost immediately and the spindles are orientated at right-angles to those in the first division so that the four telophase II nuclei are closely associated. The behaviour of the centrioles, the spindle and chromosome arrangements are similar to that of the first divisions. The second division is intranuclear but the nuclear envelope does not appear to remain intact throughout both divisions as has been suggested for *Saprolegnia* (Beakes, 1980d; Beakes and Gay, 1977; Howard and Moore, 1970).

## C. Nuclear fusion

Nuclear transfer from antheridia to oospheres in *B. lactucae* is presumed to occur at the stage when the oospheres are delimited only by a double membrane (Sargent *et al.*, 1977). Karyogamy seems to follow very soon after the antheridial nucleus enters the oosphere. They have the appearance of migrating nuclei in sporangiophores with partially contracted chromatin and microtubules radiating from their apical NAOs. Fusion seems to be initiated when the NAOs are in close proximity and to occur in the apical region near them. During fusion, the chromatin appears condensed like that of early to mid-prophase nuclei. It intermingles concurrently with the change in fusion nucleus shape from a figure eight to a sphere and then gradually disperses. Some of these stages have been described for *P. parasitica* (Sansome and Sansome, 1974) and they resemble a marine parasitic Oomycete (Schnepf *et al.*, 1978b).

During nuclear fusion and the continued development of an oospore of *B. lactucae* the supernumerary haploid nuclei of the periplasm and antheridium sometimes undergo mitosis before beginning to autolyse.

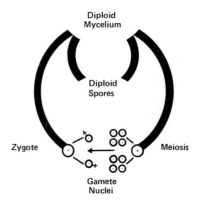

Fig. 1.   A diagram of the life cycle of a downy mildew. The diploid phase is represented by a thick line and large nuclei. The haploid state is represented by a thin line and small nuclei.

## D. Diploidy

The greater portion of the life cycle of downy mildew fungi is diploid (Fig. 1). The difference in genetic endowment between diploid and haploid inheritance has several consequences, some of which confer obvious biological advantages. At the population level in the field, a dual set of genes provides for a wider range of accommodative responses to unfavourable conditions and an increase in plasticity and potential variability of the organism (Nagl, 1978). A significant role of the diploid condition in the propagules and host-dependent phase may be to increase the adaptive ability of these biotrophs and enable them to overcome host resistance.

### E. Polyploidy

The multinucleate state of the downy mildew thallus is genetically fixed. Multinucleate cells are functionally polyploid cells (Nagl, 1978). Polyploidy can have quantitative and qualitative effects on gene expression. The coordination of activities is more easily controlled in multinucleate cells than in uninucleate tissues. Nagl (1978) has hypothesized that DNA increase is required in organisms with low DNA values during ontogenesis.

Polyploid nuclei have not been identified in any species of downy mildew. This may be a reflection of the few cultures used in these studies (Tommerup *et al.*, 1974; Sansome and Sansome, 1974). They occur in *Phytophthora megasperma* (Sansome and Brazier, 1974). The presence of polyploid nuclei would be of significance in genetical studies of disease resistance and pathogenicity.

## III. Cellular Differentiation

During the processes of infection, colonization and reproduction in a host, downy mildew fungi exhibit a sequence of distinct morphological stages. Near synchronous ontogenic development of a high percentage of propagules can be achieved and the populations used to study changes in gene expression.

### A. Vegetative development and the hyphal tip

A feature of unicellular organisms is that active growth and diversification proceed uninterrupted by cytokinesis. The various portions of the thallus have differing functions which are defined by the activities of the organelles in each region. If the thallus is homokaryotic, differentiation results from the unequal expression of the same genetic material.

Growth of hyphae is highly polarized and takes place in the tip (Fig. 2). The cytoplasm of the tip has a distinct organization. The ultrastructural organization of the apex of germ tubes and intercellular hyphae are alike. Both are similar to the papilla region in conidia just prior to germ tube emergence and to the bud of a secondary vesicle (Ingram *et al.* 1976). Comparisons between them and conidiophore apices, made with interference contrast microscopy, indicate close similarity in structure (Tommerup, unpublished). The length of each tip zone is a variable which can be correlated in 24- to 48-h-old colonies of *B. lactucae* with the rate of hyphal elongation.

The apical zone, where cell wall extension occurs, is filled with numerous small, differentially-stained cytoplasmic vesicles which fuse with the plasma membrane and there are some microtubules (Hickey and Coffee, 1977; Ingram *et al*. 1976; Sargent *et al*., 1973). The subapical zone has a full complement of cytoplasmic organelles. This region is apparently mainly employed in autosynthetic activities including the creation of new organelles. The mitochondria have well developed cristae. The dictyosomes appear to be producing cytoplasmic vesicles which are interspersed with many free and some bound ribosomes.

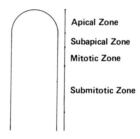

Fig. 2. A diagrammatic representation of the tip organization in a hypha based on information from light and electron microscopy.

In the mitotic zone the nuclei are distinguished by near synchronous divisions. Of all the nuclear cycle phases only the length of the mitotic phase has been examined and it is shortest in rapidly elongating hyphae (Tommerup *et al*., unpublished). The daughter nuclei may remain in the mitotic zone or enter the next basipetal zone. Germ tube tips of *B. lactucae* differ from hyphae in that there is no mitotic zone. Mitosis occurs after the fungi have established a relationship with their host plant (Ingram *et al*., 1976). This may have a nutritional basis because the synthesis of nuclear components has a high phosphate, nitrogen and energy requirement. As the hyphae of *B. lactucae* elongate the nuclei migrate towards the tip. Nuclear division appears to be initiated when the volume of cytoplasm per nucleus in the apical 50 $\mu$m reaches a critical value which is similar for slow and fast growing hyphae.

Basipetally adjacent to the mitotic zone is one where nuclear divisions are less frequent and asynchronous. In regions of the hyphae further from the tip, the nuclei divide infrequently. Functional differentiation may increase in the submitotic zone. Cytological evidence indicates that associated with the cessation of mitosis there is increased ribosomal RNA synthesis. Comparison of the nucleoli from the mitotic and sub-mitotic zones in fast-growing hyphae shows that nucleoli in the non-dividing nuclei of the sub-mitotic zone are larger, more granular and vacuolate. They are features of nucleoli that are actively synthesizing RNA (Olszweska, 1976; Jordan and Chapman, 1971). By analogy with other organisms, new stable RNA is likely to be needed preceding and during the initiation of branches, haustoria and reproductive structures. The periods between mitosis when mRNA synthesis is likely to occur are short in fast growing hyphae and some RNA may be transcribed by the nuclei adjacent to the mitotic zone and transported apically.

In the tip, the duration of a duplication cycle can be defined by the interval between two successive mitoses. Apical elongation occurs during the interval. Branches are formed at varying distances from the tip. Throughout vegetative growth, development can be defined as recurrent ontogenesis. The duplication cycle in downy mildews is a physiological feature rather than a morphological entity. It differs from the cell cycle of unicellular organisms (Mitchison, 1971, 1977) and from the duplication cycle of the apical compartment in septate moulds (Trinici, 1978).

The tip has a major morphogenetic role. Appressoria, infection pegs, primary vesicles, haustoria and conidiophores are initiated there (Ingram *et al.*, 1976). Gametangia may also be initiated only at the tip (Sansome and Sansome, 1974; Tommerup *et al.*, 1974). Each of these events follows or can be attributed to a change in the physical or chemical environment.

During spheroidal or bulbous development there is no distinct vesicular zone. The penetration peg is filled with cytoplasmic vesicles and expands into a globular primary vesicle where they are distributed throughout the cytoplasm. The process has features in common with the accumulation of small vesicles in buds of secondary vesicles and their dispersion as the structure expands (Ingram *et al.*, 1976; Sargent *et al.*, 1973). Discrete localization of cytoplasmic vesicles may not be a prerequisite for the penetration of host cells as the accumulations do not apparently precede haustorial development. The composition of the cell walls of the various different tubular, globular and bulbous structures react in the same way to histochemical stains indicating similar composition and therefore similar precursors. The cytoplasmic vesicles may be involved in wall growth. Detailed cytological and chemical studies of their origin and nature are needed to determine their morphogenetic role.

## B. Spore germination

An examination of the initial events involved in the germination of downy mildew spores provides a framework for defining when different genes are expressed and some of the possible controlling factors. One question of special interest is whether mRNA, required for germination, is synthesized during sporogenesis, spore maturation, or after germination is initiated.

Germination begins with hydration when the quiescent spores are transferred to water. Spore germination can be synchronized by washing to remove the water soluble inhibitors. How such inhibitors are held in fungal spores is not known and the location of them is uncertain (Macko et al., 1976). The inhibitor in P. tabacina is β-ionone (Leppik et al., 1972) but in other species of the Peronosporaceae they have not been identified.

The metabolic activities of conidia prior to hydration have not been determined. However, biochemical and cytological investigations of events soon afterwards provide some clues to the factors controlling the initiation of germination. Freshly harvested conidia of B. lactucae are densely packed with polysomes, few of which appear associated either in groups or with the sparsely distributed short pieces of endoplasmic reticulum (Sargent and Payne, 1974). Some protein synthesis may be initiated upon or during hydration and the spores may have all the preformed components for this. The protein synthesizing system extracted from spores of P. tabacina hydrated for about ten minutes in iced water functions in vitro (Holloman, 1971). After that time the synthesis of new ribosomes and mRNA is apparently not necessary for germination (Hollomon 1970). Whether all the mRNA essential for germination is preformed or some is synthesized soon after hydration has not been unequivocally determined.

The germination self-inhibitor may function by preventing major protein synthesis. Some synthesis occurs before complete removal of the inhibitor, but no germ tubes form (Holloman, 1971). The production of a large amount of membranous material as well as the increase in mitochondrial volume which precedes germ-tube emergence, gives some indication of the new structural proteins required during the early stages of germination (Sargent and Payne, 1974).

Synthesis of RNA may be required for germ tube elongation and appressorium formation, and this may be initiated soon after hydration (Hollomon, 1971). At that time changes in the nucleoli of B. lactucae indicate resumption of ribosomal RNA synthesis (Jordan, 1978; Sargent and Payne, 1974). The amount of heterochromatin also increases and this is another indication of changes in gene activity.

Further clarification of the events involved in spore germination may be derived from detailed cytological studies correlated with biochemical measurements of the sequence in which pathways begin functioning during and following hydration. In addition to investigations showing whether key enzymes are preformed or synthesized, identification of the energy reserves would provide more indications of the nature of the controlling systems.

## C. Appressorium and infection peg

Spore germination and germ tube elongation proceed once the water soluble inhibitor is removed. Subsequent development involves separate morphogenesis (Fig. 3). On leaves and other shoot surfaces of young plants the appressorium preferentially develops at the junction between the anticlinal walls of adjoining epidermal cells (Jones, 1978; Mence and Pegg, 1971; Preece *et al.* 1967; Maclean and Tommerup, 1979). Appressorium formation involves interactions between the germ tube, the host surface and the physical and chemical environment including other microbes present in the infection droplet. Different host–parasite combinations may be variously affected by one or more factors (Blakeman and Parbery, 1977; McCracken and Swinburne, 1979; Emmett and Parbery, 1975).

A thigmotropic mechanism possibly involving the surface structure may be involved in appressorium–plant interactions. Appressoria form on resistant or susceptible hosts, non-hosts, or freeze-killed host plants (Ingram *et al.*, 1976; Maclean and Tommerup, 1979). The potential role of microbes in appressorium initiation by downy mildew fungi has seldom been

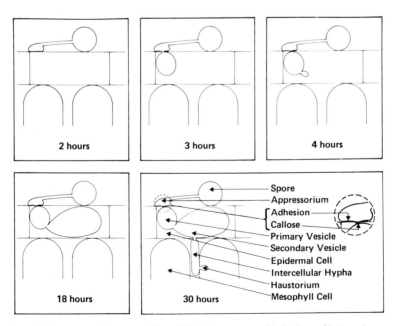

Fig. 3. A diagrammatic representation of the time-course of infection of lettuce by a compatible race of *Bremia lactucae* (after Ingram *et al.*, 1976; Sargent *et al.*, 1973; Maclean and Tommerup, 1979).

investigated (Preece *et al.*, 1967). For a *P. farinosa*–sugar beet cotyledon system there was no difference between appressorial development in the presence or absence of other microbes, indicating a lack of microbial influence for that interaction (Tommerup and Ingram, unpublished).

Initiation of appressoria of *B. lactucae* and *P. farinosa* is followed by the rapid swelling of germ tubes commencing in the sub-apical zone and extending back towards the spore but without loss of the tip organization. Immediately following the formation of microscopically visible appressoria, they can be mechanically detached from the leaf. However very soon after that the adhesion between the plant and fungal surface prevents detachment. For this step the active participation of a living host cell is unnecessary. The nature of the adhesion has not been chemically determined and there are at least three possibilities; non-specific protein adhesion, carbohydrate–carbohydrate binding and carbohydrate–protein binding. In view of the many different examples of carbohydrate-protein binding betweeen interacting cells being implicated as an early step in recognition processes (Callow, 1977; Sequeira, 1978), it is tempting to suggest that this type of binding also plays a role in adhesion of appressoria.

Information is exchanged between *B. lactucae* and living plant cells of hosts and non-hosts soon after adhesion (Ingram *et al.*, 1976). At that stage there is no differentiation for compatibility or incompatibility in the response of either component (Maclean *et al.*, 1974). The host reaction to an incompatible race is expressed by changed ultrastructure during the very early stages of contact between the fungal surfaces and host plasmalemma. The fungal response is seen a short time later.

## D. Zoospores

Chemotropism may be involved in the choice of site at which zoospores settle. The stimulus associated with photosynthesis causes zoospores of *Plasmopara viticola* and *Pseudoperonospora humuli* to select stomata and to encyst and germinate quickly (Royle and Thomas, 1971a,b, 1973; Royle, 1976). The behaviour of soil-borne downy mildew zoospores is not so well understood. By analogy with that of zoospores from other Oomycetes, a sequence of events which includes chemotropic attraction to the root surface followed by adhesion (Cameron and Carlile, 1978; Daly, 1976) mediated by terminal sugars of the mucopolysaccharide slime and sugar receptors of the zoospore, possibly on the flagellum (Hinch and Clarke, 1980) is postulated.

The ultra-structural organization of zoospores of *Sclerophthora macrospora* (Fukutomi and Akai, 1966) is similar to that of other Oomycetes (Bimpong and Hickman, 1975). During encystment of downy mildew zoospores, the sequence of structural changes may possibly also be like those of other Oomycetes (Bartnicki-Garcia and Hemmes, 1976; Hoch and Mitchell, 1972; Schnepf *et al.*, 1978a).

## E. Infection Structures

The primary infection structures may be either intercellular or intracellular with one form predominantly developed by a species. Following penetration, the structures enlarge very rapidly. When most of the spore contents have been transferred into them, a plug forms sealing off the external germ tube or appressorium (Ingram *et al.*, 1976; Pares and Greenwood, 1977; Wehtje *et al.*, 1979).

The initial intracellular structure may serve primarily as a secondary spore for *B. lactucae* in compatible and incompatible hosts, freeze-killed tissue and non-hosts (Ingram *et al.*, 1976; Maclean and Tommerup, 1979). Interaction with a living plant cell is not a prerequisite for this development. The formation of primary vesicles of several races of *B. lactucae* is apparently unaltered by a range of plant species identifying it as a morphological stage which can be used in genetic studies of host–parasite interactions.

The primary infection structures of a number of other downy mildews including *P. parasitica* (Chou, 1970), *S. sorghi* (Jones, 1978) and *P. humuli* (Pares and Greenwood, 1977) may also initially function as a type of internal spore. Those formed by *P. halstedii* in sunflower leaves may have a more host dependant relationship (Wehtje *et al.* 1979). Unlike the primary vesicle of *B. lactucae* growth of the infection vesicle of *P. halstedii* is indeterminate.

The appearance of the secondary vesicle as a nipple-like projection from the primary vesicle is the second identifiable intracellular morphological stage for *B. lactucae*. Biochemical changes, including those encompassing alterations in the relationship between the penetrated cell and primary vesicle may be detectable earlier than changes in subcellular organization and more precisely define this developmental stage. A nett increase in cytoplasmic volume and the first nuclear divisions since spore formation, take place during secondary vesicle development suggesting that the nutritional dependance of *B. lactucae* on its living penetrated cell is increased at this stage (Ingram *et al.*, 1976; Maclean and Tommerup, 1979). Additional evidence supporting this idea comes from an examination of developing secondary vesicles in highly incompatible hosts, where nuclei do not divide. One of the criteria controlling nuclear division seems to involve a relationship between nuclear number and cytoplasmic volume. There may be no stimulus for division in those secondary vesicles where the total volume of cytoplasm has not increased.

The formation of secondary vesicles is a characteristic of *B. lactucae* but unlike primary vesicles their shape and size are variable and they are influenced by the phenotype of the fungal-plant interaction. Development of secondary vesicles of races having defined virulent and avirulent genes in hosts having the appropriate genes for resistance and susceptibility represents a system for the analysis of how one genome regulates expression in the other.

Formation of the first hyphae and haustoria initiates the next ontogenetic phase. When these structures develop, $^3$H from $^3$H-leucine is transferred from plant cells to *B. lactucae* (Andrews, 1975). Haustoria can form when there is contact between a fungal wall and either the inner or outer walls of plant cells. There is some evidence that the site of initiation is influenced by a thigmotrophic response. Initiation may result when expanding fungal walls of *P. farinosa* (Tommerup and Ingram, unpublished), *P. pisi* (Hickey and Coffee, 1977; 1978) and *B. lactucae* (Ingram et al., 1976) become adpressed to host walls. An exchange of information apparently precedes haustorial entry into a cell. Haustorial initiation has features in common with the early fungal–plant cell interactions during appressorial penetration. A physiological requirement of the fungus for intracellular contact may influence gene expression for haustorium formation.

Young haustoria are metabolically very active as shown by the density and structure of the cytoplasmic organelles. There is little information on the chemical nature and role of any exchange occurring between host and parasite (Ingram et al., 1976; Takahashi et al., 1977). Haustorial function may not always depend on the presence of a fungal nucleus. The active role of a haustorium in informational exchange with the host may be of short duration, possibly only two or three days for *B. lactucae*. In *B. lactucae* and *P. farinosa*, the nuclei are not present during early haustorial expansion and they may move out during haustorial senescence as in *B. lactucae* (Tommerup and Ingram, unpublished). Haustoria of other species might be anucleate (Pares and Greenwood, 1977; Takahashi et al., 1977).

By the time vegetative colonies are competent to sporulate they may no longer have only the nuclear types that were present in the initial propagule. That hyphal anastomosis and nuclear interchange occur in downy mildews remains to be established, although hyphae apparently linking two colonies have been observed occasionally in each of several species (Tommerup and Ingram, unpublished). Hyphal anastomosis is uncommon in Oomycetes (Burnett, 1974), however it has been described in *Phytophthora* (Stephenson et al., 1974). Heterokaryosis has been postulated for *B. lactucae* (Section IV) and may be perpetuated by single spores.

### F. Parasite–host interaction

Characterization of the differences and similarities between resistant and susceptible interactions indicates some of the functions of the genes that determine the responses. Adhesion may constitute the first step in the infection process by downy mildews. Prior to penetration no distinction in the behaviour of either fungus or plant for compatible or incompatible combinations has been found (Kroeber et al., 1979; Maclean et al., 1974, 1979; Riggle, 1977). Contact by the fungus with the host plasmalemma may greatly contribute to determination of the interaction type. The possible nature of the factors involved, including products of fungal and host genes have been discussed recently (Callow, 1977; Dixon and Lamb, 1980; Sequeira, 1978).

Interactions between lettuce cultivars and B. lactucae can be characterized broadly into susceptible, resistant and intermediate groups, resistance being expressed as restriction of growth and sporulation (Crute and Johnson, 1976; Maclean and Tommerup, 1979; Viranyi and Block, 1976). Within each category, the phenotypes of both components vary cytologically indicating different physiological responses. Some of them correlate with the reacting gene pairs (Maclean and Tommerup, 1979). More examples need to be described to determine whether the phenotype for each A gene of some races varies only within narrow limits. For a few of the R/A gene pairs, changes in the cellular environment, implicating different gene expression in the host, fungus or both can result in a modified interaction phenotype (Crute and Norwood, 1978; Maclean and Tommerup, 1979). When isogenic lines of the host and parasite are available it will be possible to determine those that are primarily due to single genes.

## IV. Asexual Reproduction

The diurnal periodicity of asexual reproduction in downy mildews is an ecological adaptation closely linked to changes in temperature and relative humidity favouring sporulation and these occur at night (Yarwood, 1937). Certain combinations of temperature, light, darkness and relative humidity have been shown to stimulate and others to repress sporulation (Cohen, 1976; Cohen and Eyal, 1977; Cruickshank, 1963; Pegg and Mence, 1970; Raffray and Sequiera, 1971). Synchronous production of the successive developmental phases of single cycle sporogenesis can be induced by manipulation of the environment. In optimum conditions the initiation of a sporulation cycle depends on other complex interacting factors including the physiological state of the colony and host (Tommerup and Ingram, unpublished).

Colonies apparently need to produce a certain amount of growth before they are able to sporulate. This is reached by colonies of B. lactucae grown from single conidia under standard conditions in six days (Sargent et al.,

1973). When grown at a series of lower temperatures, the time taken to reach a total hyphal length similar to standard colonies progressively increased and was closely correlated with the time of onset of sporulation (Tommerup and Ingram, unpublished). Before a colony was competent to sporulate no exposure to conducive conditions induced sporulation.

The change from vegetative growth to the initiation of the first sporulation cycle can be seen in *B. lactucae* when the sporangiophore primordia form during a light phase. They develop at the apex of hyphae in the substomatal cavity. At the base of a stomata the primordia narrow and elongation ceases at the outer surface. It recommences after the onset of a dark phase. During the first hour there are indications that the metabolic activity of the primordia and subtending hyphal tips increases. An accumulation of cytoplasmic vesicles re-appears at the apex. The rate of nuclear division increases reaching a plateau by two to three hours. The initial developmental stages of *P. parasitica* are probably similar (Davison, 1968a). Subsequently, the basic morphological sequences (Fig. 4) follow a general pattern (Davison 1968b,c; Fried and Stuteville, 1977; Kajiwara and Iwata, 1957).

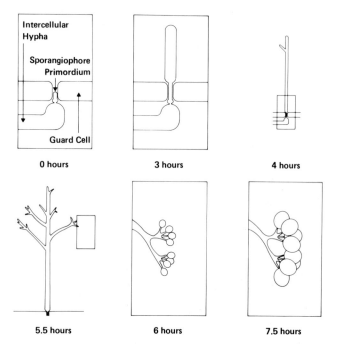

Fig. 4.   A diagrammatic representation of the time-course of sporulation of *Bremia lactucae* from the onset of a dark cycle until morphologically mature spores have formed.

Spore production begins with a change from polar to spheroidal growth. Initially one nucleus migrates into each spore of *B. lactucae* and three synchronous mitoses occur. The fourth is not always synchronous and may take place in only some nuclei, which are in one region of a spore. Incubation at 5°C maintains synchrony for the fourth division. After the third division there may sometimes be further migration of nuclei from the sporangiophore into the spores. Some spores are probably homokaryotic in heterokaryotic colonies, however the possibility of them being hetero-karyotic exists. The spores are unable to germinate immediately their morphological development is completed, but they can do so after a further 4 to 6 h. Physiological maturation may be a common feature (Fried and Stuteville, 1977; Thirumalachar *et al.*, 1953).

In the elongating sporangiophores of *B. lacutcae* the nuclei appear to migrate in mid-prophase. The NAO with microtubules extending into the relatively long pyriform nucleus is at the apex, the nucleolus towards the base and the chromatin appears partially condensed. Nuclei at late prophase, metaphase and anaphase are almost stationary, however soon after telophase they again have a prominent apical NAO and migrate.

Sporangiophore and spore development appear to be a continuous process (Davison, 1968b). Information from the sequence of cytological changes and from the differential physiological effects of environmental factors indicates that six morphogenetic stages are involved. They are primordial initiation, sporangiophore elongation and branching, spore initiation, expansion, and maturation. The existence of synchrony during sporulation cycles at optimum conditions, and the similar time taken by different downy mildews for the morphological and cytological steps, implies a close physiological and there-fore chemical basis for asexual reproduction. This suggests that common factors control sporulation in the group.

The direct effect of environmental factors on sporulation cannot be deter-mined when the fungi are growing in plants. Metabolic inhibitors that have been used, likewise do not have differential effects for fungal and plant tissues. These complications limit the biochemical implications of such studies. None-theless more information about induction and the subsequent phases would be obtained from detailed ultrastructural and chemical studies of changes which must precede the morphogenetic events that have been defined by light microscopy.

## V. Sexual Reproduction

The sexual breeding systems of most downy mildews are virtually unknown. Recent investigations of the genetics of sexual reproduction in *B. lactucae* provides a model for other members of this group. *B. lactucae* is capable of regular and predictable production of oospores in large numbers. Heterothallism has been demonstrated for many isolates with each expressing only one mating type (Michelmore and Ingram, 1980). Only two mating types have been identified in a collection of isolates differing in pathogenicity and geographical origin. In heterothallic strains, reciprocal translocation heterozygosity might function as a suppressor of segregation of mating types during vegetative growth as has been postulated for *Phytophthora* (Sansome, 1980). The presence of chromosomal structural hybridity may be essential for the establishment and maintenance of stability.

Some isolates produce oospores when cultured alone (Tommerup *et al.*, 1974). Low percentages of the colonies formed by single spores from them reproduce sexually and so exhibit a homothallic capacity (Tommerup and Ingram, unpublished; Michelmore and Ingram, personal communication). Heterokaryons that are possibly transitional may result from the nondisjunction or rearrangement at mitosis of chromosomal determinants trisomic for compatability type (Michelmore and Ingram, personal communication). The mechanism has been indicated by genetical and cytological analysis for trisomic homothallic isolates of *P. drechsleri* having different degrees of stability (Mortimer *et al.*, 1977; Sansome, 1980).

A higher incidence of mitotic crossing over may be a feature of old colonies of *B. lactucae*. An analysis by culture of single spore isolates of the progeny from each sporulation cycle of strains which rarely formed oospores, showed that only progeny from late cycles produced them (Tommerup and Ingram, unpublished). The physiological state of the host and fungus associated with old lesions may have an influence on stability. Extreme physiological conditions can differentially induce the depression of mating type recognition factors in diploid yeasts (Crandall and Caulton, 1975).

The breeding mechanisms so far identified in *B. lactucae* may be general for other downy mildews. *P. parasitica* is apparently heterothallic, however some colonies from single spores produce oospores (de Bruyn, 1937; McMeekin, 1960). Some graminaceous downy mildews also appear to be heterothallic (Michelmore and Williams, personal communication).

Where many oospores occur sporulation is diminished suggesting that induction of sexual reproduction causes repression of asexual reproduction. This phenomenon has been correlated with seasons (Berry and Davis, 1957; McKay, 1939) and stages of lesion development (Pegg and Mence, 1970) and in others found to be a regular occurrence (Ingram *et al.*, 1975; Fletcher, 1976) although the incidence is related to pathogen race (Ingram *et al.*, 1976a). Relatively high numbers of oospores have been correlated with early chlorosis or necrosis. Oospore production may, however, induce chlorosis rather than the reverse (Michelmore and Ingram, personal communication) or there may be no primary causal relationship. Environmental conditions interacting with a particular balance of host–parasite relations may be conducive to instability resulting in higher frequencies of more than one mating type.

Both homothallic and heterothallic isolates of *B. lactucae* may be bisexual. Each mating type can be either a nuclear donor or acceptor. The first sex structures produced by some cultures are of only one type and subsequently both types form with no loss in cross compatability (Tommerup and Ingram, unpublished). Other cultures initially produce both types. Very infrequent cultures may have a greater capacity to produce only one type as one such culture has been observed. Environmental conditions appear to modify the initial expression of sexuality. Difficulties in interpretation arise since only phenotype behaviour is characterized by single spore isolates. Studies using mononucleate propagules, oospores or zoospores, would avoid some of the complications and extend the information on the possible range of genotype behaviour and on the basic mechanisms regulating sex expression in downy mildews. The control of sexuality may be similar to that in *Phytophthora* (Sansome, 1980) or like some monoecious plants where expression of sex genes is complex and influenced by modifying genes, and the genetic background (Robinson *et al.*, 1976).

Sexual reproduction in the downy mildews potentially provides for a rearrangement of nuclear and cytoplasmic genes in the developing oosphere. The structural and functional specializations of the process can be subdivided into a sequence of stages: induction, differentiation of gametangia, nuclear transfer, zygote formation, and development of a dormant oospore. Within each stage there are a number of morphological and physiological changes suggesting that there is a corresponding sequence of regulatory controls.

The fact that incompatibilities occur within downy mildew species indicates that there are recognition systems. For the initiation of sexual reproduction both mating types must be present in close proximity. The involvement of phaeromes in the induction phase can be inferred from this behaviour and from the apparent chemotropic response of the hyphae. Compatible hyphae interact, resulting in intertwining (Michelmore and Ingram, personal communication).

The gametangia of homothallic and heterothallic matings of *B. lactucae* are alike. Gametangial differentiation is coordinated in a way which suggests it is also stimulated and synchronized by specific metabolites. Plasmogamy may be subject to mechanisms determining the kinds of nuclei that will fuse. It occurs between two morphologically distinct cells which are delimited by septa (Fig. 5) (McMeekin, 1960; Whitehead, 1958).

Functionally, the oogonia and antheridia are different from each other as well as from the vegetative state. An antheridium is perigynously adpressed to each oogonium (Sargent *et al.*, 1977; Tommerup *et al.*, 1974). Adhesion

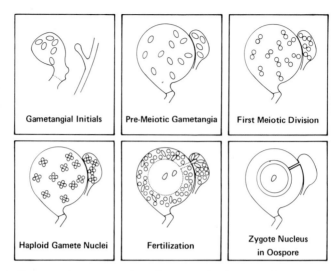

Fig. 5. A diagrammatic representation of oospore development of *Bremia lactucae* (after Sargent *et al.*, 1977; Tommerup *et al.*, 1974).

between their walls follows contact. Subsequently, mitoses and meiosis is not only synchronous within each organ, there is near synchrony between them suggesting that the preceding cell contact established specific interactions. This behaviour may follow some type of recognition between cell-surface ligand molecules. Thinning of both gametangial walls in the regions where the fertilization tube develops takes place during meiosis. The mitochondria and vesicles appear to be functionally involved. Subsequent steps in the fertilization process must be coordinated so that nuclear transfer from the antheridium to the oosphere takes place, before wall deposition between the double membranes disrupts the passage between the fertilization tube and oosphere. Some cultures of *B. lactucae* have a very high abortion rate indicating an incompatibility factor. In some, abortion apparently results from a loss in developmental synchrony between the antheridia and oogonia. This mechanism is different from that of steroid insufficiency in *P. cactorum* (Elliot and Sansome, 1977).

From delimitation, the mononucleate oosphere of *B. lactucae* is functionally unlike the multinucleate periplasm which probably has a role in oospore development. The delimiting double membrane appears to have pores. Oospore wall deposition can proceed even if fertilization fails indicating that its regulation may be maternally inherited. Other events may also be mediated by oogonial mRNA formed prior to fertilization. During fertilization there is no apparent loss of antheridial cytoplasm, indicating that little of it is transferred into the oosphere. Extranuclear inheritance is therefore predominantly female. Oospores of *S. macrospora* are multinucleate (White-head, 1958). Multinucleation may follow mitosis of the zygote nucleus (San-some, 1966).

Development of dormant oospores appears to be a sequence of well ordered morphological and structural events culminating in a zygote nucleus in a lipid-dense protoplast surrounded by a thick wall (Sargent *et al.*, 1977). The changes in organelle composition and wall structure parallel those in other Oomycetes (Beakes, 1980a, b, c; Hegnauer and Hohl, 1978) where no changes in ultrastructure are apparent during oospore dormancy. A similar ultrastructural study of dormant downy mildew oospores has not yet been made and there are no detailed chemical or physiological studies.

Dormancy seems to be an obligate and not an optional developmental pathway in oospores. The developmental events are indicative of a structure having a very low metabolic activity and therefore likely to be resistant to water logging, desiccation, freezing, thawing, and high temperatures. Oospores are unable to germinate immediately their morphological develop-ment is completed and a process of physiological maturation follows (Berry and Davis, 1957; Morgan, 1978; Pande, 1972; Pratt and Janake, 1978). Oospores are generally considered to enable downy mildew fungi to exist in the absence of living host plants. How long they remain viable and what factors affect survival are largely undefined.

Stimulation to germinate has been correlated with complexities of physical and chemical factors and the roots of host and some non-host plants (Morgan, 1978; Pratt, 1978; Safeeulla, 1976). Germination of mature oospores is ap-parently suppressed during incubation in decayed leaf tissue (Morgan, 1978). Methods for the near synchronous production of high numbers of oospores coupled with those for induction of germination may yield repeatedly high germination frequencies if maturation is a synchronous event. Genetic, physiological, and infection studies of oospores have been hampered by lack of this basic information. Contaminant-free oospores can be produced (Tommerup *et al.*, 1974) which overcomes the problems caused by myco-parasitic fungi (Pratt, 1978).

Following activation and prior to germ tube emergence the large globules are no longer evident, the inner wall layers appear to be solubilized and the oospore volume increases (Berry and Davis, 1957; McKay 1939; Morgan, 1978). This particular pattern of changes is similar to that described by electron microscopy for oospore germination in other families (Beakes, 1980 a, b; Ruben and Stanghellini, 1978).

## VI. Conclusions

The downy mildews produce during their life cycles an array of developmentally linked structures. The availability of mutants for use in biochemical and cytological studies may help to indicate where further subdivisions in the developmental processes occur. They may also be valuable in establishing some of the causal relationships for the morphological changes that are induced by environmental alterations. The regulation of genetic activities controlling ontogeny in the group might be expected to have features in common. By virtue of the specialization, additional controls would be necessary to provide differential growth and function. Analysis of the physiology, biochemistry and cytology of the various phases of asexual and sexual development have been the main research themes of downy mildew genetics. The results provide a basis for experimental genetics.

## References

Andrews, J. H. (1975). *Can. J. Bot.* **53**, 1103–1115.
Bartnicki-Garcia, S. and Hemmes, D. E. (1976). *In* "The Fungal Spore, Form and Function". (D. J. Weber and W. M. Hess Eds) 593–639. Wiley, New York and London.
Beakes, G. W. (1980a). *Can. J. Bot.* **58**, 182–194.
Beakes, G. W. (1980b). *Can. J. Bot.* **58**, 195–208.
Beakes, G. W. (1980c). *Can. J. Bot.* **58**, 209–227.
Beakes, G. W. (1980d). *Can. J. Bot.* **58**, 228–240.
Beakes, G. W. and Gay, J. L. (1977). *Trans. Br. Mycol. Soc.* **69**, 459–471.
Berry, S. Z. and Davis, G. N. (1957). *Pl. Dis. Reptr.* **41**, 3–6.
Bimpong, C. E. and Hickman, C. J. (1975). *Can. J. Bot.* **53**, 1310–1327.
Blakeman, J. P. and Parbery, D. G. (1977). *Physiol. Pl. Path.* **11**, 313–325.
Burnett, J. H. (1974). "Mycogenetics". Wiley, New York and London.
Callow, J. A. (1977). *Adv. Bot. Res.* **4**, 1–49.
Cameron, J. N. and Carlile, M. J. (1978). *Nature* **271**, 448–449.
Chou, C. K. (1970). *Ann. Bot.* **34**, 189–204.
Cohen, Y. (1976). *Aust. J. Biol. Sci.* **29**, 281–289.
Cohen, Y. and Eyal, H. (1977). *Physiol. Pl. Path.* **10**, 93–103.
Crandall, M. A. and Caulton, J. H. (1975). *In* "Methods in Cell Biology". (D. M. Prescott Ed.) Vol. **12**, 185–207. Academic Press, London and New York.
Cruickshank, I. A. M. (1963). *Aust. J. Biol. Sci.* **16**, 88–98.
Crute, I. R. and Johnson, A. G. (1976). *Ann. Appl. Biol.* **83**, 125–137.
Crute, I. R. and Norwood, J. M. (1978). *Ann. Appl. Biol.* **89**, 467–474.
Daly, J. M. (1976). *In* "Encyclopaedia of Plant Physiology". (R. Heitefuss and P. H. Williams Eds.) Vol. **4**, 25–50. Springer-Verlag, Berlin and New York.
Davison, E. M. (1968a). *Ann. Bot.* **32**, 613–621.
Davison, E. M. (1968b). *Ann. Bot.* **32**, 623–631.
Davison, E. M. (1968c). *Ann. Bot.* **32**, 633–647.
de Bruyn, H. L. G. (1937). *Genetica* **29**, 553–558.
Dixon, R. A. and Lamb, C. J. (1980). *Nature* **283**, 135–136.
Elliott, C. G. and Sansome, E. (1977). *J. Gen. Microbiol.* **98**, 141–145.
Emmett, R. W. and Parbery, D. G. (1975). *Ann. Rev. Phytopath.* **13**, 147–167.
Fincham, J. R. S., Day, P. R. and Radford, A. (1979). "Fungal Genetics". Blackwell, Oxford.
Fletcher, J. T. (1976). *Ann. Appl. Biol.* **84**, 294–298.

Fried, P. M. and Stuteville, D. L. (1977). *Phytopathology* **67**, 890–894.
Fukutomi, M. and Akai, S. (1966). *Trans. Mycol. Soc. Japan* **7**, 199–202.
Heath, I. B. (1978). *In* "Nuclear Division in the Fungi". (I. B. Heath Ed.) 89–176. Academic Press, London and New York.
Hegnauer, H. and Hohl, H. (1978). *Exp. Mycol.* **2**, 216–233.
Hickey, E. L. and Coffee, M. D. (1977). *Can. J. Bot.* **55**, 2845–2858.
Hickey, E. L. and Coffee, M. D. (1978). *Protoplasma* **97**, 201–220.
Hinch, J. and Clarke, A. E. (1980). *Physiol. Pl. Path.* **16**, 303–308.
Hoch, H. C. and Mitchell, J. E. (1972). *Protoplasma* **75**, 113–138.
Hollomon, D. W. (1970). *J. Gen. Microbiol.* **62**, 75–87.
Hollomon, D. W. (1971). *Arch. Biochem. Biophys.* **145**, 643–649.
Howard, K. L. and Moore, R. T. (1970). *Bot. Gaz.* **131**, 311–336.
Ingram, D. S., Sargent, J. A. and Tommerup, I. C. (1976). *In* "Biochemical Aspects of Host–Parasite Relationships". (J. Friend and D. R. Threlfall Eds) 43–78. Academic Press, London and New York.
Ingram, D. S., Tommerup, I. C. and Dixon, G. R. (1975). *Trans. Br. Mycol. Soc.* **64**, 149–153.
Ingram, D. S., Tommerup, I. C. and Searle, L. M. (1976a) *Ann. Appl. Biol.* **84**, 299–302.
Jones, B. L. (1978). *Phytopathology* **68**, 732–735.
Jordan, E. G. (1978). "The Nucleolus". Oxford University Press, Oxford.
Jordan, E. G. and Chapman, J. M. (1971). *J. Exp. Bot.* **22**, 627–634.
Kajiwara, T. and Iwata, Y. (1957). *Ann. Phytopath. Soc. Japan*, **22**, 201–203.
Kroeber, H., Oezel, M. and Petzold, H. (1979). *Phytopath. Z.* **94**, 16–44.
Kwiatkowska, M. and Maszewski, J. (1979). *Protoplasma* **98**, 363–367.
Leppik, R. A., Hollomon, D. W. and Bottomley, W. (1972). *Phytochemistry* **11**, 2055–2063.
Macko, V., Staples, R. C., Yaniv, Z. and Granados, R. R. (1976). *In* "The Fungal Spore, Form and Function". (D. J. Weber and W. M. Hess Eds) 73–98. Wiley, New York and London.
McCracken, A. R. and Swinburne, T. R. (1979). *Physiol. Pl. Path.* **15**, 331–340.
McKay, R. (1939). *J. R. Hort. Soc.* **64**, 272–285.
Maclean, D. J., Sargent, J. A., Tommerup, I. C. and Ingram, D. S. (1974). *Nature* **249**, 186–187.
Maclean, D. J. and Tommerup, I. C. (1979). *Physiol. Pl. Path.* **14**, 291–312.
McMeekin, D. (1960). *Phytopathology* **50**, 93–97.
Mence, M. J. and Pegg, G. F. (1971). *Ann. Appl. Biol.* **67**, 297–308.
Michelmore, R. W. and Ingram, D. S. (1980). *Trans. Br. Mycol. Soc.* **75**, 47–56.
Mitchison, J. M. (1971). "The Biology of the Cell Cycle." Cambridge University Press, London.
Mitchison, J. M. (1977). *In* "Mitosis, Facts and Questions". (M. Little, N. Paweletz, C. Petezelt, H. Ponstingl, D. Schroeter and H. P. Zimmerman Eds) 1–13. Springer-Verlag, Berlin, Heidelberg and New York.
Morgan, W. M. (1978). *Trans. Br. Mycol. Soc.* **71**, 337–340.
Mortimer, A. M., Shaw, D. S. and Sansome, E. (1977). *Archiv. Microbiol.* **111**, 255–259.
Nagl, W. (1978). "Endopolyploidy and Polyteny in Differentiation and Evolution". North-Holland Press, Amsterdam, New York and Oxford.
Olszewska, M. J. (1976). *Histochemistry* **49**, 157–175.
Pande, A. (1972). *Mycologia* **64**, 426–430.
Pares, R. D. and Greenwood, A. D. (1977). *Aust. J. Bot.* **25**, 585–589.
Pegg, G. F. and Mence, M. J. (1970). *Ann. Appl. Biol.* **66**, 417–428.
Pratt, R. G. and Janke, G. D. (1978). *Phytopathology* **68**, 1600–1605.
Preece, T. F., Barnes, G. and Bayley, J. M. (1967). *Pl. Path.* **16**, 117–118.
Raffray, J. B. and Sequeira, L. (1971). *Can. J. Bot.* **49**, 237–239.
Riggle, J. H. (1977). *Can. J. Bot.* **55**, 153–157.

142     I. C. TOMMERUP

Robinson, R. W., Munger, H. M., Whitaker, T. W. and Bohn, G. W. (1976). *Hort. Science* **11**, 554–567.
Royle, D. J. (1976). *In* "Microbiology of Aerial Plant Surfaces". (C. H. Dickinson and T. F. Preece Eds) 569–605. Academic Press, London and New York.
Royle, D. J. and Thomas, G. G. (1971a). *Physiol. Pl. Path.* **1**, 329–343.
Royle, D. J. and Thomas, G. G. (1971b). *Physiol. Pl. Path.* **1**, 345–349.
Royle, D. J. and Thomas, G. G. (1973). *Physiol. Pl. Path.* **3**, 405–417.
Ruben, D. M. and Stanghellini, M. E. (1978). *Am. J. Bot.* **65**, 491–501.
Safeeulla, K. M. (1976). "Biology and Control of the Downy Mildews of Pearl Millet, Sorghum, and Finger Millet." Wesley Press, Mysore.
Sansome, E. (1966). *In* "Chromosomes Today". (C. D. Darlington and K. R. Lewis Eds) Vol. 1, 77–83. Oliver and Boyd, Edinburgh.
Sansome, E. (1980). *Trans. Br. Mycol. Soc.* **74**, 175–185.
Sansome, E. and Brazier, C. M. (1974). *Trans. Br. Mycol. Soc.* **63**, 461–467.
Sansome, E. and Sansome, F. W. (1974). *Trans. Br. Mycol. Soc.* **62**, 323–332.
Sargent, J. A., Ingram, D. S. and Tommerup, I. C. (1977). *Proc. R. Soc. London, Ser. B.* **198**, 129–138.
Sargent, J. A. and Payne, H. L. (1974). *Trans. Br. Mycol. Soc.* **63**, 509–518.
Sargent, J. A., Tommerup, I. C. and Ingram, D. S. (1973). *Physiol. Pl. Path.* **3**, 231–239.
Schnepf, E., Deichgraeber, G. and Drebes, G. (1978a). *Can. J. Bot.* **56**, 1309–1314.
Schnepf, E., Deichgraeber, G. and Drebes, G. (1978b). *Can. J. Bot.* **56**, 1315–1325.
Sequeira, L. (1978). *Ann. Rev. Phytopathol.* **16**, 453–481.
Shaw, C. G., (1975). *Trop. Agric. Res. Ser.* **8**, 47–55.
Stephenson, L. W., Erwin, D. C. and Leary, J. V. (1974). *Phytopathology* **64**, 149–150.
Takahashi, K., Inaba, T. and Kajiwara, T. (1977). *Physiol. Pl. Path.* **11**, 255–259.
Thirumalachar, M. J., Shaw, C. G. and Narasimhan, M. J. (1953). *Bull. Torrey Bot. Club* **80**, 299–307.
Tommerup, I. C., Ingram, D. S. and Sargent, J. A. (1974). *Trans. Br. Mycol. Soc.* **62**, 145–150.
Trinici, A. P. J. (1978). *In* "The Filamentous Fungi". (J. E. Smith and D. S. Berry Eds) Vol. 3, 132–163. Edward Arnold, London.
Virányi, F. and Blok, I. (1976). *Nether. J. Pl. Path.* **82**, 251–254.
Wehtje, G., Littlefield, L. J. and Zimmer, D. E. (1979). *Can. J. Bot.* **57**, 315–323.
Whitehead, M. D. (1958). *Phytopathology* **48**, 485–493.
Yarwood, C. E. (1937). *J. Agric. Res.* **5**, 365–373.

Chapter 8

# Physiology and Biochemistry of Host–Parasite Interaction

## D. S. INGRAM

*The Botany School, Downing Street, Cambridge, UK*

## I. Introduction

Biochemical studies of the growth and survival of a pathogen and of the changes it induces in its host can lead, ultimately, to a better understanding of epidemiology, disease development and control. With a few exceptions, such studies on diseases caused by downy mildew fungi (Peronosporaceae) lag far behind those for diseases caused by other major groups of biotrophs (Brian, 1967; Scott, 1972; Heitefuss and Williams, 1976; Cooke, 1977). This neglect, which probably has its origins in the relative lack of importance of downy mildew fungi as the cause of disease in cereals in the Western Hemisphere, is unfortunate for two major reasons. First, because the class Oomycetes, which includes the downy mildews, is phylogenetically, physiologically and genetically unique (Lé John, 1971; Dick and Win-Tin, 1973) and its biotrophic members may, therefore, have evolved relationships with their hosts which differ significantly from those evolved by other fungi. Secondly, because the downy mildew diseases of graminaceous crops are now recognized to be of considerable social and economic importance in the countries of the "third world", and may also threaten the maize and related crops of north America (Frederiksen and Renfro, 1977).

In the following paragraphs I summarize some of the research on the biochemistry of host–parasite relationships published to date, and draw attention to some of the more promising areas for future study.

## II. Prerequisites for Interaction Studies

Ideal prerequisites for meaningful studies of the biochemistry of host–parasite interaction are: (a) a clear understanding of the genetic control of virulence and avirulence in the parasite and of susceptibility and resistance in the host; (b) precise histological and cytological descriptions of spore germination, infection and the establishment and development of the interaction; and (c) the availability of methods for growing the parasite, alone and in combination with its host, under controlled conditions. Unfortunately, these criteria have not yet been fully satisfied for any downy mildew disease, although some progress has been made with the combination of *Bremia lactucae* and lettuce. Attempts to define the genetics of this interaction are well advanced (see Chapters 9 and 20), there have been extensive histological and cytological studies of infection and disease development (see Chapters 7 and 10), and host and parasite may be handled with ease using seedlings or detached cotyledons (Maclean and Tommerup, 1979). There is still a need, however, to define the interaction further and to establish and define other model systems, especially among the downy mildew diseases of the Gramineae.

## III. The Biotrophic Status of Downy Mildew Fungi and the Need for Comparative Studies

Yarwood (1956) and Brian (1967) drew attention to the Peronosporales as a unique group with regard to parasitism, for within a relatively small number of closely related genera in this order are found representatives of the whole range of host–parasite relationships from the most extreme forms of necrotrophy to the most specialized forms of biotrophy (Table 1).

The genus *Pythium* occupies one end of the spectrum. Some species are pure saprotrophs but many others, such as *P. ultimum*, exhibit all the characteristics of extreme necrotrophy. The middle of the spectrum is occupied by

TABLE 1

Characters correlated with parasitism in the Peronosporales

| Character | NECROTROPHY ← Pythium spp. e.g. P. ultimum | Phytophthora spp. e.g. P. palmivora | INTERMEDIATE Phytophthora spp. e.g. P. infestans | → BIOTROPHY Members of the Peronosporaceae (downy mildews) |
|---|---|---|---|---|
| Parasitism | Weak | Weak/Strong | Strong | Strong |
| Mycelium | Inter- and intracellular | Inter- and intracellular | Mainly intercellular | Intercellular |
| Haustoria | Absent | ?Rudimentary | Rudimentary, specialized | Well developed, specialized |
| Host reaction | Very rapid necrosis | Rapid necrosis | Necrosis immediately follows sporulation | Necrosis follows sporulation, or is delayed, according to species |
| Tissue maceration | Marked | Marked | Limited | Very limited or absent |
| Morphological disturbance | None | None | None | Often marked, especially in systemic infections |
| Host range | Very wide | Wide | Limited | Limited |
| Behaviour in culture | Grow well | Grow well | Fastidious | Most species cannot be cultured |
| Competitive saprophytic ability | Good | Poor | None | None |
| Sporangiophores | Unspecialized | Unspecialized or specialized, indeterminate, evanescent | Specialized, indeterminate, evanescent | Specialized, determinate, evanescent |
| Position of sporangia | Aquatic | Aquatic or aerial | Aerial | Aerial |
| Germination of sporangia or conidia | Zoospores | Principally zoospores | Zoospores/direct | Zoospores or Zoospores/direct or direct, according to species |

the genus *Phytophthora*. Many *Phytophthora* species, such as *P. palmivora*, while retaining most of the features of necrotrophy, appear more specialized than the parasitic *Pythiums* and in some cases may even undergo a short period of biotrophic growth during the first few hours after infection. Also within the genus a small number of species, exemplified by *Phytophthora infestans*, are truly intermediate in their parasitism between necrotrophy and biotrophy: they have a narrow host range, and following infection, there is a biotrophic phase of growth lasting several hours or days with minimal tissue damage, invagination of the membranes of living cells by hyphae and rudimentary haustoria and sporulation on green tissue; this is eventually followed by death of host cells, limited degradation of cell walls and other components and, probably, a necrotrophic phase of growth.

Lastly, at the further end of the spectrum, are the downy mildews (Peronosporaceae), usually classed as biotrophs. All species are physiologically obligate parasites with a limited host range. Relationships with living hosts are usually relatively long-lived, and there is little evidence of tissue damage in the short term. The mycelium is predominantly intercellular and bears complex haustoria which invaginate plasma membranes of active cells. There is usually evidence of stimulation of host metabolism, especially during the early stages of interaction, and often abnormal growth of infected organs suggestive of a derangement of phyto-hormone levels. Sporulation is on green tissue, although infected tissues do eventually die.

Recently Whipps and Cooke (1978) have suggested that this biotrophy may not always be as complete as that of the rusts and powdery mildews, however, and that within the Peronosporaceae a further spectrum of relationships between hosts and parasites may exist. Thus, for example, in Brassicas infected by *Peronospora parasitica* the biotrophic phase of growth is not static and may be replaced, as tissues die slowly, by a necrotrophic phase with limited degradation of cellular components and sporulation occurring on the dead and dying tissues, whereas in sugar beet infected with *P. farinosa* there is no clear evidence of necrotrophy at any stage of disease development.

Correlated with the increasing specialization of host–parasite relationships within the Peronosporales as a whole is evidence of increasing adaptation of asexual reproduction to a terrestrial rather than an aquatic environment (Table 1).

It is dangerous to draw conclusions concerning evolutionary trends as a result of examination of living species only, but in the absence of a good fossil record some speculation is inevitable. There has been considerable discussion recently concerning the evolutionary origins of biotrophy, and it has been suggested that in many cases biotrophic fungi hay have evolved from necrotrophic forms through a series not unlike the one described above for the order Peronosporales, with a progressive loss of the ability to produce extracellular cell wall degrading enzymes, and a consequent loss of the ability to grow and compete as a saprophyte, being replaced by an increased dependence upon a living host for a supply of carbohydrates and other factors. Alternative theories suggest that biotrophic parasitic associations may be evolutionarily primitive with necrotrophs and saprotrophs having evolved from biotrophs during and following the emergence of land flora. Whatever the origins of biotrophy and necrotrophy, it is possible that each of the existing groups within the Peronosporales represents a step in the progression from one to the other, preserved because of its success in a particular ecological niche. This being so, the Peronosporales as a whole is an important group of fungi for investigation of the evolution and origin of biotrophy through comparative studies of the biochemistry of host–parasite interaction. The Peronosporaceae would be particularly important in such studies, especially if there is also a further spectrum of relationships within this group and if some of its members exhibit a shift from one mode of nutrition to another during the development of an infection. Comparative studies might concentrate on: the relative capacities of different genera and species to produce cell wall-degrading enzymes and how the activity of such enzymes is controlled; the capacities of different genera and species to produce toxins and to derange the phyto-hormone levels of host tissues; the structure and physiological potential of haustoria and other intracellular infection structures; the biosynthetic capacities of spores and sporelings, particularly their ability to synthesize the nucleic acids and proteins required for growth; and the degree of dependence of the parasites upon their hosts for other metabolic processes.

## IV. The Germination of Spores

Studies of the germination and biosynthetic capacity of spores and sporelings can provide valuable insights into the nature of relationships between hosts and parasites.

In common with the asexual spores of most obligate biotrophs, the conidia and encysted zoospores of the Peronosporaceae germinate readily in water to produce a germ tube. This, however, has a limited potential for growth and normally dies unless a relationship with a living host plant is established within a few hours of activation. The growth of germ-tubes may be enhanced by exogenous nutrients (Shepherd, 1972, on *Peronospora tabacina*), although this is not always the case (Pegg and Mence, 1970, on *Peronospora pisi*). Nutrient stimulated growth of germ-tubes is however, usually short lived, and an inadequate supply of simple chemical substances is unlikely to explain the dependence of downy mildew fungi on living hosts. The biochemistry of protein synthesis by the sporelings was, therefore, studied by Hollomon (1969, 1970, 1973).

*P. tabacina* was used as a model. Actidione inhibited germination of conidia, indicating a requirement for the systhesis of new protein for this process. However, although inhibitors of RNA synthesis prevented the incorporation of $^3$H uridine into the RNA of germinating spores, they had no effect on germination itself. Hollomon concluded that the differentiation of a germ-tube did not involve the synthesis of RNA and that the protein required for this process was synthesized on stable templates of mRNA, present within the dormant conidium and presumably formed during its development while still connected to an intercellular mycelium. Further biochemical analysis led Hollomon to propose that in *P. tabacina* the initiation of germination leads to release of mRNA from the stable templates and formation of initiation complexes with free small sub-units from particulate-bound ribosomes, and that attachment of these initiation complexes to the large sub-units then leads to protein synthesis.

While the features of germination in *P. tabacina* described by Hollomon have much in common with those of the rusts (Staples and Yaniv, 1976), it is not yet known to what extent they are common to other downy mildew fungi, although the circumstantial evidence of Mason (1973) suggests that in *B. lactucae* similar systems may be involved.

Many aspects of the synthesis of proteins by asexual spores of *P. tabacina* still remain to be investigated, especially: the mechanism of synthesis of the stable template of mRNA during spore formation and the nature of the complex itself; the mechanisms by which protein synthesis and germination are controlled; the mechanisms whereby fungal mRNA synthesis is activated following the establishment of a relationship with a compatible host; and the extent to which the biosynthetic capacity of the spores of *P. tabacina* compares with that of *P. infestans* and of other downy mildew fungi, including those which may be grown in culture.

A second feature of asexual spore germination studied in *P. tabacina* is the role of endogenous inhibitors. The existence of a water soluble inhibitor of germination was first recorded by Shepherd and Mandryck in 1962. This inhibitor, designated by the trivial name "quiesone", was purified and characterized by Leppik *et al.* (1972) as 5-isobutyroxy-β-ionine, a compound related chemically to abscisic acid and active at very low concentrations. Hollomon (1973) has suggested that quiesone may inhibit protein synthesis by preventing the re-attachment of the initiation complexes to the large, particulate-bound ribosomal sub-units.

Other downy mildew fungi are known to contain water soluble inhibitors of germination in their spores, e.g. *B. lactucae* (Mason, 1973), although none has been characterized. It would be interesting to know to what extent they are active against the spores of other members of the Peronosporaceae; lack of specificity within the group might indicate not only a common mode of action but also a common sequence of events leading to the synthesis of the protein required for germination.

Another approach to studies of the biochemistry of asexual spores has centred on their energy supply. Since they are capable of germinating in water this probably comes largely from stored reserves, although Andrews (1975) has shown that the conidia of *B. lactucae* may take up sugar released from the host into the infection droplet. The main energy reserve of downy mildew spores is in the form of lipid. The cytochemical studies of Duddridge and Sargent (1978) have shown that the major lipolytic enzyme within the quiescent conidium of *B. lactucae* is lipase and that this is largely restricted to the nuclear envelope and endomembrane system. As the spore germinates and the lipid reserves are mobilized there is apparently a shift towards esterase for lipolysis. In a separate study Sargent and Payne (1974) concluded that the inactivation of spores of *B. lactucae* by high temperatures occurs because the conversion of lipid reserves to the intermediates, needed for a supply of energy and the synthesis of membrane systems, is prevented.

Clearly there is a need for further cytochemical and biochemical studies of the energy relations of downy mildew spores and of the pathways which lead to lipid breakdown and interconversion during germination.

Oospores, the sexual spores of downy mildew fungi, can survive for long periods in the absence of a host. Nothing is known of the biochemical basis of this dormancy, however, nor of the processes involved in germination. This is unfortunate because an understanding of the mechanisms controlling oospore dormancy and germination is essential to a proper assessment of their role in epidemiology and variation. Morgan (1978) has shown that the germination of oospores of *Bremia lactucae* is stimulated by the presence of germinating lettuce seeds. Studies of the biochemistry of oospores and their germination must, be regarded as a research priority.

## V. Cytochemical Studies of Infection

The most important phase in the life history of a downy mildew fungus occurs during the period immediately following the germination of its spores upon the surface of a potential host plant. During this time a series of characteristic specialized infection structures is normally elaborated, the function of which is to bring the fungus from the plant surface, where its spores germinate, to a site suitable for haustorium formation. Consideration of the electron microscopic details of such infection structures (see Chapter 10) raises a number of important questions concerning the early stages of the development of an infection, which are common to all interactions involving biotrophic fungi. These include: what controls, physical or chemical, operate at the leaf surface to dictate the mode and site of infection; what substances are secreted by the fungus in advance of penetration to stimulate host metabolism and proliferation of its membranes; what substances are taken up by the fungus prior to penetration and how do these influence its metabolism; what enzymes are involved in the penetration of the cell walls of the host and how is their activity restricted to the immediate vicinity of the penetration peg; what is the role of wall-degrading enzymes in intercellular growth and the supply of carbohydrates to the parasite; do invaginated host plasma membranes have unusual biophysical properties to facilitate molecular exchange between host and parasite; how do host and parasite recognize one another and how does this lead to accommodation or rejection of infection structures; and what is the true role of intracellular infection structures, especially haustoria,—are they simply sites for the uptake of nutrients and other substances or are they also involved in the secretion of the molecules which lead to the re-orientation of host metabolism following infection. There are no clear answers to any of these questions for the downy mildews (Ingram et al., 1976; Heitefuss and Williams, 1976). A small number of attempts to address the problems cytochemically have been made, however, and are worthy of mention.

Andrews (1975) used the technique of microautoradiography to study the uptake of glucose and the amino acid leucine by B. lactucae during the formation of infection structures in lettuce cotyledons (see Chapter 10). The pathogen accumulated the label from $^3$H-glucose before penetration, and thereafter at all stages in the development of infection structures. In contrast, the label from $^3$H-leucine did not move into the fungus before penetration or during the early stages of infection structure formation, although it did so after the secondary vesicle had formed and during the development of haustoria. These results suggest a change in the ability of B. lactucae to take up certain substances from its host at different stages of infection structure development. Electron microscope autoradiography was also used by Takahashi et al. (1977) to study the distribution of label from $^{14}CO_2$ in cucumber leaves infected by Pseudoperonospora cubensis. Label was located in both host and fungal structures, although evidence for the mode and site of transfer from one to the other was inconclusive.

Problems concerning the interpretation of both of the above studies arise through lack of biochemical information on the chemical form in which the labels moved from hosts to parasites and failure of the authors to account for label incorporated into soluble substances. If such problems could be solved the techniques of microautoradiography could rapidly advance our understanding of the molecular exchange between biotrophs and their hosts.

Changes in the levels of lipolytic enzymes during infection were examined by electron microscope cytochemistry by Duddridge and Sargent (1978) using B. lactucae and lettuce. The cytoplasmic vesicles which accumulated in the appressorium before and during infection were devoid of lipolytic activity, but products of lipolysis occurred in both fungal and host cell walls adjacent to that part of the cuticle which was being penetrated, suggesting enzymatic breakdown of this barrier by enzymes from the fungus. In the presence of taurocholate, which activates lipases and slightly inhibits esterases, and quinine hydrochloride, which inhibits lipase, distinct cytochemical patterns relating to lipolytic activity were demonstrated in fungal structures. Beyond suggesting that these patterns may have resulted from the exchange of substances which occurred following contact of the pathogen with its host, interpretation was not possible. The work of Duddridge and Sargent is interesting for two reasons. First because it indicates that cytochemical studies of enzyme activity during infection by downy mildew fungi provide valuable information and secondly because it underlines the extreme difficulty of interpretation of the results of such studies in the absence of parallel biochemical investigations.

Finally, it is pertinent to mention some histological studies which relate to the nature of the interface between the haustoria of downy mildew fungi and host cells. First, following a study of haustoria of various species, Kajiwara (1971) concluded that in downy mildew fungi such structures come into direct contact with host cytoplasm. If this interpretation is correct it could have important implications for molecular exchange between the host and parasite. Subsequent studies, however, have failed to confirm Kajiwara's conclusions (Pares and Greenwood, 1979; Hickey and Coffey, 1977). Secondly, Hickey and Coffey (1978) have conducted a cytochemical investigation involving enzymic digestions, chemical extractions and specific staining methods of the host–haustorium interface in pea infected by *P. pisi*. The penetration matrix between the host wall and fungal hyphae in systemic infections and the extrahaustorial matrix had a proteinaceous component, possibly glycoprotein, while the extrahaustorial matrix had cellulose as an additional constituent. Complex carbohydrates were shown to be present in both matrices. Hydrolysis with B-1,3 glucanase revealed that B-1,3 glucans are important in the construction of the fungal cell wall but did not affect the structural integrity of the two matrices. Staining suggested differences between the extrahaustorial membrane and the normal host plasmalemma (cf. Pares and Greenwood, 1979). Although these studies do not throw any direct light on the role of haustoria in host–parasite interaction, they are important in constituting a first step in the chemical characterization of the interfaces involved.

## VI. Metabolic Changes in Infected Tissues

There are many marked shifts in the metabolic processes of plant tissues following infection by biotrophic parasites. These include changes in respiration, photosynthesis, nucleic acid and protein synthesis and phenol metabolism. There may also be changes in the translocation and accumulation of nutrients and in the levels of endogenous growth substances. All these changes are inter-related and are initiated by the secretion of toxins, enzymes and other molecules by the parasite and by the removal of metabolites from the cells of the host. Some are associated with parasite nutrition, others with resistance to infection, while others, although correlated with infection, appear to play no central role in the interaction itself. In the case of diseases caused by downy mildew fungi only changes in carbohydrate metabolism and in the levels of growth substances have received any serious attention, and even then most of the findings have been of a superficial nature. Such studies are discussed below.

## A. Carbohydrate metabolism

Much of the research concerning changes in carbohydrate metabolism within tissues infected by members of the Peronosporaceae has been made using seedlings or detached cotyledons and should, therefore, be interpreted with caution. Lack of experimental evidence has necessitated some extrapolation from experiments made with members of the Albuginaceae.

*Respiration.* A characteristic feature of the infection of plant tissues by rusts (Uredinales) and powdery mildew fungi (Erysiphaceae) is a two- to four-fold increase in the rate of respiration (Scott, 1972; Heitefuss and Williams, 1976). Most evidence suggests that in these cases the respiration of the host is shifted from a system that is predominantly channelled through glycolysis and the Krebs cycle to one that is dependent upon the pentose phosphate pathway (Cooke, 1977). Increases in the Krebs cycle have also been reported, but the relative contributions of this pathway and the pentose phosphate pathway are not clear.

A number of reports indicate that the respiration rates of tissues infected by members of the Peronosporaceae and Albuginaceae also rise dramatically (Sempio, 1959 with *B. lactucae* on lettuce; Thornton and Cooke, 1974 and Heitefuss and Fuchs, 1960 with *P. parasitica* on cabbage; Williams and Pound, 1964 and Black *et al.*, 1968 with *Albugo candida* on radish). However, the limited evidence so far suggests that such increases do not reflect any significant change in the pentose phosphate pathway. When radish cotyledons were infected with *A. candida*, respiration increased two- to three-fold, but the $C_1/C_6$ ratio together with evidence from inhibition and feeding experiments suggested that this was due to stimulation of existing pathways without any involvement of the pentose phosphate pathway (Williams and Pound, 1964). Accumulation of acyclic polyhydric alcohols, which is often associated with increases in the activity of the pentose phosphate pathway in tissues infected by rust fungi, was not detected in members of the Cruciferae infected by *P. parasitica* (Thornton and Cooke, 1974; Long and Cooke, 1974), nor in *Senecio squalidus* infected by *Albugo tragopogonis* (Long and Cooke, 1974).

*Photosynthesis.* Extensive research has indicated that the overall activity of photosynthetic pathways declines in leaves infected by rusts and powdery mildews, and this is accompanied by a decrease in the chlorophyll content of the tissue (Daly, 1976; Cooke, 1977). (Reports of increases in photosynthesis during the early stages of infection may refer to artefacts created by the abnormally high concentrations of $CO_2$ used in some experiments.) Around the sites of infection, however, green islands of chlorophyll are retained, and within them photosynthesis continues. They thus provide regions within which synthetic processes are maintained and the movement of metabolites from host to parasite may continue after the rest of the leaf has become non-functional. Enhanced fixation of $CO_2$ in the dark also occurs in green islands, but this appears to be due to the ability of the fungus itself to dark-fix. The similarity between green islands and the sites of exogenous application of cytokinins to plant tissues has been widely noted, although there is no clear evidence that these or other plant growth substances are actually responsible for the green-island phenomenon. Studies of the effects of downy mildew and related fungi on photosynthesis have been superficial, so meaningful comparisons with the rusts and powdery mildews are difficult. The following information is, however, pertinent.

Black *et al.* (1968) found by infra-red $CO_2$ analysis that the photosynthetic rate of cotyledons of radish infected with *A. candida* fell, and that this preceded the rise in respiration rate reported by Williams and Pound (1964). In another study Harding *et al.* (1968) examined the pattern of pigment retention during green island development following infection of *Brassica juncea* cotyledons with *A. candida*, the photosynthetic capacity of green island tissue and the ultrastructure of chloroplasts within green islands. They found that label from glycine-2-$^{14}$C was incorporated into chlorophylls *a* and *b* in both infected and non-infected tissue. Both infected and non-infected tissues fixed $^{14}CO_2$ in the light, but at 4 days after infection green islands fixed five times more $^{14}CO_2$ in the light than did non-infected tissue, while both types of tissue fixed about equal amounts in the dark. Photosynthesis per mole of chlorophyll fell at the same rate in green island and non-infected tissues. The maintenance of chlorophyll and continued photosynthetic activity in green island tissue was paralleled by delayed breakdown of chloroplasts. Electron microscopy showed that these retained their structural integrity in green islands for longer than in non-infected tissues. The authors drew attention to the similarity in *B. juncea* between green islands and tissues treated with kinetin.

The formation of green islands is not normally a characteristic of infection by true downy mildew fungi since the mycelium is usually distributed evenly throughout the infected tissue and clearly defined pustules are not formed. The few studies of the effects of downy mildew fungi on phototosynthesis have all been confined to crude assessments of chlorophyll levels during infection. Mason (1973) found that in detached lettuce cotyledons infected with *B. lactucae* the levels of chlorophylls *a* and *b* did not begin to decline until about six days after infection, this corresponding to the onset of sporulation. After this time the decline was more rapid in infected tissue than in non-infected tissue. Changes in chloroplast ultrastructure were not detected until 18 days after detachment in infected cotyledons and 25 days in non-infected cotyledons. Safeeulla (1976) reported decreased levels of chlorophyll following infection of members of the Gramineae with downy mildew fungi. Thornton and Cooke (1974) detected no differences in the levels of chlorophylls *a* and *b* in non-infected and *P. parasitica*-infected cabbage cotyledons during a 7 day experimental period and argued that the equivalent of green islands may be present in tissue infected with *P. parasitica* as microscopic areas of retained greenness associated with localized haustorium formation.

*Accumulation of metabolites.* A feature of infection by the rusts and powdery mildews is the accumulation of metabolites around the sites of infection. This is the result of decreased export from such areas and increased import into them from the rest of the leaf or plant. Carbohydrates accumulate as insoluble or soluble polysaccharides within the host cells and as acyclic polyhydric alcohols, trehalose and glycogen with the fungal mycelium, the latter compounds facilitating unidirectional movement of carbon from host to fungus (Cooke, 1977). Lipids may also accumulate.

Metabolite accumulation also occurs in infections caused by fungi related to the downy mildews (Williams and Pound, 1964; Harding *et al.*, 1968; Coffey, 1975 (*Albugo* spp. on members of the Cruciferae)), but whether this is due to increased fixation at infection sites, reduced export of material from them or increased import into them is not known. A clue is that when potato leaves infected with *Phytophthora infestans* were fed with $^{14}CO_2$ the radioactive products apparently accumulated in infected tissue as a result of retention of assimilates, and in infected regions carbohydrates continued to be translocated to other parts of the plant (Farrell, 1971). The accumulation of metabolites around the sites of infection of true downy mildew fungi has not been investigated.

It has already been noted that polyols are not formed in tissues infected by downy mildew and related fungi. It has been suggested, therefore, that the unidirectional flow of carbohydrates between host and parasite is facilitated by hydrolysis of host sucrose and uptake of hexoses, followed by accumulation of trehalose, glycogen and lipid within hyphae (Long and

Cooke, 1974; Cooke, 1977). It has been shown by Long *et al.* (1975) that infection of *S. squalidus* by *A. tragopogonis* results in an increase in the activity of acid invertase, localized at the sites of infection. It was concluded that sucrose from the host is first hydrolysed to glucose and fructose and then absorbed by the parasite. It was also suggested that invertase may play a key role in the provision of substrate for the accumulation of starch at infection sites; where there is a surplus of soluble carbohydrate, particularly sucrose, hydrolysis by invertase might provide hexose for starch synthesis within chloroplasts. Invertase may thus mediate a system by which the excess soluble carbohydrate at infection sites is converted to osmotically inactive polysaccharides. This conclusion is complicated, however, by the more recent suggestion of Whipps and Cooke (1978) that starch does not accumulate in *S. squalidus* infected by *A. tragopogonis*. It is not known whether invertase plays any part in the accumulation or transfer of carbohydrates in tissues infected by members of the Peronosporaceae, but in view of its implication in infection by members of the closely related family Albuginaceae, as well as by members of the Uredinales (Long *et al.*, 1975) and a member of the Ustilaginales (Billet *et al.*, 1977) it may reasonably be supposed that it does.

*Conclusions.* It is clear from the above that there is an urgent need for gathering of basic information concerning the effects of downy mildew fungi on respiration, photosynthesis and the translocation, accumulation and transfer of carbohydrates in infected host tissues. In the absence of such information, nothing but the most tentative conclusions may be drawn from the experimental evidence so far available. One particularly interesting aspect of carbohydrate metabolism that has so far received no attention whatsoever, is the relationship between the $C_4$ pathway of carbon assimilation and infection by the graminaceous downy mildews. This should be regarded as a research priority, especially in view of the remarkable correlation between the host ranges of such fungi and plants exhibiting the $C_4$ syndrome (see Chapter 18).

## B. Growth substances

A frequent, although not universal, accompaniment to infection of plants by biotrophic fungi is evidence of derangement of endogenous phytohormone levels (Brian, 1967, 1972). Such changes have been variously implicated in the stimulation of host metabolism and the redirection of translocation patterns. It has also been suggested that they could play a more fundamental role in mediating the metabolic integration of host and parasite.

Evidence of phytohormone disturbance caused by downy mildew diseases includes elongation of stem internodes of sugar cane infected by *Sclerospora sacchari*, loss of apical dominance and flower distortion in pearl millet infected by *Sclerospora graminicola*, leaf thickening and distortion in sugar beet infected by *P. farinosa* and stem stunting in sunflower infected by *Plasmopara halstedii*. Frequently morphological disturbance only occurs following systemic infection of the host and is not associated with infection of mature organs; such is the case with Brassicas and *P. parasitica* and with graminaceous downy mildews. In some instances, e.g. *B. lactucae* and lettuce and most other downy mildew diseases which do not normally involve systemic infection, morphological disturbances apparently never occur. This cannot, however, be taken as conclusive evidence that phytohormone levels are not altered in the infected tissues, but may simply reflect an inability of the tissues to respond visibly.

In only two instances have the hormonal disturbances resulting from infection by downy mildew fungi been investigated in any detail. In the first, the late Jennifer Bailey (Bailey and Ingram, unpublished results) analysed the structural changes occurring in sugar beet leaves infected by *P. farinosa*. Infection caused stunted leaf growth, i.e. a decrease in leaf area, length, fresh weight and dry weight, accompanied by general chlorosis. A marked increase in the thickness of infected leaves was the result of an increase in the size and number of cells in the mesophyll and some parts of the vascular region, accompanied by the complete disappearance of intercellular spaces and loss of distinction between the palisade and spongy mesophylls. Studies of the growth of infected and non-infected leaves revealed that the degree of morphological disturbance of infected leaves was correlated with the stage of development of the lamina at the time of infection. Detailed structural studies of this kind are an important prerequisite for investigation of the biochemical basis of changes in phytohormone levels.

In the second example structural studies were not made, but the biochemical basis of the inhibition of growth which follows infection of sunflower by *P. halstedii* were partially elucidated. Cohen and Sackston (1973, 1974) showed that inoculation of apical buds of sunflower led to inhibition of stem elongation. Reduced amounts of auxin and gibberellin-like substances were found in infected plants (Viswanathan and Sackston, 1970), and a positive correlation was established between the inhibition of stem growth and the ability of stem slices to remove indoleacetic acid from test solutions (Cohen and Sackston, 1973, 1974). It was suggested that diseased leaves produce one or more activators which are translocated to the upper stem of the sunflower plant and enhance the ability of its cells to take up IAA. Chromatography and fluorescence microscopy (Cohen and Ibrahim, 1975; Cohen, 1975) suggested that infection causes large increases in scopoletin, peroxidase activity and, during the early stages of infection, chlorogenic acid and related compounds, but it was not possible to establish any correlation between these phenomena and growth stunting.

It is clear that the downy mildew diseases represent a virtually unexplored and potentially highly rewarding area for biochemical studies of hormonal disturbances following infection and the role of such changes in pathogenesis.

## VII. Resistance and Tolerance

There have been several histo-cytological studies of the interaction between downy mildew fungi and incompatible hosts carrying specific genes for resistance. In most cases host cell death either before, during or following penetration has been implicated (e.g. Russel, 1969 with *P. farinosa* on sugar beet; Kröber and Petzold, 1972, 1977 with *P. farinosa* on spinach; Wehtje and Zimmer, 1978 and Wehtje *et al.*, 1979 with *P. halstedii* on sunflower; Greenhalgh and Dickinson, 1975, Dickinson and Greenhalgh, 1977 and Kluczewski and Lucas, Personal communication: with *P. parasitica* on *Brassica* spp.; Maclean *et al.*, 1974, Maclean and Tommerup, 1979, Crute and Dickinson, 1976 and Crute and Davis, 1977 with *B. lactucae* on lettuce). The interaction involving *B. lactucae* and lettuce has received more attention than most and the results obtained from its study illustrate the range of responses that has been recorded. First, Maclean *et al.* (1974) showed that the inter-action between a lettuce cultivar carrying the resistance factor R8 and an incompatible isolate of *B. lactucae* resulted in rapid death of the initially invaded cell followed by a slower death of the fungus. Later Maclean and Tommerup (1979) investigated in detail, using cytological and physiological techniques for detecting cell death, the range of interactions between lettuce cultivars carrying different factors for resistance and a range of isolates of *B. lactucae*. Two broad classes of incompatible interaction were identified. The first is exemplified in hosts carrying the R3 or R8 factors for resistance, and infection resulted in rapid death of host cells followed by very limited growth of the invading fungus without nuclear division. Cells adjacent to the one penetrated initially did not die. The second is exemplified in hosts carrying the R7 resistance factor, and infection is characterized by a delay in the onset of nuclear division in the fungus and the eventual necrosis of host cells one or two days later, although the fungus continued to grow and to sporulate sparsely. Thus necrosis in host organs exhibiting this type of resistance was comparatively extensive. These findings are consistent with those of Crute and Dickinson (1976), Crute and Norwood (1978) and Viranyi and Blok (1976). It is possible that a more extensive survey would reveal a range of interactions intermediate between those described as well as instances where incompatibility does not involve death of host cells (cf. Riggle and Dunleavy, 1974; Riggle, 1977).

Death of host cells has been variously associated with the resistance on non-hosts to downy mildew fungi (Maclean and Tommerup, 1979) and with a form of resistance mediated by systemic fungicides (see Chapter 20).

Nothing is known of the biochemical basis of specific resistance to downy mildew fungi. Some have attempted to implicate the rapid death of host cells described above in the process (Maclean and Tommerup, 1979), although there is no clear evidence to support this conclusion or to distinguish between the three possibilities that cell death may be: (a) an integral component of resistance; (b) a consequence of resistance; or (c) an event which is associated with, but irrelevant to, resistance (Ingram, 1978). These possiblities are not mutually exclusive and each could occur in different gene for gene combinations, even within a single host species–parasite species combination. This might explain the multiplicity of histological manifestations of incompatibility that have been described.

Whatever the role of cell death in specific resistance, it can only be one of a number of factors associated with this phenomenon which requires investigation. It is not yet possible to distinguish between susceptibility or resistance as being the active response of a host to a downy mildew fungus (Daly, 1976), and there have been no attempts to identify the mechanisms whereby mutual recognition occurs in compatible or incompatible combinations, nor to identify the ways in which such recognition is translated to bring about the metabolic changes resulting in active accommodation or rejection of the invading pathogen. Answers to these questions will only be obtained if the most sophisticated cytological, cytochemical and micro-biochemical methods are employed and if experiments are made using only hosts and parasites which are strictly defined, structurally and genetically. It would be prudent to limit further investigation of incompatibility in interactions involving downy mildew fungi until such defined material is available (see Section II), by which time the more extensive studies involving rusts and powdery mildews may have provided the insights which will lead to a more rapid and economical elucidation of some of the outstanding questions.

One form of resistance to downy mildew fungi that has been investigated biochemically is of a more general type than the specific resistance discussed above. Two examples will be quoted. In the first Greenhalgh and Dickinson (1975), working with the commercial cultivars of *Brassica oleracea*, noted a form of resistance to *P. parasitica* that was not race specific. This resistance took the form of necrosis of host tissues associated with limitation of mycelial spread and much reduced sporulation of the pathogen. Subsequent investigation revealed a correlation between high levels of flavour volatiles (e.g. allylisothiocyanate) released by tissue damage, and limitation of fungal growth in both wild and cultivated *Brassica* lines (Greenhalgh and Mitchell, 1976). It was suggested that in cultivated Brassicas breeding has resulted in reduced levels of flavour volatiles with a consequent reduction in their general resistance to *P. parasitica*.

The second example concerns lignification of cell walls following infection of radish by *P. parisitica*. It has often been suggested that the deposition of lignin in host cell walls may have a role to play in non-specific limitation of the growth of biotrophic fungi. Recently Ohguchi and Asada (1975) have gone some way towards defining the pathways and enzymes involved in lignin biosynthesis in radish following infection by *P. parasitica*.

In both of these cases the systems described can only be regarded as components of a much more complex biochemical milieu responsible for general resistance. They are, nevertheless, important in representing serious attempts to define resistance in biochemical terms.

Finally, it is important to draw attention to the phenomenon of tolerance of hosts to infection by downy mildew fungi. Simple observation of diseased plants, particularly wild species, leads to speculation that in many cases hosts are able to support high levels of infection by downy mildew fungi without any appreciable loss of productivity; a notable example in the U.K. is oilseed rape infected by *Peronospora parasitica*. Whether tolerance of this kind results from metabolic compensation or from some other process is not known. Identification and proper biochemical description of tolerance to downy mildew fungi among wild and cultivated hosts could have important implications for disease control. Whether such identification and description is yet possible, however, given the present inadequate understanding of the biochemical basis of pathogenesis in normally susceptible plants (see Section IV), is debatable.

## VIII. Growth in Culture

Biochemical studies of diseases caused by obligate parasites have always been hampered by lack of suitable systems for growing the interacting organisms under defined conditions. Host tissue culture has sometimes been considered as a means of overcoming this problem for downy mildew fungi, and a number of dual culture systems have now been devised (Ingram, 1976, 1977, 1980; Table 2). These have a clear potential role to play in the maintenance of contaminant-free clones of inoculum, for use in experiments and as a safe means of international transport of isolates. They may also be of value in the elucidation of some aspects of host–parasite interaction, including mechanisms of resistance, the roles of growth regulators and toxins in disease development and the roles of chemical and physical factors controlling sporulation. It should be remembered, however, that cultured tissues may be physiologically and genetically very different from those of intact plants (Ingram and Tommerup, 1973; Ingram and Helgeson, 1980) and results obtained using them should be interpreted with caution.

TABLE 2

Downy mildew fungi grown in association with host tissue cultures

| Fungus | Host | References |
|---|---|---|
| *Bremia lactucae* Regel[a] | *Lactuca sativa* L. (lettuce) | Mason, 1973 |
| *Peronospora farinosa* (Fr.) Fr. f.sp. *betae* | *Beta vulgaris* L. (sugar beet) | Ingram and Joachim, 1971; Ingram and Tommerup, 1973 |
| *Peronospora parasitica* (Pers. ex Fr.) Fr. | *Brassica* spp. (cabbage, turnip etc.) | Nakamura, 1965; Ingram, 1969 |
| *Peronospora tabacina* Adams | *Nicotiana tabacum* L. (tobacco) | Izard, Lacharpagne and Schiltz, 1964 |
| *Plasmopara viticola* (Berk. & Curt.) Bebl. & de Toni | *Vitis vinifera* L. (vine) | Morel, 1944, 1948 |
| *Pseudoperonospora humuli* (Miy. & Tak.) Wilson | *Humulus lupulus* L. (hop) | Griffin and Coley-Smith, 1968 |
| *Sclerophthora macrospora* (Sacc.) Thirum. Shaw & Naras | *Eleusine coracana* (L.) Gaertn. (finger millet) | Safeeulla, 1976 |
| *Sclerospora graminicola* (Sacc.) Schroet. | *Pennisetum typhoideum* (Burm.) Stapf e: C.E. Hubb. (pearl millet) | Tiwari and Arya, 1967, 1969; Arya and Tiwari, 1969; Safeeulla, 1976 |
| *Sclerospora sacchari* T. Miyake (≡ *Peronosclerospora sacchari* (T. Miyake) C.G. Shaw) | *Saccharum officinarum* L. (sugar cane) | Wen-Huei Chen, Ming-Chin Liu and Chia-Yu Chao, 1979 |
| *Sclerospora sorghi* Weston & Uppal (≡ *Peronosclerospora sorghi* (Weston & Uppal) C.G. Shaw) | *Sorghum vulgare* Pers. | Safeeulla, 1976 |

[a]The tissues became infected, but the fungus did not proliferate.

Biochemical studies of the downy mildew diseases might also be facilitated if some means could be devised for growing the causal fungi in axenic culture. So far there have been only three reports of the growth of downy mildew fungi in this way. In the first Guttenberg and Schmöller (1958) reported that *P. parasitica* made limited growth from spores on a culture medium containing beer-wort extract. In the second Tiwari and Arya (1969) noted that *S. graminicola* grew out from callus tissues of pearl millet and became established on the tissue culture medium. In the third *Sclerophthora macrospora* was said to have been grown from spores with such facility on a simple fungal culture medium that nutritional studies relating to growth and sporulation were possible (Tokura, 1975), although there was some doubt as to the true identity of the fungus. Strenuous efforts must now be made to confirm these findings and, if possible, to establish protocols for the axenic growth of other downy mildew fungi.

## IX. Conclusions

Research into the biochemistry of host–parasite interaction in plant diseases caused by downy mildew fungi lags far behind that for diseases caused by other major groups of biotrophs. Important prerequisites for such studies, particularly genetical and histo-cytological descriptions of interactions and the availability of methods for growing parasites alone and in combination with their hosts, are largely lacking, although progress is being made to redress this situation in the case of *B. lactucae* and lettuce. Greater effort is still needed, however, to identify and define further model systems, especially among the downy mildew diseases of the Gramineae.

A consideration of host–parasite relationships within the Order Peronosporales suggests that comparative biochemical studies, both between major groups within the Order as a whole and within the Family Peronosporaceae, could lead to a better understanding of the status of the downy mildews as biotrophs as well as the evolutionary pathways which have led to this mode of nutrition.

A survey of existing studies of the interaction between downy mildew fungi and their hosts, from spore germination through infection and the development of an interaction has revealed a superficial and inadequate understanding of the processes involved and underlines the urgent need for extensive basic biochemical research. Among the topics worthy of immediate consideration are: dormancy and germination of spores, especially oospores; the processes of infection and the function of infection structures; the effects of infection on respiration, photosynthesis (especially the link between graminaceous downy mildews and the $C_4$ syndrome) and other metabolic systems; the role of hormonal disturbances in pathogenesis; and the basis of systemic *vs.* local lesion infection. Studies of the effects of infection on the biochemistry of host productivity could lead to a better understanding of tolerance, a phenomenon which has great potential for disease control. It may be prudent to limit research on the biochemical basis of resistance and specificity, however, until better defined systems are available.

## References

Andrews, J. H. (1975). *Can. J. Bot* **53**, 1103–1115.
Arya, H. C. and Tiwari, M. M. (1969). *Indian Phytopath.* **22**, 446–452.
Billett, E. E., Billett, M. A. and Burnett, J. H. (1977). *Phytochemistry* **16**, 1163–1166.

Black, L. L., Gordon, D. T. and Williams, P. H. (1968). *Phytopathology* **58**, 173–178.
Brian, P. W. (1967). *Proc. R. Soc. (Lond.)* B 168, 101–118.
Brian, P. W. (1972). *Proc. R. Soc. (Lond.)* B 200, 231–243.
Coffey, M. D. (1975). *Can. J. Bot.* **53**, 1285–1299.
Cohen, Y. (1975). *Physiol. Pl. Path.* **7**, 9–15.
Cohen, Y. and Ibrahim, R. K. (1975). *Can. J. Bot.* **53**, 2625–2630.
Cohen, Y. and Sackston, W. E. (1973 *Can. J. Bot.* **51**, 15–22.
Cohen, Y. and Sackston, W. E. (1974). *Can. J. Bot.* **52**, 861–866.
Cooke, R. (1977). "The Biology of Symbiotic Fungi". John Wiley and Sons, London.
Crute, I. R. and Davis, A. A. (1977). *Trans. Brit. Mycol. Soc.* **69**, 405–410.
Crute, I. R. and Dickinson, C. H. (1976). *Ann. Appl. Biol.* **82**, 433–450.
Crute, I. R. and Norwood, J. M. (1978). *Ann. Appl. Biol.* **89**, 467–474.
Daly, J. M. (1976). *In* "Physiological Plant Pathology" (R. Heitefuss and P. H. Williams, Eds), 27–50 and 450–479. Springer-Verlag, Berlin.
Dick, M. V. and Win-Tin (1973). *Biol. Rev.* **48**, 133–158.
Dickinson, C. H. and Greenhalgh, J. R. (1977). *Trans. Brit. Mycol. Soc.* **69**, 111–116.
Duddridge, J. A. and Sargent, J. A. (1978). *Physiol. Pl. Path.* **12**, 289–296.
Farrell, G. M. (1971). *Physiol. Pl. Path.* **1**, 457–467.
Frederiksen, R. A. and Renfro, B. L. (1977). *A. Rev. Phytopathol.* **15**, 249–275.
Greenhalgh, J. R. and Dickinson, C. H. (1975). *Phytopath. Z.* **84**, 131–141.
Greenhalgh, J. R. and Mitchell, N. D. (1976). *New Phytol.* **77**, 391–398.
Griffin, M. J. and Coley-Smith, J. R. (1968). *J. Gen. Microbiol.* **53**, 231–236.
Guttenburg, H. von and Schmoller, H. (1958). *Arch. Mikrobiol.* **30**, 268–279.
Harding, H., Williams, P. H. and MacNabola, S. S. (1968). *Can. J. Bot.* **46**, 1229–1234.
Heitefuss, R. and Fuchs, W. H. (1960). *Phytopath. Z.* **37**, 348–378.
Heitefuss, R. and Williams, P. H. (1976). "Physiological Plant Pathology". Springer-Verlag, Berlin.
Hickey, E. L. and Coffey, M. D. (1977). *Can. J. Bot.* **55**, 2845–2858.
Hickey, E. L. and Coffey, M. D. (1978). *Protoplasma* **97**, 201–220.
Hollomon, D. W. (1969). *J. Gen. Microbiol.* **55**, 267–274.
Hollomon, D. W. (1970). *J. Gen. Microbiol.* **62**, 75–87.
Hollomon, D. W. (1973). *J. Gen. Microbiol.* **78**, 1–13.
Ingram, D. S. (1969). *J. Gen. Microbiol.* **58**, 391–401.
Ingram, D. S. (1976). *In* "Physiological Plant Pathology" (R. Heitefuss and P. H. Williams Eds), 743–760. Springer-Verlag, Berlin.
Ingram, D. S. (1977). *In* "Plant Tissues and Cell Culture" (2nd edn; H. E. Street, Ed), 463–500. Blackwell Scientific Publications, Oxford.
Ingram, D. S. (1978). *Ann. Appl. Biol.* **89**, 291–295.
Ingram, D. S. (1980). *In* "Tissue Culture Methods for Plant Pathologists" (D. S. Ingram and J. P. Helgeson Eds), 139–144. Blackwell Scientific Publications, Oxford.
Ingram, D. S. and Joachim, I. (1971). *J. Gen. Microbiol.* **69**, 211–220.
Ingram, D. S., Sargent, J. A. and Tommerup, I. C. (1976). *In* "Biochemical Aspects of Plant–Parasite Relationships" (J. Friend and D. R. Threlfall Eds), 43–77. Academic Press, London and New York.
Ingram, D. S. and Tommerup, I. C. (1973). *In* "Fungal Pathogenicity and the Plant's Response" (R. J. W. Byrde and C. V. Cutting, Eds), 121–137. Academic Press, London and New York.
Izard, C., Lacharpague, J. and Schiltz, P. (1964). *Annls. Div. Etud. Equip.* S.E.I.T.A. 1 (Sect. 2), 95–99.
Kajiwara, T. (1971). *In* "Morphological and Biochemical Events in Plant–Parasite Interaction" (A. Akai and S. Ouchi, Eds). 255–277. Machizuki Publ. 6, Japan.

Kröber, H., Özel, M. and Petzold, H. (1979). *Phytopath Z.* **94**, 16–44.
Kröber, H. and Petzold, H. (1972). *Phytopath. Z.* **74**, 296–313.
Lé John, H. B. (1971). *Nature*, Lond. **231**, 164–168.
Leppik, R., Hollomon, D. W. and Bottomley, W. (1972). *Phytochemistry* **11**, 2055–2063.
Long, D. E. and Cooke, R. C. (1974). *New Phytol.* **73**, 889–899.
Long, D. E., Fung, A. K., MacGee, E. E. M., Cooke, R. C. and Lewis, D. H. (1975). *New Phytol.* **74**, 173–182.
Maclean, D. J., Sargent, J. A., Tommerup, I. C. and Ingram, D. S. (1974). *Nature*, Lond. **249**, 186–187.
Maclean, D. J. and Tommerup, I. C. (1979). *Physiol. Pl. Path.* **14**, 291–312.
Mason, P. A. (1973). Ph.D. Thesis, University of Cambridge.
Morel, G. (1944). *C. r. hebd. Séanc. Acad. Sci.*, Paris **218**, 50–52.
Morel, G (1948). *Ann. Épiphyt. (Ser. Pathologie Végétale)* **14**, 1–112.
Morgan, W. M. (1978). *Trans. Brit. Mycol. Soc.* **71**, 337–346.
Nakamura, H. (1965). *In* "Proc. Intern. Conf. Plant Tissue Culture Penn. State Univ. 1963" (P. R. White and A. R. Grove, Eds), 535–540. McCutcheon Publ. Co., Berkeley, California.
Ohguchi, T. and Asada, Y. (1975). *Physiol. Pl. Path.* **5**, 183–192.
Pares, R. D. and Greenwood, A. D. (1979). *New Phytol.* **83**, 473–477.
Pegg, G. F. and Mence, M. J. (1970). *Ann. Appl. Biol.* **66**, 417–428.
Riggle, J. H. (1977). *Can. J. Bot.* **55**, 153–157.
Riggle, J. H. and Dunleavy, J. M. (1974). *Phytopathology* **64**, 522–526.
Russel, G. H. (1969). *Brit. Sugar Beet Rev.* **38**, 27–35.
Safeeulla, K. M. (1976). "Biology and Control of The Downy Mildews of Pearl Millet, Sorhum and Finger Millet". University of Mysore, Mysore, India.
Sargent, J. A. and Payne, H. L. (1974). *Trans. Brit. Mycol. Soc.* **63**, 509–518.
Scott, K. J. (1972). *Biol. Rev.* **47**, 537–572.
Sempio, C. (1959). *In* "Plant Pathology, An Advanced Treatise, Vol. I" (J. G. Horsfall and A. E. Dimond, Eds), 277–312. Academic Press, New York and London.
Shepherd, C. J. (1962). *Aust. J. Biol. Sci.* **15**, 483–508.
Shepherd, C. J. and Mandryck, M. (1962). *Trans. Brit. Mycol. Soc.* **45**, 233–244.
Staples, R. C. and Yaniv, Z. (1976). *In* "Physiological Plant Pathology" (R. Heitefuss and P. H. Williams, Eds), 86–103. Springer-Verlag, Berlin.
Takahashi, K., Inaba, T. and Kajiwara, T. (1977). *Physiol. Pl. Path.* **11**, 255–259.
Thornton, J. D. and Cooke, R. C. (1974). *Physiol. Pl. Path.* **4**, 117–125.
Tiwari, M. M. and Arya, H. C. (1967). *Indian Phytopathology* **20**, 356–368.
Tiwari, M. M. and Arya, H. C. (1969). *Science, N.Y.* **163**, 291–293.
Tokura, R. (1975). *Trop. Agric. Res. Ser. (Tokyo)* **8**, 57–60.
Viranyi, F. and Blok, I. (1976). *Netherlands J. Plant Path.* **82**, 251–254.
Viswanathan, M. A. and Sackston, W. E. (1970). *Proc. Can. Phytopath. Soc.* **36**, 28 (abstr.).
Wehtje, G., Littlefield, L. J. and Zimmer, D. E. (1979). *Can J. Bot.* **57**, 315–323.
Wehtje, G. and Zimmer, D. E. (1978). *Phytopathology* **68**, 1568–1571.
Wen-Huei Chen, Ming-Chin Liu and Chia-Yu Chao (1979). *Can. J. Bot.* **57**, 528–533.
Whipps, J. M. and Cooke, R. C. (1978). *New Phytol.* **81**, 307–319.
Williams, P. H. and Pound, G. S. (1964). *Phytopathology* **54**, 446–451.
Yarwood, C. E. (1956). *A. Rev. Pl. Physiol.* **7**, 115–142.

Chapter 9

# Sexual and Asexual Sporulation in the Downy Mildews

## R. W. MICHELMORE

*The Botany School, Downing Street, Cambridge, UK*

## I. Introduction

Sporulation may be the first macroscopic indication of infection by a downy mildew and is often employed as a measure of successful colonization. Also, the numbers and types of spores formed by one generation determine the spatial and temporal distribution of future generations. Both the incidence and intensity of sporulation may be influenced by genetic and environmental factors, effective to varying extents during the period of interaction between host and parasite; it is therefore important to consider the process in detail. This chapter emphasizes sexual sporulation and its relationship to asexual

100 µm

sporulation; epidemiological aspects are considered in other chapters. This chapter is speculative in parts due to the paucity of data, but it indicates the range of phenomena which may be involved in the sporulation of downy mildew-causing fungi.

## II. Genetic Determinants of Sexual Reproduction

Recently many isolates of *Bremia lactucae* have been shown to be heterothallic with two distinct compatibility types, designated $B_1$ and $B_2$ (Michelmore and Ingram, 1980). Sexual reproduction, resulting in the formation of oospores inside the host tissue, occurred only when mycelia of opposite compatibility types became established in the same zone of host tissue (Fig. 1). Observations with the light microscope and the scanning electron microscope revealed that physical contact between vegetative hyphae, presumably of opposite compatibility types, resulted in the production of sexual hyphae and then gametangia (Figs 2 and 3; Michelmore and Ingram, 1981). In view of the probable complexity of the compatibility type locus in the Peronosporales, the genetic determinants of each compatibility type are unlikely to be strictly allelic.

Some isolates of *B. lactucae* regularly produce oospores when cultured alone. One such isolate was analysed by deriving a large number of subcultures from single sporangia,* the majority of which produced oospores when cultured alone, and so exhibited a form of self-fertility. This was a stable character of an isolate when large numbers of sporangia were used for subculturing; but when single sporangia of a self-fertile derivative were subcultured, self-sterile segregants of one or other compatibility type were detected. This indicated the existence of heterokaryosis in *B. lactucae* with respect to compatibility type. Observations suggested that sexual reproduction in such isolates occurred between the self-sterile $B_1$ component and the other components of the population (Michelmore, 1979).

The sexual system of *B. lactucae* has many similarities to that of *Phytophthora drechsleri* (Pythiaceae). The stability of the compatibility types in heterothallic *Phytophthora* spp. is thought to be due to a reciprocal translocation involving the chromosome segment containing the determinants of

---

* In this chapter the term "sporangia" is used for the asexual spores of all downy mildew-causing fungi regardless of the mode of germination.

---

Fig. 1.   A cleared cotyledon of lettuce ten days after inoculation at opposite ends with conidia of two compatibility types of *Bremia lactucae*. The points of inoculation are arrowed. A band of dark oospore-groups is visible across the centre of the cotyledon. (Michelmore and Ingram, 1980 reproduced with the permission of the British Mycological Society.)

compatibility type. The ability of some isolates of *P. drechsleri*, a predominantly heterothallic species, to form oospores in single culture is considered to be a form of secondary homothallism as genetic and cytological studies indicated that the chromosomes containing the compatibility type determinants are trisomic in these isolates. Heterokaryosis appeared to be transient, occuring during the segregation of heterothallic components created by somatic recombination events within the translocation complex (Mortimer *et al.*, 1977; Sansome, 1980). Cytological studies of *B. lactucae* demonstrated that multiple associations of four chromosomes, indicative of structural hybridity, occurred at meiosis in oogonia produced by heterothallic isolates. Furthermore, a ring of four chromosomes with an extra chromosome attached was observed during sexual reproduction of a self-fertile isolate cultured alone (Sansome and Michelmore, unpublished). This suggests that the chromosomal determinants of compatibility type are trisomic in this isolate and that self-fertility in *B. lactucae* is a form of secondary homothallism.

Little is known of the sexual systems of other downy mildew-causing fungi.

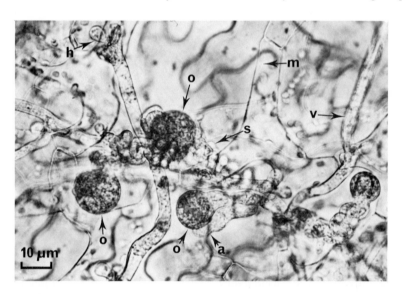

Fig. 2.   A light micrograph of developing gametangia of *Bremia lactucae* in an unfixed cotyledon of lettuce, vacuum infiltrated with water. The cotyledon had been inoculated four days previously with a 1:1 mixture of two isolates of opposite compatibility type. (Michelmore and Ingram, 1981. Reproduced with the permission of the British Mycological Society.)

Abbreviations for Figs 2–5. a = antheridium, c = sporangiophore, h = haustorium, l = lower epidermis, m = plant mesophyll cell, o = oogonium, s = sexual hypha, t = stoma, v = vegetative hypha.

The existence of both heterothallic and homothallic isolates has been demonstrated for *Peronospora parasitica* (De Bruyn, 1937). Oospore production by *P. trifoliorum* has been observed in one monosporangial isolate (Hodgden and Stuteville, 1977). Recently *Sclerospora graminicola* has been shown to exhibit heterothallism (Michelmore, Pawar and Williams, unpublished).

Sexuality in the heterothallic species of the Pythiaceae is complex. The heterothallic species of *Phytophthora* are bisexual; oogonia and antheridia are produced by mycelia of both compatibility types (Huguenin, 1973). Sexuality in the Pythiaceae can be considered as several distinct characters: (i) the tendency to reproduce sexually in single culture; (ii) the compatibility type when crossed with heterothallic type isolates; (iii) the ability to proliferate

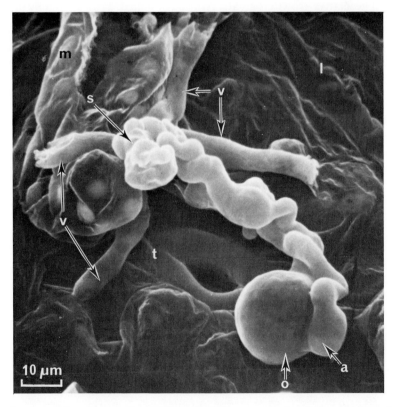

Fig. 3. A scanning electron micrograph of developing gametangia of *Bremia lactucae* in a cotyledon of lettuce 4 days after inoculation with conidia of compatibility types $B_1$ and $B_2$ in a 1:1 ratio. The material was fixed in glutaraldehyde, part fixed in osmium tetroxide, dehydrated, critical point dried, dry fractured and then splutter-coated with gold prior to observation. (Michelmore and Ingram, 1981, reproduced with the permission of the British Mycological Society.)

sexual initials when interacting with mycelia of opposite compatibility type; (iv) the proportion of oogonia which develop mature oospores. Each character, where data are available, appears to be determined independently of the others (Brasier, 1972; Pratt and Green, 1973; Shepherd and Cunningham 1978). There are no data concerning these characters in the Peronosporaceae; however future studies may reveal similar phenomena.

## III. Growth of Mycelium and Sporulation

*B. lactucae*, like many downy mildews, usually forms angular lesions, bounded by the vascular tissue of the host. At the onset of sporulation the rate of vegetative growth is greatly reduced or halted, as evidenced by optical measurements of vegetative mycelium and of the number of substomatal

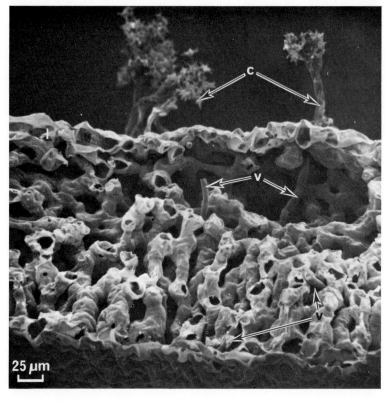

Fig. 4.   A scanning electron micrograph of a transverse section of a sporulating angular lesion of *Bremia lactucae*, following natural infection in the field. The material was fixed in glutaraldehyde, post fixed in osmium tetroxide, dehydrated, cryofractured, critical point dried and then splutter-coated with gold prior to observation.

cavities containing hyphae (Dickinson and Crute, 1974), the synchrony of the development of the gametangia and the proximity of haustoria to the tips of hyphae only after the onset of sporulation (Michelmore, 1979). Figure 4 shows a transverse section of a sporulating angular lesion of *B. lactucae*. The density of mycelium is much lower than that found in a lesion of the rust *Puccinia graminis* f. sp. *tritici*, where more hyphae than host cells were observed in section (Plotnikova *et al.*, 1979). This difference may reflect the different extents to which the two pathogens alter the pattern of nutrient translocation in their hosts. The relationship between the density of vegetative hyphae and sporulation during systemic colonization by downy mildew-causing fungi has not been studied.

When lettuce cotyledons were inoculated with *B. lactucae*, sporulation did not occur before the vegetative mycelium had ramified throughout the host tissue. The numbers of sporangia produced were correlated with the amount of vegetative mycelium present (Dickinson and Crute, 1974; Michelmore, 1979). The probability of hyphae of different compatibility types coming into contact and producing gametangia will also depend upon the density of vegetative hyphae at sporulation, provided that both compatibility types are present in the same limited zone of host tissue. When a secondarily homo-thallic isolate of *B. lactucae* was cultured alone, increased growth of vegetative hyphae appeared to increase sexual reproduction; this may have been due to the increased number of nuclear divisions enhancing the segregation of the heterothallic components (Michelmore, 1979).

When a heterothallic fungus colonizes the plant systemically, hyphae of different compatibility types can come into contact in the later stages of colonization, following infection of spatially separated parts of the host plant. The varied patterns of sporulation in graminaceous downy mildews may reflect the progressive colonization of the plant apex by mycelia of different compatibility types. Asexual sporulation of *Sclerospora graminicola* and *Peronosclerospora sorghi* is usually observed after the early stages of systemic colonization with oospore production often, but not always, occurring in leaves developing later.

## IV. Control of Sexual *vs.* Asexual Sporulation

Both sexual and asexual sporulation of *B. lactucae* occurred simultaneously over the same temperature range (5–22°C) regardless of the temperature of incubation which influenced the length of the vegetative growth phase (Michelmore, 1979). There was no evidence for a progression from asexual to sexual reproduction as the infected tissues "matured" or aged. Simultaneous sexual and asexual sporulation has also been observed in

*Peronospora parasitica* (McMeekin, 1960) and *Sclerospora graminicola* (Michelmore, Pawar and Williams, unpublished).

The data obtained for *B. lactucae* strongly suggest that the formation of oospores and sporangia are antagonistic processes. This is in agreement with the observations of McKay (1939) on *Peronospora destructor*, but disagrees with the later observations of Berry and Davis (1957) on *P. destructor* and of McMeekin (1960) on *P. parasitica*. Heavy oospore production and asexual sporulation were not observed in the same zone of host tissue in graminaceous downy mildews (Safeeulla, 1976). Production of sporangia by *B. lactucae* was obviously absent from zones of heavy oospore production (*c.* $5 \times 10^3$ cm$^{-2}$ of cotyledon), but occasionally, at lower levels of sexual reproduction, both sexual and asexual sporulation were observed in close proximity, apparently involving the same vegetative hyphae (Fig. 5). At the onset of sporulation there may be a developmental switch which prevents asexual reproduction during sexual differentiation; alternatively asexual sporulation may be suppressed by greater nutritional demands of sexual reproduction which takes priority.

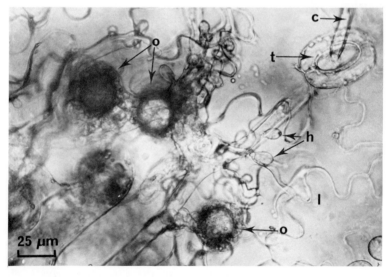

Fig. 5.  A light micrograph of a cotyledon of lettuce following infection by a conidia of *Bremia lactucae* of two compatibility types. Prior to observation the cotyledon was vacuum-infiltrated with water. The base of a sporangiophore can be seen emerging from a stoma in close proximity to several oogonia.

## V. Sporulation and Symptom Development

The appearance of chlorosis and necrosis of host tissue infected by downy mildew-causing fungi appears to be dependent upon both the extent to which the transfer of nutrients from host to fungus stresses the host and the extent to which the host can resist attack by necrotrophic secondary pathogens. Chlorosis usually occurs at the onset of sporulation when the transfer of nutrients becomes greater and the disturbance of the host's metabolism increases (Mason, 1973; Inaba and Kajiwara, 1975). In some cases chlorosis can occur before sporulation, espcially when asexual reproduction is inhibited by environmental factors (cf. Section VII). Under damp conditions, the weakened host often becomes colonized by secondary bacterial pathogens and a soft necrotic lesion develops. Under other conditions, secondary fungal pathogens may be more important than bacteria; for example, sporulation of *Peronosclerospora sorghi* is often associated with *Colletotrichum* sp. (Doggett, 1970) or *Gleocercospora sorghi* (Meenaksi and Ramalingam, unpublished) which result in dark, dry lesions.

Oospore production is associated with chlorotic and necrotic host tissue in several downy mildews. This has led to the suggestion that oospore formation is induced by senescence or necrosis of the host tissue in *Peronospora parisitica* (McMeekin, 1960), *B. lactucae* (Fletcher, 1976) and *Peronosclerospora sorghi* (Rotem *et al.*, 1978). The formation of large numbers of oospores by *B. lactucae*, however, resulted in enhanced host senescence and necrosis compared with lettuce tissue which had been inoculated with similar numbers of sporangia of only one compatibility type and on which, therefore, only asexual sporulation occurred. Thus, contrary to the above suggestion, sexual reproduction appeared to be the cause rather than the consequence of host senescence. Environmental conditions conducive to host fitness favoured the production of both oospores and sporangia; the onset of sporulation, sexual or asexual, preceded chlorosis (Michelmore, 1979).

## VI. Factors Affecting the Numbers of Spores Produced

The numbers of spores produced (sexually or asexually) depends upon the density of hyphae at sporulation (cf. Section III) and the stability of the host–parasite relationship (cf. Section V). Both these parameters will be influenced by many environmental and genetic factors. The consequences of these influences varies during the different phases of interaction between host and parasite.

## A. During establishment

In some downy mildews distinct physiologic races have been identified and the primary determinant of establishment is the interaction between the resistance phenotype of the host and the virulence phenotype of the fungus. The expression or effectiveness of some resistance genes, however, may be influenced by environmental factors, particularly temperature (Crute and Norwood, 1978).

There are environmental conditions which are not conducive to infection. Furthermore, many factors, such as the age of the host tissue or the presence of antagonistic micro-organisms, can influence the establishment of the bio-trophic relationship between a potentially compatible pathogen and its host; only a proportion of attempted infections may result in the successful development of vegetative mycelia. The larger the number of sporangia of *B. lactucae* which successfully established mycelia in an infection, the greater was the density of hyphae at sporulation (Dickinson and Crute, 1974; Michelmore, 1979) and also the greater was the probability that mycelia of different compatibility types would become established together, if both were present in the inoculum.

## B. During colonization

Low light intensities or darkness slightly lengthened the phase of vegetative growth of *B. lactucae* and greatly reduced the density of mycelium at sporulation; these effects, however, could be overcome by exogenously supplied sucrose which increased the rate of mycelial growth. When the host was senescing or stressed, sporulation occurred at low mycelial densities (Michelmore, 1979). Similarly, the hyphal density of *Pseudoperonospora cubensis* at the onset of asexual sporulation was much lower in infected leaves incubated with low light intensities or in the dark than in those incubated at high light intensities (Cohen and Rotem, 1971; Inaba and Kajiwara, 1975). These observations suggest that there may be some form of nutritional signal which induces a developmental change from vegetative to reproductive growth.

The differing patterns of symptom expression of a systemic downy mildew reflect the varying extents of colonization of the plant apex (Jones, 1978); there may be only restricted mycelial spread in mature host tissues. The relative growth rates of host and parasite may determine the extent of colonization; environmental factors probably affect the growth rates of each differently, resulting in differences in distributions and numbers of spores produced.

Different combinations of host cultivar and pathogen isolate, may result in different rates of mycelial growth or in sporulation at different hyphal densities, but there are no pertinent data on this for any downy mildew.

## C. During sporulation

The length of time over which sporulation occurs prior to the breakdown of the biotrophic relationship can again be influenced by genetic and environmental factors. The compatibility of the fungal and host phenotypes at sporulation may vary. Two heterothallic isolates of *B. lactucae*, inoculated singly, differed in their compatibility with a lettuce cultivar considered to be susceptible to all isolates of *B. lactucae*. Both isolates started to sporulate at the same time after inoculation and produced similar numbers of sporangia during the first 12 h dark period. Subsequently, however, cotyledons inoculated with one isolate decayed more rapidly and approximately half the number of sporangia were produced on them compared with cotyledons inoculated with the other isolate (Michelmore, 1979). Any environmental factor which increases the stress on the host or decreases stability of the host–parasite relationship will tend to shorten the duration of sporulation. Incubation of lettuce cotyledons infected with *B. lactucae* at 15°C resulted in more rapid sporulation, but the cotyledons decayed earlier and the total production of sporangia was lower than in infected cotyledons incubated at 10°C (Michelmore, 1979). Similar phenomena have been noted for *Pseudoperonospora cubensis* (Cohen and Rotem, 1971) and *Peronospora tabacina* (Rotem and Cohen, 1970). All of these phenomena may have been due to increased stability of the biotrophic relationships at lower temperatures or due to increased activities of biotic antagonists at higher temperatures. Sporulation of *B. lactucae* was completed earlier under septic rather than aseptic conditions; under aseptic conditions sporulation ended only after the cotyledons had become completely chlorotic (Michelmore, 1979).

From the data obtained in various studies it is not possible to separate the influences of one environmental factor on the different stages of interaction between host and parasite. Light regimes, which favour photosynthesis of the host during the periods of colonization and sporulation, tend to increase both the intensity and the length of sporulation, providing that they do not directly inhibit spore production. This has been shown for asexual sporulation of *Pseudoperonospora cubensis* (Cohen and Rotem, 1971) and for sexual sporulation of *B. lactucae* (Michelmore, 1979) and *Peronospora trifoliorum* (Hodgden and Stuteville, 1977).

## VII. Environmental Constraints on Sporulation

Asexual sporulation, producing short-lived sporangia, occurs outside the host plant and is therefore sensitive to the external environment. In consequence there appear to be further constraints superimposed on those influences outlined above. Continuous light has been shown to inhibit, partially or totally, the production of sporangia by *Peronospora tabacina* (Cruickshank, 1963; Jadot, 1966), *P. pisi* (Pegg and Mence, 1970), *Pseudoperonospora cubensis* (Kajiwara and Iwata, 1959) *Peronosclerospora sorghi* (Schmitt and Freytag, 1974) and *B. lactucae* (Raffray and Sequeira, 1971).

High relative humidity is also necessary for asexual sporulation of these fungi and critical studies on *Pseudoperonospora cubensis* and *Peronospora tabacina* have demonstrated the complexity of the situation (Cohen, 1976; Cohen and Eyal, 1977; Cohen *et al.*, 1978). A minimum period of darkness appears to be necessary for maximum sporulation to occur. During the first part of this period the relative humidity can be low, but a high relative humidity, near to 100%, is later required. Sporulation is more sensitive to inhibition by blue light than by other wavelengths. The effect of light is temperature sensitive: temperatures of 20°C and above, with very low levels of light inhibited sporulation; greater irradiation was needed to give similar degrees of inhibition at lower temperatures, until at 10°C and below, irradiation tended to promote sporulation. Indirect irradiation made by irradiating only the surrounding tissue was also inhibitory at 20°C. These effects appear to be independent of photosynthesis, as sporulation is markedly inhibited by low levels and short periods of light; also the presence of chemical inhibitors of photosynthesis do not alter the pattern of inhibition. The synthesis of an antisporulant compound has been postulated, which accumulates only at temperatures greater than 15°C. Although a high relative humidity is necessary for asexual sporulation of many downy mildews, free water on the leaf surface appears to be inhibitory (Cohen and Sherman, 1977; Rotem *et al.*, 1978; Safeeulla, personal communication).

As oospore production is not directly affected by environmental conditions external to the leaf, and occurs over a period of several days rather than during a single night, similar constraints on sexual reproduction are not expected. No constraints were found in an extensive survey of environmental influences on oospore production in *B. lactucae* (Michelmore, 1979). It is interesting that in the powdery mildew *Erysiphe cichoracearum*, where conidia are not sensitive to light or a dry environment, the production of such spores was not inhibited by light or low relative humidity (Cole, 1971).

## VIII. Qualitative Aspects of Sporulation

In addition to quantitative variation, sporulation may also vary qualitatively due to genetic and environmental influences. When *B. lactucae* was transferred from one host species to another, the size of the sporangia altered (Schweizer, 1920). The aerial morphology of *Peronospora polygoni* was dependent upon the habitat of its host *Polygonum aviculare*. Gustavsson (1959) considered that a single observation of a new downy mildew on a known host plant was likely to be an aberrant form produced as a result of abnormal environmental conditions rather than a new species. The size of sporangia of *P. pisi* was related to the temperature during sporulation (Pegg and Mence, 1970). Temperature was the only recorded environmental factor which affects the size of sporangia of *Peronospora tabacina* (Smith, 1970). Later however, continuous irradiation during sporulation, particularly by blue light, was shown to result in deformed sporangiophores; sporangia, if produced, were also deformed or much smaller, being one fifth to one tenth the size of those produced in darkness (Cohen, 1976). Intermittent light had similar effects on *P. tabacina* and *Pseudoperonospora cubensis* (Cohen *et al.*, 1978). Continuous darkness during colonization prior to sporulation resulted in sporangiophores with decreased branching and malformed sporangia of *P. cubensis* (Inaba and Kajiwara, 1975). The morphology of sporangiophores of *Sclerospora graminicola* and *Peronosclerospora sorghi* is influenced by both temperature and moisture (Safeeulla and Thirumalachar, 1956).

Oospore size and morphology have been employed as taxonomic characters at species level for many Oomycetes. The stability of oospore size, however, has been questioned. The diameters of oospores and oogonia of several *Pythium* species changed over a 30 month period of subculture (Hendrix and Campbell, 1974). Oospore size of *Phytophthora cinnamomi* depended upon the culture media used and also upon whether selfing or outcrossing with an isolate of opposite compatibility type had occurred; it appeared that the smaller the number of oospores formed, the larger was the average diameter (Zentmyer *et al.*, 1979). Clayton and Stevenson (1943) demonstrated that oospore diameter of *Peronospora* spp. found on tobacco varied greatly and that the two species *P. nicotianae* and *P. tabacina*, previously separated on oospore size, were probably one species, *P. tabacina*. Measurements of oospore diameter, in *B. lactucae* also revealed a large size range; when low numbers of oospores were formed, the mean oospore diameter was significantly greater (Michelmore, 1979). Oospore diameter, therefore, should not be regarded as a definitive taxonomic character.

Oosporogenesis involves a complex sequence of events leading to oospore formation and maturation, during which the contents of the gametangia are extensively reorganized several times. Abortion during the development of oogonia may occur in a number of ways and at a number of stages. In *B. lactucae*, the likelihood of abortion was markedly increased by the build up of large numbers of bacteria in the intercellular spaces, coinciding with the breakdown of the biotrophic relationship (Michelmore, 1979); different forms of abortion were observed in the absence of bacteria (Sargent *et al.*, 1977) and may have been due to genetic factors. Large numbers of aborted oospores have been observed in *Sclerospora graminicola* and *Peronosclerospora sorghi*, which may have been the result of the rapid drying of the infected host leaf which often occurs during sexual sporulation (Safeeulla, personal communication).

## IX. Oospore Production in the Field

One of the characteristics of sexual reproduction by many downy mildews is the infrequent and sometimes localized formation of oospores, often in particular parts of the host plant. In some species at least, this could be a consequence of heterothallism. Until recently, oospore production by *B. lactucae* had been only rarely observed and even the existence of sexual reproduction was a matter of controversy. Following the demonstration of heterothallism, sexual reproduction has been observed in any part of the plant where mycelia of both compatibility types have become established, including the roots of seedlings (Michelmore, 1979). Oospores of *Sclerospora graminicola* have also been observed in the roots of seedlings of *Pennisetum typhoides* (Safeeulla, 1976; Michelmore *et al.*, unpublished). Therefore, there appears, to be no physiological barrier to frequent oospore production by these fungi. In view of this, the significance of the natural occurrence of oospores of other downy mildews needs to be reconsidered. Previous failures to find oospores may have been due to the presence of only one compatibility type or due to zones of sexual reproduction being overlooked because of extensive decay and the absence of asexual sporulation.

In a survey of isolates of *B. lactucae*, there was a predominance of the $B_2$ compatibility type (Michelmore and Ingram, 1980a). Many heterothallic *Phytophthora* species have both compatibility types present in approximately equal proportions for only a limited part of their species range; this may be due to the physiological adaptation of some $A_2$ strains to produce oospores, in the absence of $A_1$ strains, in response to the presence of antagonistic *Trichoderma* species and certain bacteria (Brasier, 1975). Although downy mildews are frequently associated with necrotrophic antagonists, there are no data for nonspecific stimulation of sexual reproduction.

In the heterothallic species of *Phytophthora* there is also the possibility of stimulation of sexual reproduction between the $A_1$ and $A_2$ compatibility types of different species (Savage *et al.*, 1968); the frequency of this occurring naturally has yet to be determined. Inter-specific stimulation between most downy mildews is unlikely, as there is little, if any, overlap in their hosts ranges. Both *Albugo candida* (Albuginaceae) and *Peronospora parasitica* are commonly found together, however, and cross-stimulation of sexual reproduction has been suggested (Sansome and Sansome, 1974). There is some overlap in the host ranges of the fungi causing graminaceous downy mildews and therefore interspecific stimulation may take place; interspecific hybridization, if it occurs, would partly explain the confused taxonomy of some of these fungi.

## X. Concluding Discussion

From the observations on *Bremia lactucae*, the primary determinant of sexual reproduction appears to be genetic rather than environmental. Asexual sporulation occurs in the absence of heavy oospore production and a change for asexual to sexual reproduction is not induced by environmental changes. The frequency of oospore formation, however, depends not only upon the occurrence of each compatibility type or of secondarily homothallic isolates in the population, but also upon the level of infection and the pattern of fungal growth in the host which may vary with different hosts and with different environmental conditions. Environmental factors can also markedly influence the numbers of mature spores formed during both sexual and asexual reproduction.

Many more downy mildews need to be investigated for the presence of heterothallism. While not all downy mildews may be heterothallic, it is interesting that all the biotrophic pathogens and many hemibiotrophs so far studied have heterothallic sexual systems, for example, *Erysiphe graminis* (Powers and Moseman, 1956), *Puccinia graminis* (Craigie, 1927), *Ustilago maydis* (Holliday, 1961), and *Phytophthora infestans* (Smoot *et al.*, 1958). During the evolution of the Peronosporaceae, the selection pressures for variation via sexual reproduction in a biotrophic population may have been greater than the selection pressures for the more frequent formation of resistant survival spores in an inbreeding, homothallic population. The secondary homothallism exhibited by *B. lactucae* and *Peronospora parasitica* possibly provides a compromise, in which some members of an outbreeding population are functionally homothallic. In a survey of 39 isolates of *B. lactucae* collected at random from a wide range of sources, 7 appeared to be self-fertile, suggesting that secondary homothallism is common (Michelmore, 1979).

There are many similarities between the sexual systems of the hetero-thallic *Phytophthora* species and *Bremia lactucae*. Much may be gained by extrapolating from members of the Pythiaceae where studies on the complexities of the inheritance of compatibility types and pathogenicity are easier, first because they can be grown in axenic culture and secondly because of the relative ease with which the oospores of some species can be germinated. Progress may soon be made in understanding the chemical interactions involved in sexual reproduction.

Studies on asexual and sexual reproduction are both fundamental to the understanding of the epidemiology and variability of the downy mildews. From the data available, the fungi causing the downy mildews appear to be no less variable than many other plant pathogens; therefore a study of sexuality is an essential part of any control programme.

## Acknowledgements

I wish to thank Dr D. S. Ingram, and also Professor K. M. Safeeulla and colleagues at the Downy Mildew Research Laboratory, Mysore, for advice and discussions. The financial support of the Agricultural Research Council, the Lucy Ernst Trust and the Royal Society – Indian National Science Academy exchange of scientists programme is gratefully acknowledged.

## References

Berry, S. Z. and Davis, G. N. (1957). *Pl. Dis. Rep.* **41**, 3–6.
Brasier, C. M. (1972). *Trans. Br. Mycol. Soc.* **58**, 237–251.
Brasier, C. M. (1975). *New Phytol.* **74**, 195–198.
Bruyn, H. L. G. De (1937). *Genetica* **19**, 553–558.
Clayton, E. E. and Stevenson, J. A. (1943). *Phytopathology* **33**, 101–113.
Cohen, Y. (1976). *Aust. J. Biol. Sci.* **29**, 281–289.
Cohen, Y. and Eyal, H. (1977). *Physiol. Pl. Path.* **10**, 93–103.
Cohen, Y., Levi, Y. and Eyal, H. (1978). *Can. J. Bot.* **56**, 2538–2543.
Cohen, Y. and Rotem, J. (1971). *Phytopathology* **61**, 265–268.
Cohen, Y. and Sherman, Y. (1977). *Phytopathology* **67**, 515–521.
Cole, J. S. (1971). *In* "Ecology of Leaf Surface Micro-organisms" (T. F. Preece and C. H. Dickinson, Eds.) 323–337 Academic Press, New York and London.
Craigie, J. H. (1927). *Nature* **120**, 765–767.
Cruickshank, I. A. M. (1963). *Aust. J. Biol. Sci.* **16**, 88–98.
Crute, I. R. and Norwood, J. M. (1978). *Ann. Appl. Biol.* **89**, 467–474.
Dickinson, C. H. and Crute, I. R. (1974). *Ann. Appl. Biol.* **76**, 49–61.
Doggett, H. (1970). *Indian Phytopath.* **23**, 350–355.
Fletcher, J. T. (1976). *Ann. Appl. Biol.* **84**, 294–298.
Gustavsson, A. (1959). *Op. Bot. Soc. Bot. Lund.* **32**, 5–61.
Hendrix, F. F. and Campbell, W. A. (1974). *Mycologia* **66**, 681–684.

Hodgden, L. D. and Stuteville, D. L. (1977). *Proc. Am. Phytopathol. Soc.* **4**, 166–167.
Holliday, R. (1961). *Genet. Res.* **2**, 204–230.
Huguenin, B. (1973). *Cah. O.R.S.T.O.M. Ser. Biol.* **20**, 59–61.
Inaba, T. and Kajiwara, T. (1975). *Bull. Nat. Inst. Agric. Sci., Tokyo* C 29, 65–139 (Japanese, English summary).
Jadot, R. (1966). *Parasitica* **22**, 55–63.
Jones, B. L. (1978). *Phytopathology* **68**, 732–735.
Kajiwara, T. and Iwata, Y. (1959). *Ann. Phytopath. Soc. Japan* **24**, 109–113.
Mason, P. A. (1973). Ph.D. Thesis, Camb. Univ.
McKay, R. (1939). *Jl. R. Hort. Soc.* **64**, 272–285.
McMeekin, D. (1960). *Phytopathology* **50**, 93–97.
Michelmore, R. W. (1979). Ph.D. Thesis, Camb. Univ.
Michelmore, R. W. and Ingram, D. S. (1980). *Trans. Br. Mycol. Soc.* **75**. 47–56
Michelmore, R. W. and Ingram, D. S. (1981). *Trans. Br. Mycol. Soc.* (in press).
Mortimer, A. M., Shaw, D. S. and Sansome, E. R. (1977). *Arch. Microbiol.* **111**, 255–259.
Pegg, G. F. and Mence, M. J. (1970). *Ann. Appl. Biol.* **66**, 417–428.
Plotnikova, T. M., Littlefield, L. J. and Miller, J. D. (1979). *Physiol. Pl. Path.* **14**, 37–39.
Powers, H. R. and Moseman, J. G. (1956). *Phytopathology* **46**, 23.
Pratt, R. G. and Green, R. J. (1973. *Can. J. Bot.* **51**, 429–436.
Raffray, J. B. and Sequeira, L. (1971). *Can. J. Bot.* **49**, 237–239.
Rotem, J. and Cohen, Y. (1970). *Phytopathology* **60**, 54–57.
Rotem, J., Cohen, Y. and Bashi, E. (1978). *A. Rev. Phytopath.* **16**, 83–101.
Safeeulla, K. M. (1976). "Biology and Control of the Downy Mildews of Pearl Millet, Sorghum and Finger Millet". Wesley Press, Mysore.
Safeeulla, K. M. and Thirumalachar, M. J. (1956). *Phytopath. Z.* **26**, 41–48.
Sansome, E. R. (1980). *Trans. Biol. Mycol. Soc.* **74**, 175–185.
Sansome, E. R. and Sansome, F. W. (1974). *Trans. Br. Mycol. Soc.* **62**, 323–332.
Sargent, J. A., Ingram, D. S. and Tommerup, I. C. (1977). *Proc. R. Soc. Series B.* **198**, 129–138.
Savage, E. J., Clayton, C. W., Hunter, J. H., Brenneman, J. A., Laviola, C. and Gallegly, M. E. (1968). *Phytopathology* **58**, 1004–1021.
Schweizer, J. (1920). *Thurganischen Naturforschenden Gelleschaft Mitteilungen* 23, 15–60.
Schmitt, C. G. and Freytag, R. E. (1974). *Pl. Dis. Reptr.* **58**, 825–829.
Shepherd, C. J. and Cunningham, R. B. (1978). *Aust. J. Bot.* **26**, 139–151.
Smith, A. (1970). *Trans. Br. Mycol. Soc.* **55**, 59–66.
Smoot, J. J., Gough, F. J., Lamey, H. A., Eichenmuller, J. J. and Gallegly, M. E. (1958). *Phytopathology* **48**, 165–171.
Zentmyer, G. A., Klure, L. J. and Pond, E. C. (1979). *Mycologia* **71**, 55–67.

Chapter 10

# The Fine Structure of the Downy Mildews

## J. A. SARGENT

*ARC Weed Research Organization, Yarnton, Oxford, England*

## I. Introduction

Electron microscopy, particularly when used in association with physiological, biochemical and genetical studies, can provide the pathologist with information which indicates a means of understanding the complex relationship which exists between a parasite and its host. Among the downy mildews fine structure studies have been reported for *Peronospora manshurica* (Peyton and Bowen, 1963; Riggle and Dunleavy, 1974; Riggle, 1977), *P. parasitica* (Chou, 1970), *P. spinaciae* (Kajiwara, 1973), *P. farinosa* (Kröber and Petzgold, 1972; Kröber, *et al.*, 1979), *P. pisi* (Hickey and Coffee, 1977, 1978), *P. tabacina* (Kröber and Petzgold, 1972), *Pseudoperonospora humuli* (Royle and Thomas, 1971, 1973; Pares and Greenwood, 1979), *P. cubensis*

(Cohen *et al.*, 1974), *Plasmopora holstedii* (Wehtje and Zimmer, 1978; Wehtje *et al.*, 1979) and *P. viticola* (Royle and Thomas, 1973). In addition a number of reviews have highlighted differences as well as similarities in the fine structure of diverse groups of fungi (Bracker, 1967; Bracker and Littlefield, 1973; Aist, 1976). However, in only one study (Hickey and Coffee, 1978) has an attempt been made to integrate data derived from a number of techniques selected to examine in depth the relationship between parasite and host. Success in this field depends not only on techniques which provide good preservation of cell organelles but also upon cytochemical and micro-autoradiographic methods designed to indicate physiological and bio-chemical pathways at the subcellular level. This chapter attempts to illustrate the benefits of this approach by describing the results of studies aimed at elucidating the subcellular relationships between *Bremia lactucae* a biotroph, and lettuce, its host.

## II. The Fine Structure of *Bremia lactucae*

Under conditions of high humidity diseased leaves of lettuce produce abundant conidia of *B. lactucae*. These are the principal propagules of the pathogen. Under favourable conditions they germinate and if in contact with a compatible host the germling will penetrate and become established. Subsequently further vegetative conidia are formed or, in certain circum-stances, oospores capable of remaining viable throughout the winter will be produced.

### A. The conidium

The conidium (Fig. 1) is a single ovoid cell approximately $18 \times 16\,\mu$m (Sargent and Payne, 1974). At the proximal end which carries part of the sterigma on which the spore was borne traces of a cuticle remain (Fig. 2). The cell wall is uniform in thickness except at the distal papilla where it is thinner (Fig. 3). Each spore contains about 17 nuclei with prominent nucleoli. Mitochondria are abundant and in the quiescent spore are largely isodiametric. The endomembrane system is poorly organized: endoplasmic reticulum profiles are sparse, short and smooth, and the dictyosomes, although dilated peripherally, are compact (Fig. 4). Ribosomes are very abundant but since few of them are associated into polysomes they give an even, granular appearance to the ground plasm. The energy reserves of the conidium are in the form of lipid droplets which line the plasmalemma. In the quiescent spore they are strongly osmiophilic.

Conidia are unable to germinate until a water-soluble inhibitor is washed from them (Mason, 1973). When free of this they will, under a favourable temperature regime, germinate by producing a germ tube from the papilla. Rupture of the original spore wall is preceded by a well defined succession of events. First the reserves become mobilized as the lipid droplets move deeper into the cytoplasm and their affinity for osmium decreases (Fig. 5). Secondly the endomembrane system becomes activated with the generation of rough endoplasmic reticulum and a proliferation of membrane-bound vesicles from the dictyosomes (Fig. 6). Thirdly these Golgi vesicles migrate towards the papilla. Concomitant with these changes, the mitochondria and nuclei elongate and generally become orientated along the axis of the spore (Fig. 7). Microtubules also appear orientated in a similar way. These changes, which transform the cytoplasm into a highly active and polarized condition, occur very rapidly and can be completed in just over an hour following removal of the inhibitor. The vesicles which accumulate within the papilla stain differentially, suggesting that each has a specific role to play. The first of those roles is the softening of the condium wall to permit growth of the germ tube. Vesicles discharge their contents through the plasmalemma at the papilla, involving presumably, the secretion of wall degrading enzymes. Their membranes fuse to form a lomasome which, following the initiation of germ tube growth, remains as an annulus (Fig. 8). Under favourable conditions the germ tube grows at a very rapid rate (Fig. 9). A second and major role of the vesicles is the transport of wall precursors to the growing tip of the germ tube. Their production is maintained and they continue to accumulate at the tip of the germ tube (Fig. 10). There they not only discharge into the developing wall but at the same time contribute to the proliferation of the plasmalemma as their membranes fuse with and become incorporated into it. Germination occurs throughout the temperature range 0–21°C and growth of the germ tube is maximal at 15°C. At temperatures above 21°C germination is inhibited and if conidia are held for a prolonged period at a high temperature organelles suffer irreversible damage leading to spore death. Electron microscopy has revealed that the lesion resulting from high temperature stress occurs at an early stage in the events leading to germination. Within conidia washed free of inhibitor and maintained at 28°C the dispersal of the lipid droplets is the only detectable pregermination event to occur (Figs 11 and 12). However the lipid fails to lose its affinity for electron stains. Presumably at high temperatures metabolism of the reserves is blocked thus making unavailable intermediates necessary for activation of the endomembrane system. Provided the temperature is not raised excessively or it is lowered sufficiently quickly the lesion is repaired and germination proceeds. Prolonged stress results in progressive loss of integrity of membranes: mitochondria swell and become disorganized, double membranes of the endomembrane system and nuclear envelope become dilated and vacuolation is extensive (Fig. 13). Clearly conidia are unable to repair all but the mildest stress-induced damage to their organelles.


## B. The germ tube and appressorium

The germ tube is capable of fast and extensive growth; presumably the supply of energy reserves imposes the principal limit to growth. Vacuolation, which begins in the germinating spore, progresses along the elongating germ tube thus concentrating the cytoplasm at the developing end. On contacting the host the tip of the germ tube swells to produce an appressorium which makes close contact with the epidermal surface. Penetration of the host then occurs by enzymatic dissolution of the host cuticle and cell wall and the entry of a penetration peg (Sargent *et al.*, 1973). However, electron microscopy has revealed that even before penetration peg formation commences the host is sensitive to the presence of the fungus (Fig. 14). Directly beneath the appressorium the host cell wall stains very readily indicating the movement into the wall, or the formation within it, of compounds possessing abundant osmiophilic groups. At the same time host cytoplasm accumulates beneath the appressorium and membrane-bound crystals develop within it. Clearly the organisms are interacting vigorously at this early stage and the production of the membrane-bound crystals may be a means by which certain material of fungal origin is effectively de-activated by the host.


## C. The penetration peg

Digestion of the host cuticle and cell wall involves certain of the fungal Golgi vesicles in further specific roles. The mixed vesicles occupy a strategic position in the appressorium adjacent to the area of contact of the two organisms. It seems likely that the variety of enzymes required to solubilize the components of cuticle and cell wall are distributed in different vesicles. Whether or not that proves to be the case there can be no doubt about the crucial involvement of the vesicles in the penetration process. Their discharge at the point of penetration peg formation is so rapid that the plasmalemma is unable to incorporate their membranes fast enough to avoid the formation of a distinctive lomasome (Figs 15 and 16). The host cuticle is not only dissolved rapidly beneath the appressorium but it becomes swollen and thrown into folds around the developing pore (Duddridge and Sargent, 1978). All components of the cell wall are digested as the penetration peg develops. Neither the wall microfibrils nor the cuticle at the edge of the penetration pore are turned inwards, supporting the view that penetration is achieved solely as a result of enzymatic digestion and that no pressure from the fungus is involved. The fungal wall of the penetration peg is evidently deposited in a fluid form. Sections through the peg show it ramifying between the free ends of host-wall microfibrils (Fig. 17). During development of the penetration peg a pad of callose forms in the host cytoplasm directly beneath the peg. However, this is not necessarily a specific response of the host to the invading pathogen; callose is frequently induced in cells by wounding.

## D. Lipolytic activity in the germinating spore and germ tube

The subcellular distribution of lipolytic enzymes during germination of the conidium and digestion of the host cuticle has been examined using a technique which distinguishes lipases responsible for the hydrolysis of long-chain fatty acid esters and esterases which attack the shorter chains (Duddridge and Sargent, 1978). In the quiescent spore, lipase is abundant within the cisternae enclosed by the double membrane systems of the nuclear envelope, the endoplasmic reticulum and the forming face of the dictyosomes, and on certain membranous regions surrounding small vacuoles (Fig. 18). During activation of the conidium when the endomembrane system becomes elaborated further sites of lipase activity are distinguishable within the newly formed endoplasmic reticulum. Possibly the enzyme complex is involved in the synthesis of new membrane. As germination proceeds and the cytoplasm is swept into the germ tube, lipase activity is lost from these membrane systems (Fig. 19) but esterase activity develops, principally in association with the lipid droplets (Fig. 20). Presumably extensive hydrolysis of lipid is required at this stage when either the pool of available substrates for cell component synthesis has become exhausted or shorter chain products are required to synthesize compounds involved specifically in the extension of the germ tube or the preparation of the germling for penetration of the host. Although the cytoplasmic vesicles at the tip of the germ tube and appressorium might be expected to contain enzymes responsible for digestion of the host cuticle and epidermal wall, no lypolytic activity has been detected within them. Lipase is present, however, in the appressorial wall in contact with the host particularly in the region adjacent to the zone of cuticle undergoing digestion (Figs 21 and 22). It seems likely, therefore, that the enzyme is synthesized from precursors discharged from the cytoplasmic vesicles in response to the presence of host cuticle.

## E. The primary vesicle

Following penetration, the pathogen swells rapidly within the epidermal cell to form a primary vesicle (Fig. 23). In common with other invading fungi

(Pares and Greenwood, 1979) *Bremia* does not pierce the plasmalemma of the
host cell. This membrane becomes invaginated as the primary vesicle enlarges
but it is not stretched tightly. The way in which rupture of the plasmalemma
is avoided has been revealed by an examination of sections through the
penetration region (Sargent *et al.*, 1973). As the primary vesicle begins to
expand the plasmalemma immediately adjacent to the penetration pore
becomes considerably folded (Fig. 24). Evidently in response to the presence
of a developing penetration peg the host plasmalemma expands, infolding
upon itself as it does so, and subsequently unfolds to accomodate the
enlarging primary vesicle. Proliferation of host plasmalemma in the presence
of the fungus is not confined to the penetration area. Whorls of convoluted
membrane occur elsewhere in the invaded cell and even in adjacent cells in
areas close to a primary vesicle (Fig. 25). The nucleus of an adjacent cell is
also frequently observed in a position close to the fungus (Figs 26 and 27).
It is not know why the nucleus is located in this position. Possibly material
of fungal origin diffusing into those areas reduces cytoplasmic streaming. The
transfer of cytoplasm from the appressorium to the primary vesicle is
terminated by the deposition of a callose plug within the penetration peg
(Fig. 27). Although the appressorium can absorb carbohydrate secreted by the
host (Andrews, 1975) the steps which lead up to and include the development
of the primary vesicle are probably dependent upon reserves from the spore
and they can be regarded as the transfer of fungal cytoplasm to a site from
which, in a compatible host, establishment can begin. The callose which
formed in the host just prior to fungal penetration becomes dispersed around
the primary vesicle as the latter expands and further callose, generated by
the host, is deposited around the neck of the vesicle. Growth of the fungal wall
is apparently dependent upon a supply of precursors supplied through the
plasmalemma by fusing Golgi vesicles but the aggregation of cytoplasmic
vesicles which typically occurs at the growing tip of the germling ceases
immediately penetration has been achieved, a step in the closely controlled
programme of events which occurs as the host and pathogen interact.

Abbreviations: A = appressorium; An = antheridium; FT = fertilization tube; GT = germ
tube; H = haustorium; IH = intercellular hypha; O = oogonium; Os = oospore; Pp = periplasm; PV = primary vesicle; c = cuticle; ca = callose; ch = chloroplast; cm = crystalline microbody; cv = cytoplasmic vesicle; d = dictyosome; e = endoplasmic reticulum fw = fungal cell
wall; hp = host plasmalemma; hw = host cell wall; l = lipid droplet; lo = lomosome; m =
mitochondria; mt = microtubule; n = nucleus; p = papilla; s = sterigma; v = vacuole. Bar scale
= 1 μm.

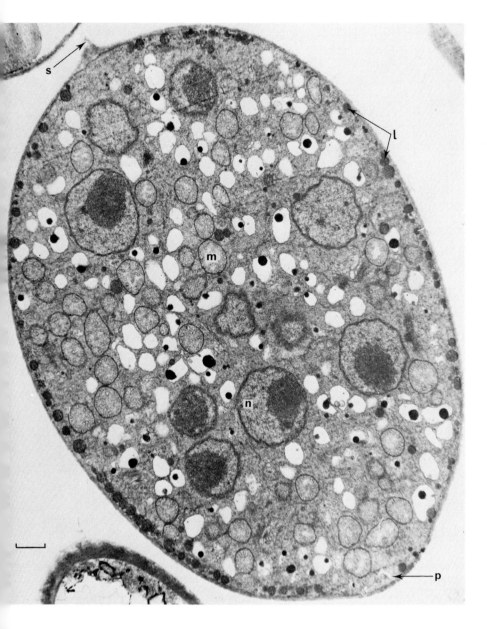

Fig. 1. A median longitudinal section through a freshly collected quiescent conidium of *Bremia lactucae*. Nuclei and mitochondria are spherical and evenly distributed between the proximal fractured sterigma and the distal papilla. Densely staining lipid droplets are confined to the periphery of the cytoplasm.

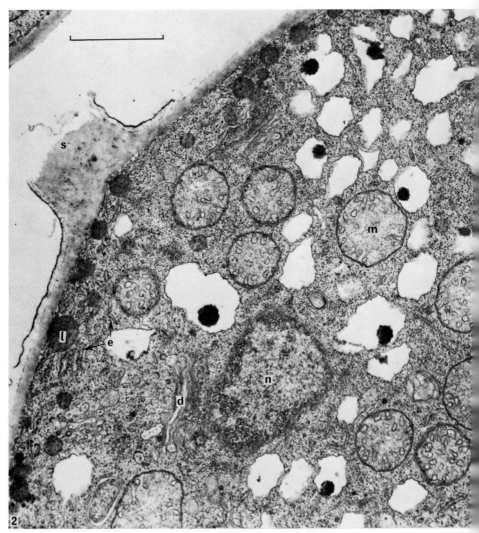

Fig. 2. A section through the proximal end of a fresh, quiescent conidium. Close to the sterigma, the wall carries a cuticle-like layer. The ground plasma is densely packed with ribosomes but endoplasmic reticulum profiles are short and sparse. The dictyosomes are relatively quiescent.

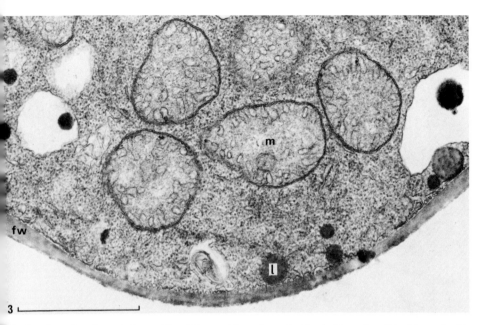

Fig. 3. A section through the papilla of the conidium shown in Fig. 1. The cell wall is thinner in this region but the cytoplasm resembles that elsewhere in the spore.

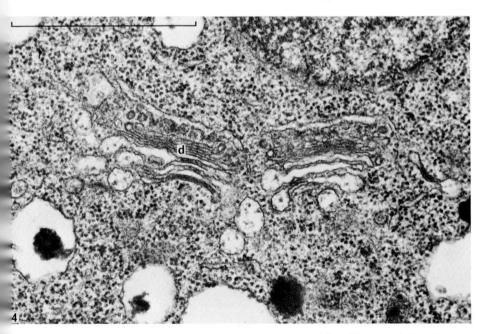

Fig. 4. A section through two dictyosomes within a fresh quiescent conidium. They are relatively inactive.

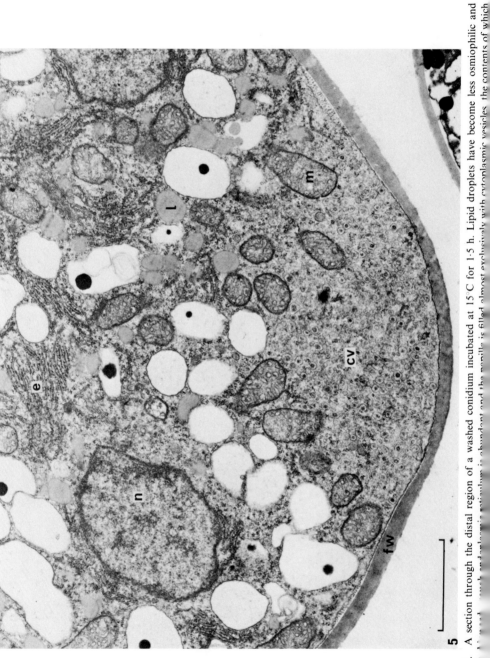

Fig. 5. A section through the distal region of a washed conidium incubated at 15 °C for 1·5 h. Lipid droplets have become less osmiophilic and ... and ... rough endoplasmic reticulum is abundant and the conidium is filled almost exclusively with cytoplasmic vesicles, the contents of which

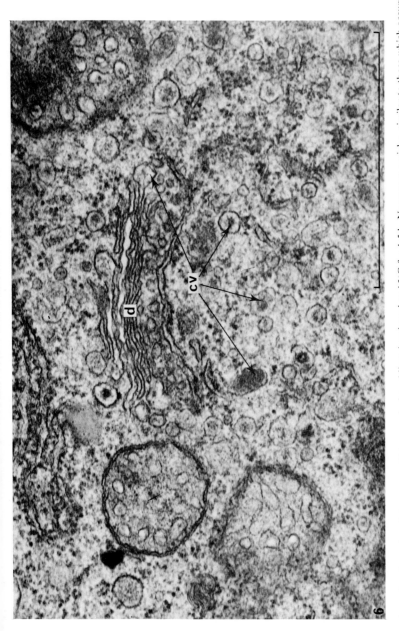

Fig. 6. A section through a dictyosome within a washed conidium incubated at 15 C for 1·5 h. Numerous vesicles, similar to those which accumulate within the papilla appear to arise from this structure and its associated endoplasmic reticulum.

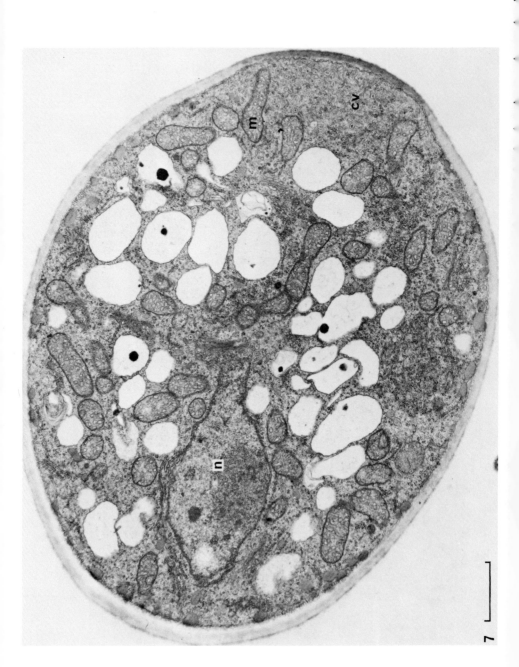

7

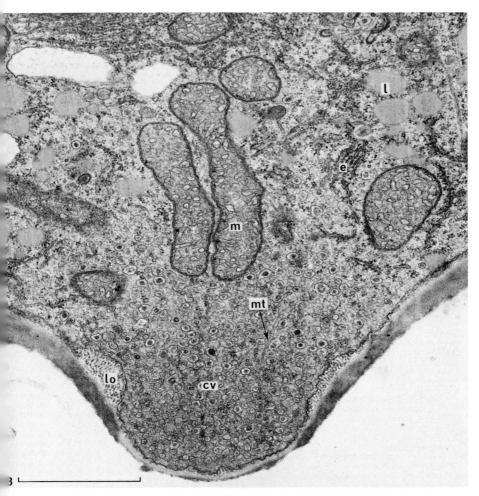

Fig. 8. A section through the distal end of a washed conidium incubated at 15 C for 1·5 h. The conidial wall has ruptured to permit the outgrowth of a germ tube. Lomasomes are adjacent to the ruptured wall and microtubules are orientated towards the germ pore.

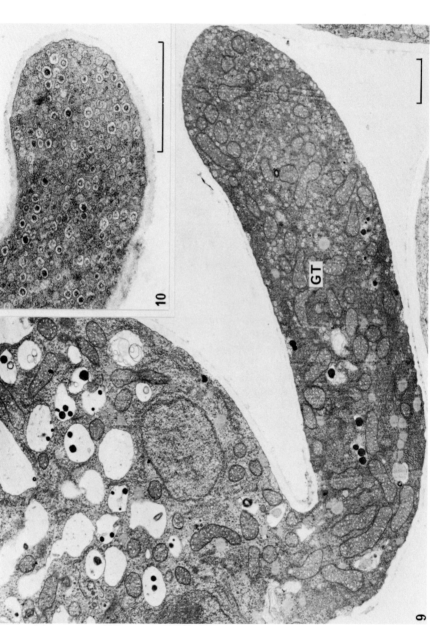

Fig. 9.   A longitudinal section through a young germ tube emerging from a conidium incubated at 15°C for 2h. Cytoplasm within the germ tube is dense and cytoplasmic vesicles are more abundant towards the tip of the germ tube which is out of the plane of section.

Fig. 10.   A longitudinal section through the tip of a germ-tube similar to that in Fig. 9. The tip is tightly packed with cytoplasmic vesicles similar

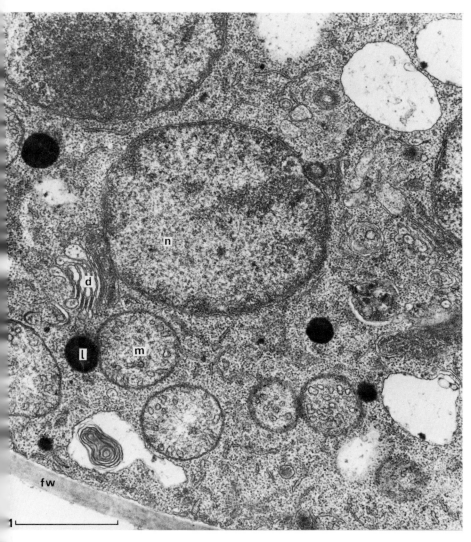

Fig. 11.  A section through part of a washed conidium incubated at 28°C for 3 h. The structure is similar
that of a freshly collected conidium except for the position of the osmiophilic lipid droplets which have
ved inwards.

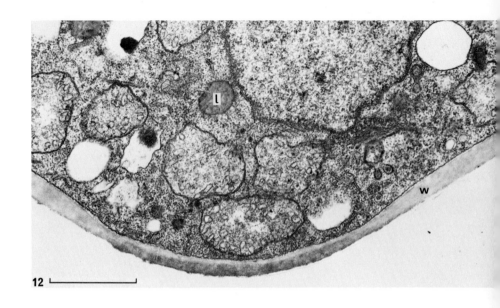

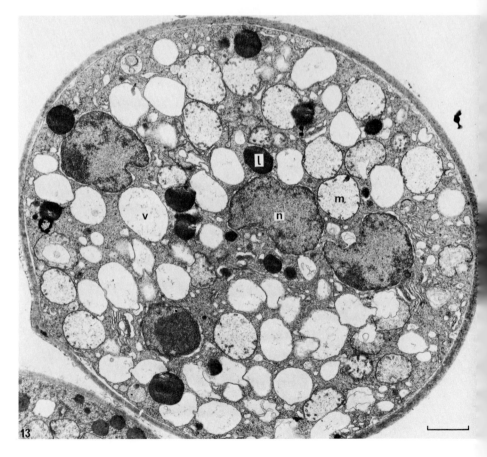

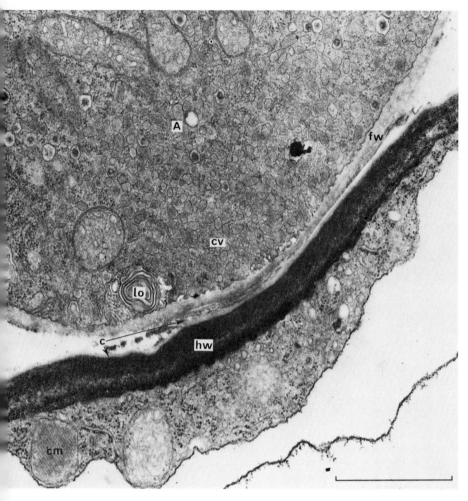

ჳ. 14.   A section through the zone of contact between an appressorium of *B. lactucae* and the epidermis
a compatible lettuce cotyledon at a stage before a penetration peg has formed. Fungal cytoplasmic
ᵢicles are discharging through the plasmalemma with the formation of lomasomes and the presence of
: fungus has caused the host cell's cuticle to swell, its wall to become highly osmiophilic, its cytoplasm
accumulate beneath the area of contact and become vesiculate, and the induction of a membrane bound
stalline microbody.

---

ჳ. 12.   A section through the papilla of a washed conidium incubated at 28°C for 24 h. No changes
structure normally associated with activation of the conidium have occurred except for the inward
vement of the lipid droplets.

ჳ. 13.   A section through a washed conidium incubated at 28°C for 24 h and showing signs of structural
ჳradation which occur with prolonged exposure to this temperature. Lipid reserves have apparently
ᵢlesced without loss of affinity for electron stains, mitochondria have become swollen and their cristae
ᵢorganized, nuclear envelopes and the endomembrane system have dilated and vacuolation has become
ᵢnsive.

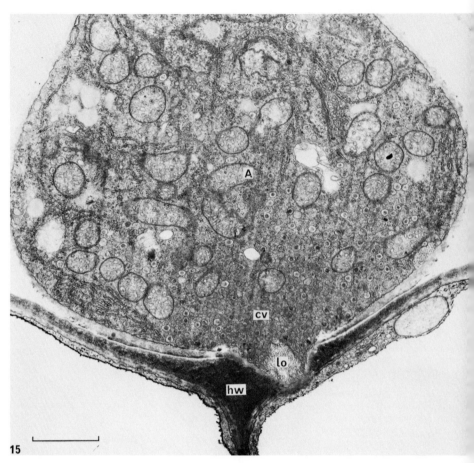

**15**

Fig. 15. A section through an appressorium in contact with the surface of a compatible lettuce cotyledon. Fungal cytoplasmic vesicles have accumulated close to the area of contact and a penetration peg is developing through the host cell wall. This is accompanied by the discharge of vesicles into the region at such a rate that their membranes form a large lomasome.

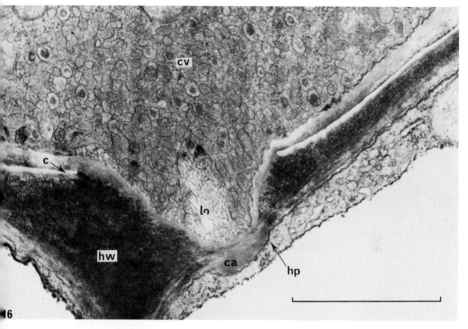

6

Fig. 16. A further enlarged serial section of the zone of contact shown in Fig. 15. The penetration peg which contains a large lomasome has almost completely penetrated the host cell wall and a callose pad has developed beneath it. The cuticle has been cleanly digested without distortion and the ends of the digested wall microfibrils remain in a position they would have occupied before invasion by the fungus.

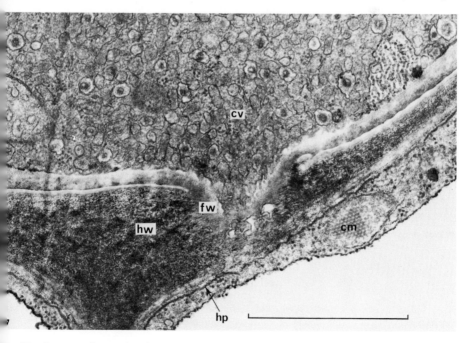

Fig. 17. A non-median section through a penetration peg showing the way in which the newly formed fungal wall surrounding the penetration peg appears to flow between the remaining ends of the digested wall microfibrils. A fungal-induced microbody is present in the host cytoplasm.

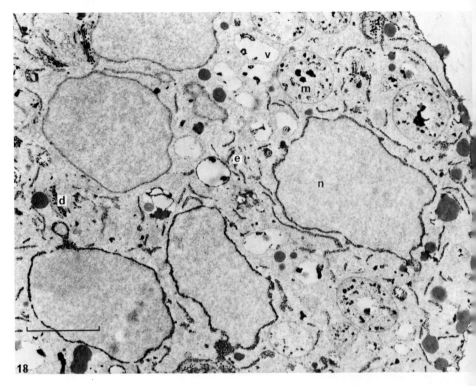

Fig. 18. A section through a quiescent conidium stained to reveal sites of lipase activity. Labelled areas include nuclear envelopes, endoplasmic reticulum, the forming face of the dictyosomes and membrane inclusions within the vacuoles. Labelling within the mitochondria is thought to be an artifact.

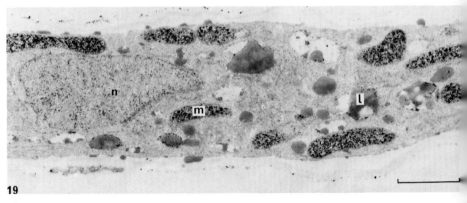

Fig. 19. A section through the distal region of a germ tube stained to reveal sites of lipase activity. Apar from heavy labelling in the mitochondria, which is thought to be artifactual, and a slight deposit over th nuclei, only the lipid droplets show signs of significant activity.

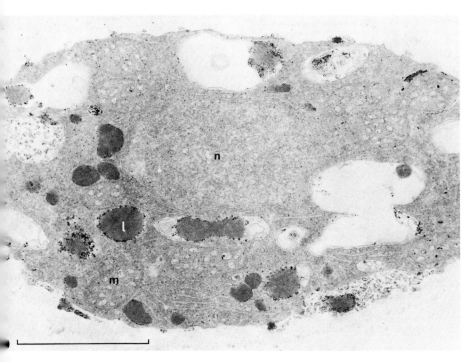

g. 20. A section through a germ tube stained to reveal sites of esterase activity. Only the lipid is ;nificantly labelled.

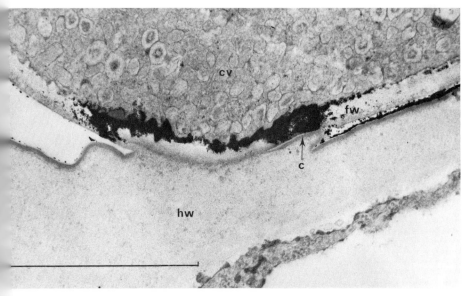

. 21. A section through the zone of contact between appressorium and host stained to reveal sites ipase activity. No fungal vesicles are labelled but activity is present within the fungal wall in contact h the host, particularly adjacent to regions of cuticle which are undergoing digestion.

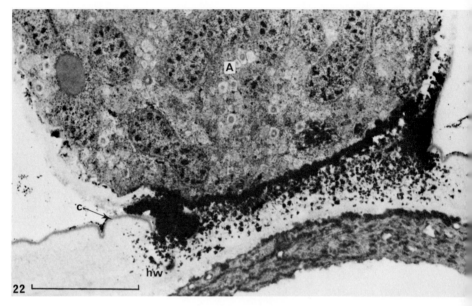

Fig. 22. A section through a zone of contact between appressorium and host at a stage slightly la
than that shown in Fig. 21. The annulus of high lipase activity has enlarged as cuticle digestion f
proceeded. Label within the fungal mitochondria is thought to be artifactual.

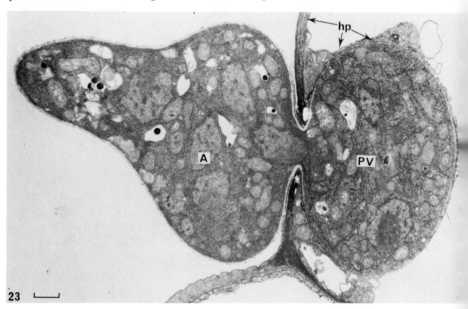

Fig. 23. A section through an appressorium and primary vesicle soon after penetration through the f
epidermal wall has been accomplished. Cytoplasm is flowing into the expanding primary vesicle. The f
cell plasmalemma is invaginated by the primary vesicle.

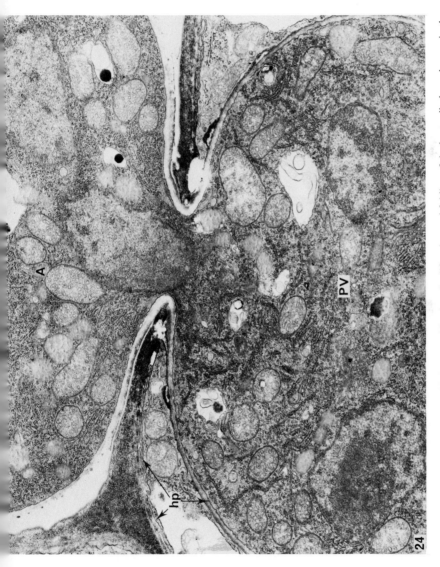

Fig. 24. An area of the section shown in Fig. 23 at higher magnification. Proliferation and folding of the host cell plasmalemma in the area of fungal penetration is evident by the multiple unit-membrane profiles around the penetration pore. Unfolding of this membrane permits its invagination without rupture by the enlarging primary vesicle.

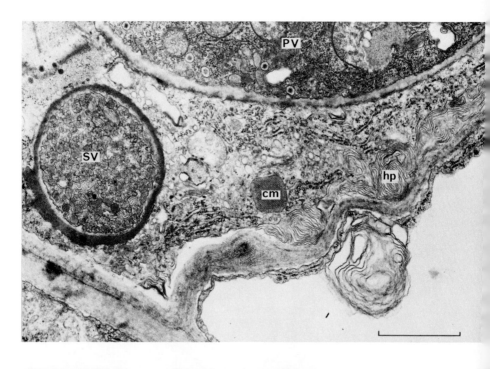

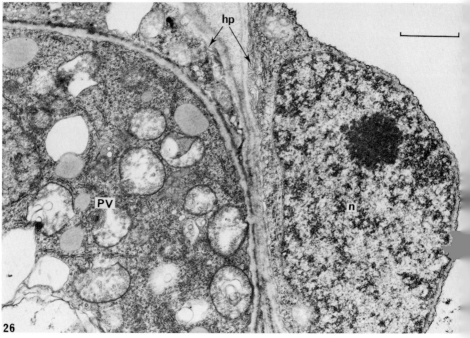

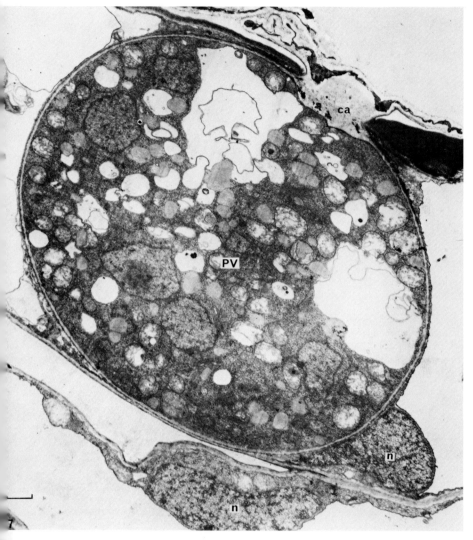

ig. 27. A section through two lettuce cells, within one of which is a mature primary vesicle. The uclei of both cells are located in positions close to the fungus. A callose plug has sealed the primary esicle from the collapsed appressorium.

ig. 25. A section through a lettuce epidermal cell containing an enlarged *Bremia* primary vesicle and developing secondary vesicle. The plasmalemma of the host cell has been induced to proliferate and a ystal-containing microbody is present. A dark osmiophilic layer surrounds the wall of the secondary sicle.

g. 26. A section through two lettuce cells, one of which contains a primary vesicle of *Bremia*. Proliferation of host plasmalemma has occurred in both infected and pathogen-free cells and the nucleus of the tter occupies a characteristic position in close proximity to the fungal structure in the adjacent cell.

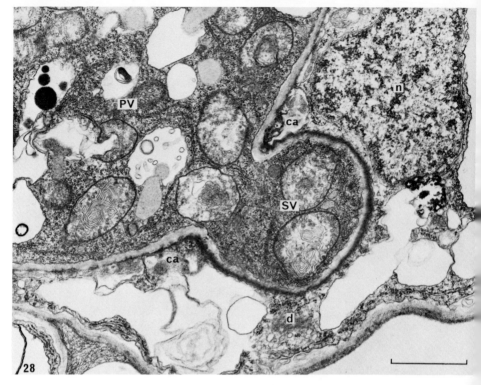

Fig. 28.   A section through a primary vesicle at the point from which a young secondary vesicle is arising. The wall of the secondary vesicle is distinguishable from that of the primary by its darkly-staining outer layer. A callose collar is already developing around the neck of the secondary vesicle, the material for which probably originates in the nearby highly active host dictyosomes. The host cell nucleus is positioned close to the fungal structures.

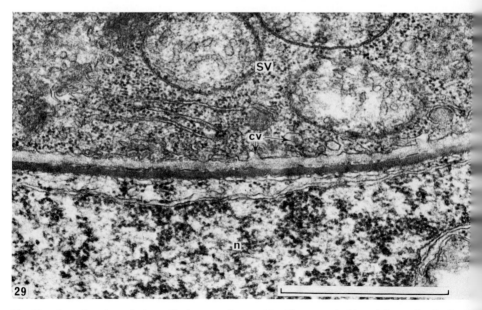

Fig. 29.   A section through the wall of a secondary vesicle. The outer darkly staining region is clearly visible and cytoplasmic vesicles are discharging into the inner layer. The host nucleus lies close to the fungus

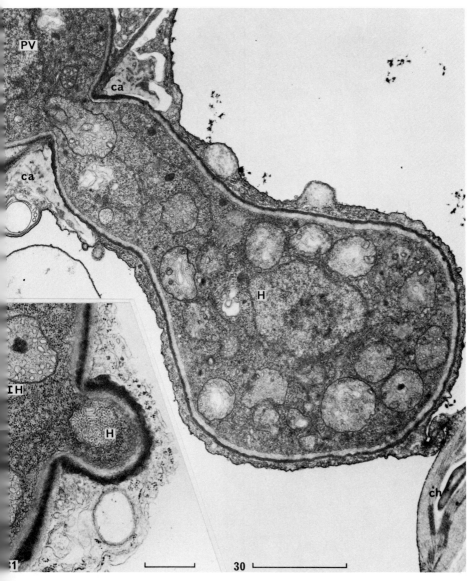

Fig. 30. A section through a young haustorium which has arisen directly from a primary vesicle in an adjacent host cell. The outer layer of its wall stains darkly and is fairly even in thickness. Apart from the callose collar secreted around the neck of the haustorium the host cytoplasm in this compatible host appears little affected by the presence of the fungus.

Fig. 31. A section through a very young haustorium developing from an intercellular hypha. Even at this stage of development the outer layer of the haustorial wall is darkly staining as is the "cement" between the hypha and the host cell wall. No clear callose collar has as yet developed at the base of the haustorium.

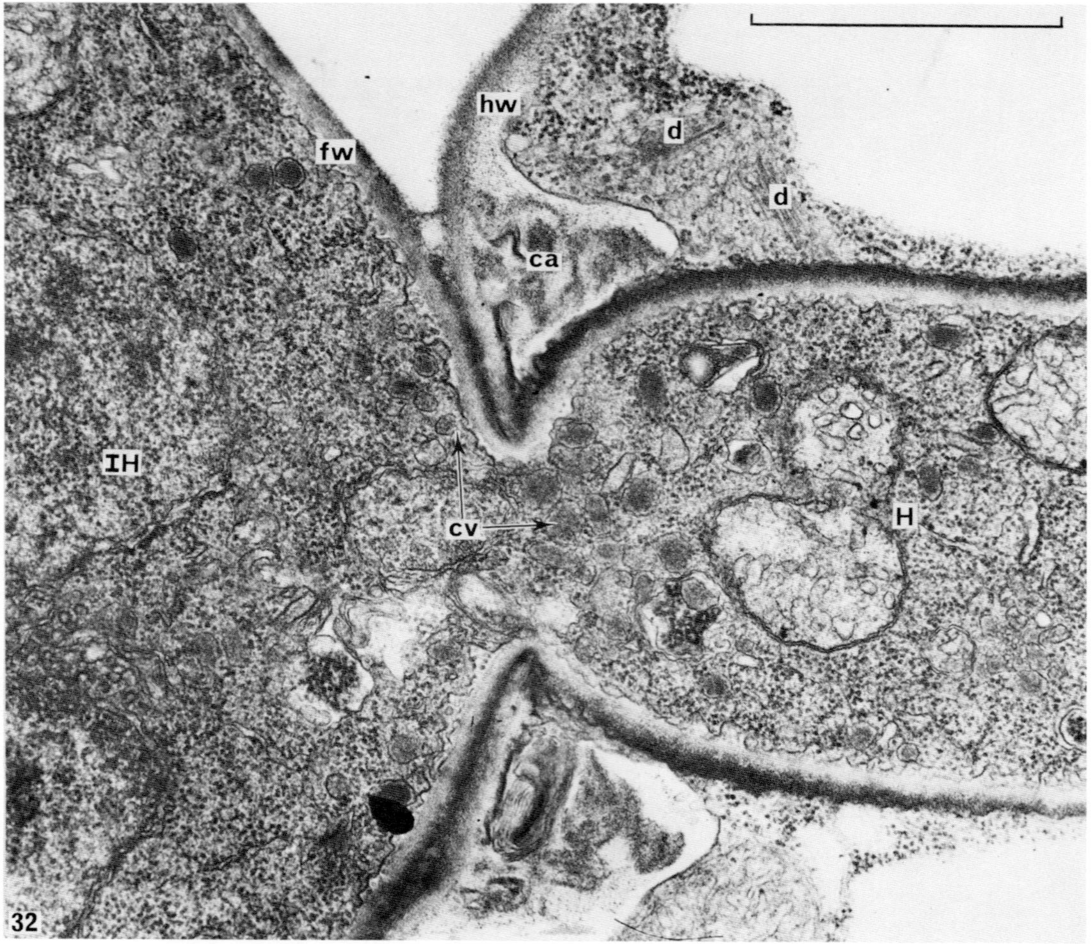

Fig. 32. A section through the base of an haustorium intermediate in development between those shown in Figs 30 and 31. Within the cytoplasm

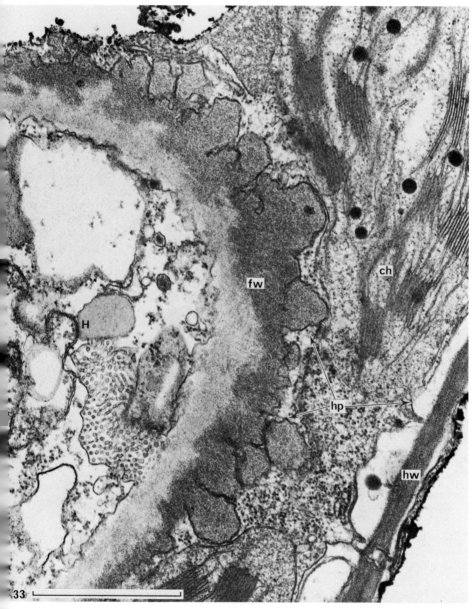

Fig. 33.  A section through the wall of a mature haustorium. Few fungal organelles remain intact but the outer layer of the wall has developed protuberances each surrounded by invaginated host plasmalemma. At this stage of development host organelle membranes lose integrity and starch is no longer present in the chloroplasts.

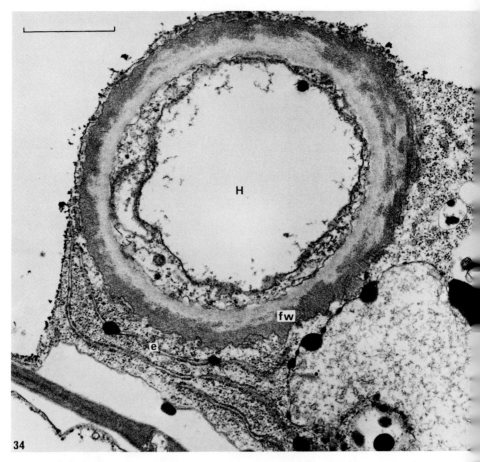

34

Fig. 34. A section through an haustorium at a stage similar to that shown in Fig. 33. External wall protuberances extend into the degenerating host plasmalemma. Long profiles of host endoplasmic reticulum are closely associated with the haustorium.

Fig. 35.　A section through an old haustorium which has become completely encased in callose.

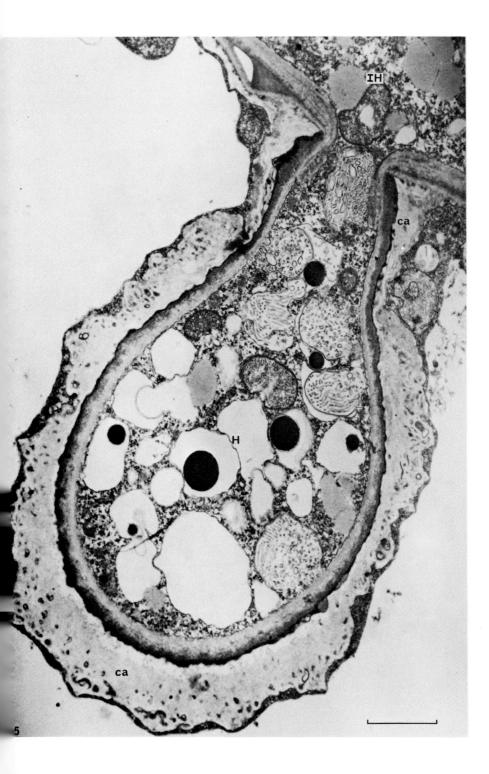

5

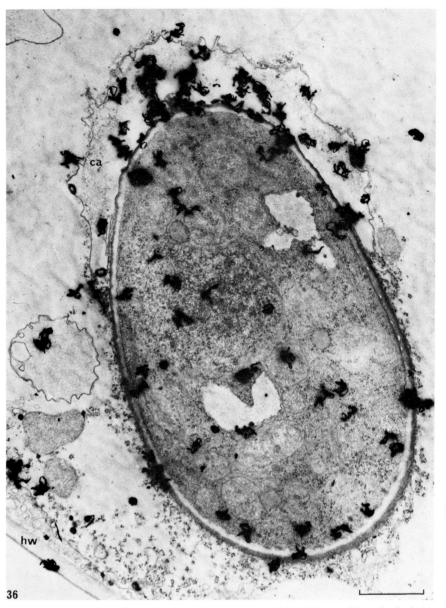

Fig. 36. An autoradiograph of a section through a secondary vesicle within an epidermal cell of a lettuce cotyledon supplied with $^3$H-glucose. Radioactivity, indicated by the silver grains, has been concentrated within the fungus and the callose collar.

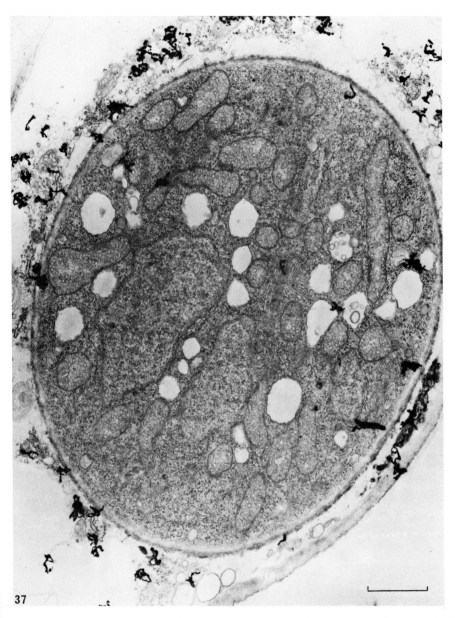

37

Fig. 37.  An autoradiograph of a section through a primary vesicle within a lettuce cotyledon supplied with ³H-leucine. Radioactivity has accumulated in the host cytoplasm surrounding the primary vesicle but little has entered the fungus.

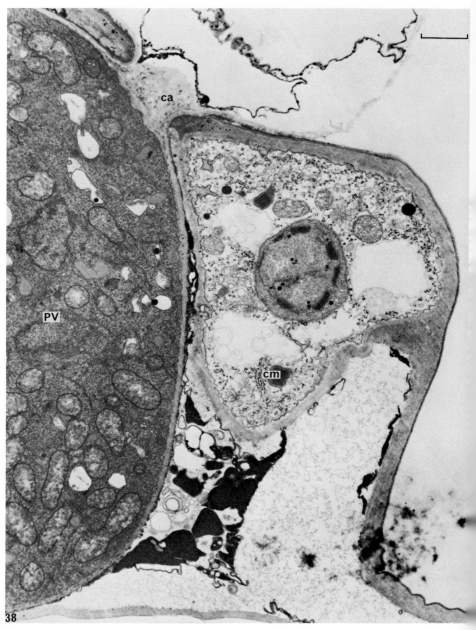

Fig. 38.   A section through a mature primary vesicle within an epidermal cell of a resistant lettuce cultivar. The fungal cytoplasm appears normal and healthy and a callose plug is present in the neck of the vesicle. However the cytoplasm within the penetrated host cell is completely disrupted. A number of crystal containing microbodies are present in the adjacent non-penetrated cell.

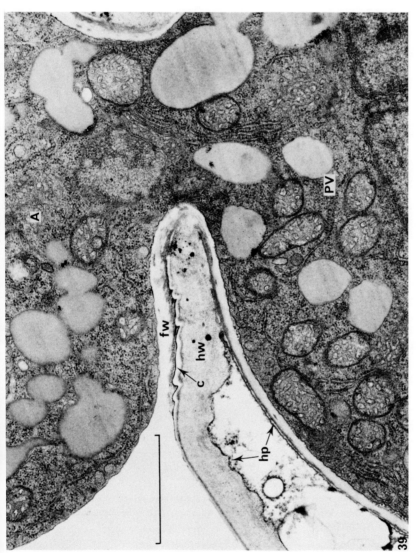

Fig. 39. A section through the penetration pore in the epidermal wall of a resistant lettuce cultivar fixed at a stage when cytoplasm is still passing into the primary vesicle from the appressorium. Severe disruption of host cell organelles has occurred but the invaginated host plasmalemma is still intact.

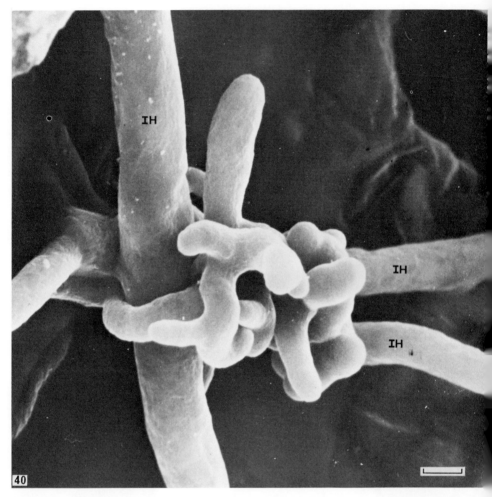

Fig. 40.   A scanning electron micrograph of an interaction between vegetative hyphae of dissimilar strains of *B. lactucae* in a lettuce cotyledon. Thin sexual hyphae have arisen from the vegetative hyphae and are intertwining.

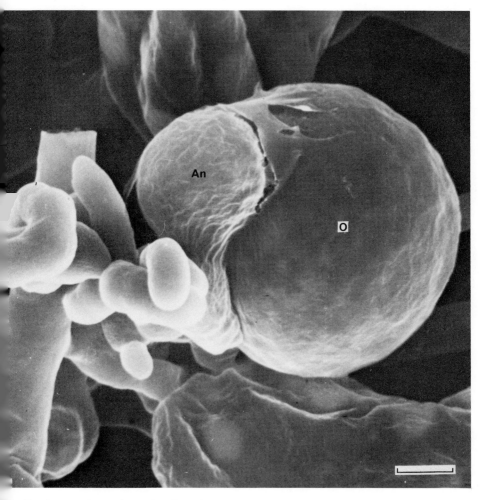

Fig. 41.    An oogonium and an antheridium which have arisen on sexual hyphae of differing strains formed in the manner shown in Fig. 40. An extracellular secretion appears to cement the gametangia together (see Figs 45 and 50).

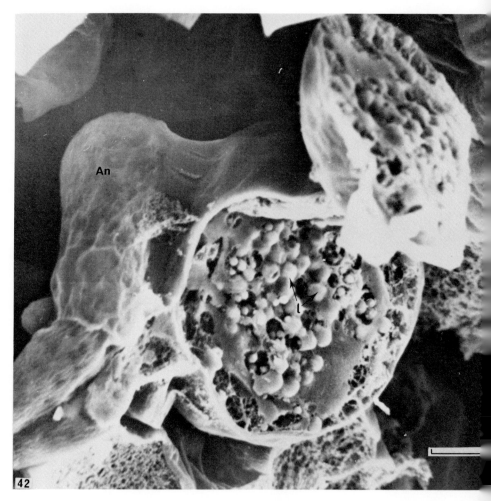

Fig. 42. A scanning electron micrograph of an oogonium fractured at a stage during condensation of the oosphere. Spherical lipid droplets are concentrated in the centre and surrounded by a sheet of membrane

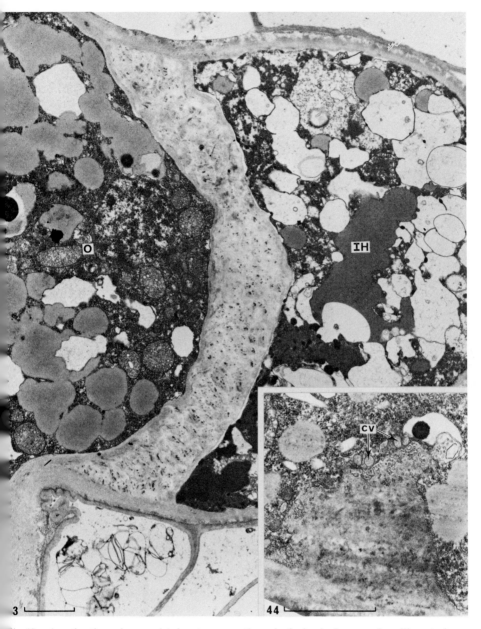

Fig. 43.   A section through a completed septum across the neck of a developing oogonium. The outer layers
of the septum are fibrilar and continuous with the walls of the hypha and oogonium. The inner core
resembles callose in structure and contains fragments of membrane.

Fig. 44.   A section through a developing septum within the neck of an oogonium. Cytoplasmic vesicles
whose contents resemble the material of the septum appear to be discharging through the crenellated
plasmalemma surrounding the septum.

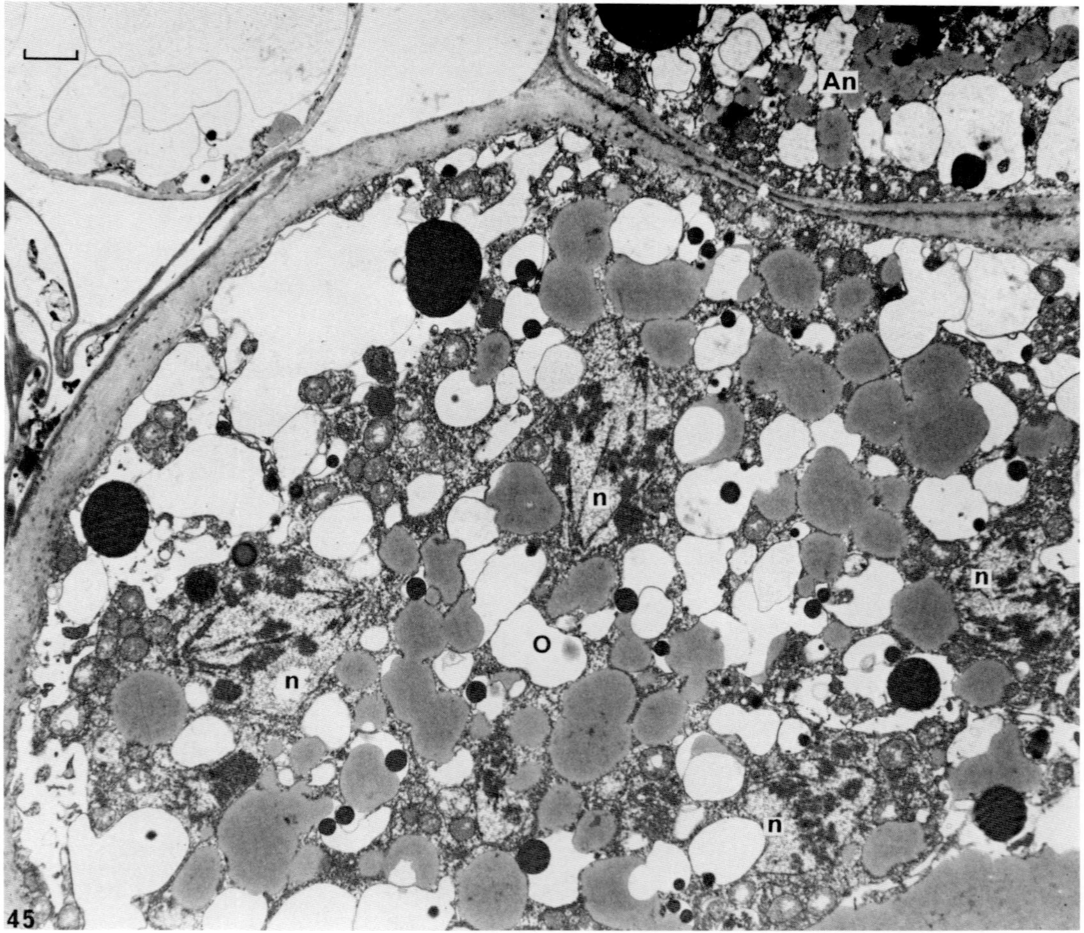

Fig. 45. A section through a developing oogonium. Four nuclei in synchronous division are clearly visible. Lipid droplets are abundant and osmiophilic

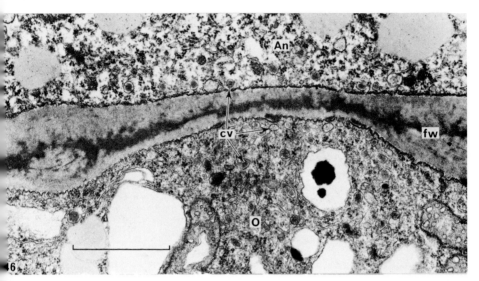

Fig. 46. A section through the region of gametangial contact just before the formation of a fertilization pore. Thinning of the cell walls in this area is accompanied by the accumulation of cytoplasmic vesicles, many of which appear to be discharging through the plasmalemma.

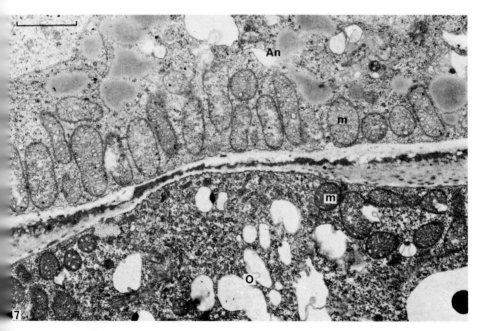

Fig. 47. A section through a region of antheridium–oogonium contact at a stage similar to that shown in Fig. 46. Mitochondria commonly accumulate in this region and are frequently in an orderly fashion, particularly in the antheridium.

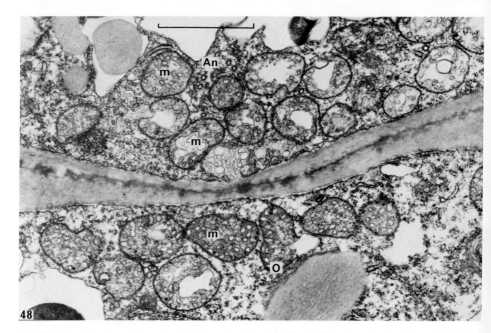

Fig. 48. A section through a region similar to that shown in Fig. 47. Here the mitochondria are becoming disorganized. The disrupted contents of one of these appears to have been discharged through the plasmalemma.

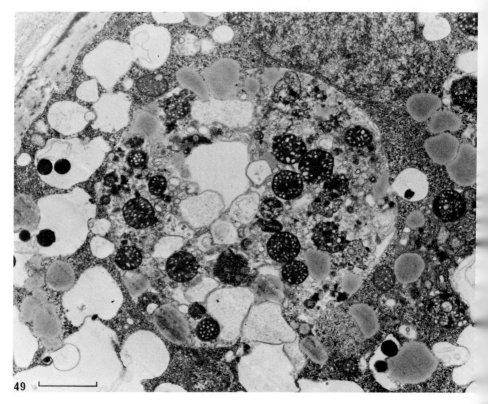

Fig. 49. A section through an area of lysis within a mature oogonium. Within this area is much disorganized membrane and the mitochondria stain darkly.

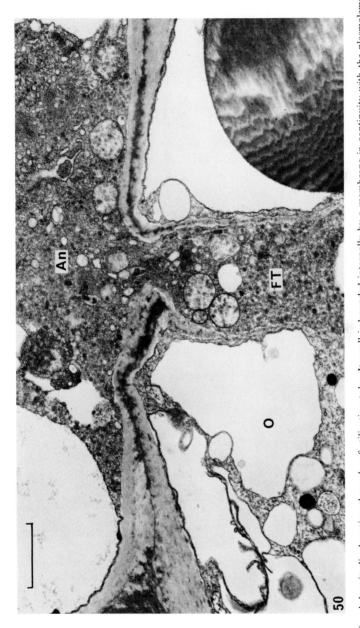

Fig. 50. A longitudinal section through a fertilization tube. Its wall is bounded internally by a membrane in continuity with the plasmalemma of the antheridium and externally by one continuous with the plasmalemma of the oogonium.

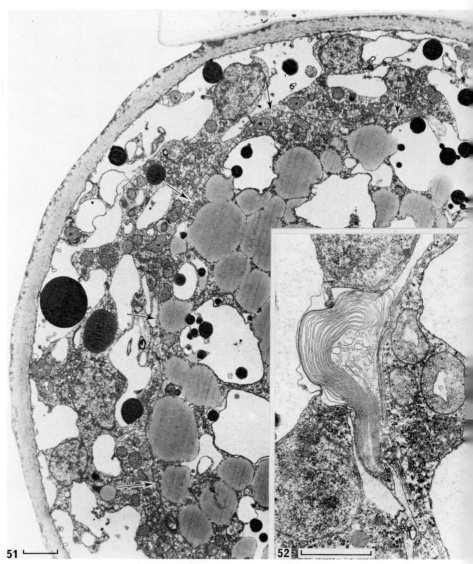

Fig. 51.   A section through an oogonium at about the time of fertilization. Most of the lipid has moved t the central region, which is delimited by a sphere of double unit membrane (arrowed).

Fig. 52.   A section through an oogonium at the stage shown in Fig. 51. In this region the newly forme sphere of double unit membrane has proliferated to form a continuous membrane whorl.

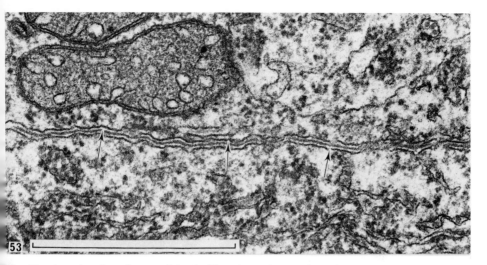

Fig. 53. A section through part of the double unit membrane sphere shown in Fig. 51. Amorphous material (arrowed) has begun to accumulate between the membranes.

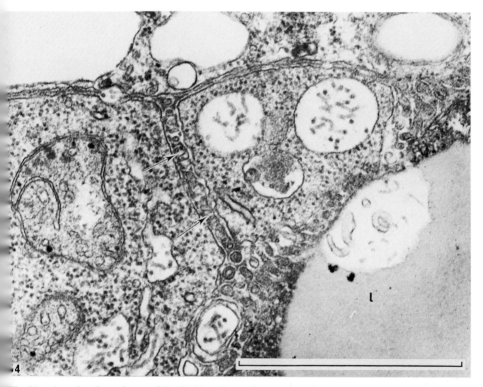

Fig. 54. A section through part of the double unit membrane sphere from which a membranous extension (arrowed) connects the space between the membranes with a region close to an eroding lipid droplet. From this region membrane-bound vesicles appear to be transported through the extension to the space between the membranes where the oospore wall will form.

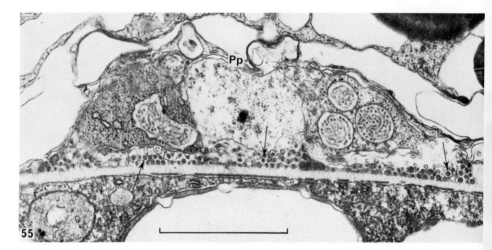

Fig. 55. A section through the developing oospore wall. Tubules (arrowed) composed of unit membrane in continuity with the outer oosphere-delimiting membrane contain darkly-staining contents.

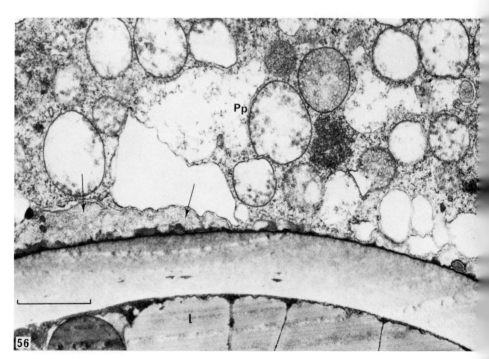

Fig. 56. A section through the maturing oospore wall and the periplasm. Apart from the smooth membrane (arrowed) close to the wall most of the organelles in the periplasm are in an advanced stage of disintegration.

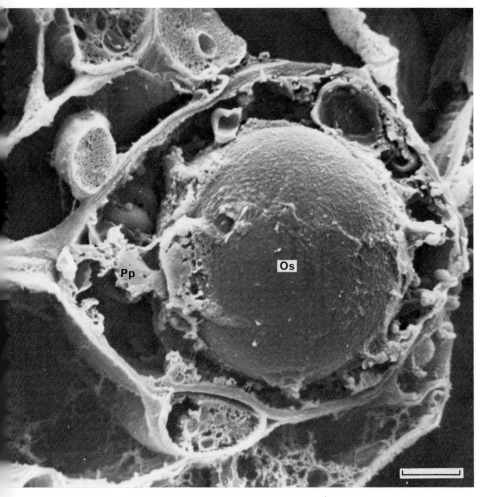

57. A scanning electron micrograph of a fractured oogonium containing a mature oospore.

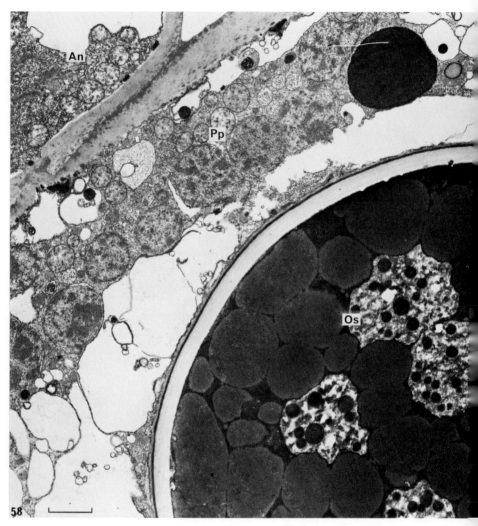

Fig. 58.  A section through a maturing oospore and its surrounding periplasm. The latter is degenerating while the contents of the spore become more condensed and highly osmiophilic.

## F. The secondary vesicle

Establishment of the pathogen in a compatible host begins with the production of a secondary vesicle. This structure arises as a projection from the primary vesicle (Fig. 28) and develops within the same host cell frequently enlarging to an extent that the cell is almost completely filled. Virtually all the cytoplasm in the primary vesicle migrates into it but no callose plug forms to separate the two structures. A major structural development associated with the formation of the secondary vesicle is the presence of a dark osmiophilic layer at the host-pathogen interface, immediately beneath the invaginated host plasmalemma and continuous with the fungal cell wall (Fig. 29). A similar layer, or encapsulation, surrounds haustoria of the downy mildews and other groups of fungi (Bracker and Littlefield, 1973; Littlefield and Bracker, 1972; Rijo and Sargent, 1974). Whether it originates from the host or the pathogen is not known but its presence around the haustoria which subsequently develop suggests an involvement in nutrient uptake by the pathogen.

Additional events within the secondary vesicle herald a new phase in the relationship between host and pathogen. The nuclei which were transported from the spore begin to divide and the volume of cytoplasm increases. Formation of the secondary vesicle marks, it seems, the first phase of establishment of the fungus within a compatible host.

## G. The haustorium

Thereafter establishment is consolidated by the development of intercellular hyphae which grow from the secondary vesicle, branch and ramify throughout the leaf, and haustoria which arise commonly from a hypha or occasionally directly from the secondary vesicle or even the primary vesicle (Figs 30 and 31). From whichever part of the fungal body they develop, cell wall penetration into an adjacent lettuce cell occurs in a similar way to that by which the germling gains entry to the epidermis. Solubilization of the host cell wall is preceded by proliferation of the host plasmalemma which prepares it for subsequent invagination as the club-shaped haustorium enlarges. A dark layer, similar to that around the secondary vesicle occurs between the fungal wall and the host plasmalemma. This layer fully covers the haustorium and extends to the edge of the penetration pore (Fig. 32). No neck-band of the type present on the haustoria of rusts is present. The young haustorium is densely packed with cytoplasm and contains up to three nuclei. In this form it has traditionally been considered to be the principal part of the fungal body involved in nutrient absorption. Within a compatible host, however, the host cytoplasm shows, at this stage, few signs of disturbance apart from the slow deposition of callose around the neck of the haustorium.

Over a period of time changes occur in the structure of both the haustorium and the host cytoplasm with which it is in contact. In common with all the older parts of the fungal body the haustorium becomes vacuolate with the cytoplasmic organelles dispersing or degenerating (Fig. 33). At the same time outward projections arise on the wall (Ingram, et al., 1976). Each of these is surrounded by the host plasmalemma whose area, and consequently the interface between the organisms, is thereby increased dramatically. The haustorium wall is consequently transformed into a structure resembling the wall of a transfer cell in higher plants and it is tempting to speculate on the enhanced opportunity for absorption which this change might afford the haustorium. Similar projections from the haustorial matrix of *Peronospora pisi* have been reported by Hickey and Coffee (1977). At the same time as the haustorium wall is transformed alterations, mostly of a degradative nature, occur in the host cytoplasm: starch is lost from the chloroplast, additional vacuoles develop and osmiophilic droplets form particularly on the tonoplast. Progressive degradation of the cytoplasm might well increase the pool of nutrients available for absorption by the haustorium. Commonly profiles of host endoplasmic reticulum occupy positions close to the mature haustorium (Fig. 34). In doing so they may possibly enhance the transport of nutrients to the fungus. However in the absence of knowledge about the absorptive capacity of the haustorium throughout its life, the significance of structural changes in both it and the host cytoplasm which surrounds it must remain speculative. Throughout the development of the haustorium, callus is deposited around it from vesicles derived from host dictyosomes. This deposition begins as a collar around the neck of the haustorium and spreads distally. Ultimately the haustorium becomes completely encased in callus (Fig. 35).

### H. Nutrient absorption from the host

In any consideration of the supply of nutrients by one organism to another a distinction must be drawn between the dependence of a biotrophic parasite upon its host for materials necessary for completion of its life cycle and the opportunity it might exploit when readily absorbable materials are available. It has been demonstrated that the *Bremia* sporeling will absorb both glucose and leucine when these compounds are available in the germination medium (Andrews, 1975). In addition tritium enters the fungus when sporelings are in contact with lettuce cotyledons supplied with $^3$H-glucose. Presumably the glucose or a metabolite is secreted to the surface of the cotyledon in a form readily available for uptake by the sporeling. In addition radioactivity from $^{14}$C-glucose supplied to lettuce cotyledons accumulates in the fungus soon after penetration (Fig. 36) but the parasite does not become labelled with tritium from $^3$H-leucine supplied to the cotyledon before it has reached the stage in development when haustoria have formed (Fig. 37). Thus it may well be that an adequate, balanced supply of nutrients can be absorbed from a compatible host only through mature haustoria. A very rapid transport of carbon to the hyphae and haustoria of *Pseudoperonospora cubensis* soon after it was supplied as $CO_2$ to the host plant, cucumber, was reported by Takahashi *et al.* (1977).


## J. Incompatibility

Not all strains of *B. lactucae* are compatible with the many varieties of lettuce. Some strains fail to become established within the host in which they usually elicit a hypersensitive response. With these, penetration of an epidermal cell is achieved and a primary vesicle is produced normally. However, the development of a secondary vesicle is much retarded and although an inter-cellular hypha usually forms, seldom do more than 2 or 3 haustoria develop before growth of the fungus stops. No nuclear divisions occur. Host hyper-sensitivity becomes manifest by melanization of the penetrated epidermal cell, and of each cell in which haustoria subsequently develop. Electron microscopy has revealed that during penetration of the epidermal cell invagination of the host plasmalemma occurs in the same way as it does during a compatible reaction (Maclean *et al.*, 1974). However all the remaining membrane systems within the cell are very rapidly degraded to the extent that no host organelles remain recognizable by the time penetration is complete (Fig. 38). Only the invaginated portion of the host plasmalemma retains a degree of integrity (Fig. 39). What protects the membrane from an agent which passes through it and yet is so destructive of other host cell membranes is not known. The presence of the fungus does not affect adjacent host cells apart from inducing the development of numerous unit membrane-bound crystals resembling those induced by compatible strains of *Bremia* in epidermal cells prior to entry. The fungal cytoplasm within the primary vesicle appears healthy and normal, confirming that the events which give rise to this hypersensitive reaction conform with the traditional view of the mechanism which attributes the failure of pathogen establishment to the death of penetrated host cells. Resistance to infection by downy mildews does not always result from such a dramatic effect on host tissues. *Peronospora farinosa* and *P. tabacina* respectively cause cell death in resistant varieties of spinach and tobacco but not all invaded cells die (Kröber and Petzgold, 1972; Kröber *et al.*, 1979). Moreover, *Peronospora manshurica* can invade resistant varieties of soybean without host cell death but growth of the fungus is very slow compared with that in susceptible varieties (Riggle and Dunleavy, 1974).


## K. Sexual reproduction

Although conidia of *Bremia* can be stored in a viable condition at sub-zero temperatures it is improbable that the organism overwinters in this form. The oospore is the likely propagule by which infection is transmitted through the soil to a developing crop in the following season. Recent work has demonstrated that *B. lactucae* is heterothallic (Michelmore and Ingram, in press). In an examination of approximately 40 isolated strains only 2 compatible types were identified. Scanning electron microscopy has shown that interactions between heterothallic isolates occur in areas where vegetative hyphae of opposite compatible types approach one another (Michelmore, 1979). Characteristically thin sexual hyphae arise from vegetative hyphae in these areas and intertwine (Fig. 40). Subsequently oogonia and antheridia are borne on the tips of opposite types of sexual hyphae (Fig. 41). Cytoplasm fills the gametangia as they swell and the oogonium accumulates a large quantity of lipid (Fig. 42). Each becomes partitioned from the rest of the fungal body by a septum composed largely of callus deposited from cyto-plasmic vesicles (Figs 43 and 44) (Sargent *et al.*, 1977). Each antheridium approaches an oogonium to which it becomes attached perigynously with the aid of an extracellular secretion. Maturation of the gametangia involves the synchronous reduction-division of the many nuclei contained within them (Fig. 45). Dissolution of the cell walls to permit the formation of a fertilization pore between an oogonium and antheridium is accompanied by the discharge of cytoplasmic vesicles through the plasmalemma of both gametangia at that point (Fig. 46). Presumably this is the means by which wall hydrolysing enzymes are deposited in this area. At the same time mitochondria accumulate along the interfacial walls particularly in the antheridium (Fig. 47). They may be drawn passively into this position by a transcellular transport system responsible for directing the movement of cytoplasmic vesicles but the discharge of contents of degraded mitochondria which occurs through the plasmalemma at the fertilization pore area may play an additional role in wall dissolution (Fig. 48). At this stage of development certain areas of lysis have been recognized in the oogonium (Fig. 49). Possibly they result from a premature initiation of the mechanism responsible for the extensive degrada-tion of cytoplasm around the oosphere which subsequently develops. The movement of antheridial cytoplasm towards the centre of the oogonium following fertilization-pore formation is directed by a fertilization tube which originates as a double membrane. Each membrane is continuous with the plasmalemma of either oogonium or antheridium (Fig. 50). Concomitant with the formation of these membranes and following the migration of the bulk of oogonial lipid to the centre of the cell a sphere of double membrane is laid down within the oogonium delimiting the future oosphere (Fig. 51). It is continuous with the membranes of the fertilization tube, and the presence of occasional whorls of membrane along its profile indicates a lack of precise control in the programming of its production (Fig. 52). Soon after the fertilization-tube membranes are formed they become separated from each other as wall material, continuous with that of the gametangia walls, is deposited between them.

Following zygote formation the fertilization tube is closed by the completion of the sphere of membrane and there follows a prolonged phase of oosphere wall development between the membranes of the sphere (Fig. 53). Initially the lipid within the sphere appears to contribute largely to wall synthesis (Fig. 54). Membrane-lined channels containing small vesicles connect outer lipid droplets with the developing wall and the droplets bear eroded areas associated with membranous figures containing coiled structures. Further wall development appears to be linked with degradation of the periplasm surrounding the oosphere. While organelles within this region of the cytoplasm progressively disintegrate, a network of tubules develops in continuity with the unit membrane surrounding the oosphere wall. The contents of these tubules are darkly staining and it seems probable that they are involved in the transport of wall precursors synthesized from the products of periplasm autolysis (Fig. 55). The existence of such a secretion mechanism is supported by the presence of smooth endoplasmic reticulum close to the wall even at a late stage of periplasm degradation (Fig. 56). Thickening of the oosphere wall is accompanied by condensation of the cytoplasm within the oosphere and changes within the lipid droplets which render them more osmiophilic (Figs 57 and 58). The zygote nucleus occupies a position roughly in the centre of the oosphere. Thus the oosphere undergoes a structural transformation to become an oospore capable of resisting the stresses likely to be experienced within the surface layers of soil and in decaying litter throughout the winter months. As yet little is known about the physiological or biochemical control of oospore dormancy and likewise, the structural changes which occur during germination have yet to be investigated.

This account has deliberately dealt almost exclusively with the subcellular organization of only a single downy mildew pathogen, *Bremia lactucae*. Many questions concerning the changes in function which accompany alterations in its fine structure remain to be answered and progress in this direction must await the systematic application of cytochemical techniques. Nevertheless, the experimental approach described in this chapter has already yielded much useful information on both the development of this particular biotroph and its relationship with its host. As a model it has much to commend it in the investigation of parasitic relationships.

## Acknowledgements

I am indebted to Dr J. H. Andrews for providing the micrographs reproduced in Figs 36 and 37 and to Dr R. W. Michelmore for those which appear as Figs 40, 41, 42 and 57. Dr D. S. Ingram has been a close collaborator in the work described here on *Bremia lactucae* and I acknowledge with gratitude his encouragement in the preparation of this chapter.

## References

Aist, J. R. (1976). *In* "Encyclopaedia of Plant Physiology. Vol 4. Physiological Plant Pathology" (R. Heitefuss and P. H. Williams, Eds) 197–221. Springer Verlag, Berlin.
Andrews, J. H. (1975). *Can. J. Bot.* **53**, 1103–1115.
Bracker, C. E. (1967). *A. Rev. Phytopath.* **5**, 343–374.
Bracker, C. E. and Littlefield, L. J. (1973). *In* "Fungal Pathogenicity and The Plant's Response" (R. J. W. Byrde and C. J. Cutting, Eds) 159–313. Academic Press, London and New York.
Chou, C. K. (1970). *Ann. Bot.* **34**, 189–204.
Cohen, Y., Perl, M., Rotem, J., Exal, H. and Cohen, J. (1974). *Can. J. Bot.* **52**, 447–450.
Duddridge, J. A. and Sargent, J. A. (1978). *Physiol. Pl. Path.* **12**, 289–296.
Hickey, E. L. and Coffee, M. D. (1977). *Can. J. Bot.* **55**, 2845–2858.
Hickey, E. L. and Coffee, M. D. (1978). *Protoplasma* **97**, 201–220.
Ingram, D. S., Sargent, J. A. and Tommerup, I. C. (1976). *In* "Biochemical Aspects of Plant–Parasite Relationships" (J. Friend and D. R. Threlfall, Eds) 43–78. Academic Press, London and New York.
Kajiwara, T. (1973). *Shokubutsu Byogai Kenkyu* **8**, 167–178.
Kröber, H., Özel, M. and Petzgold, H. (1979). *Phytopath. Z.* **94**, 16–44.
Kröber, H. and Petzgold, H. (1972). *Phytopath. Z.* **74**, 296–313.
Maclean, D. J., Sargent, J. A., Tommerup, E. C. and Ingram, D. S. (1974). *Nature* **249**, 186–187.
Mason, P. A. (1973). Ph.D. Thesis, University of Cambridge.
Michelmore, R. W. (1979). Ph.D. Thesis, University of Cambridge.
Pares, R. D. and Greenwood, A. D. (1979). *New Phytol.* **83**, 473–477.
Peyton, G. A. and Bowen, C. C. (1963). *Am. J. Bot.* **50**, 787–797.
Riggle, J. H. (1977). *Can. J. Bot.* **55**, 153–157.
Riggle, J. H. and Dunleavy, J. M. (1974). *Phytopathology* **64**, 522–526.
Rijo, L. and Sargent, J. A. (1974). *Can. J. Bot.* **52**, 1363–1367.
Royle, D. J. and Thomas, G. G. (1971). *Physiol. Pl. Path.* **1**, 345–349.
Royle, D. J. and Thomas, G. G. (1973). *Physiol. Pl. Path.* **3**, 405–417.
Sargent, J. A., Ingram, D. S. and Tommerup, I. C. (1977). *Proc. R. Soc. Lond. B.* **198**, 129–138.
Sargent, J. A. and Payne, H. L. (1974). *Trans. Br. Mycol. Soc.* **63**, 509–518.
Sargent, J. A., Tommerup, I. C. and Ingram, D. S. (1973). *Physiol. Pl. Path.* **3**, 231–239.
Takahashi, K., Inaba, T. and Katiwara, T. (1977). *Physiol. Pl. Path.* **11**, 255–259.
Wehtje, G., Littlefield, L. J. and Zimmer, D. E. (1979). *Can. J. Bot.* **57**, 315–323.
Wehtje, G. and Zimmer, D. E. (1978). *Phytopathology* **68**, 1568–1571.

Chapter 11

# The Host Specificity of Peronosporaceous Fungi and the Genetics of the Relationship between Host and Parasite

I. R. CRUTE

*National Vegetable Research Station, Wellesbourne, Warwick, UK*

# I. Introduction

Previous publications which have considered the host specificity of downy mildew fungi have done so to emphasize the relevance of the phenomenon to the taxonomy and nomenclature of the group (Yerkes and Shaw, 1959; Shaw, 1970; Waterhouse, 1973). In this treatment of the subject an attempt will be made to broaden the discussion of host specificity beyond taxonomic considerations to include the genetic and mechanistic basis for specificity and its evolutionary significance. No attempt will be made to consider tissue specificity. The discussions will deal primarily with the Peronosporaceae, a family considered to comprise the genera: *Basidiophora* Rose & Cornu;

*Bremia* Regal; *Bremiella* Wilson; *Peronosclerospora* (Ito) C. G. Shaw; *Peronospora* Cornu; *Plasmopara* Shroeter; *Pseudoperonospora* Rostowze; *Sclerophthora* (Sacc.) Thirum.; and *Sclerospora* (Schroet.) de Bary.

The host specificity and variation for pathogenicity in downy mildew fungi is considered at five levels although it is recognized that related phenomena are likely to be involved at all these levels:

1.  Peronosporaceous fungi are specific to living plant tissue.
2.  Different morphologically recognized genera of the Peronosporaceae exhibit a degree of specificity to certain host families.
3.  A single morphologically recognized pathogen species is usually specific to hosts belonging to a single host order, family, sub-family or tribe.
4.  Within a pathogen species, different pathotypes can frequently be identified which are specific to hosts of a particular genus or other species aggregate. Such pathotypes have sometimes been named as *formae speciales*.
5.  Specificity is further expressed in pathotypes of fungi which are pathogenic only on certain genotypes of the same or closely related host species. At this level, pathogen variation has frequently been described by categorizing the population into named physiological races.

## II. Specificity to Living Host Tissue

Under natural circumstances, downy mildew fungi obtain their nutrition by biotrophic processes; they grow and reproduce solely in living plant tissue. Dependence upon the living processes of the host is reflected in the fact that only two successful, but unrepeated attempts to grow these fungi axenically have been reported (Pioth, 1957; Guttenberg and Schmoller, 1958). The exception, however, is *Sclerophthora macrospora* (Sacc.) Thirum. Shaw & Naras., the least discriminating of downy mildew species in terms of host range, which is readily cultured on synthetic media (Tokura, 1975).

Several downy mildew fungi have been grown successfully in association with host tissue cultures and this subject has been considered fully in recent reviews by Ingram (1973, 1976) and Ingram and Tommerup (1973). On occasion, mycelia have been observed to spread from the host tissue on to the culture medium.

Generally, however, when the hyphal connections with the host have been severed, the isolated mycelium has not survived although Tiwara and Arya (1969) reported axenic growth of *Sclerospora graminicola* (Sacc.) Schroet., after mycelium from pearl millet (*Pennisetum typhoides* Stapf. and Hubb.) tissue cultures had grown on to the culture medium.

It can be concluded that saprotrophic nutrition is atypical of these fungi and unlikely to occur in nature. There are nevertheless indications that the chemicals utilized by downy mildew fungi in their growth processes when in association with their hosts are similar to those utilized by saprotrophic species (Sempio, 1942; Thornton and Cooke, 1974; Andrews, 1975). In addition, media in which extracts of host tissues are incorporated have proved unsuccessful as substrates for axenic growth (Murphy and McKay, 1926; McKay, 1939; Yarwood, 1943). Hence it is unlikely that the observed obligate biotrophy can be explained in simple nutritional terms. One possible explanation hinted at by Brian (1967) is that an intimate association with living host cells is required to "switch on" or derepress the appropriate fungal enzyme systems which enable it to utilize the substrates necessary for normal growth and development. Haustoria may function in maintaining this association and enable the continual exchange of information-carrying molecules.

It appears that the spores of peronosporaceous fungi (zoospores or conidia) have the capability to utilize stored or exogenously applied nutrients only to facilitate germination and host penetration. In terms of the above hypothesis, only that part of the fungal genome coding for enzymes necessary for these processes is "switched-on". Germination proceeds in at least two species (*Peronospora tabacina* Adam. and *Bremia lactucae* Regel.) only after the removal from the spore of a water-soluble autoinhibitor which allows protein synthesis to proceed (Shepherd, 1962; Shepherd and Mandryk, 1962, 1963; Holloman, 1969, 1970, 1973; Leppik et al., 1972; Mason, 1973). If spores of *Bremia lactucae* germinate *in vitro* or if host penetration does not occur, the total storage substance of the spore, probably lipid (Sargent and Payne, 1974; Duddridge and Sargent, 1978), is converted into the cell wall material of the germ tube until an almost empty shell remains (Crute and Dickinson, 1976). No nuclear division or increase in cytoplasmic volume occurs until the fungus becomes associated with a compatible host (Maclean et al., 1974; Maclean and Tommerup, 1979). Clearly a marked change in fungal metabolism must occur once this association has been formed (Ingram et al., 1976). However, Andrews (1975) has demonstrated that exogenous nutrients are utilized by germinating spores of *Bremia lactucae* which suggests that an influence on germ-tube extension is to be expected. However, evidence that nutrient-stimulated germ-tube extension occurs in downy mildew fungi is conflicting (McKay, 1939; Yarwood, 1943; Powlesland, 1963; Pegg and Mence, 1970; Mason, 1973).

Elongation of germ tubes *in vitro* under the influence of exogenous nutrients could be interpreted as saprotrophic development and more studies on early stages of development following spore germination are needed to reveal the processes which are most likely to determine the apparent absolute dependence of the parasites on their hosts. Indeed, the stimulus provided by a particular host to enable further development of peronosporaceous hyphae produced from germinating spores may be the key to host specificity at all levels. It could be that when this stimulus is absent or disturbed, the characteristic and possibly non-specific, non-host and resistant-host responses ensue. These are discussed in Section IX.

## III. Specificity of Fungal Genera to Higher Host Taxa

Of the 287 or so families of flowering plants recognized in standard taxonomic treatments, only 15% (i.e. less than 45 families) contain genera which are known hosts for peronosporaceous fungi. Both monocotyledonous and dicotyledonous families are represented, and the families attacked range from those considered to be among the most primitive (e.g. Ranunculaceae) to those thought to be the most advanced (e.g. Gramineae and Compositae). In many instances, a single host family is only host to a single peronosporaceous genus. Alternatively, a single fungus genus may occur only on hosts belonging to one family.

There are, however, exceptions to these trends although incorrect identification of the fungus genus may account for some records. When more than one fungus genus does parasitize a host family, it is frequently only one genus which is common and well represented on several host genera within the family; the other fungus genus or genera usually assume minor importance in comparison. In other instances, the different fungal genera may be restricted to host genera within different aggregates (tribes or sub-families). Sometimes fungal genera may share common hosts.

Gross analysis of host range may lead to tentative conclusions about the phylogeny of, and relationships between the downy mildew genera (see Section X) and the way in which host and pathogen co-evolution, and hence specificity, has developed but the relevance of such conjecture depends entirely on the validity of the host taxonomy. Conversely, it could be suggested that studies of pathogen specificity may provide a good guide to the relationships between host taxa. However the morphological characters of the sporophores and the means of conidiosporangial germination which are used to separate peronosporaceous genera could be characters under simple genetic control and may not, in fact, represent fundamental differences.

## IV. Specificity of Fungal Species to Higher Host Taxa

Any consideration of the host specificity is complicated by taxonomic preferences in deciding what constitutes a peronosporaceous species. Some workers have favoured the erection of many species from closely related hosts based on small differences in the size of the asexual propagules although it is now more common to consider the fungal species in a broader context (Yerkes and Shaw, 1959). Nonetheless, regardless of the taxonomic approach there appear to be no authentic records of fungal collections capable of parasitizing hosts from different families or orders. If a broad concept of the fungal species is accepted, it is usual to record a single species on a single higher host taxa (order, family, tribe, etc.).

Exceptions to this practice occur when distinct morphological criteria result in the taxonomic division of species occurring on the same host family. Such is the case with the downy mildew fungi on the family Gramineae where taxonomy and nomenclature have become more complicated.

The graminicolous downy mildew genera illustrate a clear phylogenetic relationship (Shaw, 1978) which shows a trend from the ability to be cultured in artificial media, zoospore production, and wide host range on the one hand to obligate biotrophy, direct germination of propagules and a high degree of host specificity on the other.

Such a clear relationship between genera is not apparent in downy mildews on other families, but exceptions to the trend of one fungal species per higher host taxa outside the Gramineae usually involve species of *Plasmopara* sharing hosts in common with species of other genera. This suggests that the zoosporic genus *Plasmopara* may be a common ancestor of other Peronosporaceae.

Regardless of the way in which it is considered, the specificity of fungi to higher host taxa is indicative of host and pathogen co-evolution during which time geographical and ecological isolation on a microscale have probably led to the observed speciation. There is no evidence that reproductive isolation has played a role.

In the situation when different peronosporaceous species or genera are known to occur on common hosts and if hybridization can be achieved, it offers the potential for investigating the genetic basis for some of the morphological and physiological characters which separate these taxonomic groups. This could provide an important insight into the molecular basis for biotrophy and the determination of specificity.

## V. Variation within Fungal Species for Specificity to Lower Host Taxa

The fact that a peronosporaceous fungus has been identified morphologically and on the basis of the higher host taxa from which it was collected as a particular species, does not mean that all collections of that species share the same pathogenicity characteristics. It is usual for a fungus collected from one species within a particular host genus to prove pathogenic only on species within the same or closely related genera, or sometimes on host-species aggregates below the genus level or occasionally on just a single species. This level of specificity to lower host taxa has often been described by the erection of *formae speciales*.

There is however a danger that the recognition and naming of *formae speciales* and their consideration as an extension of the fungal taxonomy may be a misleading over-simplification. The identification of *formae speciales* may simply summarize a dynamic situation in the pathogen population. The major component of the population may conform to certain narrow host range characteristics, while it is possible that components capable of varying degrees of more extensive pathogenicity may also occur at lower frequencies. Relative frequencies of the components of such a pathogen population will be determined by fitness and in most instances, it is likely that the component highly adapted to its environment (i.e. a specific host) will be most fit at the expense of flexibility (this is particularly true in an agricultural context). However, the balance could change if, for example, the presence of a particular specific host became less certain or subject to greater temporal and spatial fluctuations. There is some evidence that pathotypes with differing degrees of host specificity may occur within the population of a single peronosporaceous species (Palti, 1974).

It seems reasonable to take the view that the different species and genera of the Peronosporaceae identified on the Gramineae reflect a dynamic relationship between forms: those with a broad host range (identified as *Sclerophthora macrospora*) and those with an intermediate or narrow host range (identified as *Sclerospora graminicola* and *Peronoscleropora* species). It is only the apparent morphological correlation which occurs with their physiology that has brought about the categorization of three genera and numerous species. It is not beyond question that this correlation is complete. At this level of consideration the boundary between taxonomy and population genetics begins to become blurred. The genetic basis for the differences in host-range and morphological characteristics could be investigated where different pathotypes and morphological forms are known to parasitize a common host.

## VI. Variation within Fungal Species for Specificity to Genotypes within Lower Host Taxa

The fact that a downy mildew species or *forma specialis* is known to be pathogenic on a particular host species or collection of related species does not mean that all collections of that fungus will be pathogenic on all host genotypes within the species or species aggregate. This has led to the identification and classification of so-called physiological races in several downy mildew fungi of economically important crop plants. Although such races have only been identified in peronosporaceous species parasitizing crops, there is no reason to assume that races do not naturally occur in populations of Peronosporaceae attacking non-crop plants.

The existence of physiological races and the ability to identify them is determined by the presence of race-specific resistance factors in the host gene-pool. Assuming that a gene-for-gene type of relationship exists between host and pathogen (Person, 1959), the number of races which can be identified is related to the number of host specific resistance factors. The theoretical number of host genotypes and pathogen races is $2^n$ where $n$ is the number of race specific factors identifiable.

Crute and Johnson (1976) proposed an gene-for-gene relationship between *Lactuca sativa* L. cultivars and *Bremia lactucae*. This was substantiated in later studies (Johnson *et al.*, 1977, 1978) and 11 specific resistance factors with their matching virulence factors are considered to control the observed interrelationship.

Fewer specific resistance factors control the response of spinach cultivars to *Peronospora farinosa* f. sp. *spinaciae* Byford, but Eenink (1976b) has proposed that a gene-for-gene relationship involving three interacting resistance and virulence factors explains the patterns of resistance and susceptibility observed. Where a gene-for-gene relationship has not been worked out, one must be suspected in cases where differential resistance has been demonstrated: (Geesman, 1950; Lehman, 1953, 1958; Grabe and Dunleavy, 1959; Natti *et al.*, 1967; Dunleavy, 1970, 1971, 1977; Dunleavy and Hartwig, 1970; Stuteville, 1973; Titatarn and Exconde, 1974; Zimmer, 1974; Ilescu and Pirvu, 1975; Hubbeling, 1975; Viranyi, 1978).

Robinson (1976) suggested that vertical or differential resistance is most likely to be found in annual crops, particularly in temperate regions, since the phenomenon is only advantageous to a host when there is a discontinuity in the epidemic due to the absence of a host or unfavourable environmental conditions. Since most crop hosts of peronosporaceous fungi, in which specific virulence has been recorded, are cultivated as annuals, this would tend to support Robinson's argument. It may be relevant that differential resistance to downy mildew fungi on the perennial crops of hops and vines has not been demonstrated. However, Robinson's other suggestion, that vertical resistance should be more frequently encountered in outbreeding species, is less readily supported by the evidence from downy mildews since at least three of the species in which vertical resistance is common (lettuce, peas and soyabeans) are strongly inbreeding. It is possible, however, that although a species is commonly inbreeding, sufficient outbreeding may occur

244    I. R. CRUTE

in natural populations to make genes which confer differential resistance advantageous within any heterogeneous plant population which results.


## VII. The Genetics of Host Specificity


In discussing the genetics of host specificity in peronosporaceous fungi it is only possible to confine our considerations to the levels of specificity between races of pathogens and genotypes of related host species. Even within the confines of race-cultivar specificity, the discussion can only be one-sided since no genetical studies of peronosporaceous fungi have been made.

In all instances where response to downy mildew fungi has been shown to be race specific, and inheritance studies have been carried out, resistance has been found to be under simple genetic control. thus in *Brassica oleracea* Natti *et al.* (1967) identified two independent dominant genes which controlled the host reaction to races of *Peronospora parasitica*. At least three single dominant genes control reaction to races of *Plasmopara halstedii* ($Pl_1$, $Pl_2$ and $Pl_3$) in sunflowers (Zimmer and Kinman, 1972; Zimmer, 1974; Fick and Zimmer, 1975; Fernandez Martinez and Domingues Gimenez, 1978) and to races of *Peronospora farinosa* f. sp. *spinaciae* Byford in spinach (Smith, 1950; Smith *et al.*, 1962; Eenink, 1974, 1976a, b). The most extensive work on the genetics of resistance to a downy mildew species has been done with the lettuce —*Bremia lactucae* system (Jagger, 1924; Jagger and Whitaker, 1940; Ventura *et al.*, 1971; Sequeira and Raffray, 1971; Zink, 1973; Globerson *et al.*, 1974; Crute and Johnson, 1976; Bannerot and Boulidard, 1976; Boulidard and Bannerot, 1975; Johnson *et al.*, 1977, 1978; Norwood and Crute, 1980). Of the 11 resistance factors identified by Crute and Johnson (1976) and Johnson *et al.* (1977, 1978), there are no data on the mode of inheritance of three (*R1*, *R9* and *R11*), although circumstantial evidence suggests that *R9* and *R11* are probably single dominant genes. However, evidence that *R2–6*, *R8* and *R10* are single dominant genes is good and these have been given the appropriate *Dm* gene designation. Segregation of resistance factor *R7* suggested that it behaved as a complementary pair of independent dominant genes (Globerson *et al.*, 1974; Johnson *et al.*, 1977) which were designated *Dm-7/1* and *Dm-7/2*. Since segregation ratios have also been obtained for *R7* which suggest a single dominant gene, it is possible that the ratios which gave rise to the complementary gene interpretations resulted from incomplete expression in the heterozygous condition (Crute and Norwood, 1978).

In addition to cultivated lettuce, Netzer *et al.* (1976) studied resistance to downy mildew in the wild species *Lactuca saligna* L. In the F2 and F3 generations of a cross between the resistant wild parent and susceptible

*Lactuca sativa* cultivar they obtained segregation ratios which they were unable to interpret. However, in later generations, ratios indicating the presence of a single dominant gene were obtained and resistance was stabilized (D. Netzer and D. Globerson, personal communication). Crute and Norwood (1979) tested, with appropriate indicator isolates of *B. lactucae*, both the original *L. saligna* parent and a selected resistant F4 line from the above cross. These tests revealed that while the original parent carried novel resistance unattributable to *R1–11*, the F4 line appeared to carry solely *R3*. Hence, the *L. saligna* source carries a combination of resistance factors one of which is identical to a factor already located in cultivated genotypes. In Israel the isolate used for selection purposes resulted in the retention of *R3* but the loss of one or more additional resistance factors during selection.

There are cases where inheritance of resistance to downy mildew fungi has been shown to be simple but where differential reactions to it have not been reported. This may reflect cases of simple inherited non-specific or horizontal resistance; or alternatively situations where races capable of overcoming the resistance have not yet been identified (Ward and Tims, 1952; Russell, 1969; Bernard and Cremens, 1971; Marani *et al.*, 1972; van Vliet and Meijsing, 1974, 1977).

Genes for resistance to downy mildew fungi frequently exhibit genetic linkage with reaction to other parasites. Eenink (1974, 1976a) found tight linkage between two genes conditioning specific resistance to *Peronospora farinosa* f. sp. *spinaciae* in spinach and also noted that these were linked to a gene for resistance to cucumber virus 1. Resistance in cucumber to a powdery mildew (*Sphaerotheca fuliginea* Poll.) was found to be controlled by a gene which was linked to the locus that determined the reaction to downy mildew (*Pseudoperonospora cubensis*) (van Vliet and Meijsing, 1977). Fick and Zimmer (1975) found that gene $Pl_2$ for resistance to *Plasmopara halstedii* in sunflower was linked to gene *R1* contributing resistance to rust (*Puccinia helianthi* Schw.). In lettuce there is evidence for linkage between genes *Dm2*, *Dm3* and *Dm6* (Norwood and Crute, 1980) while it is probable that the latter gene may also be linked to a gene contributing to resistance to lettuce root aphid (*Pemphagus bursarius* L.) (Crute and Dunn, 1980). Gene *Dm8* is linked to a gene for reaction to turnip mosaic virus (Zink and Duffus, 1969, 1970, 1973, 1975) and possibly powdery mildew (*Erysiphe cichoracearum* D.C. ex Merat) (Schnathorst *et al.*, 1958).

Linkage between loci concerned with host response to parasites may prove to be of some significance in the elucidation of any underlying common molecular basis for parasite specificity. Genes controlling similar characters are likely to become associated spatially on chromosomes during the evolutionary process possibly as a result of gene duplication and differentiation. It is possible that a particular type of protein is coded for by the loci determining responses of parasites.

If, as seems probable, peronosporaceous species are diploid for most of their life cycle (Dick and Win Tin, 1973; Tommerup et al., 1974; Sansome and Sansome, 1974) and outbreeding occurs, then heterozygosity can be expected. Should specific virulence determinants be recessive, as indicated by the preliminary experiments of Michelmore (personal communication) with Bremia lactucae and, as seems to be the case with other groups of plant pathogens (Day, 1974), then alleles for virulence could be masked within a non-virulent genotype. Were mitotic recombination to occur, however, it would be possible to produce "virulent" nuclei within the coenocytic mycelium which could be transferred to the next generation of asexual propagules. In such instances, when the population of propagules is exposed to a resistant host, those carrying the necessary virulence would survive and reproduce. Crute and Norwood (1980) have suggested that for Bremia lactucae the sexual process is the most likely means whereby new specific virulence combinations are produced and that heterozygosity and somatic recombination could readily account for the observed race changes that they and others have studied in this fungus (Channon et al., 1965; Dixon and Dodson, 1973; Boulidard and Bannerot, 1975).

If heterokaryosis is assumed to occur in the Peronosporaceae, then further complications in the interpretation of selection experiments and evolution of races may arise in some genera. Species of Bremia, Peronospora and Peronosclerospora do not produce uninucleate zoospores and hence single infections result from a multinucleate spore. In such instances, the presence of the dominant allele at a locus determining specific virulence in only one of several nuclei could determine the host reaction to the spore or, alternatively, some critical proportion of particular alleles may determine what host reaction the spore precipitates.

## VIII. Reasons for Studying Variation for Specific Virulence in Populations of Peronosporaceous Fungi

The practical reasons for studying variation of specific virulence in plant pathogens relate, primarily to the breeding and exploitation of resistant cultivars.

When selecting breeding material it is important to know which pathotypes to use in the screening process, how the resistance is expressed and inherited and whether it is likely to prove adequate and durable. It is therefore necessary to study variation in the pathogen population with respect to the same resistance factors as those being utilized in the breeding programme, i.e. the differential hosts and sources of resistance being exploited should carry the same resistance genes. In order to be efficient in exploiting host genotypes produced from breeding programmes, quantitative data are required on the frequency of occurrence of particular specific virulence factors and their combinations within the pathogen population in the geographical area where such resistance is to be employed. In addition, it is important to ascertain the capacity of the pathogen for change under different selection pressures and the rapidity with which this may occur. A further requirement is that the resistance genotypes of cultivars already in use be identified since this enables information on variation for specific virulence to be used in the recommendation of cultivars likely to prove resistant in a particular area.

Studies of variation for specific virulence in peronosporaceous species on differential hosts has usually only entailed the identification and labelling of physiological races. Few studies have attempted to record either the geographical distribution or the relative frequency of occurrence of such races within the pathogen population and how this relates to the frequency and distribution of specific virulence determinants.

In studies with other major fungal pathogens where differential host response is commonly encountered, the taxonomic attitude to the subject of pathogen variation is being superseded by a population genetics approach (Wolfe et al., 1976; Shattock et al., 1977; Browder and Eversmeyer, 1977). It is important to know not how many different races occur, but how frequently such races are found, where they are located and how frequencies relate to, and change under, different selection pressures.

In this context, the work of Dixon and Wright (1978), Dixon (1978) and Wellving and Crute (1978) with *Bremia lactucae* has started to develop the ideas that categorization of physiological races is less important than knowing how frequently certain specific virulence determinants and their combinations occur within the population. By analysing the results of surveys in terms of virulence factor frequencies it was clearly demonstrated that combinations of resistance factors *1–10* in lettuce were unlikely to provide adequate control of *Bremia lactucae* in the UK since virulence to overcome this resistance was frequently encountered and widely distributed (Crute et al., unpublished data).

## IX. The Expression and Mechanisms of Specificity

It appears that spore germination either by the production of zoospores or directly by means of a germ tube is a non-specific process in peronosporaceous species. It proceeds normally on any plant surface or inert material provided the physical environment is suitable (Arens, 1937; Iwata, 1939; Wang, 1949; Royle and Thomas, 1973; Crute and Dickinson, 1976; Viranyi and Henstra, 1976; Riggle and Dunleavy, 1974; Riggle, 1977). There is no evidence that inhibition of spore germination by any host contributes to specificity but it must be said that evidence to the contrary is scant. In those species where spores germinate directly, there is also little evidence that germ tube extension is in any way affected although Viranyi and Hestra (1976) recorded that germ tubes from *Bremia lactucae* spores on a resistant lettuce cultivar were longer and grew in an irregular way compared with those on a susceptible cultivar.

Determination of specificity can occur before or after penetration has proceeded normally. Iwata (1939) studied the behaviour of *Pseudoperonospora cubensis* on 73 different higher plant species representing 32 families. The fungus was shown to penetrate the leaves of 49 of the species examined from 22 families. All members of the Cucurbitaceae were penetrated normally. Wang (1949) reported the normal penetration of several non-host cruciferae by isolates of *Peronospora parasitica* while Arens (1937) stated that *Bremia lactucae* from lettuce penetrated only this species and *Senecio cruentis* D.C. Three other species of compositae (all non-hosts of the genus *Bremia*) and four non-compositae species were not penetrated. Verhoeff (1960) found that *Bremia lactucae* from lettuce penetrated resistant lettuce cultivars but not four compositae species known to be common hosts to forms of the same fungal species. However, the work of Maclean *et al.* (1974), Crute and Dickinson (1976), Viranyi and Blok (1976), Viranyi and Henstra (1976) and Maclean and Tommerup (1979) indicated that the determination of specificity of *Bremia lactucae* from lettuce both on some non-hosts and on lettuce cultivars carrying specific resistance factors occurs after penetration. In contrast, Wehtje and Zimmer (1968) and Wehtje *et al.* (1979) concluded that in sunflower lines, carrying the $Pl_2$ gene for specific resistance to *Plasmopara halstedii*, incompatibility results solely from contact between a zoospore and a host cell and that no penetration occurs.

More attention has been paid to the events associated with compatibility or incompatibility which occur after penetration of a potential host. A common feature associated with the expression of incompatibility to peronosporaceous fungi, in common with other groups of biotrophic parasites, is host cell death (the so-called hypersensitive response) which may occur at different rates and to differing extents depending upon the host–parasite association in question (Salmon and Ware, 1928; Hoerner, 1940; Foster, 1947; Wang, 1949; Russell, 1969; Krober and Petzold, 1972, 1977; Greenhalgh and Dickinson, 1975; Dickinson and Greenhalgh, 1977; Riggle, 1977). The histology and symptoms associated with resistant plant response to lettuce isolates of *Bremia lactucae* have received more attention than for any other downy mildew fungus. Both non-hosts and lettuce cultivars carrying different specific resistance factors have been studied in this connection (Maclean *et al.*, 1974; Crute and Dickinson, 1976; Viranyi and Blok, 1976; Crute and Davis, 1977; Crute and Norwood, 1978; Maclean and Tommerup, 1979) and the findings and implications of this work have been summarized by Crute and Dixon (1981).

In the work cited above, the compatible association is characterized by minimal host cell disruption and any necrosis is considerably delayed, usually until after copious sporulation has occurred. While the almost universal association of incompatibility and host cell death is clear, the role of cell death *per se* in the determination of specificity or the way in which it is induced is a matter for debate and conjecture.

An hypothesis on the mechanistic basis for specificity needs to be able to accommodate variation in the characteristics of the response associated with a single gene in different genetic backgrounds and environments; and differences between the responses associated with different genes. Similarly, the variation in non-host responses should continually be considered since the concept of incompatibility being the unique reaction, in the so-called "quadratic check" diagram can, if taken in isolation, be misleading. Considered more broadly, compatibility is the exception and while it has been postulated that specific resistance genes are superimposed upon the normally compatible genotype of the host species, this does not readily explain the relative degrees of incompatibility which occur beyond the confines of the normal host. Intriguing observations such as Yarwood's (1977) discovery that beans will support the growth of *Pseudoperonospora cubensis* (normally non-pathogenic on beans) if first infected by *Uromyces phaseoli* (Pers.) Wint. and fungicide-induced incompatibility reactions (Crute *et al.*, 1977; Crute and Jagger, 1979) also need to be accommodated.

It is clear that more data are needed before it will be possible to interpret whether cell death is a non-specific response purely symptomatic of the degree of affinity between organisms (e.g. resulting from non-depression of the fungal genome; see Section II) or whether it and the other responses associated with it, such as phytoalexin production, are truly determinants of specificity rather than non-specific wound responses. Indeed it may be true that superficially similar phenomena could result from totally different processes but at this stage arguing from the particular to the general may be dangerous.

## X. The Evolutionary Significance of Specificity

It is probable that fungi first emerged from an aquatic environment either loosely associated with or as true symbionts of higher plants. Kevan *et al.* (1975) suggested that ancestors of present day Phycomycetes are present in the fossil record as early as the Devonian period. There is no way of knowing whether or not the ancestors of existing plant pathogens were specific to their associated higher plant species. However, it is likely that the Peronosporales share a common ancestor and hence, considering the total host range of the group, evolution has probably been in the direction of the specificity seen today.

Similarly, zoospore production, an essentially aquatic character, has probably given way in the more highly evolved genera to direct germination which is presumed to have survival value on land, particularly since it was evolved at the expense of the ability to initiate several infections from a single propagule. However, directly germinating genera still require free water for the process and it is possible that the advantage of direct germination may result from the potential variation imparted by a multinucleate and possibly heterokaryotic source of infection rather than its lack of physical dependence on water. Considering that the directly germinating genus *Peronospora* has a wider host range than any other genus of the Peronosporaceae, the character of direct germination may have differentiated before the characteristic morphological features of the sporangiophores which distinguish present day genera. Host range considerations would imply that unless sporangiophore characters are under simple genetic and readily mutable control, morphological differentiation probably occurred prior to the development of extensive specificity such that the ancestral form of each genus would be pathogenic on at least the range of host families represented in the present day host range of the genus.

The evolution of specificity can only be considered in terms of the coevolution of host and parasite. In general terms, the host plant will evolve to fit a niche while retaining a degree of flexibility in the gene pool to buffer it against environmental changes (including its parasites). The principal environmental influence on an obligately biotrophic plant pathogen is its host and it will evolve to exploit it to the full. If the host species is either continuously present or its presence at regular intervals is guaranteed, e.g. in agriculture, this process of adaptation will continue at the expense of flexibility until any individual is capable of parasitising only a very restricted host range. In the agricultural context, fungal forms on crop species will become equally specialized to host genotypes if the host provides a uniform environment. If the host population carries a range of specific resistance factors, however, survival of the pathogen will be best served by retaining the flexibility of carrying a range of specific virulence determinants.

## References

Andrews, J. H. (1975). *Can. J. Bot.* **53**, 1103–1115.
Arens, L. (1937). *Botanica (Sao Paulo)* **1**, 39–54.
Bannerot, H. and Boulidard, L. (1976). *Proc. Eucarpia Meeting on Leafy Vegetables*, IVT, Wageningen, Netherlands, 53–54.
Bernard, R. L. and Cremens, C. R. (1971). *J. Hered.* **62**, 359–362.

Boulidard, L. and Bannerot, H. (1975). *Report Stn de Genetique et d'amelioration des Plantes, Versailles* for 1972 to 1974, M1–M18.
Brian, P. W. (1967). *Proc. R. Soc. Lond.* Series B, **168**, 101–118.
Browder, L. E. and Eversmeyer, M. G. (1977). *Phytopathology* **67**, 766–771.
Channon, A. G., Webb, M. J. W. and Watts, L. E. (1965). *Ann. Appl. Biol.* **56**, 389–397.
Crute, I. R. (1978). *PANS.* **24**, 519–520.
Crute, I. R. and Davis, A. A. (1977). *Trans. Br. Mycol. Soc.* **69**, 405–410.
Crute, I. R. and Dickinson, C. H. (1976). *Ann. Appl. Biol.* **82**, 433–450.
Crute, I. R. and Dixon, G. R. (1981). *In* "The Downy Mildews". Academic Press, London and New York.
Crute, I. R. and Dunn, J. A. *Euphytica* **29**, 483–488.
Crute, I. R. and Johnson, A. G. (1976). *Ann. Appl. Biol.* **83**, 125–137.
Crute, I. R. and Norwood, J. M. (1978). *Ann. Appl. Biol.* **89**, 467–474.
Crute, I. R. and Jagger, B. M. (1979). *Rep. Natn. Veg. Res. Stn.* for 1978, 77 pp.
Crute, I. R. and Norwood, J. M. (1979). *Rep. Natn. Veg. Res. Stn.* for 1978, 77 pp.
Crute, I. R. and Norwood, J. M. (1980). *Ann. Appl. Biol.* **94**, 275–278.
Crute, I. R., Wolfmann, S. H. and Davis, A. A. (1977). *Ann. Appl. Biol.* **85**, 147–152.
Day, P. R. (1974). "Genetics of Host–Parasite Interaction". W. H. Freeman and Co., San Francisco.
Dick, M. W. and Win Tin (1973). *Biol. Rev.* **48**, 1973.
Dickinson, C. H. and Greenhalgh, J. R. (1977). *Trans. Br. Mycol. Soc.* **69**, 111–116.
Dixon, G. R. (1978). *In* "Plant Disease Epidemiology" P. R. Scott and A. Bainbridge, Eds) 71–78. Blackwell, Oxford.
Dixon, G. R. and Doodson, J. K. (1973). *Hort. Res.* **13**, 89–95.
Dixon, G. R. and Wright, I. R. (1978). *Ann. Appl. Biol.* **88**, 187–294.
Duddridge, J. and Sargent, J. A. (1978). *Phys. Pl. Path.* **12**, 289–276.
Dunleavy, J. M. (1970). *Crop. Sci.* **10**, 507–509.
Dunleavy, J. M. (1971). *Am. J. Bot.* **58**, 209–211.
Dunleavy, J. M. (1977). *Pl. Dis. Reptr.* **61**, 661–663.
Dunleavy, J. M. and Hartwig, E. E. (1970). *Pl. Dis. Reptr.* **54**, 901–902.
Eenink, A. H. (1974). *Euphytica* **23**, 485–488.
Eenink, A. H. (1976a). *Euphytica* **25**, 713–715.
Eenink, A. H. (1976b). *Proc. Eucarpia Meeting on Leafy Vegetables*, IVT, Wageningen, 53–54.
Fernandez Martinez, J. and Dominguez Gimenez, J. (1978). *Rev. Pl. Path.* Abstr. **58**, 69. 1979.
Fick, G. N. and Zimmer, D. E. (1975). *Crop Sci.* **15**, 777–779.
Foster, H. H. (1947). *Phytopathology* **37**, 433.
Geeseman, G. E. (1950). *Agron. J.* **42**, 257–258.
Globerson, D., Netzer, D. and Tjallingii, F. (1974). *Euphytica* **23**, 54–60.
Grabe, D. F. and Dunleavy, J. (1959). *Phytopathology* **49**, 791–793.
Greenhalgh, J. R. and Dickinson, C. H. (1975). *Phytopath. Z.* **84**, 131–141.
Guttenburg, H. von and Schmoller, H. (1958). *Arch. Mikrobiol.* **30**, 268–279.
Hoerner, G. R. (1940). *J. Agric. Res.* **61**, 331–334.
Hollomon, D. W. (1969). *J.Gen. Microbiol.* **55**, 267–274.
Hollomon, D. W. (1970). *J. Gen. Microbiol.* **62**, 75–87.
Hollomon, D. W. (1973). *J. Gen. Microbiol.* **78**, 1–13.
Hubbeling, N. (1975). *Med. Fac. Landbouww. Rijksuniv. Gent.* **40**, 539–543.
Ilescu, H. and Pirvu, N. (1975). *Rev. Pl. Path.* Abstr. **56**, 816, 1977.
Ingram, D. S. (1973). *In* "Plant Tissue and Cell Cultures". (H. E. Street, Ed.) 392. Blackwell.
Ingram, D. S. (1976). *In* "Encyclopedia of Plant Physiology Vol. 4; Physiological Plant Pathology" (R. Heitefuss and P. H. Williams, Eds) 740–760. Springer-Verlag, Berlin.
Ingram, D. S. and Tommerup, I. C. (1973). *In* "Fungal Pathogenicity and the Plant's Response" (R. J. W. Byrde and C. V. Cutting, Eds) 121–137. Academic Press, London and New York.

Ingram, D. S., Sargent, J. A. and Tommerup, I. C. (1976). *In* "Biochemical Aspects of Plant Parasite Relationships" (J. Friend and D. R. Threlfall, Eds) 43–78.
Iwata, Y. (1939). *Ann. Phytopathol. Soc. Japan* **8**, 124–144.
Jagger, I. C. (1924). *Phytopathology* **14**, 122.
Jagger, I. C. and Whitaker, T. W. (1940). *Phytopathology* **30**, 427–433.
Johnson, A. G., Crute, I. R. and Gordon, P. L. (1977). *Ann. Appl. Biol.* **86**, 87–103.
Johnson, A. G., Laxton, S. A., Crute, I. R., Gordon, P. L. and Norwood, J. M. (1978). *Ann. Appl. Biol.* **89**, 257–264.
Kevan, P. G., Chaloner, W. G. and Savile, D. B. O. (1975). *Palaentology* **18**, 391–417.
Kröber, H. and Petzold, H. (1972). *Phytopath. Z.* **74**, 296–313.
Kröber, H. and Petzold, H. (1977). *In* "Current Problems in Plant Pathology" (Z. Kiraly, Ed.) 49–56. Akademiae Kidu, Budapest.
Lehman, S. G. (1953). *Phytopathology* **43**, 460–461.
Lehman, S. G. (1958). *Phytopathology* **48**, 83–86.
Leppik, R. A., Hollomon, D. W. and Bottomley, W. (1972). *Phytochemistry* **11**, 2055–2063.
Maclean, D. J., Sargent, J. A., Tommerup, I. C. and Ingram, D. S. (1974). *Nature* **249**, 186–187.
Maclean, D. J. and Tommerup, I. C. (1979). *Physiol. Pl. Path.* **14**, 265–280.
Marani, A., Fishler, G. and Amirar, A. (1972). *Euphytica* **21**, 97–105.
Mason, P. H. (1973). Ph.D. Thesis, University of Cambridge.
McKay, R. (1939). *J. Royal Hortic. Soc.* **64**, 272–285.
Murphy, P. A. and McKay, R. (1926). *Proc. R. Dublin Soc. Science* **18**, 237–261.
Natti, J. J., Dickson, M. K. and Atkin, D. D. (1967). *Phytopathology* **57**, 144–147.
Netzer, D., Globerson, D. and Sacks, J. (1976). *Hort. Sci.* **11**, 612–613.
Norwood, J. M. and Crute, I. R. (1980). *Ann. Appl. Biol.* **94**, 127–135.
Palti, J. (1974). *Phytoparasitica* **2**, 109–115.
Pegg, G. F. and Mence, M. J. (1970). *Ann. Appl. Biol.* **66**, 417–428.
Person, C. C. (1959). *Can. J. Bot.* **37**, 1101–1130.
Pioth, L. (1957). *Z. PflZucht.* **37**, 127–158.
Powlesland, R. (1954). *Trans. Br. Mycol. Soc.* **37**, 362–371.
Riggle, J. H. (1977). *Can. J. Bot.* **55**, 153–157.
Riggle, J. H. and Dunleavy, J. M. (1974). *Phytopathology* **64**, 522–526.
Robinson, R. A. (1976). "Plant Pathosystems". Springer Verlag, Berlin, Heidelberg and New York.
Royle, D. J. and Thomas, G. G. (1973). *Physiol. Pl. Path.* **3**, 405–417.
Russell, G. E. (1969). *I.I.R.B.* **4**, 1–10.
Salmon, E. S. and Ware, W. M. (1928). *Ann. Appl. Biol.* **15**, 352–370.
Sansome, E. and Sansome, F. W. (1974). *Trans. Br. Mycol. Soc.* **62**, 323–332.
Sargent, J. A. and Payne, H. L. (1974). *Trans. Br. Mycol. Soc.* **63**, 509–518.
Schnathorst, W. C., Grogan, R. G. and Bardin, R. (1958). *Phytopathology* **48**, 538–543.
Sempio, C. (1942). *Riv. Biol.* **34**, 3–7.
Sequeira, L. and Raffray, J. B. (1971). *Phytopathology* **61**, 578–579.
Shattock, R. C., Janssen, B. D., Whitbread, R. and Shaw, D. S. (1977). *Ann. Appl. Biol.* **86**, 249–260.
Shaw, C. G. (1970). *Indian Phytopath.* **23**, 364–370.
Shaw, C. G. (1978). *Mycologia* **70**, 594–604.
Shepherd, C. J. (1962). *Aust. J. Biol. Sci.* **15**, 483–508.
Shepherd, C. J. and Mandryk, M. (1962). *Trans. Br. Mycol. Soc.* **45**, 233–244.
Shepherd, C. J. and Mandryk, M. (1963). *Aust. J. Biol. Sci.* **16**, 77–87.
Smith, P. G. (1950). *Phytopathology* **40**, 65–68.
Smith, P. G., Luhn, C. H. and Webb, R. E. (1962). *Phytopathology* **52**, 365.
Stuteville, D. L. (1973). Abstr. No. 0715. 2nd Int. Congress of Plant Pathology, Minneapolis.

Thornton, J. D. and Cooke, R. C. (1974). *Physiol. Pl. Path.* **4**, 117–126.
Titatarn, S. and Exconde, O. R. (1974). *Philipp. Agric. J.* **58**, 90–104.
Tiwari, M. M. and Arya, H. C. (1969). *Science, NY* **163**, 291–293.
Tokura, R. (1975). *Trop. Agric. Res., Tokyo* **8**, 57–60.
Tommerup, I. C., Ingram, D. S. and Sargent, J. A. (1974). *Trans. Br. Mycol. Soc.* **62**, 145–150.
van Vliet, G. J. A. and Meijsing, W. D. (1974). *Euphytica* **23**, 251–256.
van Vliet, G. J. A. and Meijsing, W. D. (1977). *Euphytica* **26**, 793–796.
Ventura, J., Netzer, D. and Globerson, D. (1971). *J. Amer. Soc. Hort. Sci.* **96**, 103–104.
Verhoeff, K. (1960). *Tijdschr. Pftentenz.* **66**, 133–203.
Viranyi, F. (1978). *Phytopath. Z.* **91**, 362–364.
Viranyi, F. and Blok, I. (1976). *Neth. J. Pl. Path.* **82**, 251–254.
Viranyi, F. and Henstra, S. (1976). *Acta Phytopath. Academiae Scientarum Hungaricae* **11**, 173–182.
Wang, T. M. (1949). *Phytopathology* **39**, 541–547.
Ward, W. A. and Tims, E. C. (1952). *Phytopathology* **42**, 22.
Waterhouse, G. M. (1973). *In* "The fungi: an advanced treatise". Vol. IVB. (G. C. Ainsworth, F. K. Sparrow and A. S. Sussman, Eds) 165–183. Academic Press, New York and London.
Wehtje, G., Littlefield, L. J. and Zimmer, D. E. (1979). *Can. J. Bot.* **57**, 315–323.
Wehtje, G. and Zimmer D. E. (1978). *Phytopathology* **68**, 1568–1571.
Wellving, A. and Crute, I. E. (1978). *Ann. Appl. Biol.* **81**, 251–256.
Wolfe, M. S., Barrett, J. A., Shattock, R. C., Shaw, D. S. and Whitbread, R. (1976). *Ann. Appl. Biol.* **82**, 369–374.
Yarwood, C. E. (1943). *Hilgardia* **14**, 595–691.
Yarwood, C. E. (1977). *Phytopathology* **67**, 1021–1022.
Yerkes, W. D. and Shaw, C. G. (1959). *Phytopathology* **49**, 499–507.
Zimmer, D. E. (1974). *Phytopathology* **64**, 1465–1467.
Zimmer, D. E. and Kinman, M. L. (1972). *Crop. Sci.* **12**, 749–751.
Zink, F. W. (1973). *J. Amer. Soc. Hort. Sci.* **98**, 193–195.
Zink, F. W. and Duffus, J. E. (1969). *J. Amer. Soc. Hort. Sci.* **94**, 403–407.
Zink, F. W. and Duffus, J. E. (1970). *J. Amer. Soc. Hort. Sci.* **95**, 420–424.
Zink, F. W. and Duffus, J. E. (1973). *J. Amer. Soc. Hort. Sci.* **98**, 49–50.
Zink, F. W. and Duffus, J. E. (1975). *Phytopathology* **65**, 243–245.

Chapter 12

# Breeding for Resistance to Downy Mildews

PETER MATTHEWS

*John Innes Institute, Norwich, Norfolk, England*

## I. Introduction

In this chapter breeding for resistance is described in 15 crops, both temperate and tropical, which are attacked by downy mildew from the genera *Bremia, Peronospora, Pseudoperonospora, Plasmopara, Sclerophthora* and *Sclerospora*.

Variation in the crop is outlined and sources of resistance are identified. Inheritance of resistance, where this is known, is described. For further details of the origins, development and floral biology of the crops covered in this chapter see Simmonds (1976).

A prerequisite in breeding for disease resistance is a knowledge of variation in the pathogen and such variation in races, aggressiveness and pathotypes is described where pertinent. For detailed general discussions of the underlying genetic principles involved in the interplay of host and pathogen and its importance in plant breeding the reader is referred to Day (1974), Robinson (1976), and Van der Plank (1978).

A number of the crops covered present special breeding problems related to their floral biology and breeding systems, examples being dioecy in grapes, hops and spinach; monoecy in cucurbits and maize; incompatibility in cabbages and sunflowers; and marked protogyny in millet.

Resistance to downy mildews in the crops surveyed is determined by genetic systems which require different management and the development of different strategies. Race-specific dominant genes are being used in lettuce; major genes in cucumber, grapes, maize, melons, millet, peas, soybean, spinach, sorghum and tobacco; and polygenic resistance in grapes, hops, maize, onions and sugar beet. Both major gene and polygene resistance is being utilized in some crops, e.g. grapes and maize. Methods range from simple crossing and backcrossing to incorporate dominant gene resistance (lettuce, soybean) to the complex and sophisticated development of balanced populations utilizing polygenic resistance derived from as many as 90 varieties/sub-populations (maize). In other crops resistance has been or is being introduced from wild species. Interspecific crossing is the basis of resistance breeding in grapes, onions, tobacco, squashes and sunflower. In nearly every crop a compromise has to be reached between resistance and agronomic characters. For an up-to-date and realistic approach to the problems and principles of crop improvement see Simmonds (1979); and for a general account of plant breeding for disease resistance see Russell (1978).

## II. *Bremia*

### A. Lettuce (*Lactuca sativa*)

Control of downy mildew of lettuce (*Bremia lactucae*), based on resistance has not been entirely effective because of the occurrence of many races and pathotypes (Crute and Johnson, 1976a).

Resistance is determined by a simple Mendelian system either by a single dominant gene or a pair of complementary dominant genes (Jagger and Whitaker, 1940; Ventura *et al.*, 1971; Sequeira and Raffay, 1971; Zink, 1973; Globerson *et al.*, 1974; Bannerot and Boulidard; 1976b; Crute and Johnson, 1976b; Crute and Gregory, 1976; Johnson *et al.*, 1977; Crute and Norwood, 1978; Johnson *et al.*, 1978).

Information on race/variety reactions, sources of resistance in present day varieties and varieties with identical resistance genes are tabulated by Crute and Johnson (1976c). Available data is interpreted in terms of a gene-for-gene relationship (Crute and Johnson, 1976b, c). Ten genes are postulated for resistance (*R* genes) and 10 for virulence in *Bremia* (*V* genes). Adopting the more conventional system for proven dominant resistance, *R* genes have been re-designated *Dm1*, *Dm2*, *Dm3* etc. (Johnson *et al.*, 1977, 1978). Hypersensitivity resistance in which 1–10 cells are killed is determined by *Dm1*, *Dm2*, *Dm3*, *Dm4*, *Dm5*, *Dm8*, *Dm9*, *Dm10* and "incomplete resistance" where 10–100 cells are killed, by *Dm6* and *Dm7*. *Dm2*, *Dm3*, *Dm4*, *Dm8* and probably *Dm6* are genes of major effect; *Dm7/1* and *Dm7/2* a pair of complementary dominant genes (Johnson *et al.*, 1972, 1978). Lists of varieties containing known *Dm* genes are to be found in Johnson *et al.* (1977). The widely used resistant varieties Calman, Valverde and other US varieties derive from PI* 104864, as do Avoncrisp and Avondefiance all of which contain *Dm8*. *Dm7/1* and *Dm7/2*, acting as complementary dominant genes, and *Dm6* condition "incomplete resistance" in both seedling and mature plant (Crute and Norwood, 1978). The incomplete reaction is hypostatic to genes giving more complete resistance. *Dm7/1* and *Dm7/2* resistance is incomplete in outdoor crisp types and modified in other backgrounds. *Dm6* is completely expressed at 20°C. Incomplete resistance associated with race specific resistance leads to marked reduction in sporophore production which is also associated with polygenic horizontal resistance. One of the pair of complementary genes *Dm7/1* may be effective on its own (Bannerot and Boulidard, 1976b; Johnson *et al.*, 1978). There is evidence that genes *Dm6* and *Dm8* are linked (30%) (Eenink, 1976a; Johnson *et al.*, 1978). No evidence has been

---

* PI prefixes indicate US Department of Agriculture plant introductions.

found for linkage between resistance and horticulturally useful characters (Jagger and Whitaker, 1940; Ventura *et al.*, 1971).

*R11* a new dominant factor has recently been identified from the cross *Hilde* × *L. serriola* (Bannerot and Boulidard, 1976b; Johnson *et al.*, 1978). Another recessive type of resistance has been found amongst collections of wild *L. saligna* in Israel (Netzer and Globerson, 1976; Netzer *et al.*, 1976).

Breeding has relied upon the use of race specific dominant gene resistance from which varieties have been developed by crossing and selfing, back-crossing is rarely used; selection is for resistance to indigenous (local) races (Jagger, 1931; Jagger and Whitaker, 1940; Thompson *et al.*, 1941; Bohn and Whitaker, 1951; Whitaker *et al.*, 1958; Sequeira and Raffay, 1971; Ventura *et al.*, 1971; Zinkernagel, 1975; Bannerot and Boulidard, 1976b; Dawson, 1976; Eenink, 1976; Handke, 1976; Welling and Crute, 1978). Over the past 50 years 120 varieties have been discovered or bred with race-specific domi-nant gene resistance (Crute and Johnson, 1976c). Resistance can be traced to nine sources: Blondine, Meikoningin, Gotte à forcer types, *L. serriola* PI 167150, Grand Rapids, Romaine blonde d'hiver, *L. serriola* PI 91532 = PI 104854, Bourguignonne grosse blonde d'hiver and Sucrine.

The occurrence of 10 specific virulence genes in *Bremia* means that $2^{10}$ races are theoretically possible. The demonstration of such a large race potential and a gene-for-gene relationship has far reaching consequences for downy mildew resistance breeding in lettuce. There are already examples of resistance breaking down (Jagger, 1931; Handke, 1976; Crute and Davis, 1976). Also, new races appear to arise spontaneously by mutation (Bannerot and Boulidard, 1976a). More recently isolates have been found which attack all known resistance genes (Crute and Davis, 1976; Lebeda *et al.*, 1980). Six races have been identified to date in the UK (Crute and Dickinson, 1976), 6 possibly 7 in West Germany (Zinkernagel, 1975; Handke, 1976), 6 in Holland (Eenink, 1976) and 3 in Israel (Ventura *et al.*, 1971); 16 from a sample of 104 Swedish isolates (Wellving and Crute, 1978), and 4 in Czechoslovakia (Lebeda *et al.*, 1980).

New strategies for the deployment of existing *Dm* genes and the develop-ment of horizontal resistance have been proposed. Bannerot and Boulidard (1976b) note that to accumulate all genes in one variety would be difficult, if not impossible, but suggest the incorporation of several genes at one time in new varieties, breeding for tolerance and for horizontal resistance. Handke (1976) suggests that all resistant varieties will eventually be attacked and proposes that susceptible varieties should be grown, and as consequence all except one race would disappear. Resistant varieties should then be re-introduced, which would have high levels of resistance. Resistance alleles should be closely linked.

The strategies proposed by Crute and Davis (1976) and Crute and Johnson (1976a, b) are more comprehensive. They comment that resistance associated with hypersensitivity is ephemeral and that the progressive addition of *Dm* genes does not give lasting control. They suggest the matching of *Dm* and *V* genes in different areas, and that *Dm* genes may be longer lasting in geographically distant areas. Alternative strategies proposed are the use of disruptive selection, multilines, location and utilization of new *Dm* genes, and the location and utilization of general resistance. In disruptive selection varieties containing different *Dm* genes, either singly or in pairs, would be rotated; this would require a knowledge of local races and selection of varieties with the least number of superfluous genes. In the multiline system, varieties would be bred agronomically uniform but heterogeneous for *Dm* genes. New sources of resistance have been found in wild *Lactuca* species *L. serriola* (4), *L. virosa* (7), *L. saligna* (4) (Crute and Johnson, 1976b). These species are closely related to each other and to *L. sativa* and are capable of crossing amongst themselves (Thompson *et al.*, 1941). These resistance factors are being transferred to *L. sativa*. General resistance cannot be identified in the presence of hypersensitive *Dm* genes. Selection for general resistance in the absence of *Dm* genes is difficult, it is only expressed in the adult plant and is often dependent on environment and inoculum concentration. Crute and Johnson (1976a) doubt that useful levels of general resistance will be found, and if found, are unsure of its economic value.

## III. *Peronospora*

### A. Cabbage, Broccoli and other Cruciferae (*Brassica oleracea*)

Resistance or immunity to *oleracea* (broccoli) strains of downy mildew (*Peronospora parasitica*) is available in *Brassica* species (Natti, 1958). Commercial varieties of *B. oleracea* are susceptible, with the exception of vars. *fimbriata* (Kale), whilst all other horticultural *Brassica* varieties are either resistant or immune. Cotyledon inoculations with five broccoli isolates from New York State indicated moderate resistance in PI 189028 and high resistance in PI 231210; field tests of broccoli varieties revealed variety differences, but none with high resistance.

Two races are known, the predominant one in New York State affecting *B. oleracea*, designated race 1 (Natti *et al.*, 1967). Resistance as found in PI 189028—a broccoli type— is determined by a single dominant gene, and in the cross PI 189028 × Waltham 29 (waxless) is dependent on foliage wax. Race 2 attacks PI 189028; resistance found in PI 245015 is determined by a single dominant gene inherited independently of race 1 resistance. Of 230 plant introductions screened, 16 had resistance to race 2, some also to race 1. Both types of resistance are used in breeding. However Natti *et al.* (1976)

suggest that because of the likely development of new races, a search should be made for field resistance in other *Brassica* species.

Resistance in *B. oleracea* cv. January King is expressed as necrosis of cotyledon mesophyll cells which inhibits growth of hyphae in the host (Greenhalgh and Dickinson, 1975). A similar type of hypersensitive resistance was reported in two populations of "wild" (naturalized) cabbages of *B. oleracea* (Greenhalgh and Dickinson, 1976). An allyl-isothiocyanate volatile is associated with resistance but this is undesirable in commercial cultivars, hence there has been progressive selection to reduce its level with a consequent progressive decrease in resistance (Greenhalgh and Mitchell, 1976).

## B. Onion (*Allium cepa*)

Downy mildew (*Peronospora destructor*) is one of the most destructive diseases of the onion seed crop, and also causes heavy losses of the bulb crop. Resistance breeding stems from work carried out in California (Jones *et al.*, 1939; Jones and Mann, 1963). Three sources of resistance have been identified in *Allium cepa*. In 13-53, a male sterile selection from Italian Red, seed stems are immune and leaves highly resistant; in 13-20-3, another selection from Italian Red, seed stalks are immune but foliage is only slightly resistant, however it has better bulb quality. The third source of resistance was found in an F1 hybrid between Red 21 and two inbred lines of Stockton Yellow Flat (50-6 and 50-6-1) which has seed stalk immunity.

Breeding is usually by mass or individual selection at diploid level, interspecific crossing is rarely used. *Allium fistulosum* is resistant to downy mildew and other fungal pathogens but is unexploited because of hybrid sterility. The use of male sterility in the resistant Italian Red 13-53 led to the production of hybrid seed. Emphasis is now on developing hybrid onion varieties. Calred and Lord Howe are derived from Italian Red 13-53 and have resistant stem and moderately resistant leaves. F1 populations between 13-53 and susceptible varieties possess intermediate resistance. Resistance appears to be polygenically rather than monogenically determined. *A. cepa* × *A. fistulosum* hybrids are reported to be extremely susceptible, suggesting that *A. fistulosum* resistance is recessive.

## C. Pea (*Pisum sativum*)

Downy mildew of peas (*Peronospora viciae*) is now regarded as being of increasing importance in the UK especially during early establishment (Gent, 1966; Gane *et al.*, 1971). Control using resistant varieties has been suggested as the best long term strategy to reduce losses (Olofsson, 1966; Allard, 1970;

Cousin, 1974). In the UK field assessments of varietal resistance are routinely carried out by the Processors and Growers Research Organization (PGRO) and the National Institute of Agricultural Botany (NIAB) whose respective annual reports and publications should be consulted for details. Similar reports are available for peas grown in Eire (Ryan, 1967, 1968, 1971) and France (Cousin, 1974).

More than six hundred lines of the John Innes *Pisum* Germplasm Collection have been screened for resistance using an oospore inoculum obtained from Ireland (Matthews and Dow, 1971, 1976). Reactions were assessed on a 1–5 scale, only symptomless lines (rated 1) were regarded as being resistant; 47 lines were of this type. The majority were round-seeded, including a number of commercially available dried peas. There were few resistant vining peas, exceptions being Sharpes 777 and Midfreezer, otherwise the principal sources of resistance are to be found in accessions of *P. arvense*, *P. abyssinicum*, *P. humile* and *P. thebaicum* and in primitive cultivars from Afghanistan. In a Russian survey of 85 varieties high resistance was found in Laga and Puali (Kirik *et al.*, 1975).

Resistance in JI 85 (Lamprecht 1402—Afghanistan) is determined by a single dominant gene and in JI 411 (Cobri) by two recessive genes, in other lines resistance is polygenically determined (Matthews and Dow, 1971).

Cousin (1974) reported the high resistance found in Clause 50, Clauserva, Cobri, Starcovert, and Starnairn to be determined by the same recessive gene, associated with which are genes determining horizontal resistance.

Resistance, derived from lines carrying all known genetic types is being introduced into a leafless pea breeding programme at the John Innes Institute (Dow, 1978, 1979). Six resistant lines have been used: JI 85—Lamprecht 1402; JI 114—Hangildin 5618; JI 393—Sharpes 777; JI 423—Dik Trom; JI 470—Zem; JI 493—Neptune. Selection is for combined resistance to both downy mildew and pod spot (*Ascochyta pisi*, race 3). Resistant lines are crossed to the best advanced leafless and semi-leafless selections to produce large F2 populations and screened first for resistance to pod spot and then for downy mildew resistance; leaflet, stipule and tendril characters are selected later. Selections are backcrossed to advanced leafless selections.

There is now clear evidence for the existence of at least 8 races in *Peronospora viciae*; 5 have been reported in Holland (Hubbeling, 1975; Ester and Gerlagh, 1979); 4 in Germany (Heydendorff and Hoffman, 1978); and 2 in Ireland (Dow, 1980). There is no known resistance to race 8 (Ester and Gerlagh, 1979). In the light of these reports breeding programmes need to be revised and new sources of resistance identified. An inoculum pool containing races 2, 4 and 5 is being used in the John Innes programme (Dow, 1980).

## D. Soybean (*Glycine max*)

Breeding for resistance to downy mildew (*Peronospora manshurica*) is not the primary aim in the development of new varieties. General breeding objectives are first and foremost to obtain high yielding varieties which are high in oil and protein content, then resistance to frost shattering, followed by disease, nematode and insect resistance. Breeding relies on pure line (pedigree) selection, bulk populations, and especially back-crossing; objectives and breeding methods are discussed and described in detail by Brim (1973).

The soybean crop is genetically very vulnerable, resting on a very narrow genetic base. Most currently grown varieties are related to 11 lines of oriental origin; in particular to Mandarin, which occurs in the pedigrees of 84% of varieties grown in the northern and central soya growing states of the US (Committee on Genetic Vulnerability of Major Crops, 1972). Breeding is complicated by the presence of numerous races; 24 have so far been described (Geeseman, 1950a; Lehmann, 1953, 1958; Graber and Dunleavy, 1959; Dunleavy, 1970a, 1971; Buibago *et al.*, 1970).

In screening 72 varieties to races 2, 5, 7–18 inclusive, Mendola was found to be immune to all races, Kanrich either immune or resistant (races 2 and 9), and Kanro either immune, resistant or heterozygous in response (races 10 and 14) (Dunleavy, 1970a). From a field survey over five years of 300 lines and varieties, resistance to race 10 and other unidentified races was located in 7 accessions which were never infected: Pine Dell Perfection, PI 166140, PI 171443, PI 174885, PI 183930, PI 200527, PI 201422 (Dunleavy and Hartwig, 1970; Hartwig, 1973). Glasshouse tests showed these were either immune or resistant to races 2, 8, 10, 12, 13, 16, 17, 18, 19. In addition Mendola and Kanrich were apparently resistant to all known races.

Resistance to three races in, Chief, Dunfield and Mukdeen is determined by a single partially dominant major gene (Geeseman, 1950b). In other crosses, genes of complementary effect ($M_1$ and $M_2$) were postulated resulting in progenies which segregated for resistance. A third factor $Mi_R$ hypostatic to $Mi_1$ and $Mi_2$ was identified in the variety Richland.

Kanrich resistance has been used in breeding since 1962 by Dr C. F. Williams, Colombia, Missouri; it is being transferred by back-crossing. This material was used by Bernard and Cremeens (1971) who selected from back-cross populations of Chippewa and Wayne. Initial selection was based on screening with mixed inoculum of races 1 and 2 (Dunleavy, 1970). Heterozygotes were intermediate in resistance. Resistance was subject to both genotypic and environmental variation. In these populations resistance behaved as a dominant character. Dunleavy (1970) concluded that a single gene, *Rpm*, determined resistance. The susceptible varieties Beeson, Carsoy,

Chippewa, Harosoy, Merit and Wayne carry *Rpm*. Bernard and Cremeens (1971) identified *Rpm* in Pursell Perfection, and noted that *Rpm* had no effect on seed yield under epiphytotic conditions.

## E. Spinach (*Spinacea oleracea*)

Early attempts were made to select plants resistant to *Peronospora farinosa* from large populations of seedlings derived from existing susceptible varieties (Smith, 1950). Presumptive resistance was found in Nobel, Prickly Winter and Viroflay; however derived progenies all proved to be susceptible. Further testing of USDA plant introductions led to the identification of resistance (immunity) in PI 140467 originating from Iran. Data from F1s and back-crosses showed that immunity found in PI 140467 is determined by a single dominant gene.

Infection of Califlay, a variety carrying immunity derived from PI 140467, in California, indicated the presence of a second race (Zink and Smith, 1958). Race 2 was shown to occur in California and possibly the Netherlands. Califlay is immune to race 1 but susceptible to race 2; all USDA breeding stocks were found to be immune to both races. These breeding lines were derived from PI 140462, whereas the immunity of Califlay originates from PI 140469 (Smith *et al.*, 1962). The gene for resistance in Califlay was designated $M_1$, and $M_2$ the gene in USDA lines conferring resistance to races 1 and 2. Reimann-Philipp (1976) concluded that $M_1$ and $M_2$ are alleles of the same gene, $M_2$ being pleotropic; there was no clear evidence of any epistatic effects. He attempted to show linkage of $M_1$ and $M_2$ using trisomic analysis. Trisomics were identified carrying $M_1$ and $M_2$ resistance; and he showed that the same chromosome was involved i.e. chromosome 1, by crossing a fully susceptible tetraploid to diploid Califlay ($M_1$ donor) and diploid W. Remona ($M_2$ donor).

In Holland resistance to race 2 was screened in progenies of the cross (Wintra × Nores) × Eerste Oogst and shown to be controlled by a single dominant gene which was weakly linked to the gene for resistance to cucumber mosaic virus (Eenink, 1974). Crosses between the susceptible Eerste Oogst and Nores, resistant to races 1 and 2, showed resistance to be controlled by two closely linked dominant genes, 4·6% recombination (Eenink, 1976b). Combined resistance to races 1 and 2 lasted for 20 years until the appearance of a third race (Eenink, 1976c). Six of 237 selections/lines in the IVT spinach gene bank were found resistant to race 3, no lines were resistant to all races. Resistance genes have been re-designated $R_1$, $R_2$ and $R_3$.

The methodology for breeding F1 hybrids using X-ray induced markers is described by Handke and Reimann-Philipp (1976). Also, they report segregation of downy mildew resistance and tight linkage with cucumber mosaic virus resistance in the F2 of Virginia Savoy × Frutemona.


## F. Sugar beet (*Beta vulgaris*)

Losses due to downy mildew (*Peronospora farinosa* f, sp. *betae*) may approach 40% of sugar yield. The disease is difficult to control using fungicides so there is reliance on resistant varieties and/or avoidance of susceptible varieties (Leach, 1945; Russell, 1978). The first half of his century saw mass selection of diploids, multiples and multigerm varieties based on 20–50 stocks. Later selections were derived from polyploids produced in the late 1930s, leading to development of monogerm triploids with use of male sterility.

A survey of field resistance of many varieties in France indicated the presence of resistance in sugar beet varieties (Singalovsky, 1937). First attempts at selecting for resistance were carried out in California in the 1940s (Leach, 1945). Significant differences in resistance were detected between varieties and considerable variation in varietal response between years was noted, however three levels of susceptibility were recognized. The varieties Hartmann and Eagle Hill were significantly less susceptible. Since varieties were extremely heterozygous Leach's strategy was to select within commercially acceptable varieties. Seedling inoculation and selection was practised under glasshouse conditions; with the assumption that seedling and field resistance are correlated. Subsequent testing of resistant selections under epidemic field conditions confirmed this assumption. He also observed a correlation between a tendency to low bolting and resistance to downy mildew. Highly resistant selections were made from Hartmann and US 33.

A strategy of selecting from seedling glasshouse populations was continued by McFarlane (1952) who retested progenies of seedling selections for field resistance, and noted an apparent lack of correlation between glasshouse and field performance in selections derived from the susceptible SL 453 crossed with the moderately resistant SL 4201. The advantages of glasshouse selection were dependability and ability to screen large numbers at a low cost. He noted that resistance could be more readily attained by selecting a self-fertile, line, than in crosses or selection which could not be selfed. He was the first to suggest the possibility of developing highly resistant hybrids using cytoplasmic male sterility which had recently been reported (Owen, 1945). He also found that the original monogerm selections described by Savitsky (1950) possessed intermediate resistance.

Sugar beet breeding in UK started in 1940 at the Plant Breeding Institute, Cambridge; breeding rationale and trialling procedures are reviewed by Campbell and Russell (1965). Downy mildew was considered a minor, though potentially serious disease and beet varieties/selections were screened for resistance. Inoculation techniques were poor and there was a lack of correlation between glasshouse and field performances (Russell, 1968, 1969). Varieties were found to differ in resistance at different stages of development —although a few stocks showed similar resistance responses at different ages and in different environments (Russell, 1969). Before 1969, downy mildew resistance was not included in breeders' indices. Field trials to evaluate resistance of varieties and selections were set up at the NIAB regional trials centre at Trawscoed, Wales. These were complemented with trials of few varieties at many locations (Willey, 1969). Differences were detected between varieties, and responses varied between years.

Glasshouse tests did not usually agree with field tests, neither did seedling and mature plant responses. Russell (1969) concluded that glasshouse tests were of little value in selecting for resistance. He noted that most multigerm/ monogerm varieties currently in use had acceptable levels of resistance. On the other hand seedling selection at high temperature and high humidity in the glasshouse can be useful (Brown, 1977). With natural infection reactions varied from immunity to maximum susceptibility. Using artificial inoculations he consistently differentiated between very resistant and very susceptible seedlings, but failed to pick up useful intermediate resistance. A mixed inoculum was used to minimize selection of narrow race specific resistance. Resistance was reported to be polygenically determined.

Hypersensitive resistance was discovered in FC 63-9058, and was shown to be determined by a single dominant gene in paired crosses of FC 63-9058 with susceptible "G" and SLCU 3MS lines (Russell, 1969). He warned of resistance-breaking genes of this easily manipulated resistance if used in breeding, and advocated the use of non-specific resistance common in varieties and breeding stocks. The problems affecting selection and breeding relate to variability in the expression of resistance with environment and plant age, and the presence of other pathogens affecting resistance, complex inheritance (polygenic), and unreliability of glasshouse techniques. Selection under suitable field conditions was advocated e.g. at Trawscoed. At least six components of resistance were identified by Russell (1972):

  (i) hypersensitivity as in the inbred US line FC 63-9058, determined by a single dominant gene;
  (ii) resistance to conidial germination; suggests a relationship with variability of conidial stimulating substances;
  (iii) resistance to inoculation; possibly related to differences in sugar contents of the first two true leaves;
  (iv) resistance to growth in tissues; suggests this may either be related to growth substances or fungistatic substances;
  (v) tolerance;
  (vi) resistance to sporulation; associated with low concentrations of sugar in leaves.

He noted that polygenically controlled components of resistance (ii) to (vi) could be affected by many factors—trace elements, reduced sugar content, viruses etc., and concluded that breeding material should be tested in as many environments as possible.

It has been shown that hypersensitivity resistance can be overcome (Russell, 1972). Line FC 63-9058 was deliberately exposed to natural and artificial inoculation to encourage development of new races. Two races, 0 and 1, are now recognized; race 1 overcoming resistance in FC 63-9058 (Russell, 1978).

Selection at the Plant Breeding Institute, Cambridge, is based on exposure of seedlings to artificially induced epidemics throughout the growing season. One generation of selection is sufficient to give a significant increase in resistance. Such field resistance is more strongly expressed in polyploids (Russell, 1978).

### G. Tobacco (*Nicotiana tabacum*)

For a survey of the early development of blue mould-resistance (*Peronospora tabacina*), and a listing of some currently grown resistant cultivars see Lucas (1975). Early searches for resistance were confined to accessions of *N. tabacum* from Central and South America, a few lines of which were found to have low levels of resistance. This was not sufficient for breeding purposes (Clayton, 1945); and was linked with impaired yield and poor quality (Lucas, 1975). No spontaneous resistant mutants have been found in more than one million *N. tabacum* seedlings screened by Wark in Australia (Wuttke, 1972). The lack of useful resistance in *N. tabacum* led to a survey of other *Nicotiana* species. Sources of resistance were found in *N. amplexicaulis, cavieola, debneyi, exigua, excelsior, goodspeedii, gossei, hesperis, ingulba, longiflora, maritima, megalosiphon, occidentalis, obliqua, plumbaginifolia, repanda, rosulata, rotundifolia, simulans, suaveolens, velutina* (Smith-White *et al.*, 1936; Clayton, 1945; Hill and Mandryk, 1962; Wark, 1964; surveys summarized by Burk and Heggestad, 1966). High resistance is common in Australian species. Either immunity or high resistance was found in seedling tests of *N. cavieola, debneyi, exigua, megalosiphon, obliqua, occidentalis, repanda, rosulata,* and *simulans* (Hill and Mandryk, 1962).

Resistance is markedly affected by host vigour, and increases with age of plant (Clayton, 1945, 1968), is affected by temperature (Krober and Mass-feller, 1962), and inoculation concentration (Hill, 1966a). There is reported to be a positive correlation between resistance and nicotine content (Wittman and Barbetta, 1971).

Selection may be carried out either at the cotyledon stage or in the field. A cotyledon test has been devised which permits rapid screening and is extensively used in breeding (Schiltz and Izard, 1962). It assumes that cotyledon and mature plant resistance are correlated. When done under controlled conditions in the glasshouse, it can be completed within 21 days. Watering seedlings with $MgSO_4$ enhances the resistance reaction (Bartolucci et al., 1971). Field scanning for resistance is a prerequisite for varietal status.

Four genes inherited in a dominant polygenic manner have been reported to determine N. debneyi resistance (Dudek, 1964). Resistance may involve seven chromosomes; segregation data of N. debneyi resistance transferred to N. tabacum indicated that resistance was determined by at least three major genes; in addition a major gene from N. tabacum enabled four levels of resistance to be identified (Clayton, 1968).

In a 7 × 7 diallel containing four resistant and three susceptible varieties, debneyi resistance evaluated at the seedling stage, behaved as a dominant (hypersensitive) character with a predominance of additive variation, in this study resistance was located on three separate chromosomes (Dean, 1974). In Hicks breeding lines debneyi resistance behaves as a monogenic dominant (Lea, 1961; Sato, 1971). In some selections resistance may be enhanced by one or more factors (Sficas, 1966; Schweppenhauser, 1974). Resistance in N. goodspeedii is apparently determined by a single dominant gene (Lucas, 1975). Resistance in N. tabacum is determined by a major gene plus a number of minor genes (Wuttke, 1969).

No blue mould resistant varieties were available prior to 1960. The development of resistant varieties has involved the transfer of resistance found in Australian species, in particular N. debneyi, by interspecific crossing (Lea, 1961; Clayton et al., 1967a, b, c, 1968; Wark, 1970; Wuttke, 1972; Lucas, 1975; Evans, 1976; Wark et al., 1976). The transfer of resistance from Nicotiana species is usually achieved by the production of the sesquidiploid either by direct or indirect means. The least successful method is indirect, and produces the F1 interspecific hybrid which is then used as the female parent in a backcross with N. tabacum. The N. debneyi × N. tabacum hybrid, and interspecific crosses involving Australian species are sterile; hence the necessity for making the amphiploid. It is more usual to convert the sterile F1 to the fertile allopolyploid and produce the sesquidiploid by back crossing the alloploid with N. tabacum as the recurrent parent to obtain quality. Quite often the alloploid of distantly related species is stable and can be maintained by seed. The direct production of the sesquidiploid is more efficient and is achieved by the cross 4n (N. tabacum) × N. species (Chaplin and Mann, 1961). Here the hybrid cannot be maintained via the seed. In order to avoid cytoplasmic male sterile modifications on recurrent back-crossing, N. tabacum (2n and 4n) are used as female parents (Clayton, 1950; Burk, 1960). In certain crosses it may be necessary to make the cross between autotetraploid forms of both N. tabacum and the resistant species (Clayton, 1945). Where there is severe cross sterility, one may have to resort to the use of a bridging species (Burk and Dropkin, 1961).

Since 1963 there have been attempts to develop F1 hybrid tobaccos from

male sterile parents (Clayton, 1962; Lea, 1963; Wark, 1964). Worthwhile
levels of resistance and reasonable quality were obtained in F1 hybrids (Dean
*et al.*, 1968); subsequently it was concluded better to select superior pure
lines from segregating progeny rather than develop F1 hybrids (Dean,
1974).

Resistance breeding owes much to programmes started in North America
and Australia. Australian breeding started in the 1930s in New South Wales
(Wuttke, 1972). The programme eventually led to the development of
Resistant Hicks from a Canadian line of *N. debneyi* × *N. tabacum* (Lea, 1961).
Leaf quality of Resistant Hicks was unsatisfactory, but further selection led
to the development of the varieties Beerwah H. and Beerwah G.G. Subsequently attempts were made to introduce alternative forms of resistance from
a range of *Nicotiana* species; and two varieties, Sirogo and Sirone, with resistance from *N. goodspeedii*, were released (Wark, 1970). Three pathotypes,
APT1, APT2 and APT3 are recognized in Australia (Wark *et al.*, 1960;
Wuttke, 1972). Sirogo and Sirone are resistant to APT1 but not to the other
two Australian pathotypes. Programmes are now underway to select for field
resistance to APT2 by back-crossing to Canberra lines carrying resistance
from *N. debneyi*, *N. excelsior* and *N. velutina* (Wuttke, 1972). Resistant
varieties now account for nearly all Australian grown tobacco (Evans, 1976).

Both *N. exigua* and *N. glauca* have been used in Poland in the development
of resistant varieties; and mutation treatment with $^{60}CO$ has also shown some
promise (Kobus, 1975).

Rhodesia is free from blue mould and sends breeding lines for test to
countries where downy mildew is endemic. Anticipatory resistance breeding
is now under way in Rhodesian Flue Cured tobaccos using Sel. 39 (Commercial Fixed Hicks A2 line) with dominant *N. debneyi* resistance (Schweppenhauser, 1979).

There are numerous reports of blue mould strains, pathotypes, ecotypes,
biotypes; and variations in virulence and aggressiveness which must give some
concern to the breeder (Cruickshank, 1961a, b; Hill, 1963, 1966, b; Hill and
Green, 1965; Mandryk, 1966; Shepherd and Muller, 1967; Golenia, 1968;
Krober and Massfeller, 1962; Tuboly, 1966; Corbaz, 1970; Egerer, 1972;
O'Brien, 1970, 1973; Zalpov, 1970; Tosic, 1966; Legenkaya, 1971; Schweppenhauser, 1974).

## IV. *Pseudoperonospora*

### A. Cucumber (*Cucumis sativus*)

Breeding for disease resistance in cucumbers is reviewed by Sitterly (1972).
Pioneer work on breeding for downy mildew (*Pseudoperonospora cubensis*)

resistance in cucumbers stems from work at the Puerto Rico Agricultural Experiment Station (Roque and Adsuar, 1938, 1939). From more than 100 varieties screened only a Chinese variety (Chinese Long), introduced in 1933, was found to be very resistant. Oriental stocks have long been known to possess downy mildew resistance (Cochran, 1937). Using Chinese Long, two resistant strains, Puerto Rico 39 and 40, were distributed in 1940. These were used in breeding programmes in South Carolina which led to the development of Ashly Palmetto and Santee. The practical methodology of cucumber breeding has been outlined by Barnes (1947).

Resistance was reported in 80 varieties originating from India and China, the most resistant being Chinese Long and Puerto Rico 37 (Jenkins, 1942). In a recent survey of 14 cucumber lines only Poinsett and Chippa were found to be highly resistant and stable under high inoculum loads (Cohen, 1976a). Resistance is related to differences in degree of infection, rate of spread, and restricted conidial formation.

Inheritance of resistance in Chinese Long and Puerto Rico 37 in crosses exposed to natural infection, is apparently polygenic, F1s being intermediate (Jenkins, 1942, 1946). More recently resistance in the variety Poinsett has been shown to be determined by a single recessive gene ($p$) (van Vliet and Meysing, 1974). Seedling tests were carried out in the glasshouse and resistance was indicated by absence of spores. The gene $p$ was found to be linked with $D$ (dull green fruit colour) and $S$ a gene for powdery mildew resistance; $p$ and $S$ may be identical. Resistance in the Japanese variety Aoji hai is determined by a triple recessive system and is also apparently linked with dull green fruit colour (Shimizu et al., 1963). Resistance in the Indian variety Bangalore was analysed in F1, F2 and Bxs and was shown to be determined by several factors (Cochran, 1937).

Cochran (1937) reported breeding programmes in Australia, UK, Puerto Rico and US. Two types of resistance are known, yellow lesion type in Palmetto (Barnes and Epps, 1954) and brown lesion type as found in PI 197087. Crosses made in 1939 between two resistant varieties Puerto Rico 37 and Chinese Long and the susceptible varieties A and C followed by several generations of inbreeding led to lines with resistance approaching that of the resistant parents (Barnes et al., 1946). From this programme two susceptible varieties Cubit and Marketeer were introduced. Subsequently these were crossed with the resistant Puerto Rico 40; selection was based on field performance. From this programme came Palmetto, possessing yellow lesion type resistance; it is highly resistant to infection, and lesions are small with no sporangia (Barnes, 1948).

A joint project between Clemson College Truck Experiment Station and Asgrow Seed Corporation led to the pooling of breeding stocks (Barnes, 1961b). Stocks were crossed with PI 197087 (brown lesion resistance), the aim being to produce pickling cucumbers with resistance to anthracnose race

1, downy mildew, powdery mildew and cucumber mosaic virus. Yellow lesion resistance in Palmetto was shown to be controlled by a large number of genes, F1s being intermediate. Brown lesion resistance (PI 197087) was also multigenic, F1s with both yellow and brown types of resistance were found to be slightly better than intermediate. Since this programme involved selecting widely divergent characters and three or more diseases, recombination crosses were made between selected plants in F3–F5 carrying desirable complementary characters. From this programme came the slicing variety Ashley and a number of multiple disease resistant pickling varieties e.g. Chipper, Poinsett.

The occurrence of male sterility led to the development of hybrid cucumbers (Peterson and Weigle, 1958; Barnes, 1961a). In one type, derived from line 307 of complex oriental/domestic variety origins, staminate flowers abort; this condition is controlled by a cytoplasmic factor and two mendelian factors, one dominant, the other recessive. Male sterile lines, and gynomonoecious steriles were developed with good resistance to anthracnose race 1, powdery mildew and downy mildew. Subsequent breeding has led to the development of hybrid slicing and pickling cucumbers with multiple resistance to as many as 6 diseases (Barnes, 1961b, 1966). Gy3 and Gy54 (gynoecious lines) and SC10 (pollinator) have been released. PI 197087 is the source of much downy mildew resistance in this programme.

The evolution of new races of *Pseudoperonospora cubensis* was suggested by Ellis (1951) and indicated by the appearance of downy mildew in 1950 on the previously resistant Palmetto (Epps and Barnes, 1952). Two biological forms, a *V* isolate from Watermelon and *H* from cucumber and their reactions on cantaloupe melons and cucumber have been reported (Hughes and Van Haltern, 1952). Genes derived from Puerto Rican sources are apparently resistant to Israeli forms of *P. cubensis* (Cohen, 1976b).

## B. Melon (*Cucumis melo*)

Work at the Texas Experiment Station with four tolerant West Indian varieties indicated that resistance behaves as a partially dominant character (Ivanoff, 1944). Although segregation occurred in F2 families resistance was hard to assess. Downy mildew and aphid resistance were frequently associated; and they showed some correlation with leaf characters.

Resistance in cantaloupe melons may be expressed as immunity (Seminole), resistant (Edisto), moderately resistant (Georgia 47), and tolerant (Smith's Perfect); a typical susceptible variety being Hales Best (Sitterly, 1972). Both Edisto and Seminole also carry resistance to powdery mildew.

Sitterly (1972), quoting unpublished notes of Audous (1970), comments that "susceptibility is usually dominant in the F1; and that hybrids are often more susceptible than genetic parentage indicates".

## C. Squash (*Cucumis pepo*)

Interspecific hybrids have been developed using *C. lundelliana* as a bridging species with *C. moschata, C. mixta, C. pepo* and *C. maxima* to give inter-breeding populations (Rhodes, 1959). From these Sitterly (1971) produced *C. pepo* breeding lines combining resistance to downy mildew and squash mosaic virus.

The development of hybrids in cantaloupe melons and squashes has been advanced by the incorporation of gynoecy (Sitterly, 1972).

## D. Hop (*Humulus lupulus*)

Downy mildew (*Pseudoperonospora humuli*) of the hop is the most important disease limiting hop production in UK and presents serious problems in Belgium, Czechoslovakia, West Germany, Yugoslavia and more recently in East Germany, Poland and USA.

Up to date surveys of resistance breeding, selection and manipulation of the breeding system in hops are described by Royle (1977) and Neve (1977). Control in the USA is based on regulating rootstock infection; in the UK the fungus is active throughout the season. Hence breeding in Europe is for resistance in both rootstock and herbaceous parts of the plant, whilst in the USA emphasis is biased towards rootstock resistance. As the hop is dioecious there are problems in breeding; plants are highly heterozygous, and male plants produce no cones yet the crop consists of the female cones. For-tunately disease resistance can be assessed in male plants as well as a limited number of other characters; where these cannot be determined male parents have been assessed by progeny testing. Aims have been to combine high yield, high α-acid content and multiple disease resistance to *Verticillium, Pseudo-peronospora* and *Sphaerotheca*. Additional problems arise with crosses (i.e. back-crosses) which tend to reduce heterozygosity; there is a loss of vigour accompanied by the expression of highly undesirable recessive genes (multi-bines, pagoda dwarfs, miniature dwarfs, albinos, etc.). It is suggested that some of these difficulties could be overcome by developing monoecious strains; for example tetraploid *XXXY* plants where both male and female flowers are fertile.

Of the varieties grown in the 1920s only Fuggle was found to be resistant. All parts of this cultivar are resistant and much subsequent breeding is derived from it. Resistance showed signs of breaking down in the 1930s, this was attributed either to the evolution of new races, increased virulence or climatic conditions. However, since then resistance has been maintained and there is no evidence of race differentiation.

Selection relies heavily on artificial inoculations of seedlings, detached leaves, and potted rootstocks, with a final evaluation for field resistance. Resistance can be missed if selection follows maximum (optimum) disease expression. Also, seedling and detached leaf responses may not reflect mature plant response, as all are factors influencing selection efficiency. The situation may be further complicated by differential responses of organs and tissues. Little is known of the mechanisms of resistance or inheritance. In herbaceous tissues resistance is expressed after pathogen entry; whilst in the Zattler varieties, leaf resistance is expressed as incomplete hypersensitivity; inheritance appears to be polygenic. Breeding based on Fuggle was started in the US in 1957, from this programme have come diploid Cascade (resistant as rootstock, cone susceptible), and the triploids Columbia and Williamette (moderate resistance as rootstock and resistant to cone infection) (Horner, 1964). In Germany breeding started in 1926 but has utilized alternative sources of resistance in wild hops. Typical of early selections are Hüller Anfang and Hüller Stort and Hüller Festschritt; these were unsatisfactory. Later selections were better such as Hüller Bitterer and Hallertauer Gold. The best known of the more recent selections from this programme are 7K491, 2L118 and 2L183. These varieties were introduced to form the basis of resistance breeding in the UK (Dept of Hop Research, Wye College). Resistant male plants were produced from the German varieties and crossed with UK varieties, α-acid being introduced from Northern Brewer. This programme led to the release of Wye Northdown and of Wye Challenger which is virtually immune and requires no fungicidal sprays.

Three male germplasm clones GP2, GP3 and GP4, have recently been released (Horner et al., 1974). GP2 and GP4 originate from open pollinated females 2L118 and 7K491 resistant clones. They transmit resistance to Ps. humuli to a large proportion (69–84%) of their progeny in crosses with the highly susceptible variety, Yakima Cluster.

## V. Plasmopara

### A. Grape (Vitis vinifera)

Resistance to grape downy mildew (Plasmopara viticola) was first reported by Demaree et al. (1937) who screened approximately 300 varieties and species and concluded that the most likely sources of resistance were Vitis aestivalis, V. labrusca, V. lincecomii and V. riparia. European varieties (V. vinifera) were more susceptible than American. They also noted various degrees of resistance. The aestivalis-labrusca group of varieties had high resistance, lincecomii outstanding resistance; some resistance was found in vinifera varieties.

Subsequent breeding has involved intercrossing American and European varieties followed by continuous back-crossing to European varieties to transfer high resistance from the American species. Development of resistance by breeding up to 1962, particularly in Germany, was described by Husfeld (1962); who reported five resistance classes.

A survey of Portuguese varieties indicated that none were resistant, but differences in susceptibility were noted indicating possibilities for breeding and selecting varieties with "economic" levels of resistance (Coutinho, 1950). Later work utilized interspecific crossing followed by clonal selection (Coutinho, 1964). Results from 200 crosses are reported from 1945–1964. Selections were made in F1, F2 and back-crosses. Mass sowings of seedlings were exposed to natural infection followed by two rounds of inoculation followed again by exposure to natural infection. Resistance may be expressed as a hypersensitive reaction; either as "limited patch resistance", mycelial growth limited beyond necroses; or as "ring-spot resistance", in which necrotic tissues are marginal to mycelial growth. Phasic resistance may appear in leaves, and is related to leaf age and climate. Economic resistance has been found in selections showing limited mycelial development or hypersensitive resistance. Out of 110 F2 resistant clones produced and tested for production and quality, five were promising (CI.-C6; CI.-C19; CI.-C22; CI.-C67 and CI.-C76). X-rays and neutrons have been used in attempts to induce downy mildew resistant mutations (Coutinho, 1972). A few resistant plants were obtained from $10^6$ seeds treated. Resistance was of "limited patch" or "ring-spot" types. Three promising selections were obtained (Vital No. 8, Fernao Pires No. 703 and Fernao Pires No. 777).

Breeding work in Freiburg, Germany, started in 1937 (Becker and Zimmermann, 1976). The first crosses attempted to transfer resistance from *V. riparia* to Gamay (Oberlin, 595); these were unsuccessful. From 1954 onward resistance in French hybrids (Leyvre-Villard) was used in crosses with European vines. Only seedlings which showed resistance following artificial inoculation were selected. Resistant seedling hybrids derived from crosses with European vines transmitted resistance easily, provided they were used as female parents. European vines were found to provide, in various proportions, both complementary resistance genes and others. Crosses with Gewurtztraminer produced fewer resistant seedlings than *Pinot* × hybrid crosses. Interspecific crosses are characteristically robust, vigorous and resistant to both downy mildew and *Botrytis*. Wines produced from hybrids are not different in nature and in certain crosses may be better than those from European vines. Hybrids have not been selected *per se*, but because of their inherent resistance.

Different types of resistance and ways of increasing resistance were reported and discussed by Boubals (1959). Complete immunity is rare in the *Vitaceae*, only being found in *Cissus discolor*. Resistance to germ-tube penetration, apparently related to hairy or warty surfaces occurs in *V. lincecomii*, *V. piasezkii* and cultivars Pinot meunier and Berger. In the genera *Tetrastigma* and *Cissus* resistance in some species is conditioned by a hypersensitive reaction associated with stomatal and mycelial death, this form of resistance is determined by a single dominant gene. All *V. vinifera* varieties are homozygous recessive for this gene (i.e. susceptible). In other species, and interspecific hybrids, resistance is expressed as a restriction in mycelial growth in the host and is polygenically determined. There may be partial dominance of resistance or susceptibility, depending on the *vinifera* variety used. The number of genetic factors involved varies between 1 and 4. *V. riparia* was found to be the best species to cross with *V. vinifera*, giving more seedlings with higher levels of resistance in F2 than other species. Foliage of late, as compared to early maturing types, is susceptible. There is no correlation between downy mildew and powdery mildew resistance. Some variation in *Plasmopara viticola* isolates was noted by Boubals (1959).

## B. Sunflower (*Helianthus annuus*)

Sunflower downy mildew (*Plasmopara halstedii*) had a wide host range, many *Helianthus* species are susceptible (Orellana, 1970). Interspecific hybridization has been used to transfer resistance from *H. tuberosus* (2n = 102) into selected *H. annuus* (2n = 34) (Leclercq *et al.*, 1970). Aneuploid analysis showed resistance to be controlled by a dominant gene or genetic system carried by the extra chromosome, apparently derived from *H. tuberosus*. The origin of diploid resistant selections as found in HIR 34, has not been established; it is assumed that they must have arisen either by translocation or substitution. Resistance in HIR 34.1.3.4 analysed in F2 and back-crosses, is determined by a single gene (Belkov and Shopov, 1975); in Romanian material resistance was also found to be determined by a single dominant gene (Vranceanu and Stoenescu, 1970). Three resistance genes $Pl_1$, $Pl_2$ and $Pl_3$ have now been identified (Zimmer, 1974). HIR 34 originating from cv. Armavir 9343 × *H. tuberosus* carries $Pl_1$. In American material resistance is controlled by two dominant genes $Pl_2$ ($H_1$) *and* $Pl_3$ ($H_2$), the presence of either one conferring resistance (Vear and Leclercq, 1971). Both $Pl_2$ and $Pl_3$ are different from $Pl_1$ and are inherited independently of the male sterile gene *msl* and the tillering gene *b*. There is evidence of linkage between $Pl_1$ and $R_1$ resistance to race 1 *Puccinia helianthi* (Vranceanu and Stoenescu, 1970). Resistance to the US form of *Plasmopara* traces back to 953–88, a selection from a natural cross cv. Sunrise × wild *H. annuus*, which probably carried the $Pl_2$ gene (Zimmer, 1974). All three genes control resistance to the European race but $Pl_1$ and $Pl_3$ are ineffective against the North American race.

A rapid method of estimating resistance has been described in Russia where interspecific crosses are being used to transfer resistance from *H. tuberosus* (Pancenko, 1965; Pustovoit, 1966a). Assessments based on natural infection in Canada proved unreliable (Putt, 1964) but screening by hypocotyl inoculation in green lines can give reliable results (Sackston and Govern, 1966). Field tests have been used to screen varieties for resistance in Italy (Monotti and Zazzerini, 1974) and in Iran (Rahmani and Madjidiah-Gassemi, 1975).

There is clear evidence that pathogenic races of *Plasmopara halstedii* exist; many lines selected for resistance in Europe are susceptible in US; whereas all US selected lines are resistant in European trials (Zimmer, 1974). There is also evidence indicating the occurrence of pathogenic races in eastern Europe (Savulescu, 1941; Savulescu and Vanky, 1956; Yagodkina, 1956). Russian downy mildew is thought to be similar to the US Red River form since no resistance has been found in *H. annuus* (Pustovit, 1966b).

Resistance to US downy mildew (Red River race) occurs in three lines selected from 953-88-3 × Armavirsky 2497:—HA61, HA62 and HA63 (Zimmer and Kinman, 1972). Resistance to the Red River race in HA61 is controlled by $Pl_2$; also HA61 is resistant to the European race, this resistance being determined by $Pl_3$ which is ineffective against the Red River race. Resistance in HA61 and other sources is being used in crosses with cytoplasmic male sterile lines (Leclercq, 1969). The rapid development of downy mildew resistant hybrids is envisaged with the possibility of obtaining hybrids containing all three resistance genes. It is assumed that a new pathotype is less likely to arise when all three genes are used (Vear and Leclercq, 1971; Zimmer and Kinman, 1972).

## VI. *Sclerophthora* and *Sclerospora*

### A. Maize (*Zea mays*)

There are eight downy mildews which attack maize, they may also be important pathogens of other tropical *Gramineae* including millet, sorghum and sugar cane. Economically the most important diseases are brown stripe downy mildew (*Sclerophthora rayssiae*), Philippine downy mildew (*Sclerospora philippinensis*), sorghum downy mildew (*Sclerospora sorghi, S. andropogonis-sorghi, S. maydis*) and sugar cane downy mildew (*Sclerospora sacchari*). Immunity to downy mildew is unknown in maize. Resistance occurs in native varieties in Indonesia, Philippines, Taiwan and Vietnam. Resistance genes from native varieties are being incorporated into introduced varieties with high combining ability; resistant inbreds are being used in the parentage of double crosses. Mochizuki (1975) warns that new varieties may be resistant or tolerant under mild conditions but susceptible under severe conditions. Host resistance was seen by Frederiksen and Renfro (1977) as the "most efficient, effective and economical means of controlling downy mildew diseases". Little attention has been paid to the occurrence of physiological races in breeding programmes, although gross differences in virulence have been reported in *S. sorghi* (Pupipat, 1975); and the reactions of seven isolates on five varieties gave six virulence patterns which indicated the existence of physiological races in *S. philippinensis* (Titatarn and Exconde, 1974).

Maize breeding in recent years has placed emphasis on the development of improved populations (synthetics, composites), rather than on improved inbreds. Breeding is based on crosses between individuals and at the same time selfing; crossing plants to testers coupled with selfing individual plants; taking seed of the best test crosses to produce hybrids between two slightly inbred lines. The ensuing hybrids lack uniformity but development is shorter taking from two to four seasons as opposed to five.

The reader is referred to Frederiksen and Renfro (1977) for a review of tropical downy mildews affecting maize, sugar cane and millet and to Mochizuki (1975) for a review of the inheritance of host resistance in maize to the economically important downy mildews, and to Payak (1975) for a discussion of downy mildews in India.

Handoo *et al.* (1970) detected in the cross Pem (R) × Venz (S) a single dominant gene for resistance; whilst in Pem (R) × PH 3 (S) resistance was controlled by two dominant complementary genes. In all crosses other than these resistance was found to be polygenically determined, an estimated 11 genes were apparently involved.

They concluded that reciprocal recurrent selection would be the best breeding strategy (see Comstock *et al.*, 1949). The procedure they suggest would be to develop synthetics and composites by recurrent selection, and mobilize these populations to develop hybrids in order to minimize nongenetic effects. Yield, not disease resistance, would be the primary criterion in deciding which synthetic/hybrid was to be grown. Since additive genetic variation was more important, mass selection, recurrent selection, SI selection or any combination of these would be the most effective methods for building up resistance genes.

Singh and Asani (1975a, b) concluded that breeding, because of the importance of general combining ability, should be based on mass selection, recurrent selection and full sib family selection methods. They also suggested that genotype evaluation should be carried out at more than one location with plots of more than one row to minimize error.

## B. Maize—Brown stripe downy mildew (*Sclerophthora rayssiae*)

Sources of resistant germplasm to *Sclerophthora rayssiae* and *Sclerospora sacchari* are listed by Lal (1975). Breeding started in 1969 from 19 elite composites. In 1971, 22 populations were artificially inoculated and screened for resistance, not only to downy mildew but also to *Erwinia carotovora* and *Cephalosporium acremonium*. Resistant plants to all three diseases with good agronomic qualities were selfed, S1 lines taken to S3 and populations reconstructed; top crosses were made using a tester parent. In all, 102 lines were selected.

Biometrical studies of the inheritance of resistances are in broad agreement, indicating that resistance is polygenic with partial or complete dominance; additive gene action being more important than non-additive; epistasis could also be important (Asani and Bhusan, 1970; Handoo *et al.*, 1970; Singh and Asani, 1975a, b).

Asani and Bhusan (1970) noted that Cuba 24 was resistant in natural infections but susceptible under epiphytotic conditions; suggesting that this variety is resistant to foliar (natural) infections but susceptible to root infection (epiphytotic—inoculated directly into soil). They concluded that considerable genetic advances could be achieved through mass selection or bi-parental selection in various populations.

## C. Maize—Philippine downy mildew (*Sclerospora philippinensis*)

There is a wide range of resistant germplasm available (Payak, 1975; Carangal, 1975). Three major corn breeding programmes have been established: in the Philippines (Aday, 1975), the Inter Asian Corn Programme, Taiwan (Carangal, 1975) and in Thailand (Sriwatanapongse, 1975).

Gomez *et al.* (1963) showed that resistance to infection was partially dominant, that few loci were involved and that resistance was influenced by seedling vigour. Mochizuki *et al.* (1974) analysed resistance in a $9 \times 9$ F1 diallel and found that resistance was controlled by dominant genes with over dominance; considerable gene interaction was also detected. However in one inbred line genes for susceptibility were dominant. Resistance in Mexican varieties was shown to be quantitative with partial dominance and additive in inheritance (Aday, 1975), who also reported significant location × genotype, and season × genotype interactions.

Resistance to infection in local Philippine varieties has been analysed in a $13 \times 13$ F1 diallel using artificial inoculation (Yamada and Aday, 1977). Resistance behaved as a partially dominant character and they concluded that it was possible to introduce resistance genes from local varieties.

The Philippine downy mildew resistance programme started in 1953, from which have been developed six Philippine downy mildew resistant synthetics (Aday, 1975; Exconde, 1975). Between 1953 and 1963 resistant inbred lines were selected, and later, between 1964 and 1968, local varieties and introduced-germplasm were used to produce high yielding open-pollinated varieties. Local varieties Munies White Flint and MIT Sel 2, were found to be highly resistant; yellow flint varieties were not as resistant as White flint types. Between 1967 and 1973 local resistant varieties, MIT S-2, Ph 9DMR, A206 DMR, Aroman White Flint and Tiniguib were hybridized with introduced high yielding yellow flints and composites. Advanced generation hybrids were screened for resistance and the survivors from ten populations selected for improvement. Survivors were S1 progeny tested for resistance, selections chain-crossed; original and re-combined populations were then re-screened at three sites. Eight populations were selected for improvement (Philippine DMR 1, 2, 3, 4, 5, 6, 8 and 10) and bulk seed populations of these were reselected. F1 and F2 plants for seed production were grown in mildew-free conditions. Bulk F2 seeds underwent yield trials and resistance screening; survivors were crossed, sib-mated and bulked and high yield downy mildew resistant lines selected. Emphasis is now being placed on intra-population improvement in composite populations. Varieties and composites are grouped according to desirable characters and resistance crossing is practised within groups (plant-to-plant sibbing). Each group is then treated as a base population, further improvement being made through recurrent selection. Present aims are for high downy mildew resistance coupled with higher yields, combined with resistance to leaf blight, rust stalk rot, corn borer and weevil, as well as early maturity, good standing and high protein quality.

The Inter Asian Corn Programme has been responsible for the Co-operative development of Taiwan Composites No. 1 and No. 2 (Carangal, 1975).Cooperative breeding started in 1972 using Caribbean Composite (CIMMYT) as a base population in crosses with resistant Philippine inbreds (Ph 9 DMR, 2, Munies Aroman 206, MIT VAR, Tiniguib, Aroman White Flint). Equal amounts of each cross were bulked and taken through three generations with no selection for resistance. Populations were then exposed to infection in downy mildew nurseries. Survivors were self-pollinated and equal amounts of seed from each ear bulked to produce Caribbean DMR Composite. This base population was then planted in Taiwan, and downy mildew-free plants selfed. S1 lines were further evaluated in Taiwan and Thailand. The best 40 lines were selected for yield and resistance, chain crossed, and then random mated with recurrent selection on a regional basis. No. 1 was based on 42 inbreds and has good resistance in several countries. No. 2 is broader based and is derived from 92 diverse collections exposed to infection in both Taiwan and Philippines; it carries moderate resistance to downy mildew.

In Thailand the aim has been to develop downy mildew resistant Opaque-2 varieties (Sriwatanapongse, 1975). Resistance was introduced from Ph DMR 1 and P'₁ DMR 5 in crosses with seven élite populations. These were then crossed to Thai Opaque-2 Composite 1, taken to F2 when Opaque-2 types were selected. In the F4 full hard endosperm families were selected in each population, and bulked to form Thai Opaque-2 Composite No. 3. This was exposed to downy mildew in a nursery, and reciprocal full sibs taken from mildew-free plants, only ears with good hard endosperms were saved. Back-up populations were developed by inter-population selection using improved normal populations, opaque populations from CIMMYT crosses with normal downy mildew resistant varieties, and Thai Opaque-2 Composite No. 3.

Resistance to Java downy mildew (*Sclerospora philippinensis*) is polygenic with additive variation and no dominance (Hakim and Dahlan, 1974).

### D. Maize—sorghum downy mildew (*Sclerospora sorghi, Sclerospora maydis*)

In the US breeding programme at the Texas Agricultural Experiment Station varieties are screened in field nurseries and conidia are used in glasshouse tests which have been found to be cheaper and more reliable than field tests (Frederiksen *et al.*, 1973; Frederiksen and Ullstrup, 1975; Frederiksen and Renfro, 1977).

Two types of resistance are recognized, which may be to both soil-borne oospores and air-borne conidia or alternatively only to soil-borne oospores. Resistance was analysed in two sets of diallels involving Texas lines, and Texas with selected Southern lines (Frederiksen *et al.*, 1973). In the first, diallel resistance ranged from full dominance to intermediate; in the second, with two exceptions, resistance was dominant. In the two exceptions either resistance was associated with an additional susceptibility gene, or susceptibility was dominant. It was concluded that there appears to be one major locus involved, resistance being dominant in some crosses, recessive in others, with more than one pair of alleles being involved. Alternatively there is the possibility of there being two genes, one showing dominance or partial dominance and one being recessive; modifying genes may also be involved. Two studies in Thailand found resistance to be polygenically determined with additive variation or with some dominance (Jinahyon, 1973); and in Indonesia resistance to *S. maydis* is also reported to be polygenic and additive with partial dominance (Hakim and Maesum, 1973).

Breeding for resistance has been largely concentrated in the US and is based on two sets of inbreds (Frederiksen *et al.*, 1973; Frederiksen and Ullstrup, 1975).

Other breeding programmes have been established in India, Nepal and Thailand. The Indian programme is based on resistance found in Ph DMR 1 and Ph DMR 5 from which double-top-cross hybrids were produced (Payak, 1975). In Nepal resistance in local lines and varieties has been utilized (Senanarong, 1975). Breeding has involved selecting resistant lines, high yielding lines, S1 line selection, full sib selection, back crossing, varietal crosses, selection among and within inbred lines, in order to produce synthetics and composites. The Thailand programme is based on polygenic resistance found in Philippine material (Jinahyon, 1975). Breeding relies on recurrent mass selection based on S1 and full sib progenies.

### E. Maize—sugar cane downy mildew (*Sclerospora sacchari*)

Since 1969, germplasm in India has been evaluated and resistance identified (Payak, 1975; Lal, 1975). Resistance is determined by a single dominant gene, *Dmr*, located in the short arm of chromosome two (Chang and Cheng, 1968, 1969).

### F. Sorghum (*Sorghum bicolor*)

World collections have been screened for resistance to *Sclerospora sorghi* in India, Nigeria and US (Futrell and Webster, 1966; Miller *et al.*, 1968; King and Webster, 1970). Results are in general agreement. The best source of resistance is to be found in a Cafforum race; Cafforums, Milos and Hegari sorghums in general have good resistance. High levels of resistance have been identified in sweet sorghums (Zumno *et al.*, 1969). Sources of resistance are listed and related to country of origin by Wall and Ross (1970). 14% of entries in the world collection were found to carry resistance, and a high proportion (47%) found in Cafforum races; most resistant accessions were collected from South Africa.

There is evidence for both conidial and oospore resistance (Frederiksen, 1974; Nider *et al.*, 1974). Resistance is dominant, but may involve more than one gene (Frederiksen and Ullstrup, 1975). In two hybrid forage sorghums, Sudax 12a and Sx131, resistance is polygenic and partially dominant (Nider *et al.*, 1974).

Selection may be accomplished either by using artificial conidial inoculations in glasshouses, or by field screening to oospore infection. Glasshouse tests are cheaper and more reliable than field tests, and have been successfully used to identify resistance under field conditions (Jones, 1970, 1971; Craig,

1976). Resistance to one phase of the disease does not necessarily hold for other phases (Miller *et al.*, 1968).

Downy mildew is now an important disease of high yielding varieties although good sources of resistance are available, there are indications that in some varieties this is breaking down. Cytoplasmic male sterile lines widely used in the development of hybrid varieties, are highly susceptible and hybrid seed production may be impaired (Safeeula, 1977). Hybrid sorghums particularly in the US, are vulnerable because of the almost universal use of two sources of male sterility emphasizing again the necessity for maintaining genetic diversity in a breeding programme.

Breeding for resistance to downy mildew (*Sclerospora sorghi*) has been rapid; within only 10 years breeders in Texas and elsewhere developed resistant silage and grain hybrids (Wall and Ross, 1970; Rao and House, 1972; Frederiksen *et al.*, 1973; Frederiksen and Ullstrup, 1975).

## G. Pearl Millet (*Pennisetum typhoides*)

Breeding and selection of pearl millets for resistance to downy mildew (*Sclerospora graminicola*) has recently been reviewed by Nene and Singh (1976). The production of a "sick plot" i.e. adding oospore material to soil every year is considered desirable for screening germplasm and breeding material. There is evidence for resistance to both conidial and oospore infection (Frederiksen, 1974). Resistance is based on percentage plants infected though variation in severity may occur (King and Webster, 1970). There is much knowledge of sources of resistance (Appadurai *et al.*, 1975; Chahal *et al.*, 1975); although good stable sources have still to be found.

Inbred resistant lines, have been developed from the India Co-ordinated Millets Improvement Programme; typical of these are J-104, HB-5, 126 $D_2$ × J-1270, MS-628 × 7140-6, PHB-14, PHB-10, $18D_2B$, 111-B; four tolerant populations have been developed from Nigerian sources, Maiwa A, Maiwa B, Senegal Dwarf synthetic, and Cassidy Dwarf (Nene and Singh, 1976).

Inheritance of resistance in IP 1246 and IP 2287 in crosses with K560 is determined by two dominant genes (Nene and Singh, 1976); Appadurai *et al.*, (1975) report resistance determined by one or two dominant genes.

There is some evidence for race specialization based on differential responses of resistant varieties when grown in different regions (Girard, 1975; Safeeula, 1977). The vulnerability of recently introduced high yielding varieties is attributed to the unilateral use of cytoplasmic male sterility derived from Tift 23A in hybrid seed production (Safeeula, 1977).

282          P. MATTHEWS

# References

Aday, B. A. (1975). *Trop. Agric. Res. Ser.* **8**, 207–219.
Allard, C. (1970). *Annls. Phytopath.* **2**, 87–115.
Appadurai, R., Parambaramani, C. and Natarajan, U. S. (1975). *Indian J. Agric. Sci.* **45**, 179–180.
Asani, V. L. and Bhusan, B. (1970). *Indian Phytopath.* **23**, 220–230.
Bannerot, H. and Boulidard, L. (1976a). *Proc. Eucarpia Meet. Leafy Vegetables, Wageningen* 107–108.
Bannerot, H. and Boulidard, L. (1976b). *Proc. Eucarpia Meet. Leafy Vegetables, Wageningen* 86–87.
Barnes, W. C. (1947). *Proc. Amer. Soc. Hort. Sci.* **49**, 227–230.
Barnes, W. C. (1948). *Proc. Amer. Soc. Hort. Sci.* **51**, 437–441.
Barnes, W. C. (1961a). *Proc. Amer. Soc. Hort. Sci.* **77**, 415–416.
Barnes, W. C. (1961b). *Proc. Amer. Soc. Hort. Sci.* **77**, 417–423.
Barnes, W. C. (1966). *Proc. Amer. Soc. Hort. Sci.* **89**, 390–393.
Barnes, W. C., Clayton, C. N. and Jenkins, J. M. (1946). *Proc. Amer. Soc. Hort. Sci.* **47**, 357–360.
Barnes, W. C. and Epps, W. M. (1954). *Pl. Dis. Reptr.* **38**, 620.
Bartolucci, A., Becherelli, A., Fiaschi, A., Montanari, L. and Testa, F. (1971). *Tabacco. Roma* **75**, 1–8.
Becker, N. Y. and Zimmermann, H. (1976). *Wein-Wissenschaft* **4**, 238–258.
Belkov, V. and Shopov, T. (1975). *Rasteniev"dni Nauki* **12**, 103–107.
Bernard, R. L. and Cremeens, C. R. (1971). *J. Hered.* **62**, 359–362.
Bohn, G. W. and Whitaker, T. W. (1951). *U.S. Dept. Agr. Res. Serv. Circ.* **881**, pp. 1–27.
Boubals, D. (1959). *Annls. Amél. Pl.* **9**, 1–233.
Brim, C. A. (1973). *In* "Soybeans: Improvement, Production and Uses". (B. E. Caldwell, R. W. Howell, J. W. Judd and H. W. Johnson, Eds) 155–186. American Society of Agronomy, Wisconsin, US.
Brown, S. J. (1977). *Ann. Appl. Biol.* **86**, 261–266.
Buibago, G., Montes, M. R. and Orozco, S. H. (1970). *Acta Agron. Palmira* **20**, 1–7.
Burk, L. G. (1960). *J. Hered.* **51**, 27–31.
Burk, L. G. and Dropkin, V. H. (1961). *Pl. Dis. Reptr.* **45**, 734–735.
Burk, L. G. and Heggestad, H. E. (1966). *Econ. Bot.* **20**, 76–88.
Campbell, C. K. G. and Russell, G. E. (1965). *Rep. Pl. Breed. Inst. 1963–4*, 6–32.
Carangal, V. R. (1975). *Trop. Agric. Res. Ser.* **8**, 195–205.
Chahal, S. S., Gill, K. S., Phul, P. S. and Aujla, S. S. (1975). *Crop Improvement* **2**, 65–70.
Chang, S. C. and Cheng, C. E. (1968). *Rep. Corn Res. Centre* (Potsu, Chiayi, Taiwan) **6**, 1–6.
Chang, S. C. and Cheng, C. E. (1969). *Rep. Corn Res. Centre* (Potsu, Chiayi, Taiwan) **7**, 1–6.
Chaplin, J. F. and Mann, T. J. (1961). *Bull. N. Carol. Agric. Expt. Sta. Tech.* **145**.
Clayton, E. E. (1945). *J. Agric. Res.* **70**, 79–87.
Clayton, E. E. (1950). *J. Hered.* **41**, 171–175.
Clayton, E. E. (1962). *Bull. Inf. Coresta* **2**, 25–30.
Clayton, E. E. (1968). *Tob. Sci.* **12**, 112–120.
Clayton, E. E., Heggestad, H. E., Grosso, J. J. and Burk, L. G. (1967a). *Tob. Sci.* **11**, 91–102.
Clayton, E. E., Heggestad, H. E., Grosso, J. J. and Burk, L. G. (1967b). *Tob. Sci.* **11**, 102–106.
Clayton, E. E., Heggestad, H. E., Grosso, J. J. and Burk, L. G. (1967c). *Tob. Sci.* **11**, 107–112.
Clayton, E. E., Heggestad, H. E., Grosso, J. J. and Burk, L. G. (1968). *Tob. Sci.* **12**, 112–120.
Cochran, F. D. (1937). *Proc. Amer. Soc. Hort. Sci.* **35**, 541–543.
Cohen, Y. (1976a). *Phytoparasitica* **4**, 25–31.

Cohen, Y. (1976b). *Phytoparasitica* **4**, 209.

Committee on Genetic Vulnerability of Major Crops (1972). National Academy of Sciences, Washington D.C.

Comstock, R. E., Robinson, H. F. and Harvey, P. H. (1949). *Agron. J.* **41**, 360–367.

Corbaz, R. (1970). *Phytopath. Z.* **67**, 21–26.

Cousin, R. (1974). "Le Pois" Institut National de la Recherche Agronomique, Paris.

Coutinho, M. P. (1950). *Ann. Junto Nac. Vin.* **2**, 13–135.

Coutinho, M. P. (1964). *Vitis* **4**, 341–346.

Coutinho, M. P. (1972). *Genet. Iber.* **24**, 77–92.

Craig, J. (1976). *Pl. Dis. Reptr.* **60**, 350–352.

Cruickshank, I. A. M. (1961a). *Aust. J. Biol. Sci.* **14**, 45–65.

Cruickshank, I. A. M. (1961b). *Aust. J. Biol. Sci.* **14**, 198–207.

Crute, I. R. and Davis, A. A. (1976). *Ann. Appl. Biol.* **83**, 173–175.

Crute, I. R. and Dickinson, C. H. (1976). *Ann. Appl. Biol.* **82**, 433–450.

Crute, I. R. and Gregory, A. (1976). *Rep. Natn. Veg. Res. Stn.* 1975.

Crute, I. R. and Johnson, A. G. (1976a). *Ann. Appl. Biol.* **84**, 287–290.

Crute, I. R. and Johnson, A. G. (1976b). *Proc. Eucarpia Meet. Leafy Vegetables, Wageningen* 88–94.

Crute, I. R. and Johnson, A. G. (1976c). *Ann. Appl. Biol.* **83**, 125–137.

Crute, I. R. and Norwood, J. M. (1978). *Ann. Appl. Biol.* **89**, 467–474.

Dawson, R. R. (1976). *Ann. Appl. Biol.* **84**, 282–283.

Day, P. R. (1974). "Genetics of Host Parasite Interaction." Freeman, San Francisco.

Dean, C. E. (1974). *Crop Sci.* **14**, 482–484.

Dean, C. E., Heggestad, H. E. and Grosso, J. J. (1968). *Crop Sci.* **8**, 93–96.

Demaree, J. B., Dix, I. W. and Magoon, C. A. (1937). *Proc. Amer. Soc. Hort. Sci.* **35**, 451–460.

Dixon, G. R. (1976). *Ann. Appl. Biol.* **84**, 283.

Dixon, G. R. and Doodson, J. K. (1973). *Hort. Res.* **13**, 89–95.

Dow, K. P. (1978). *Rep. John Innes Institute* 1977, p. 31.

Dow, K. P. (1979). *Rep. John Innes Institute* 1978, p. 28.

Dow, K. P. (1980). *Rep. John Innes Institute* 1979, 27–28.

Dudek, M. (1964). *Centr. Lab. Przm. Tyton. Biul.* **3**, 37–45.

Dunleavy, J. M. (1970). *Crop Sci.* **10**, 507–509.

Dunleavy, J. M. (1971). *Am. J. Bot.* **58**, 209–211.

Dunleavy, J. M. and Hartwig, E. E. (1970). *Pl. Dis. Reptr.* **54**, 901–902.

Eenink, A. H. (1974). *Euphytica* **23**, 485–487.

Eenink, A. H. (1976a). *Proc. Eucarpia Meet. Leafy Vegetables, Wageningen* 78–83.

Eenink, A. H. (1976b). *Euphytica* **25**, 713–715.

Eenink, A. H. (1976c). *Proc. Eucarpia Meet. Leafy Vegetables, Wageningen* 1976, 53–54.

Egerer, A. (1972). *Ber. Inst. TabForsch.* **19**, 5–13.

Ellis, D. E. (1951). *Pl. Dis. Reptr.* **35**, 91–93.

Epps, W. M. and Barnes, W. C. (1952). *Pl. Dis. Reptr.* **36**, 14–15.

Ester, A. and Gerlagh, M. (1979). *Zaadbelangen* **33**, 146–147.

Evans, L. T. (1976). *Nature* Lond. **261**, 655–657.

Exconde, O. R. (1975). *Trop. Agric. Res. Ser.* **8**, 21–30.

Frederiksen, R. A. (1974). Annual report on development of improved high yielding Sorghum cultivars with disease and insect resistance. 175 pp. Texas A. and M. University.

Frederiksen, R. A., Bockholt, A. J., Clark, L. E., Cosper, J. W., Craig, J., Johnson, J. W., Jones, B. L., Matocha, P., Miller, F. R., Reyes, L., Rosenow, D. T., Tuleen, D. and Walker, J. H. (1973). *Texas Agric. Exp. Stn. Res. Monogr.* **2**, 1–32.

Frederiksen, R. A. and Renfro, B. L. (1977). *A. Rev. Phytopathol.* **15**, 249–275.

Frederiksen, R. A. and Ullstrup, A. J. (1975). *Trop. Agric. Res. Ser.* **8**, 39–44.

Futrell, M. C. and Webster, O. J. (1966). *Pl. Dis. Reptr.* **50**, 641–644.
Gane, A. J., King, J. M. and Gent, G. P. (1971). "Pea and Bean Growing Handbook 1971 Vol. 1—Peas." Pea Growing Research Organization, Peterborough, England.
Geeseman, G. E. (1950a). *Agron. J.* **42**, 257–258.
Geeseman, G. E. (1950b). *Agron. J.* **42**, 608–613.
Gent, G. P. (1966). *Pea Growing Research Organization. Misc. Publn.* No. 17, Peterborough, England.
Girard, J. C. (1975?). *In* "Proceedings of the consultants group meetings on downy mildew and ergot of pearl millet." (R. J. Williams, Ed.) 59–73. ICRISAT, Hyderabad, India.
Globerson, D., Netzer, D. and Tjallinghii, F. (1974). *Euphytica* **23**, 54–60.
Golenia, A. (1968). *Pr. Nauk. Inst. Ochr. RasL.* **10**, 1–103.
Gomez, A. A., Aquilizan, F. A., Payson, R. M. and Calub, A. G. (1963). *Philipp. Agric.* **47**, 113–116.
Graber, D. F. and Dunleavy, J. (1959). *Phytopathology* **49**, 791–793.
Greenhalgh, J. R. and Dickinson, C. H. (1975). *Phytopath. Z.* **84**, 131–141.
Greenhalgh, J. R. and Dickinson, C. H. (1976). *Ann. Appl. Biol.* **84**, 278–281.
Greenhalgh, J. R. and Mitchell, N. D. (1976). *New Phytol.* **77**, 391–398.
Handke, S. (1976). *Proc. Eucarpia Meet. Leafy Vegetables, Wageningen* 95–104.
Handke, S. and Reimann-Philipp, R. (1976). *Proc. Eucarpia Meet. Leafy Vegetables, Wageningen* 1976, 55–61.
Handoo, M. I., Renfro, B. L. and Payak, M. M. (1970). *Indian Phytopath.* **23**, 231–249.
Hakim, R. and Dahlan, M. (1974). *Contribution, Central Res. Inst. Agric., Indonesia* **9**, 1–9.
Hakim, R. and Maesum, D. (1973). Proc. of the Ninth Inter-Asian Corn Improvement Workshop 59–64.
Hartwig, E. E. (1973). *In* "Soybeans. Improvement, Production and Uses". (B. E. Caldwell, R. W. Howell, J. W. Judd and H. W. Johnson, Eds) 187–210. American Society of Agronomy, Wisconsin, USA.
Heydendorff, C. von and Hoffman, G. M. (1978). *Z. PflKrankh. PflPath. PflSchutz.* **85**, 561–569.
Hill, A. V. (1963). *Bull. Inf. Coresta* **3**, 6–8.
Hill, A. V. (1966a). *Bull. Inf. Coresta* **7**, 7–15.
Hill, A. V. (1966b). *Aust. J. Agric. Res.* **17**, 133–146.
Hill, A. V. and Green, S. (1965). *Aust. J. Agric. Res.* **16**, 597–615.
Hill, A. V. and Mandryk, M. (1962). *Aust. J. Exp. Agric. Anim. Husb.* **2**, 12–15.
Horner, C. E. (1964). *Rep. Hop Dis. Conf. E. Mall. Res. Sta. Wye College* 1964.
Horner, C. E., Brooks, S. N., Hannold, A. and Likens, S. T. (1974). *Crop Sci.* **14**, 341.
Hubbeling, N. (1975). *Med. Fac. Landbouww. Rijksuniv. Gent.* **40**, 539–543.
Hughes, M. B. and Van Haltern, F. (1952). *Pl. Dis. Reptr.* **36**, 365–367.
Husfeld, R. (1962). *In* "Handbuch der Pflanzenzuchtung BdVI", 721–773. Parey, Berlin.
Ivanoff, S. C. (1944). *J. Hered.* **35**, 35–39.
Jagger, I. C. (1931). *U.S. Dept. Agric. Yearbook* 1931, 348–350.
Jagger, I. C. and Whittaker, T. W. (1940). *Phytopathology* **30**, 427–433.
Jenkins, J. M. (1942). *J. Hered.* **33**, 35–38.
Jenkins, J. M. (1946). *J. Hered.* **37**, 267–271.
Jinahyon, S. (1973). *Proc. of the Ninth Inter-Asian Corn Improvement Workshop* 30–39.
Johnson, A. G., Crute, I. R. and Gordon, P. L. (1977). *Ann. Appl. Biol.* **86**, 87–103.
Johnson, A. G., Laxton, S. A., Crute, I. R., Gordon, P. L. and Norwood, J. M. (1978). *Ann. Appl. Biol.* **89**, 257–264.
Jones, B. L. (1970). *Pl. Dis. Reptr.* **54**, 603–604.
Jones, B. L. (1971). *Proc. VIIth Bien. Intern. Grain Sorghum Res. and Util. Conf.* **7**, 3–5.
Jones, H. A., Porter, D. R. and Leach, L. D. (1939). *Hilgardia* **12**, 531–550.

Jones, H. A. and Mann, L. K. (1963). "Onions and their Allies". London and New York.
King, S. B. and Webster, O. J. (1970). *Indian Phytopath.* **23**, 342–349.
Kirik, N. N., Sidorchuk, V. I. and Koshevskii, I. I. (1975). *Selektsiya i Semenovodstvo* **6**, 29–30.
Kröber, H. and Massfeller, D. (1961). *NachrBl. dt. PflSchutzdienst. Braunschweig* **13**, 81–85.
Kröber, H. and Massfeller, D. (1962). *NachrBl. dt. PflSchutzdienst. Braunschweig* **14**, 113–118.
Kobus, I. (1975). *Genet. Pol.* **16**, 1–28.
Lal, S. (1975). *Trop. Agric. Res. Ser.* **8**, 245–248.
Lea, H. W. (1961). *Bull. Inf. Coresta* **2**, 21–27.
Lea, H. W. (1963). *Bull. Inf. Coresta* **3**, 13–15.
Leach, L. D. (1945). *Hilgardia* **16**, 317–334.
Lebeda, A., Crute, I. R., Blok, I. and Norwood, J. M. (1980). *Z. Pflzucht.* **85**, 71–77.
Leclercq, P. (1969). *Annls. Amél. Pl.* **19**, 99–106.
Leclercq, P., Cauderon, Y. and Dauge, M. (1970). *Annls. Amél. Pl.* **20**, 363–373.
Legenkaya, E. I. (1971). *Rev. Pl. Path.* **50**, Abst. 3160.
Lehmann, S. G. (1953). *Phytopathology* **43**, 460–461.
Lehmann, S. G. (1958). *Phytopathology* **48**, 83–86.
Lucas, G. B. (1975). "Diseases of Tobacco" 3rd ed. Biological Consulting Associates, Raleigh, North Carolina.
Mandryk, M. (1966). *Aust. J. Agric. Res.* **17**, 39–47.
Matthews, P. and Dow, K. P. (1971). *Pisum Newsletter* **3**, 22–23.
Matthews, P. and Dow, K. P. (1976). *Ann. Appl. Biol.* **84**, 281.
McFarlane, J. S. (1952). *Proc. Am. Soc. Sug. Beet Technol.* 1952, 415–420.
Miller, F. R., Frederiksen, R. A., Ali Khan, S. T. and Rosenow, D. T. (1968). *Tex. Agric. Exp. Sta. Misc. Publ.* **890**.
Mochizuki, N. (1975). *Trop. Agric. Res. Ser.* **8**, 179–193.
Mochizuki, N., Carangal, V. R. and Aday, B. A. (1974). *JARQ* **8**, 185–187.
Monotti, M. and Zazzerini, A. (1974). *Inftore. fitopatol.* **24**, 5–14.
Natti, J. J. (1958). *Pl. Dis. Reptr.* **42**, 656–662.
Natti, J. J., Dickson, M. K. and Atkin, J. D. (1967). *Phytopathology* **57**, 144–147.
Nene, Y. L. and Singh, S. D. (1976). *PANS* **22**, 366–385.
Netzer, D. and Globerson, D. (1976). *Phytoparasitica* **4**, 210.
Netzer, D., Globerson, D. and Sacks, J. (1976). *Hort. Sci.* **11**, 612–613.
Neve, R. A. (1977). *Rep. Dep. Hop Res. Wye. Coll.* 1976, 54–59.
Nider, F., Semienchuk, P., Semienchuk, J. and Krull, C. (1974). *Revta Agron. N. E. Argent.* **11**, 93–98.
O'Brien, R. G. (1970). *Qd. J. Agric. Sci.* **27**, 137–146.
O'Brien, R. G. (1973). *Aust. Pl. Path. Soc. Newsletter* **2**, 2.
Olofsson, J. (1966). *Pl. Dis. Reptr.* **50**, 258–261.
Orellana, R. G. (1970). *Bull. Torrey Bot. Club.* **97**, 91–97.
Owen, F. V. (1945). *J. Agric. Res.* **71**, 423–440.
Pancenko, A. (1965). *Selekcija Semenovodstvo* **2**, 52–54.
Payak, M. M. (1975). *Trop. Agric. Res. Ser.* **8**, 13–19.
Peterson, C. E. and Weigle, J. L. (1958). *Bull. Mich. Agric. Exp. Sta. Quart.* **40**, 960–965.
Peyrot, J. (1973). *Coresta* **1**, 5–9.
Pupipat, U. (1975). *Trop. Agric. Res. Ser.* **8**, 63–78.
Pustovoit, G. (1966a). *In Proc. Second Int. Sunflower Conf.* 82–89, Morden, Manitoba.
Pustovoit, G. (1966b). *Genetika* **2**, 59–69.
Putt, E. E. (1964). *In Proc. First Int. Sunflower Conf.* College Station, Texas.
Rahmani, Y. and Madjidieh-Ghassemi, S. (1975). *Iran J. Pl. Path.* **11**, 96–104.
Rao, N. G. P. and House, L. R. (1972). (Eds) "Sorghum in the Seventies". Oxford and New Delhi.

Reimann-Philipp, R. (1976). *Proc. Eucarpia Meet. Leafy Vegetables, Wageningen* 1976, 62–66.
Rhodes, A. M. (1959). *Proc. Amer. Soc. Hort. Sci.* **74**, 546–552.
Robinson, R. A. (1976). "Plant Pathosystems". Springer, Berlin.
Roque, A. and Adsuar, J. (1938). *Ann. Rept. Puerto Rico Agric. Expt. Sta.* 1937–1938, 45–46.
Roque, A. and Adsuar, J. (1939). *Ann. Rept. Puerto Rico Agric. Exp. Sta.* 1938–1939, 61–62.
Royle, D. J. (1977). *Rep. Dep. Hop Res. Wye Coll.* 1976, 43–53.
Russell, G. E. (1968). *J. Agric. Sci. Camb.* **71**, 251–256.
Russell, G. E. (1969). *J. Int. Inst. Sugar Beet Res.* **4**, 1–10.
Russell, G. E. (1972). *In* "The Way Ahead in Plant Breeding". *Proc. 6th Congr. Eucarpia* 1972, 99–107.
Russell, G. E. (1978). "Plant Breeding for Pest and Disease Resistance". Butterworth, London.
Ryan, E. W. (1967). *Res. Rep. Hort. For. Div., An Foras Taluntais* 115–116.
Ryan, E. W. (1968). *Res. Rep. Hort. For. Div., An Foras Taluntais* 104–105.
Ryan, E. W. (1971). *Ir. J. Agric. Res.* **10**, 315–322.
Sackston, W. E. and Goossen, P. G. (1966). *In Proc. Second Int. Sunflower Conf.* 40–42. Morden, Manitoba.
Safeeulla, K. M. (1977). *Ann. N.Y. Acad. Sci.* **287**, 72–85.
Sato, M. (1971). *Iwata, Jap. Tab. Exp. Sta. Bull.* **3**, 21–34.
Savitsky, V. F. (1950). *Proc. 6th Gen. Meet. Am. Soc. Sugar Beet Technologists* 1950, 156–159.
Savulescu, T. (1941). *Bull. Sect. Scient. Acad. Roum.* **24**, 46–47.
Savulescu, T. and Vanky, L. (1956). *Arch. Freunde NatGesch. Mecklenb.* **2**, 236–365.
Schiltz, P. and Izard, C. (1962). *C.r. Acad. Agric. Fr.* 1962, 561–564.
Schweppenhauser, M. A. (1974). *S. Afr. J. Agric. Sci.* **70**, 349–351.
Senanarong, A. (1975). *Trop. Agric. Res. Ser.* **8**, 31–34.
Sequeira, L. and Raffray, J. B. (1971). *Phytopathology* **61**, 578–579.
Sficas, A. G. (1966). *Coresta Inform. Bull. Spec.* **3**, No. 3082.
Shepherd, C. J. and Muller, A. (1967). *Rep. Commonwl. Scient. Ind. Res. Org. Aust.* 1967, 114–116.
Shimizu, S., Kanazawa, K. and Kono, H. (1963). *Bull. Hort. Res. Sta. Hiratsuka serA.* **2**, 65–81.
Simmonds, N. W. (1976). "Evolution of Crop Plants". Longman, London.
Simmonds, N. W. (1979). "Principles of Crop Improvement". Longman, London.
Singalovsky, Z. (1937). *Annls. Epiphyt. n.s.* **3**, 551–618.
Singh, I. S. and Asani, V. L. (1975a). *Indian J. Genet. Pl. Breed.* **35**, 123–127.
Singh, I. S. and Asani, V. L. (1975b). *Indian J. Genet. Pl. Breed.* **35**, 128–130.
Sitterly, W. R. (1971). "Cucurbits. Plant Disease Control by Heredity Means". Penn. State Univ. Press.
Sitterly, W. R. (1972). *A. Rev. Phytopath.* **10**, 471–490.
Smith, P. G. (1950). *Phytopathology* **40**, 65–68.
Smith, P. G., Webb, R. E. and Luhn, C. H. (1962). *Phytopathology* **52**, 597–599.
Smith-White, S., Macindoe, S. L. and Atkinson, W. T. (1936). *J. Aust. Inst. Agric. Sci.* **2**, 26–29.
Sriwatanapongse, S. (1975). *Trop. Agric. Res. Ser.* **8**, 245–248.
Thompson, R. C., Whitaker, T. W. and Kosar, W. F. (1941). *J. Agric. Res.* **63**, 91–107.
Titatarn, S. and Exconde, O. R. (1974). *Philipp. Agric.* **58**, 90–104.
Tosic, L. (1966). *Fourth Int. Tob. Sci. Cong. Proc.* 721–732.
Tuboly, L. (1966). *Bull. Inf. Coresta* **4**, 27–34.
Tyagi, C. S., Paroda, R. S., Arora, N. D. and Singh, K. P. (1975). *Indian J. Genet. Pl. Breed.* **35**, 403–408.
van Vliet, G. J. A. and Meysing, W. D. (1974). *Euphytica* **23**, 251–255.
Van der Plank, J. E. (1978). "Genetic and Molecular Basis of Pathogenesis". Springer, Berlin.
Vear, F. and Leclercq, P. (1971). *Annls. Amél. Pl.* **21**, 251–255.
Ventura, J., Netzer, D. and Globerson, D. (1971). *J. Amer. Soc. Hort. Sci.* **96**, 103–104.

Vranceanu, V. and Stoenescu, F. (1970). *Probl. Agric.* **2**, 34–40.
Wall, J. S. and Ross, W. M. (Eds) (1970). "Sorghum Production and Utilization". Connecticut.
Wark, D. C. (1964). *Third Int. Sci. Tob. Cong. S. Rhodesia 1963* 259–269.
Wark, D. C. (1970). *Tob. Sci.* **15**, 147–150.
Wark, D. C., Hill, A. V., Mandryk, M. and Cruickshank, I. A. M. (1960). *Nature, Lond.* **187**, 710–711.
Wark, D. C., Wuttke, H. H. and Brouwer, H. M. (1976). *Tob. Int.* **178**, 127–130.
Wellving, A. and Crute, I. R. (1978). *Ann. Appl. Biol.* **89**, 251–256.
Whitaker, T. W., Bohn, G. W., Welch, J. E. and Grogan, R. G. (1958). *Proc. Amer. Hort. Soc.* **72**, 410–416.
Willey, L. A. (1969). *J. Int. Sugar Beet Res.* **4**, 175–182.
Wittman, G. and Barbetta, F. (1971). *Genet. Agric.* **25**, 153–173.
Wuttke, H. H. (1969). *Aust. J. Exp. Agr. Anim. Husb.* **9**, 545–548.
Wuttke, H. H. (1972). *Aust. Tob. Growers Bulletin* **20**, 6–10.
Yagodkina, V. P. (1956). *Bull. Nauch. Tech. Inf. Zas. Rast.* **1**, 33–34.
Yamada, M. and Aday, B. A. (1977). *Maize Genet. Coop. Newsletter* **51**, 68–70.
Zalpov, N. (1970). *Ent. Phytopath. App.* **29**, 1–5.
Zimmer, D. E. (1974). *Phytopathology* **64**, 1465–1467.
Zimmer, D. E. and Kinman, M. L. (1972). *Crop Sci.* **12**, 749–751.
Zink, F. W. (1973). *J. Amer. Soc. Hort. Sci.* **98**, 293–295.
Zink, F. W. and Smith, P. G. (1958). *Pl. Dis. Reptr.* **42**, 818.
Zinkernagel, V. (1975). *NachrBl. PflSchutzdienst.*, Stutt. **27**, 185–188.
Zumno, N., Coleman, O. H., Jones, B. L., Frederiksen, R. A. and Bockholt, A. J. (1969). *Crop Sci.* **9**, 783–784.

Chapter 13

# Control of Downy Mildews by Cultural Practices

## J. PALTI AND J. ROTEM

*Agricultural Research Organization, Bet Dagan, Israel*

## I. Introduction

Cultural control of crop diseases is largely a matter of sanitation and of manipulating the environment to the advantage of the host and to the detriment of the pathogen. Various comprehensive reviews on the subject are available, foremost among them being that by Stevens (1960), and an up-to-date survey was recently made by Palti and Rotem (1981). Essentially similar cultural practices evidently serve to obviate the effects of diseases of many kinds. We shall emphasize those most suitable for downy mildews, with due regard to (a) the characteristics of Peronosporaceae as pathogens, especially their modes and rates of reproduction, survival and transmission of inoculum, and (b) the host–pathogen relations of these mildew diseases, with special regard to the pronounced susceptibility of young tissue and to transmission by propagating material.

## II. Pathogen and Disease Characteristics Affecting the Efficacy of Cultural Control

### A. Rate of inoculum build-up

Build-up of inoculum is an integrative process dependent upon those factors which affect the various phases of the pathogen's life cycle, and subject to the genetic make-up of the species. Most, if not all, downy mildews are capable of an explosive build-up of inoculum under suitable conditions. The comparatively short moisture periods needed for infection and sporulation facilitate build-up even under conditions which would not appear particularly favourable, as will be evident from other chapters. Only under special conditions, e.g. in glasshouse culture, are some specific cultural measures (e.g. heating) able to prevent build up of inoculum entirely. In general, cultural measures can only slow down such build-up, but even this may suffice to help the crop escape extensive damage.

From the point of view of cultural control, certain characteristics of the downy mildews are of importance. For example the sporulation process is restricted to living tissue and is inhibited by necrotization, and mildews generally prefer young rather than maturing tissues for reproduction. Therefore, cultural practices aimed at inhibition of sporulation, which are crucial when there is a large amount of young, highly susceptible foliage (e.g. in nurseries), cease to be important in advanced stages of disease development, even though dense foliage in the mature crop may create a microclimate greatly favouring sporulation. As a general rule, manipulation of moisture conditions that will affect infection and sporulation of downy mildews must be expected to be more effective in climatic regimes in which moisture is a limiting factor for disease development, than in regions where moisture is abundant (Rotem and Palti, 1980).

## B. Viability of spores

Cultural practices can obviously affect fungi more readily if the spores are highly sensitive to drought than if they are more robust. Unfortunately, the susceptibility of downy mildew spores to environmental hazards has been established for only a few species. The knowledge available indicates that downy mildew spores are definitely, but not extremely, sensitive to environmental hazards. Within this group, fungi of the genus *Peronospora*, with their relatively resistant conidia (e.g. *P. tabacina* on tobacco), may be expected to react less markedly to practices aimed at reducing viability of spores, than the sporangia-bearing species of *Plasmopara* or *Pseudoperonospora* (e.g. *Pseudoperonospora cubensis* on cucumbers).

## C. Host susceptibility

The peak of susceptibility that occurs at young growth stages is of greater significance in temperate and relatively humid, than in Mediterranean or semi-arid, regions. In these hot and relatively dry areas the susceptibility of the young plants is sometimes evident only in dense stands (such as nurseries), or in fields sprinkled frequently, whereas in young, well-spaced plants the susceptibility of tissues is offset by good aeration. Continuous development of susceptible young growth renders chemical treatment difficult so that fungicides have to be applied at short intervals unless they have systemic action. This applies to numerous downy mildew hosts, such as cucumbers, vines and broccoli (Davidson *et al.*, 1962).

## III. Sanitation

Good sanitation is essential for securing disease-free sowing and planting material and involves the destruction of diseased plant material which could provide inoculum for early infections in the following season. Detailed knowledge of the methods of transmission of any given disease (e.g. by seeds or by oospores left in plant debris) is essential to determine what practises may help in controlling any particular downy mildew. Table 1 lists some major diseases in which seasonal carry-over is markedly aided by oospores.

In general, diseases caused by species of *Peronospora* (which represent the majority of downy mildew diseases) seem less likely to perennate by oospores than do those diseases caused by species of *Plasmopara*, *Bremia* or *Sclerospora*.

TABLE 1

Some major downy mildew diseases in which
oospores are involved significantly in perennation

| Downy mildew species | Crop | Reference |
|---|---|---|
| *Bremia lactucae* | lettuce | Humphreys-Jones (1971) |
| | | Dixon (1978) |
| *Plasmopara halstedii* | sunflower | Goossen and Saxton (1968) |
| *Plasmopara viticola* | vine | Anderson (1956) |
| *Peronospora farinosa* | spinach | Messiaen and Lafon (1970) |
| *Peronospora pisi* | pea | Oloffson (1966) |
| *Sclerospora graminicola* | pearl millet, *Setaria* | Shetty *et al.* (1977) |
| | | Kenneth (in this volume) |
| *Peronosclerospora sorghi* | sorghum, maize | Kenneth (in this volume) |
| | | King and Webster (1970) |
| *Sclerophthora rayssiae* | maize | Kenneth (in this volume) |
| *Sclerophthora macrospora* | rice, maize | Kenneth (in this volume) |

## A. Transmission on propagating material

Evidence for the transmission of mycelium of downy mildews in seeds has been reported for *Sclerospora graminicola* on *Pennisetum typhoides* (Shetty *et al.*, 1977) and *Peronosclerospora sorghi* on sorghum (Kenneth, see Chapter 18). Other cases of transmission by mycelium are suspected, but have not been proved. On corn seed, transmission of *P. sorghi* can be prevented by drying the seeds to 9% moisture or storing them for 40 days (Jones *et al.*, 1972). Experimental evidence to exclude mycelium transmission by seeds has been furnished for *Peronospora pisi* on peas (Hagedorn, 1974), and for *P. destructor* on onions (Rondomanski, 1971). Oospores adhering to the seeds are held to be responsible for the transmission of all mildews affecting sorghum, maize and related graminaceous crops. Whether the disease is transmitted internally in the seed, or as oospores adhering to the surface is of practical importance since internal transmission can be controlled only by heat treatment, while chemical disinfection can be used to control infection from oospores clinging to seed (Shetty *et al.*, 1977).

Downy mildews can be transmitted on infected vegetative propagating material and the most important case of transmission by seed bulbs is that of *Peronospora destructor* on onions (see Chapter 21). There is conclusive proof that such transmission in some countries may constitute the only, or at least the most important means of seasonal carry-over for this disease (Rondomanski, 1971). However, infected onion sets, kept over the winter at 27–30°C, subsequently produced healthy spring crops. Treatment with hot water (25 min at 50°C) has also been advocated to prevent transmission of

*Sclerophthora sacchari* on sugar cane sets, of *Peronospora viciae* by pea seed, and of *P. farinosa* on spinach seed (Messiaen and Lafon, 1970).

Leach (1931) has pointed out that mycelium of *Peronospora schachtii* surviving on crowns of sugar-beet stored over the winter for subsequent seed-production, infects the young leaves in spring, and that exclusion of such beets before replanting is essential.

## B. Transmission by soil and manure

The contamination of soil by oospores constitutes a major source of infection not only for succeeding crops of susceptible species in the same field (see Section VI), but also for transfer of the disease to other fields by every soil-carrying agent. Most important among the latter are wheels of tractors and of trucks driven into the field. Seasonal carry-over by soil-borne oospores (see Table 1) is of special importance in all the crops of Compositae affected by *Plasmopara halstedii* (Goossen and Saxton, 1968) and in some such crops affected by *Bremia lactucae* (Humphreys-Jones, 1971). Oospore transmission by means of manure from cattle, which had been fed sorghum infected with *Peronosclerospora sorghi*, has been reported (King and Webster, 1970).

## C. Removal of infected plants or organs, and debris management

Since most downy mildews are capable of explosive inoculum build-up from primary sources, physical removal of plants or organs that carry inoculum from season to season or display symptoms early in the season, can be of obvious value; however, this is economically feasible only where labour is cheap. Some relevant cases are described here.

The few leaves of grapevine overwintering on the canes in regions with mild winters may carry *Plasmopara viticola* into the following season, as has been observed in Mediterranean countries.

Volunteer plants, growing from seeds or vegetative organs (bulbs, corms, sets, etc.) left over in the field, constitute possible dangerous sources of downy mildew infection. This may be due to early systemic infection, e.g. of sorghum by *Peronosclerospora sorghi*, or to the fact that such plants germinate or sprout well before the main crop. These plants thus provide a source of inoculum awaiting the new crop as is the case with *Peronospora destructor* on onions (Palti *et al.*, 1972) and *P. tabacina* on tobacco (Lucas and DeLoach, 1964). Destruction of such volunteer plants is not easy when they grow in other crops, but seems well worth the effort, if it can delay mildew development by a critical few weeks.

Removal of plants infected early, i.e., roguing, has been advocated for pearl millet infected by *S. graminicola*, before oospores are formed (Kenneth, 1977) and for sugar-cane infected by *Sclerophthora sacchari*. More commonly, removal of the first leaves (or other plant organs) showing infection, or of particularly infection-prone organs, has been recommended: Romanko (1965) found that removal of all basal growth of hop, up to 4 feet in height, reduced secondary infection by *Pseudoperonospora humuli*. Similarly, removal of flowers or leaves with early mildew symptoms has been mentioned by Pirone *et al.* (1960) as a means of limiting spread of species of *Plasmopara* on such flower crops as *Anemone*, *Erigeron*.

One aspect of downy mildews that has received too little attention in recent decades concerns the persistence of mycelium in senescent or dormant vegetative tissue. Melhus (1915) showed that this dormant tissue can be a source of perennation, which certainly requires attention. Among the mildews he listed as persisting in this way on crops and allied hosts, are: *Plasmopara halstedii* on *Helianthus diversicatus* (on rhizomes); *P. effusa* on *Spinacia oleracea*; *Peronospora schachtii* on *Beta vulgaris*; *P. ficariae* on *Ranunculus* spp.; and *Peronospora viciae* on *Vicia sepium*.

The destruction of overwintering vegetative sources is also considered an important aspect of sanitation to prevent early season build-up of *P. farinosa* on sugar-beets in England (Byford, personal communication). Debris management is an important means of reducing primary inoculum of all the downy mildews that form oospores which are likely to transmit disease. Ploughing to bury leaves carrying oospores as deep as possible has long been standard advice for control of *P. viticola* (Anderson, 1956), and of *P. sorghi* on sorghum (see Chapter 18); it should also contribute towards better control of *B. lactucae* on lettuce, *P. halstedii* on sunflower, and many other mildew fungi under conditions conducive to oospore formation.

Burning debris suspected of carrying oospores is a routine measure recommended for pea crops affected by *P. viciae* and for most flower crops such as *Matthiola*, *Anemone*, *Stellaria* which are subject to downy mildew attack (Pirone *et al.*, 1960).

## IV. Planning of Crops

The planning of crops to minimize losses from downy mildews includes regional measures to keep the disease out of whole farming communities at certain seasons, and on-farm measures such as choice of topography, soil type, sowing date and proximity to other crops. Crop rotation also has to be planned properly, and will be discussed in a later section.

## A. Regional planning and proximity of susceptible crops

The basic consideration here is the limited host range of all major downy mildews on the one hand, and the extremely rapid inoculum build-up of which downy mildews are capable under favourable conditions, on the other hand.

Regional planning aims at the prevention of early season outbreaks of the mildew on crops that can be sown in succession, e.g. in early spring, late spring, summer, etc., or avoids cultivation of the same crop under cover in winter and in the open in spring. Limitation of sowing to the season in which the crop can be sown most profitably can be achieved by common consent or by law. Instances in which the practical value of this approach has been demonstrated are:

(a) Tobacco: Prohibition of early winter sowing of seedbeds of the minor crop of chilli (*Capsicum annum*), which is affected by *Peronospora tabacina*, has delayed attacks of this mildew on the major tobacco crop in Greece (Zachos, 1963).

(b) Cucurbits: In Israel restriction of winter sowing of cucumbers and melons under cover in regions, in which spring sowings of these crops are important, has served to delay attacks of *Pseudoperonospora cubensis* on these crops. or even helped them to escape such attacks altogether.

Rondomanski (1966) has pointed out the advantage of keeping seed crops of onions free from *Peronospora destructor* by growing them in regions in which no commercial onion crops are grown.

Proximity planning of crops on the farm, i.e., avoidance of young crops being sown or planted near older crops likely to be infected, is of particular importance with downy mildews. Since most of these fungi are highly infective to young foliage, the inoculum from an older crop may serve to stunt growth very much in the younger crop. It follows that special care must be taken not to sow seedbeds of crops such as brassicas, tobacco and onions anywhere near sources of inoculum. As pointed out by Rotem (1978) and Aust *et al.* (1980), abundance of inoculum from nearby sources may compensate a pathogen for deficiencies in environmental conditions and may precipitate serious attacks under conditions otherwise not very favourable to the pathogen.

Seed crops kept over the winter may also constitute important sources of disease to market crops sown in spring in the region. Byford and Hull (1967) recommend keeping seed and root crops of sugar beet as far apart as possible, in order to reduce chances of attack by *P. farinosa*, and emphasize that epidemics in root crops originate most commonly in seed crops.

## B. Choice of sowing season, soil type and topography

Sowing in seasons in which the pathogen is at a disadvantage offers an easy, but not always practical, means of control. It is of value for example, where humid seasons alternate with much drier conditions, or where there are marked changes in temperature. A case in point is that of the sorghum downy mildew whereby sowing sorghum in soil with temperatures below 20°C will prevent its infection by *Peronosclerospora sorghi*.

In Israel, early autumn sowings of onions escape *Peronospora destructor*, which generally appears only in February, and early spring crops of melons often escape *Pseudoperonospora cubensis*. Conversely, delaying sowing of tobacco from November to early January has been reported to reduce attacks of *Peronospora tabacina* in Greece (Zachos, 1963), and delaying spring planting of sunflower reduced attacks of *Plasmopara halstedii* in the USA (Zimmer, 1971).

The topography of fields or plantations can be used in avoiding downy mildews, since such mildews are so sensitive to moisture conditions. Advantage is thus taken of the micrometeorological differences between even slight slopes of varying exposition: Slopes facing the sun early in the day have higher soil and leaf temperatures, lower relative humidity, and less (or no) dew than those not so exposed. Moreover, even on slopes not exposed to early insolation, dew is limited because the downward movement of cool air at night may induce dew and fog in low-lying areas, but much less so on the slopes (Rotem, 1978). It follows that slope conditions may be expected to inhibit development of downy mildews (a) which require fairly long moist periods for infection and sporulation, rather than those which infect and sporulate quickly; and (b) which disperse by sporangia sensitive to adverse conditions rather than those with more tolerant conidia. All of these diseases may benefit from slopes under conditions that are suboptimal for their development (Palti and Rotem, 1980), but much less so under conditions in which frequent rains and/or other favourable factors can cancel the micrometeorological advantages of slopes. Several examples will be presented to illustrate these points.

In the western states of the USA, serious economic damage from *Peronospora viciae* to the pea crop was restricted to the topographically lower portions of fields; the frequency of infected plants decreased as elevations increased, and there were no infections near the top of slopes (Wilson, 1971). Similarly, Rotem (1978) found tobacco plants to be severely attacked by *Peronospora tabacina* at the bottom of a valley in the Galilee of Israel, while plants on adjoining slopes were healthy. The absence of *Plasmopara viticola* from vineyards on the slopes of the Carmel mountain range in Israel also contrasts with attacks of this mildew on vines in all other parts of the coastal plain.

The type of soil on which crops are grown has an indirect effect on downy mildew attacks. Lighter soils can often be sown earlier than heavier soils, and then facilitate escape from downy mildews before the latter can multiply for massive attacks. On the other hand, irrigated crops on sandy soils require more frequent water applications, which favour the mildews especially where irrigation is by sprinkling (see below).

Careful avoidance of topographical or edaphic hazards is especially important for seedbeds and in crops grown for seed. In these cases even marked increases in the cultivation costs will be repaid by healthy seeds or transplants.

## V. Moisture Management

High moisture conditions are needed for infection and sporulation of downy mildews. However, both processes, and especially infection, require a period of free moisture which, compared with that needed by many other pathogens, is relatively short. Any practice designed to reduce the number of hours that tissues retain a water film, may inhibit mildew development.

The role of slopes in affecting disease development chiefly through shorter wet periods has already been discussed. We shall now review the management of moisture (a) coming from natural sources such as rain or dew, (b) deriving from irrigation, and (c) in crops under cover.

To predict the efficacy of different kinds of moisture management, it has to be borne in mind that under natural conditions of high humidity throughout the daytime, little can be done to reduce the risk of downy mildew disease. However, under conditions suboptimal for the development of such diseases, with days or hours of plant wetness definitely limited, proper moisture management is likely to affect development of these diseases more decisively than any other practices.

### A. Minimizing the effects of rain, dew and guttation

The aim of all operations to minimize downy mildews in conditions which produce moisture films on leaves and fruits, is to promote the circulation of air within the crop and to hasten the drying up of the soil and plant surfaces.

Basic for this purpose is strict avoidance of excessive density of foliage. This is even more important with downy mildews than with many other leaf diseases. Some important downy mildews will, under conditions of intermittent, high humidity, chiefly attack plants in crowded seedbeds, in which the proportion of very susceptible tissue is extremely high: examples are *B. lactucae* on lettuce (Messiaen and Lafon, 1970), *P. parasitica* on Brassicae,

*P. tabacina* on tobacco and especially on chilli and eggplant, and generally also *P. destructor* on onions. It is typical of some of these diseases, that infected seedlings, when transferred from the nursery and spaced out in the field, cease temporarily to develop further infection until the rows in the field close and the density of foliage again reaches the point where circulation of air is inhibited. As for the direct-seeded crops, Bohn *et al.* (1964) have pointed out that wide spacing is beneficial in musk-melons subject to attacks by *Pseudoperonospora cubensis.*

On grapevines, removal of leaves enveloping the bunches is useful for better control of *Plasmopara viticola*, both because air circulation is improved and because better coverage of the bunches is achieved when sprays are applied.

Tall structures, especially hedges or trees serving as windbreaks, certainly interfere with air circulation and also cast shadows over part of the field. The effects of such hedges in promoting development of *Phytophthora infestans* on tomatoes in Israel have been recorded (Palti and Netzer, 1963), and their effect on downy mildew development may be assumed to be similar.

Dew is often heavier on top leaves, which are not shaded by other foliage, but these leaves will also be the first to dry. Since many of the top leaves constitute young growth highly susceptible to certain downy mildews, their wetting by dew creates a dangerous combination of circumstances. Other than covering plant beds with dense nets or canopies, which is an expensive procedure, little can be done to prevent dew formation in open fields, though periods of dew can be shortened by good air circulation.

Another means of shortening dew periods in the morning is in the choice of the direction in which row crops are sown. Palti *et al.* (1972) found that the incidence of *Peronospora destructor* on onions in Israel's coastal plain was perceptibly higher in rows running from east to west—in which dew persisted longer, than in rows running north to south—where the prevalent winds blowing from the south dried the leaves earlier.

In apparent contradiction of what has already been said, continuous coverage of leaves by a film of moisture may reduce infection. Cornford (1953) has reported that maintaining wetness continuously for two days increased the proportion of sugar-beet cotyledons and leaves that resisted attack by *Peronospora schachtii.*

Guttation, induced or increased by irrigation, has been found to play a role in the development of two species of *Pseudoperonospora*. Under dewless conditions, guttation facilitated infection of *P. cubensis* on cucumbers in Israel, though irrigation applied directly to the soil reduced the rate of guttation by about one-third, as compared with sprinkler irrigation (Duv-devani *et al.*, 1946). Similarly, in California, hops irrigated by flooding became infected by *P. humuli* during those days following irrigation, when guttation was appreciable (Sonoda and Ogawa, 1972).

## B. Irrigation

Irrigation effects on downy mildews are important chiefly under moisture conditions which are suboptimal for the pathogen, but not under highly humid or extremely dry conditions.

The use of irrigation practices to minimize damage from downy mildews while the crop grows, involves choosing the irrigation technique and optimal timing of water application.

All irrigation techniques practised up to the early 1970s, involved the intermittent application of water at various intervals, with consequent marked rises and falls in the water available to the crop. This has two consequences: (a) periodic water stress, a fact less relevant to downy mildew than to some other diseases, and (b) periodic formation of flushes of young growth, i.e., the simultaneous appearance of large amounts of soft tissue particularly susceptible to downy mildew. This is particulary dangerous in the case of grapevines. It can be avoided in part by conventional irrigation at shorter intervals, but the recently introduced irrigation by trickling, in which water is steadily supplied at a low rate, prevents the formation of such flushes.

### 1. Irrigation techniques

The more important characteristics of the various irrigation techniques, from the point of view of downy mildew control, are the following:

(a) Overhead sprinkling intermittently wets all above-ground parts of the crop, providing the free moisture required for infection and sporulation. It also wets the entire soil surface, which increases dew formation (Duvdevani et al., 1946) and causes splashing, which may convey oospores from the soil to the plant. On some soils, overhead sprinkling tends to encourage shallow growth of roots, thus causing crops to need more frequent irrigation.

(b) Flooding often wets some of the lower, but not the upper, leaves and also causes some splashing. It wets most of the soil surface, but is frequently applied at longer intervals than irrigation by sprinkling or by furrows.

(c) Irrigation by furrows intermittently wets part of the soil surface (depending on distances between rows) but not the shoot, and causes little splashing.

(d) Trickle irrigation constantly wets part of the soil surface, the proportion wetted again depending on the distance between rows, but does not wet the shoot.

The results of comparative experiments to determine the effects of the various techniques on downy mildew development have been published in relation to only a few downy mildews. Avizohar-Hershenzon and Hochberg (1969) found that development of *Plasmopara viticola* on grapevine in Israel's coastal plain was much stronger in plots irrigated by sprinkling than in those under furrow irrigation. Messiaen and Lafon (1970) also stated that onion seed crops should be irrigated by means other than sprinkling to minimize attack of *P. destructor*, and Crute (1978) made the same point for lettuce grown under cover and subject to attack by *B. lactucae*. With regard to other downy mildews, although experimental evidence may be lacking, there is no doubt that sprinkling is apt to further their development, and that other means of irrigation should be practised wherever possible.

## 2. Timing of irrigation

Proper timing of irrigation involves the choice of:
  (a) The correct stage of crop growth at which to begin irrigation.
  (b) The optimal irrigation intervals.
  (c) The best time of day to apply water.
Where crops are sown in soil wetted by rain but which ripen as the rainy season draws to a close, as in many semi-arid countries, the question arises whether the benefit of irrigation in terms of increased yields outweighs its mildew-promoting effect. This problem is encountered in the Mediterranean area in spring crops of onions, melons and peas. The farmer then has to decide, on the basis of previous rainfall, soil type, and other local conditions, whether or not to irrigate, and, if so, when to start.

Choice of optimal irrigation intervals is most important on susceptible crops under sprinkler irrigation. Crops on sandy soils often have to be sprinkled twice a week, and this may add a significant number of hours of wetness and could greatly promote downy mildew development. This has been repeatedly observed in relation to *Pseudoperonospora cubensis* on cucumbers in Israel. Reducing the number of sprinklings by lengthening intervals between them has also been found to lessen attacks of *P. humuli* on hops in California (Sonoda and Ogawa, 1972). Determination of irrigation intervals which, on the one hand, are short enough to ensure an adequate water supply to the crop and prevent water stress, but, on the other hand, do not favour the disease too much, represents one of the problems only experienced farmers can solve.

The time of day at which sprinkling should be performed depends largely on the presence or absence of dew. If there is no dew, the time of sprinkling is of secondary importance, but if dew is formed, care must be taken not to create a long, uninterrupted wet period by sprinkling in the morning, contiguous with the dew period. Sprinkling during the daytime has been proved advantageous in relation to *Plasmopara viticola* on grapevine (Avizohar-Hershenson and Hochberg, 1969) and to *Pseudoperonospora cubensis* on cucumbers. However, it is practicable only where there are no strong winds during the daytime hours.

An interesting effect of daytime sprinkling has been indicated by Cohen and Rotem (1971). Since sporangia of *P. cubensis* die if they are wetted for periods not sufficient to complete zoospore release and are then exposed to dry conditions, daily application of one short sprinkling in the daytime reduced the amount of viable inoculum in a cucumber field. This approach certainly merits further study.

## C. Moisture management in crops under cover

Conditions highly favourable for downy mildew development often prevail in unheated greenhouses covered by glass or plastic sheeting, in nurseries under glass, and in all crops grown in plastic tunnels. On the other hand, all growth under cover is to some extent protected from the influx of inoculum. Thus, disease outbreaks sometimes occur in individual houses or tunnels, but not in others in close proximity.

Ventilation and heating of greenhouses is the obvious way to manage moisture conditions in greenhouse crops, for the control of downy mildews on vegetables such as lettuce and cucumbers, and on flower crops such as roses and stocks (*Matthiola*).

In vegetables grown in low plastic tunnels, or in seedbeds of vegetables, flowers or tobacco covered by plastic sheets, proper ventilation is equally fundamental, but more difficult to achieve. Pending the development of automatic devices for erecting and retracting such tunnels quickly and cheaply, proper ventilation of such crops involves a large outlay for labour, but may be indispensable. A further source of moisture in plastic tunnels is the condensation forming on the sheets and dripping on to the plants. In Israel this has been found to contribute to the development of *Pseudoperonospora cubensis* on cucumber. Chemicals have recently become available for coating the sheets and reducing water condensation.

In hot and dry regions and seasons, glasshouses are sometimes cooled by air forced into the house through moist pads. This method often, although not invariably, creates moist conditions inside the house, and may be detrimental.

## VI. Crop Rotation

In relation to downy mildews, crop rotation can be of value where oospores may persist in soil or plant debris, where perennating mycelium persist on vegetative parts, and where volunteer plants may precipitate early attacks.

The practical value of rotation for this purpose has been stressed by Zimmer (1971) in relation to *Plasmopara halstedii* on sunflowers, and by Hagedorn (1974) for *Peronospora pisi* on peas. On the other hand, Oloffsson (1966) considers rotation of pea crops over 6 years not to be effective against *P. pisi*, since its oospores probably survive on the soil for 10–15 years.

## VII. Minimizing Susceptibility of Host Tissue

As mentioned before, downy mildew fungi usually attack young, soft tissue, and the period over which such tissue is available in amounts to facilitate destructive mildew attack is limited in most crops. Anything that can be done to shorten the danger period or to reduce the amount of highly susceptible tissue present at any one time is therefore of value in checking downy mildew attack.

Mention has already been made of the importance of avoiding excessive density in seedbeds and of irrigation practices. The effect of nutrients on disease incidence also needs to be considered.

In most crops fertilizer management, in relation to downy mildews, consists basically of properly balanced nitrogen, phosphorus and potassium application to avoid excessive soft growth. This, as a rule, is of special importance during the first one-third or one-half of the crops's growth span, e.g. in brassicas, onions, peas, spinach, clover, lucerne, and tobacco.

Where systemic infection is a threat, special care must usually be taken over a relatively brief period after germination to reduce such infection by speeding up plant growth with an adequate supply of nitrogen. However, this must not be excessive since growing the highly susceptible tobacco crop, for example, on pasture soils rich in organic matter has been found to enhance greatly the danger of basal stem infection by *Peronospora tabacina* (Paddick *et al.*, 1967; Hill, 1962).

Unbalanced nutrition has been reported variously to favour downy mildew attack on various hosts. Lack of potassium or phosphorus has been claimed to predispose *Setaria* to attack by *Sclerospora graminicola* (Kenneth, 1977), and lack of potassium to increase susceptibility of cabbage to *Peronospora parasitica* (Felton and Walker, 1946).

Most work on the effect of nutrition on the susceptibility of crops to downy mildews has been carried out under laboratory conditions, or in greenhouses and more field work is required to provide a more reliable picture of these effects.

## VIII. The Role of Cultural Practices in the Integrated Control of Downy Mildews

In the quadrangle of integrated control (chemical treatments—cultural practices—biological control—resistance breeding), biological control is not yet relevant to downy mildews. Resistance breeding has succeeded in producing cultivars of reduced susceptibility (e.g. onions, melons, grapevines) but genuine immunity has rarely been achieved.

Protectant fungicides cannot be relied upon to give adequate protection against pathogens capable of explosive epidemic development where hosts are highly susceptible and when inoculum is plentiful. The advent of systemic fungicides which are effective against downy mildews may enable protection of valuable crops to be based chiefly on these fungicides—as long as their level of activity can be maintained. One way to retain their activity is to reduce the frequency of dosage of their use, by management so as to ensure that crops are less susceptible and conditions are less favourable to downy mildews. This is where resistance breeding and avoidance of factors predisposing hosts to their mildews interact with sanitation, regional measures and in-field practices (density of stand, irrigation, etc.) to produce fields in which inoculum will not build up too rapidly and which can be protected with a minimum of chemical treatments.

## References

Anderson, H. W. (1956). "Disease of Fruit Crops." McGraw-Hill Book Co. Inc., New York.

Aust, H. J., Bashi, Esther and Rotem, J. (1980). *In* "Comparative Epidemiology" (J. Palti and J. Kranz, Eds), PUDOC, Wageningen.

Avizohar-Hershenzon, Zehara and Hochberg, N. (1969). *Annls de Phytopathologie* **1**, no. hors-series, 55–56.

Bohn, G. W., Mets, K., Breece, J. R. and Whitaker, T. W. (1964). *Pl. Dis. Reptr.* **48**, 258.

Byford, W. J. and Hull, R. (1967). *Ann. Appl. Biol.* **60**, 281–296.

Cohen, Y. and Rotem, J. (1971). *Trans. Br. Mycol. Soc.* **57**, 67–74.

Cornford, C. E. (1953). *Rep. Rothamsted Exp. Stn. 1952*, 79–92.

Crute, I. R. (1978). *Agric. Res. Coun. Rev.* **5**, 1–4.

Davidson, A. E., Vaughan, E. K. and Hikida, H. R. (1962). *Pl. Dis. Reptr.* **46**, 310–311.

Dixon, G. R. (1978). *In* "Plant Disease Epidemiology" (P. R. Scott and A. Bainbridge, Eds), 71–78. Blackwells, Oxford.

Duvdevani, S., Reichert, I. and Palti, J. (1946). *Palestine J. Bot. Rehovot Series* **5**, 127–151.

Felton, M. W. and Walker, J. C. (1946). *J. Agric. Res.* **72**, 69–81.

Goossen, P. G. and Saxton, W. E. (1968). *Can. J. Bot.* **46**, 5–10.

Hagedorn, D. J. (1974). *Pl. Dis. Reptr.* **58**, 226–229.

Hill, A. V. (1962). *Aust. J. Agric. Res.* **13**, 650–661.

Humphreys-Jones, D. R. (1971). *Pl. Soil* **35**, 187–188.

Jones, B. L., Leeper, J. C. and Frederiksen, R. A. (1972). *Phytopathology* **62**, 817–819.
Kenneth, R. (1977). *In* "Diseases, Pests and Weeds in Tropical Crops" (J. Kranz, H. Schmutterer and W. Koch, Eds), 96–99. Paul Parey, Berlin.
King, S. B. and Webster, O. J. (1970). *Indian Phytopathology* **23**, 342–349.
Leach, L. D. (1931). *Hilgardia* **6**, 203–251.
Lucas, G. B. and DeLoach, L. R. (1964). *Pl. Dis. Reptr.* **48**, 524.
Melhus, I. E. (1915). *J. Agric. Res.* **5**, 59–70.
Messiaen, C. M. and Lafon. R. (1970). "Les Maladies des Plantes Maraicheres". Institut National de Recherche Agronomique, Paris.
Oloffson, J. (1966). *Pl. Dis. Reptr.* **50**, 257–261.
Paddick, R. G., Hill, A. V. and Green, S. (1967). *Aust. J. Agric. Res.* **18**, 589–600.
Palti, J. and Netzer, D. (1963). *Phytopathologia Mediterranea* **2**, 265–274.
Palti, J. and Rotem, J. (1981). *In* "Plant Pathologist's Pocketbook". 2nd edn. Comm. Mycol. Inst., Kew, UK.
Palti, J., Brosh, S., Stettiner, M. and Zilkha, M. (1972). *Phytopathologia Mediterranea* **11**, 30–36.
Pirone, P. P., Dodge, B. O. and Rickett, H. W. (1960). "Diseases and Pests of Ornamental Plants". Ronald Press Co., New York.
Romanko, R. R. (1965). *Pl. Dis. Reptr.* **49**, 247.
Rondomanski, W. (1966). *Zeszyty Problemowe Postepow Nauk Rolniczych* **60**, 271–276. (In Polish).
Rondomanski, W. (1971). *TagBer. dt. Akad. Landw Wiss. Berl.* **115–117**, 157–171.
Rotem, J. (1978). *In* "Plant Disease—An Advance Treatise" (J. G. Horsfall and E. B. Cowling, Eds), 317–337. Academic Press, New York and London.
Rotem, J. and Palti, J. (1980). *In* "Comparative Epidemiology" (J. Palti and J. Kranz, Eds). 104–116.
Shetty, H. S., Neergard, P. and Mathur, S. B. (1977). *Proc. Indian Acad. Nat. Sci.* **43**, 201–220.
Sonoda, R. H. and Ogawa, J. M. (1972). *Hilgardia* **41**, 457–473.
Stevens, R. B. (1960). *In* "Plant Pathology" Vol. III (J. G. Horsfall and A. Dimond, Eds). Academic Press, New York and London.
Wilson, V. E. (1971). *Pl. Dis. Reptr.* **55**, 730.
Zachos, D. G. (1963). *Annls Inst. Phytopathologique Benaki, N.S.* **5**, 380–390
Zimmer, D. E. (1971). *Pl. Dis Reptr.* **55**, 11–12.

Chapter 14

# Chemical Control of Downy Mildews

## F. J. SCHWINN

*Agricultural Division, Ciba-Geigy Ltd., Basel, Switzerland*

## I. Introduction

The downy mildews are included among the most devastating of plant
diseases, one of the best known examples being downy mildew of the grape
vine, caused by *Plasmopara viticola*. This disease, after its introduction from
the USA and its first appearance in France in 1878, threatened the crop
throughout Europe for a number of years and despite considerable progress
in its control it is still the major disease of the vine. Downy mildews occur
on a wide variety of other field and speciality crops such as hops, tobacco,
sunflowers, maize, sorghum, millet and lettuce and they are therefore of
primary economic significance. As a consequence, efforts to control them
have been made since the early phase of plant protection, the actual starting
point being the accidental detection of the fungicidal effect of Bordeaux
mixture by Millardet (1885). The dimensions both of the disease problems
caused by downy mildews and of their control have grown dramatically since
these early days, and it is beyond any doubt that in spite of considerable
progress in alternative methods of control, such as breeding for resistance,
chemical control is to date, the most effective and economic control measure
to protect crops from these diseases.

The market for downy mildew fungicides represents the biggest single segment of the fungicide market. In 1978, it was estimated to be of the order of US\$450 million (Ciba Geigy internal information) from a total worldwide fungicide market of about US\$16 000 million (Anonymous, 1979). Until recently, this market was completely supplied by the broad spectrum, residual, protectant fungicides, however, during the past few years, a new generation of systemic and curative fungicides has appeared and begun to penetrate this segment.

Downy mildew pathogens cause two major types of disease. Since familiarity with them is a prerequisite for understanding the problems of their chemical control, they are briefly described here.

Type 1 is characterized by a local infection on the leaves arising from pathogens surviving on plant debris in the soil; the primary infection is soil-borne, the secondary is air-borne. These pathogens have the potential to cause epidemic outbreaks. Typical examples are: *Plasmopara viticola* on grape vine, *Bremia lactucae* on lettuce, *Peronospora tabacina* on tobacco and *Pseudoperonospora cubensis* on cucumber.

Type 2 comprises diseases on shoots and/or leaves by pathogens surviving in the soil (with subsequent infection of the hypocotyl), in seeds or rootstocks; they grow systemically in the host plant and cause disease symptoms on the foliage. Typical examples are: *Plasmopara halstedii* on sunflowers, *Peronosclerospora maydis* on maize and sorghum, and *Pseudoperonospora humuli* on hop.

The following sections will discuss the various types of downy mildew fungicides based on their chemistry, starting with copper, then moving to the dithiocarbamates and the more recent residual fungicides and finally dealing with the new systemic fungicides. Since there is a wide choice of excellent review articles and book chapters which deal mainly with the various aspects of the classical residual fungicides (McCallan, 1967; Martin, 1973; Scheinpflug *et al.*, 1977; Spencer, 1977), these will be dealt with relatively briefly here, and more emphasis will be placed on the new systemic compounds.

## II. Downy Mildew Fungicides

### A. Residual fungicides

#### 1. Copper compounds

The chemistry of the inorganic copper fungicides has been described in detail by Schlör (1970) and Martin (1973).

The first copper compound used in plant protection was copper sulphate; its fungicidal activity was discovered by Prévost in 1807 (Martin, 1973), but it was not used as a foliar fungicide, until, in 1885, Millardet accidentally detected its excellent activity against downy mildew of grape vines, when used in the form of the so-called Bordeaux mixture, which is a blue gelatinous suspension made of copper sulphate and limestone in an aqueous medium. A common recipe uses 1 kg copper sulphate crystals (bluestone) and 1·25 kg hydrated lime per 1000 litres (Martin and Worthing, 1976). However, a variety of prescriptions has been developed and ready-to-use preparations have long been available. The approximate composition of Bordeaux mixture is considered to be

$$[Cu(OH)_2]_x . CaSO_4$$

The fungicidal activity on the plant surface is associated with the slow formation of water soluble copper compounds, from the Bordeaux deposit, the ultimate toxicant being the cupric ion (Martin, 1969). The details of this process, especially of the factors leading to the formation of free copper ions, are, even today, not fully understood. The persistence of the mixture is largely due to the strong adherence of the freshly prepared precipitate to the foliage; since this property deteriorates rapidly with time, the spray must be used as soon as possible after preparation. Bordeaux mixture shows excellent spreading properties and has considerable resistance against weathering.

The introduction of Bordeaux mixture can be regarded as the starting point of modern crop protection. It was rapidly accepted by growers and was their main weapon against downy mildews and other diseases for more than 40 years. Based on Horsfall's observation on the fungicidal activity of cuprous oxide ($Cu_2O$) (Horsfall, 1932), the so-called "fixed-coppers" have been applied as foliar sprays or dusts to avoid the troublesome preparation of Bordeaux mixture and to reduce the risk of adverse effects to the crop. However, their tenacity for the leaf surface is lower. These products are prepared as ready-to-use formulations from a variety of basic copper salts, mainly from copper oxychloride ($CuCl_2 . 3Cu(OH)_2 . nH_2O$). Of lesser but still of practical importance are cuprous oxide ($Cu_2O$), basic copper carbonate or malachite ($Cu(OH)_2 . CuCO_3$), Cheshunt compound ($Cu(NH_3)_2)^{2+}$, cupric hydroxide ($Cu(OH)_2$), cupric dihydrazinium sulphate ($Cu(N_2H_5)_2 . (SO_4)_2$), tribasic copper sulphate ($CuSO_4 . 3Cu(OH)_2 . nH_2O$) and oxine-copper (Corbett, 1974).

oxine-copper

Copper compounds present few hazards to man and the environment, but they may be injurious to those sensitive varieties of plants which take up copper ions too readily. Thus, the low chemotherapeutic index is a limiting factor for their use in practice. On the other hand copper has a retarding effect on vegetative growth and a hardening effect on foliage, which is regarded as beneficial especially for the grape vine crop growing at its climatic optimum.

Copper fungicides are nowadays mainly used in admixtures with organic fungicides; they are still an important chemical weapon for the control of downy mildew, particularly on grape vines in the Mediterranean countries and also other crops including hops and certain vegetables. Common rates of use are in the order of 1–2 kg of metallic copper/ha for single copper fungicides, and of 0·5 kg of metallic copper/ha in mixtures. The economic importance of copper fungicides is well reflected by a recent estimate (Anon., 1975) which allocates them 15% (US$150 million) of the world fungicide market in 1974, with a forecast of US$180 million by 1980.

The mode of action of copper fungicides has recently been discussed by Corbett (1974) and Lyr (1977a, b). The key factor is the formation of free copper ions ($Cu^{2+}$) on the plant surface through the influence of $CO_2$ from the air and organic acids excreted by the plant or the fungal spore. The copper ion penetrates into the spores where it interferes with various enzyme reactions, both by blocking SH-groups or by chelation. The high activity against Oomycetes is associated with the water solubility of copper salts and the cell wall composition of these pathogens.

## 2. Derivatives of dithiocarbamic acid

This was the first group of organic fungicides to be discovered, and even now, more than 40 years later, is still by far the most important. The compounds are entirely residual chemicals, and have a broad spectrum of activity including the downy mildews of type 1 (as defined in this chapter) as well as other classes of pathogens. In 1974, their overall worldwide use was estimated to represent a value of US$400 million (Anon., 1975). They are the most widely used group of fungicides controlling the downy mildews. The compounds in use in this context are derivatives of the main types shown below.

(1) Thiuram disulphides    (2) Dimethyldithiocarbamates (3) Ethylenebis(dithiocarbamate)s

The most important derivative of the first type is the tetramethyl derivative, known as thiram or TMTD. It was the first commercial fungicide to emerge from this chemistry and was first introduced in 1937, in Germany. Its fungicidal properties had been described by Tisdale and Williams in 1931 (McCallan, 1967).

Thiram

Even though more than 40 years old, it is still an important broad-spectrum, seed-dressing fungicide for many crops, mainly used in admixtures with other products. It is of minor importance, however, against downy mildews.

The iron and zinc salts (common names: ferbam, ziram) are the best known representatives of the second type; they were introduced in the 1930s. Ziram is still used to a limited extent mainly in mixtures with other dithiocarbamates or copper, as a foliar fungicide against downy mildews of the grape vine and certain vegetables.

Ziram

The most important group of dithiocarbamates is the third. The zinc and the manganese salts (common names zineb and maneb, respectively) from this group, and, more recently, the complex of the two, mancozeb, are dominating the foliar- and soil-applied downy mildew fungicides market. They are applied as wettable powder formulations at dose rates in the order of 1–2 kg ai ha$^{-1}$ or 150–250 g ai hl$^{-1}$.

R = Zn, Mn
Zineb, Maneb

Mancozeb

Maneb and mancozeb are also widely used as seed dressing fungicides. Later developments of this chemistry are the zineb-ethylenebis (thiuram disulphide) mixed precipitation introduced in 1963 and propineb (zinc propylenebis) (dithiocarbamate). The main advantage of the dithiocarbamates over copper is that crops can tolerate them under a wide variety of climatic conditions, and they tend to improve the appearance of the foliage of many crops. On the other hand, they are much more susceptible to wash-off by rainfall. Therefore, they have to be applied at relatively frequent intervals of 3 to 14 days.

Recently, this group has come under suspicion due to the potential toxic properties of their common metabolite, ethylenethiourea (ETU) (Innes *et al.*, 1969; Graham *et al.*, 1975). However, the topic has not yet been fully clarified.

$$\begin{array}{c} R \\ \diagdown \\ \phantom{xx} N-\overset{\overset{\displaystyle S}{\|}}{C}-S^{-} \\ \diagup \\ R \end{array}$$

The fungicidal moiety of the dithiocarbamates is suggested above. This anion interferes with fungal metabolism in different ways, mainly by reaction with metal-containing enzymes or SH-groups (see Lukens, 1971; Martin, 1973; Corbett, 1974; Lyr, 1977a, b). It is worth mentioning in this context, that dithiocarbamates exhibit the same type of toxic activity against the plant cell provided they get in contact with it. What prevents them from being phytotoxic is the fact that they are not taken up by the plant.

## 3. Triphenyltin compounds

Organotin compounds are, in general, broad spectrum biocides. Few of them are suitable for plant protection but for the sake of completeness, two members of this group, fentin hydroxide and fentin acetate, are mentioned here.

Fentin hydroxide      Fentin acetate

They exhibit a strong protective fungitoxic action, and were the first fungicides to be found to have a curative activity; however, due to their indigenous phytotoxicity, they can be used on only a few crops mainly in mixtures with other fungicides. Their only use against downy mildews is against primary infections on hops. Triphenyltin fungicides are suggested to be inhibitors of oxidative phosphorylation (Corbett, 1974).

## 4. Polyhalogenalkylthio compounds

This group of residual broad-spectrum fungicides introduced in the 1950s, is only second in importance to the dithiocarbamates. Its estimated market value was around US$340 million in 1974 (Anon., 1975), i.e. over 30% of the total fungicide market. The two main representatives are folpet and captan, the former being the more efficient at controlling downy mildew. They are used as wettable powders, either alone or in admixture, against

Folpet                    Captan

foliar downy mildews in grape vines, hops and vegetables, and are also widely used, in combination with other fungicides, as seed-dressing agents or soil fungicides.

Captafol is another compound belonging to this group; but is only used against downy mildews of hops.

Captafol

Dichlofluanid

Dichlofluanid, another analogue of the group, was introduced in 1965. It shows a broader spectrum of activity than captafol, but maintains the high efficacy against foliar downy mildews such as in grape vines. These chemicals show a favourable persistence of activity; they are well tolerated by a large variety of host plants and they can readily be mixed with most other fungicides. They are applied at dose rates of the order of 1·5 kg ha$^{-1}$ or 120–200 g ai hl$^{-1}$

The fungitoxic moieties of the polyhalogenalkylthio molecules are the —SCCl$_3$, —SCCl$_2$CHCl$_2$ and —SCCl$_2$F groups. They react with thiol groups of fungal enzymes, which are oxidized to disulphides (Lukens, 1971). This leads to an inhibition of the energy metabolism of the pathogens; thus, their action is identical with that of many other residual fungicides (Kaars Sijpesteijn, 1970).

## 5. *Phthalonitriles*

Tetrachloroisophthalonitrile, introduced in 1963, is a highly active broad-spectrum foliar fungicide. It exhibits an excellent protective activity against downy mildew of grape vines, even though, in practice, it has been only little used to date in this crop. The use rate is 1·8 kg ai ha$^{-1}$ or 150 g ai hl$^{-1}$.

Chlorothalonil

## B. Systemic fungicides

### 1. Carbamates

Since the introduction of the polyhalogenalkylthio compounds in the 1950s until a few years ago there had been no major, new, downy mildew fungicides introduced commercially. In contrast to other areas of disease control where a new generation of systemic, highly active compounds had been introduced from 1968 onwards, the classic residual fungicides remained as the only weapon against the downy mildews. Finally, after a standstill of about 20 years, signs of a new step forward in the control of downy mildews appeared in 1974 with the introduction of prothiocarb the first systemic fungicide active against Oomycetes (Bastiaansen et al., 1974). Although its main area of use, as with its somewhat weaker but better tolerated analogue propamocarb, introduced in 1978 (Pieroh et al., 1978) is the control of soil-borne pythiaceous pathogens, it is also being proposed against lettuce and hop downy mildew. Recently, it has shown promising results against tobacco black shank (Prinsloo, 1977).

$$(CH_3)_2N(CH_2)_3NH-\underset{\underset{O}{\|}}{C}-SC_2H_5 \qquad\qquad (CH_3)_2N(CH_2)_3NH-\underset{\underset{O}{\|}}{C}-OC_3H_7$$

Prothiocarb                                        Propamocarb

Even though both compounds belong to a well-known chemical group, they differ essentially from their forerunners (cf. Section II.B.2) in showing apoplastic translocation after root-application. However, for various technical and commercial reasons, their use is limited to ornamentals and, in the case of propamocarb, certain vegetables. The recommended use rates are of the order of 0·09 to 0·15% ai. They are the first downy mildew fungicides, which, due to their systemicity, can be applied as soil drenches against foliar pathogens, such as Bremia lactucae. However, their uptake through the roots is limited. This fact is reflected in their relatively weak curative effectiveness.

The activity of prothiocarb and propamocarb lasts for 3–8 weeks, depending on the pH-value of the soil. The whole molecule seems to be responsible for the fungitoxic effect of these chemicals which can be reversed by sterols (Kerkenaar and Kaars Sijpesteijn, 1977; Papavizas et al., 1978).

### 2. [(2-cyano(methoxyimino)acetyl]-3-ethylurea

This chemical, known as cymoxanil, was announced in 1976 (Serres and Carraro, 1976). Like the carbamates, it is a fungicide with specific and selective action against the Oomycetes.

$$C_2H_5NH-C-NH-C-C=NOCH_3$$
$$\quad\quad\quad \|\quad\quad\quad \|\ \ |$$
$$\quad\quad\quad O\quad\quad\quad O\ \ CN$$

cymoxanil

It is a weak protectant, losing its activity after a few days, especially in hot weather conditions. However, its curative action up to 3 days after infection is remarkable. It is recommended for use at low rates (10–12 g ai. 100 litres$^{-1}$), in combination with standard rates of protective fungicides such as folpet, copper or mancozeb. These admixtures exhibit a synergistic effect and thus improve the level and duration of protective activity (Serres and Carraro, 1976). Cymoxanil has been introduced under a number of trade marks in the European grape vine market and it is also recommended for use against hop downy mildew. Although only locally systemic after foliar application (Serres and Carraro, 1976; Roussel, 1978), the compound is readily taken up by roots following soil application (Kluge, 1978). Cymoxanil loses its activity within 3–6 days of application by a rapid breakdown in plant tissue (Douchet *et al.*, 1977). Its low stability in soil may be a reason why its use via this route is not recommended (Douchet *et al.*, 1977). Nothing has been published to date regarding the mechanism of the fungitoxic action of this fungicide.

## 3. *Metal ethyl phosphonates*

Fosetyl-aluminium has become the best known compound of this class of chemicals (Bertrand *et al.*, 1977; Williams *et al.*, 1977).

$$\left[ \begin{matrix} C_2H_5O & H \\ \diagdown & \diagup \\ & P \\ \diagup\diagup & \diagdown \\ O & O^- \end{matrix} \right]_3 Al^{+++}$$

Fosetyl-aluminium

Only recently has the fungicidal activity of this long-known chemical class been detected. This observation opened new possibilities for the control of Oomycetes. The most remarkable property of these chemicals is a strong basipetal as well as acropetal translocation in the plant (Bertrand *et al.*, 1977). Thus, fosetyl-aluminium is the first true, fully systemic commercial fungicide to become available for plant protection. The other remarkable feature is its spectrum of fungicidal activity, which includes many but not all Oomycetes (Schwinn, 1980) nor all downy mildews. For example, it is not active against *Peronospora tabacina*. On the other hand, it is active against several pathogens belonging to the Deuteromycetes such as *Phomopsis viticola*, *Guignardia bidwellii* and *Alternaria tenuis* var. *mali* (Bertrand *et al.*, 1977; Chazalet *et al.*, 1977). From this it could be inferred that the spectrum of activity of fosetyl-aluminium is to some extent determined by factors associated with the host plant.

Another surprising aspect of this chemical is its low fungitoxicity *in vitro* (Hai *et al.*, 1979; Vegh *et al.*, 1977). The ED 50 value for the inhibition of mycelial growth of various Oomycetes is in the order of 3000 $\mu$g ml$^{-1}$ depending on pH conditions. This is a remarkable contrast to the activity in the field where dose rates of about 5–15 g ai m$^{-2}$ for soil application and 200 to 300 g ai hl$^{-1}$ (2000–3000 $\mu$g ml$^{-1}$) or about 1·5 kg ai ha$^{-1}$ for foliar applications give full disease control. Thus, the compound is relatively much more active in the field. Based on these differences, Clerjean and Beyries (1977) and Molot and Beyries (1977) have speculated on the formation of a more fungitoxic degradation product in the plant which, however, has not yet been isolated.

The mode of action of fosetyl-aluminium is not yet known; however, there are indications that it may not only act directly against the pathogen, but that it may also initiate or stimulate the defence reaction of the host plant by increasing the levels of phenolic substances and phytoalexins in the tissues (Hai *et al.*, 1979).

Fosetyl-aluminium is readily translocated from treated leaves to roots and to new growth (Bertrand *et al.*, 1977; Williams *et al.*, 1977). Upward translocation gives good protection of the new shoots, though it seems that this is obtained at the expense of the concentration retained in the sprayed older leaves. In view of this and other factors fosetyl-aluminium is mainly recommended in combination with residual fungicides, such as folpet, or mancozeb. In such combinations synergistic effects have been shown, and these allow an extension of the traditional spray intervals and offer a broader spectrum of activity (Chazalet *et al.*, 1977; Lafon *et al.*, 1977; Marais and Van der Walt, 1978).

Foliar treatment against soil- or seed-borne pathogens, on the other hand, may require high dose rates; therefore, the practicability of treatment via this route still remains to be demonstrated convincingly.

## 4. Acylalanines

This is a new class of fungicides (Schwinn *et al.*, 1977a) with high *in vitro* activity (ED 50 0·1 to 1 $\mu$g ml$^{-1}$) and protective as well as curative *in vivo* efficacy against most of the pathogenic Oomycetes. The compounds are readily absorbed and show a fast acropetal translocation in the plant (Staub *et al.*, 1978a; Cohen *et al.*, 1979), but phloem transport though less pronounced, does occur (Staub *et al.*, 1978a; Zentmyer and Ohr, 1978). These features give the fungicides an unusual versatility, leading to recommendations against airborne, seed- and soil-borne pathogens, by foliar, soil or seed application. The two leading representatives are metalaxyl and furalaxyl.

Whilst these two compounds have an identical spectrum of activity, they

metalaxyl                                    furalaxyl

differ in water solubility (furalaxyl = $0.23$ g litre$^{-1}$, metalaxyl = $7.1$ g litre$^{-1}$) and crop tolerance. Furalaxyl, the less soluble of the two, is better tolerated by the plant. Since it is very safe on a wide variety of species and cultivars, it is mainly recommended for use in ornamentals (Wiertsema and Wissink, 1977). Recently, another acylalanine, ofurace, has been announced (Kaspers and Reuff, 1979).

Metalaxyl has been introduced commercially for downy mildew control on agricultural and speciality crops such as grape vines, tobacco, hops, maize, sorghum, vegetables. It is the most effective downy mildew fungicide at present in common use; the recommended dose rates are 200–300 g ai ha$^{-1}$, or 20–30 g hl$^{-1}$, respectively; they give excellent disease control even under high disease pressure. Thanks to its excellent curative activity it can be used in extended spray schedules (Schwinn et al., 1977a, b; Urech et al., 1977). Rainfall, which favours downy mildew diseases and strongly decreases the action of conventional residual fungicides, has been shown to increase the efficiency of foliar sprays of acylalanines (Staub et al., 1978a) probably by favouring penetration into leaves and green parts of the stem.

The strong systemic activity of metalaxyl offers new possibilities for the control of type 2 downy mildews (as defined in this chapter) which so far have not been accessible to chemical control (Urech et al., 1978). According to the specific disease pattern, methods for granular (Paulus et al., 1977; Darvas et al., 1978) seed-dressing (Safeeulla and Venugopal, 1978), drench (Morton et al., 1978) or soil application (Johnson et al., 1979) have been described.

On crops such as grape vines, hops and vegetables where other diseases occur simultaneously, metalaxyl, like other specific downy mildew fungicides, is recommended to be used in mixtures with other fungicides, mainly dithio-carbamates, folpet or copper (Urech et al., 1977). Such mixtures also improve the level of downy mildew control on senescing foliage, where acylalanines are less effective than on young tissue. Futhermore, they may be considered as a useful means of reducing the risk of developing resistant pathogen strains (Dekker, 1977; Delp, 1979); a risk which cannot be judged before the mode of action of any fungicide is known. It has recently been shown that *Phytophthora infestans* has the ability to develop *in vitro* resistance but as yet no published data known which report on strains with *in vivo* resistance to acylalanines (Staub *et al.*, 1979).

Metalaxyl has little effect on the pathogen on the leaf surface. It exhibits its fungitoxic activity only after ingress of the pathogen into the host tissue (Staub *et al.*, 1978b; Crute, 1978). Recent findings indicate that metalaxyl blocks the formation of secondary haustoria and mycelial growth inside the leaf, lesion formation and sporulation. (Staub *et al.*, 1980; Bruck *et al.*, 1980). In soil-borne pathogens, metalaxyl acts also against the formation of chlamydospores and sporangia (Staub and Vandepeute, 1979; Staub and Young, 1980). See note on p. 320.

## III. Appraisal of Currently Used Fungicides

The mainstay for the chemical control of downy mildews of type 1 are still the residual fungicides (see Sections II.A.1–5). These are all non-specific, broad spectrum fungitoxicants. Provided the farmer uses them according to a strict schedule, either based on calendar spray plans or on forecasting, they provide adequate protection against this type of downy mildew under average disease pressure. They therefore have great merit but under severe disease pressure, e.g. in rainfall or under conditions of high humidity, they demonstrate severe limitations, mainly because they are purely surface fungicides. Thus they are subject to weathering factors and cannot reach the pathogen once it has penetrated the host tissue. This is a particular disadvantage in the control of pathogens with the biological and epidemiological characteristics of the downy mildews. These are: short duration of generation, rapid reproduction especially under conditions of high rainfall, adaptation to long-distance wind transport, fast developing lesions. Protectant fungicides are at a particular disadvantage in preventing downy mildew infection of flowers and early fruit stages (e.g. in grape vines) especially when conditions are favourable for the disease. Once such organs are infected they cannot be cured.

A further limitation to the usefulness of the residual chemicals is that they are insufficiently active against all the downy mildews of type 2 which, due to the absence of viable alternative control methods, continuously cause severe yield losses. A typical case is that of the graminicolous downy mildews. The recent discovery of the new systemic compounds is gradually resulting in real progress towards a high standard of control of both types of downy mildew. These chemicals though differing in various characteristics are highly active, easily translocated in the plant and exhibit a clear curative action. Their environmental impact looks favourable.

One peculiarity of these fungicides is that they are highly water soluble. Whether this feature is directly linked to their high performance, remains to be studied. Since the attractiveness of the new systemic compounds is evident (Schwinn, 1980), they are quickly penetrating the market. However, they will probably not supersede the traditional fungicides which will still be needed for several reasons:

1. In areas of low disease pressure they will still represent an economic method for disease control.
2. In crops attacked simultaneously by various pathogens, they serve to supplement the narrow spectrum of the new systemic compounds.
3. When used as mixture partners, they will help to reduce or delay the risk of a build-up of resistant pathogen strains (Delp, 1979).

It is therefore evident that a wise combination of the new and the traditional compounds will offer the best service to the farmer.

## IV. Future Trends

In spite of the remarkable progress achieved with the introduction of the new downy mildew fungicides, there is still room for further improvement. The following points may illustrate this.

The potential of the new chemicals cannot be fully exploited in terms of flexibility, when they have to be used in mixtures with standard protectants in order to broaden the spectrum. Here, novel partners for the control of pathogens other than the downy mildews are needed.

There is a clear need for the improvement of application methods. Even though a chemical like metalaxyl is applied at low rates, a major part of the spray does not reach its target, i.e. the plant surface. On the other hand, full coverage of the plant surface may no longer be necessary in view of the systemicity of the products. Also the concept of using granular and seed-dressing formulations against foliar diseases needs to be investigated and explored further.

The curative action of the new compounds makes them suitable for integrated control strategies, especially when used in conjunction with predictive systems (Bruck et al., 1979; Schwinn, 1980). In these circumstances curative fungicides offer greater flexibility, in that once an infection warning has been received there is no urgency to apply them immediately. Protectants on the other hand have to be applied strictly at the time of the warning since they have no "kick-back" action. Now that these curative chemicals are at our disposal, such systems must be studied systematically. In spite of the general attractiveness of this idea there are certainly limitations, above all in the fact that such an approach requires scientific understanding by the farmer and a high standard of equipment and organization. Therefore, this whole area needs to be carefully investigated.

A further point which needs more consideration is the use of the new fungicides in combination with natural resistance of cultivars, as explained by Fry (1977). Such a concept would allow the simultaneous exploitation of different types of selection parameters on the fungal population, and thus in all probability ensure an increased longevity of both resistant cultivars and highly active fungicides. In view of the attractiveness of the new generation of downy mildew fungicides and the legitimate interest of the farmer in having them available over a long period, such concepts are certainly worthy of study.

## References

Anonymous (1975). *Fm. Chem.* **138**, (9), 45–48.
Anonymous (1979). *Fm. Chem.* **142**, (9), 61–68.
Bastiaansen, M. G., Pieroth, E. A. and Aelbers, E. (1974). *Meded. Fac. Landbouww. Gent* **39**, 1019–1025.
Bertrand, A., Ducret, J., Debourge, J. C. and Horrière, D. (1977). *Phytiatrie Phytopharm.* **26**, 3–18.
Bruck, R. I., Fry, W. E. and Apple, A. E. (1980). *Phytopathology* **70**, 597–601.
Bruck, R. I., Fry, W. E. and Mundt, C. C. (1979). *IX. Int. Congr. Pl. Prot.* Washington, USA Abstr. p. 814.
Bruin, G. C. A. (1980). PhD Thesis, Univ. Guelph, Canada, 110 pp.
Bruin, G. C. A. and Edgington, L. V. (1979). *Amer Phytopath. Soc.* N.E. Reg. Meeting, Abstr. p. 22.
Bruin, G. C. A. and Edgington, L. V. (1980). *Phytopathology* **70**, 459–460.
Chazalet, M., Crisinel, P., Horrière, D. and Thiollière, J. (1977). *Phytiatrie Phytopharm.* **26**, 41–54.
Clerjean, M. and Beyries, A. (1977). *Phytiatrie Phytopharm.* **26**, 73–83.
Cohen, Y., Reuveni, M. and Eyal, H. (1979). *Phytopathology* **69**, 645–649.
Corbett, J. R. (1974). "The Biochemical Mode of Action of Pesticides". Academic Press, London and New York.
Crute, I. R. (1978). *3rd Int. Congr. Pl. Path. Munich* Abstr. p. 383.
Darvas, J. M., Kotzé, J. M. and Toerien, J. C. (1978). Citrus and subtrop. *Fruit J.* 6–7.
Davidse, L. C. (1980). *Acta Bot. Neerl.* **29**, 216.
Davidse, L. C. (1981). *Neth. J. Pl. Path.* **87**, 17–31.
Delp, C. J. (1979). *IX. Int. Congr. Pl. Prot.* Washington, USA Abstr. p. 725.
Douchet, J. G., Absi, M., Hay, S. J. B., Muntan, L. and Villani, A. (1977). *Brit. Crop Prot. Conf.* 535–540.
Fisher. D. J. and Hayes, A. L. (1979). Long Ashton Res. Stn., Annu. Res. Rep. p. 137.
Fry, W. E. (1977). *In* "Antifungal Compounds" (M. Siegel and H. D. Sisler, Eds), **1**, 19–50. M. Dekker Inc., New York and Basel.
Graham, S. L., Davis, K. J., Hansen, W. H. and Graham, C. H. (1975). *Food Cosmet. Toxicol.* **13**, 493–499.
Hai. V. T., Bompeix, G. and Ravise, A. (1979). *C. r. Ser. D.* **228**, 1171–1174.
Horsfall, J. G. (1932). *Bull. N.Y. Agric. Exp. Sta.* **615**, 1–26.
Innes, J. R. M., Ulland, B. M., Valerio, M. G., Petrucelli, L., Hart, E. R., Pallotta, A. J., Bates, R. R., Falk, H. L., Gart, J. J., Klein, M., Mitchell, J. and Peters, J. (1969). *J. Nat. Cancer Inst.* **42**, 1101.
Johnson, G. I., Davis, R. D. and O'Brien, R. G. (1979). *Pl. Dis. Reptr.* **63**, 212–215.

Kaspers, H. and Reuff, J. (1979). *Mitt. Biol. Bundesanst.* **191**, 240.

Kerkenaar, A. (1981). *Pestic. Biochem. Physiol.* (in press).

Kerkenaar, A. and Kaars Sijpesteijn, A. (1977). *In* "Internal Therapy of Plants" (A. Fuchs, Ed.). *Neth. J. Pl. Path.* **83**, Suppl. 1, 145–152.

Kaars Sijpesteijn, A. (1970). *Wld Rev. Pest Control* **9**, 85–93.

Kluge, E. (1978). *Arch. Phytopathol.* **14**, 115–122.

Lafon, R., Bugaret, Y. and Bulit, J. (1977). *Phytiatrie Phytopharm.* **26**, 19–40.

Lukens, R. J. (1971). "Chemistry of Fungicidal Action". Springer Verlag, Berlin, Heidelberg and New York.

Lyr, H. (1977a). *In* "Antifungal Compounds" (M. R. Siegel and H. D. Sisler, Eds), Vol. **2**, 301–332. Marcel Dekker Inc., New York and Basel.

Lyr, H. (1977b). *In* "Plant disease—an advanced treatise" (J. G. Horsfall and E. B. Cowling, Eds), Vol. **1**, 239–261. Academic Press, New York and London.

McCallan, S. E. A. (1967). *In* "Fungicides—an advanced treatise" (D. C. Torgeson, Ed.), Vol. **1**, 1–37. Academic Press, New York and London.

Malathrakis, N. E. (1980). *Proc. 5th Congr. Mediterr. Phytopath. Soc.* Abstr. p. 145–146.

Marais, P. G. and Van der Welt, H. S. (1978). *Phytophylactica* **10**, 89–91.

Martin, H. (1969). *In* "Fungicides—an advanced treatise" (D. C. Torgeson, Ed.), Vol. **2**, 101–117. Academic Press, New York and London.

Martin, H. (1973). "The scientific principles of crop protection". 6th edn. E. Arnold, London.

Martin, H. and Worthing, C. R. (1976). "Insecticide and Fungicide Handbook". Blackwell Scientific Publications, Oxford.

Millardet, A. (1885). *J. Agr. Prat.* **49**, 513–516.

Molot, P. M. and Beyris, A. (1977). *Phytiatrie Phytopharm.* **26**, 63–72.

Morton, H. V., Staub, T. and Young, T. R. (1978). *3rd Int. Congr. Pl. Path. Munich* Abstr. p. 383.

Papavizas, G. C., O'Neill, N. R. and Lewis, J. A. (1978). *Phytopathology* **68**, 1667–1671.

Pappas, A. C. (1980). *Proc. 5th Congr. Mediterr. Phytopath. Soc.* Abstr. p. 146–148.

Paulus, A. O., Nelson, J., Gafney, J. and Snyder, M. (1977). *Proc. Brit. Crop Prot. Conf.* **9**, 929–935.

Pieroh, E. A., Krass, W. and Hemmen, C. (1978). *Meded. Fac. Landbouww. Gent* **43**, 933–942.

Prinsloo, G. C. (1977). *Phytophylactica* **9**, 17–18.

Raynal, G., Ravisé, A. and Bompeix, G. (1980). *Ann. Phytopathol.* **12**, 163–175.

Reuveni, M., Eyal, H. and Cohen, Y. (1980). *Plant Dis.* **64**, 1108–1109.

Roussel, C. (1978). *Phytoma* **30** (Juillet-Août, Nr. 300), 13–14.

Safeeulla, K. M. and Venugopal, M. N. (1978). *3rd Int. Congr. Pl. Path. Munich* Abstr. p. 363.

Scheinpflug, H., Schlör, H. and Widdig, A. (1977). *In* "Chemie der Pflanzenschutz- und Schädlingsbekämpfungsmittel" (R. Wegler, Ed.) **4**, 120–238. Springer Verlag, Berlin, Heidelberg and New York.

Schlör, H. (1970). *In* "Chemie der Pflanzenschutz- und Schädlingsbekämpfungsmittel" (R. Wegler, Ed.), Vol. **2**, 44–161.

Schwinn, F. J., Staub, T. and Urech, P. A. (1977a). *Meded. Fac. Landbouww. Gent* **42**, 1181–1188.

Schwinn, F. J., Staub, T. and Urech, P. A. (1977b). *Mitt. Biol. Bundesanstalt* **178**, 145–146.

Schwinn, F. J. (1980). *Proc. Brit. Crop Prot. Conf. 1979*, 791–802.

Serres, J. M. and Carraro, G. A. (1976). *Meded. Fac. Landbouww. Gent* **41**, 645–650.

Spencer, E. Y. (1977). *In* "Antifungal Compounds" (M. R. Siegel and H. D. Sisler, Eds), Vol. **1**, 1–17. Marcel Dekker Inc., New York and Basel.

Staub, T. and Schwinn, F. J. (1980). *Proc. 5th Congr. Mediterr. Phytopath. Soc.* Abstr. p. 154–157.

Staub, T. and Vendepeute, J. (1979). *Plant Physiol.* **63** (Suppl. 5), 136.

Staub, T. and Young, T. R. (1980). *Phytopathology* **70**, 797–801.

Staub, T., Dahmen, H. and Schwinn, F. J. (1978a). *Z. PflKrankh. PflSchutz* **85**, 162–168.
Staub, T., Dahmen, H. and Schwinn, F. J. (1978b). *3rd Int. Congr. Pl. Path. Munich* Abstr. p. 366.
Staub, T., Dahmen, H., Urech, P. A. and Schwinn, F. J. (1979). *Pl. Dis. Reptr.* **63**, 385–389.
Staub, T., Dahmen, H. and Schwinn, F. J. (1980). *Z. PflKr. PflPath. PflSchutz*, **87**, 83–91.
Urech, P. A., Schwinn, F. J. and Staub, T. (1977). *Proc. Br. Crop Prot. Conf.* 623–631.
Urech, P. A., Eberle, J. and Ruess, W. (1978). *3rd Int. Congr. Pl. Path. Munich* Abstr. p. 359.
Vegh, J., Baillot, F. and Roy, J. (1977). *Phytiatrie Phytopharm.* **26**, 85–95.
Ward, E. W. B., Lazarovits, G., Stoessel, P., Barrie, S. D. and Unwin, C. H. (1980). *Phytopathology* **70**, 738–740.
Williams, D. J., Beach, B. G. W., Horrière, D. and Maréchal, G. (1977). *Proc. Brit. Crop Prot. Conf.* 565–573.
Wiertsema, W. P. and Wissink, G. H. (1977). *Meded. Fac. Landbouww. Gent* **42**, 1189–1194.
Zentmeyer, G. and Ohr, H. D. (1978). *Phytopath. News* **12**, p. 142.

*Note added in proof.*

Crute (1978), Cohen *et al.* (1979) have reported necrotic reactions on metalaxyl-treated susceptible lettuce and tomato plants, respectively, which were indistinguishable from those normally observed on untreated resistant cultivars. This may be a direct effect stimulating the host's resistance mechanisms or indirect by slowing down the fungal growth sufficiently to allow resistance, even in susceptible hosts, to be expressed. Ward *et al.* (1980) showed that the phytoalexin content of compatible host races is strongly increased by metalaxyl and Fisher and Hayes (1979) showed that acylalanines have little effect on respiration, membrane permeability or cell wall formation, whereas RNA, DNA and protein synthesis were reduced. Kerkenaar, (1981 and in press), proposes interference with RNA synthesis. Since pathogens have developed resistance to several other fungicides, above all to the benzimidazoles and the pyrimidines, this aspect needs to be reviewed for the new downy mildew fungicides, too. As yet there is no indications of the development of resistant strains for prothiocarb, propamocarb, cymoxanil and fosetyl aluminium. However, for acylalanines preliminary studies had indicated a negative correlation between *in vitro* resistance and pathogenicity or sensitivity, but *in vivo* investigations in which mutagens were used (Bruin and Edgington, 1979, 1980; Bruin, 1980; Davidse, 1980, 1981) revealed *Phytophthora* strains with both *in vitro* and *in vivo* resistance and with unchanged pathogenicity. Towards the end of 1979 the first occurrences of metalaxyl failures against cucumber downy mildew due to resistant populations were reported from Israel and Crete, (Malathrakis, 1980; Pappas, 1980; Reuveni *et al.*, 1980). During 1980 a more serious development of resistance occurred in late blight of potatoes in several Western European countries, above all in Netherlands and Ireland, and in New Zealand.

In all these cases, matalaxyl had been used singly and exclusively over a longer period of time, with a tendency to curative use and/or the extension of the spray intervals beyond those recommended. It is now clear that acylalanines have an inherent resistance risk. Reduction of this risk, called for a change in the recommendations for use. Both for theoretical and practical considerations the introduction of prepack mixtures of metalaxyl with conventional fungicides has been initiated (Schwinn, 1979; Staub and Schwinn, 1980).

Chapter 15

# Downy Mildew of Brassicas

A. G. CHANNON

*The West of Scotland Agricultural College, Auchincruive, Ayr, UK*

## I. Introduction

Among the downy mildew diseases affecting vegetable crops, the one caused by the fungus *Peronospora parasitica* (Pers. ex Fr.) Fr. on horticultural and agricultural members of the genus *Brassica* must be among the commonest. It is widespread in those regions of the world where brassica crops are grown, and although attack by *P. parasitica* at the seedling stage can lead to death of the young plant, infection at a later stage in growth results in a more balanced relationship between host and parasite. In spite of this, however, the effect of the pathogen is generally one of debilitation and reduction in performance or quality of the host plant. On some plants *P. parasitica* occurs alongside *Albugo candida* (Pers. ex Hook) O. Kuntze, the cause of white blister disease, and the effects of the individual pathogens, being somewhat similar, are not easy to distinguish (Butler and Jones, 1949).

In contrast to the powdery mildew, which one tends to associate with dry conditions (Butt, 1978), downy mildew of brassicas is most prevalent during persistent cool damp weather, when high humidity and air movement favour sporulation, dissemination and infection by the causal fungus.

## II. The Disease

### A. Host species affected

*Peronospora parasitica* occurs frequently on seedlings of cabbage (*Brassica oleracea* var. *capitata*),* Brussels sprout (*B. oleracea* var. *gemmifera*) and cauliflower (*B. oleracea* var. *botrytis* sub-var. *cauliflora*), particularly on the latter when raised over winter, under glass (Moore, 1959). The foliage of adult plants of these crops is also commonly attacked and infection may spread both in the field and in storage to the heads of cabbage (Thung, 1926a; Ramsey, 1935) and to the curds of cauliflower (Jenkins, 1964; Lund and Wyatt, 1978). The sprouts of Brussels sprouts may also be infected (Moore, 1959). Additional horticultural crops likely to be attacked include radish (*Raphanus sativus*), horseradish (*Armoracia rusticana*), watercress (*Nasturtium officinale*), stock (*Matthiola incana*), wallflower (*Cheiranthus cheiri*) and aubretia (*Aubretia* sp.) (Moore, 1959). Chinese cabbage (*B. campestris* subsp. *pekinensis*) can also be severely attacked by *P. parasitica* (Ikata, 1930; Chang *et al.*, 1963).

Among the agricultural crops, downy mildew is common on turnips (*B. campestris* subsp. *rapifera*); swedes (*B. napus* var. *napobrassica*), and occurs occasionally on rape (*B. napus* var. *napus* f. *annua*); marrow-stem kale (*B. oleracea* var. *acephala* sub-var. *medullosa*) and seakale (*Crambe maritima*) (Moore, 1959). *P. parasitica* also attacks kohlrabi (*B. oleracea* var. *gongyloides*); Chinese kale (*B. alboglabra*) (Johnston, 1963); broccoli (*B. oleracea* var. *botrytis* sub-var. *cymosa*); colza (*B. napus* var. *napus* f. *biennis*) and Chinese mustard (Anon., 1968), and other *Brassica* species including *B. fruticulosa*, *B. tournefortii*, *B. juncea* and *B. nigra* (Gäumann, 1926) and *B. chinensis* (Hiura and Kanegae, 1934).

Other species which may be attacked by *P. parasitica* are *Camelina sativa* (Darpoux, 1945), *Cheiranthus allioni* (De Bruyn, 1935a), *Coronopus didymus*

---

* Classification of common *Brassica* species according to Nieuwhof (1969).

(Langdon, 1948), *C. squamatus* (Dias and Da Camara, 1953), *Eruca sativa* (Gäumann, 1926), *Lepidium graminifolium* (Nicolas and Aggéry, 1940), *Malcolmia africana* (Thind, 1942), and *Raphanus raphanistrum, Sinapis alba* and *S. arvensis* (Gäumann, 1926). In addition to the above, Gäumann (1923) listed over 80 cruciferous species as susceptible to infection by the numerous species of *Peronospora* which he recognized but which are now generally regarded as all being *P. parasitica.*

## B. Geographical distribution

Downy mildew occurs on cultivated brassicas and on other cruciferous species in widely separated localities throughout the world. Although the list below is not exhaustive, among the countries cited by various authors are the following:

Angola (Serafim and Serafim, 1968), Argentine (Lindquist, 1939), Australia (Samuel, 1925; Anon., 1955), Austria (Glaeser, 1970), Bermuda (Waterston, 1940), Brazil (Grillo, 1937), Britain (Moore, 1959), Brunei (Herb. IMI),* Canada (Jones, 1944; Downey and Bolton, 1961), Chile (Mujica and Vergara, 1960), China (Porter, 1926; Pai, 1957), Columbia (Orjuela, 1965), Costa Rica (McGuire and Crandall, 1967), Cuba (Fernandez Roseñada, 1973), Cyprus (Herb. IMI), Czechoslovakia (Rýdl, 1968), Denmark (Gram and Rostrup, 1924), Dominica (Anon., 1972a), Egypt (Elarosi and Assawah, 1959), Ethiopia (Herb. IMI), Fiji (Anon., 1969), Finland (Herb. IMI), France (Darpoux, 1945), Germany (Neumann, 1955), Greece (Herb. IMI), Guatemala (Muller, 1950), Haiti (Anon., 1972a), Hong-Kong (Johnston, 1963), India (Thind, 1942), Iran (Ershad, 1977), Iraq (Herb. IMI), Israel (Peleg, 1953), Italy (Ciferri, 1961), Jamaica (Leather, 1967), Japan (Hiura and Kanegae, 1934), Kampuchea (Soonthronpoct, 1969), Kenya (Anon., 1957), Korea (Anon., 1972b), Libya (Herb. IMI), Malawi (Peregrine and Siddiqi, 1972), Malaysia (McIntosh, 1951), Malta (Herb. IMI), Mauritius (Orian, 1951), Mexico (Rodriguez, 1972), Morocco (Herb. IMI), Mozambique (Decarvalho, 1948), Nepal (Bhatt, 1966), Netherlands (Thung, 1926b), New Zealand (Jafar, 1963), Northern Ireland (McKee, 1971), Norway (Ramsfjell, 1960), Panama (McGuire and Crandall, 1967), Pakistan (Perwaiz *et al.*, 1969),Papua New Guinea (Herb. IMI), Philippines (Ocfemia, 1925), Poland (Zarzycka, 1970), Portugal (Da Costa and Da Camara, 1954), Puerto Rico (Anon., 1972a), Roumania (Savulescu, 1960), Sabah (Anon., 1962), American Samoa (Firman, 1975), South Africa (Doidge *et al.*, 1953), Spain (Gonzalez Fragroso, 1924), Sri Lanka (Park, 1932), Sweden (Nilsson, 1949), Switzerland (Gäumann, 1923), Taiwan (Lo, 1961), Tanzania (Herb. IMI), Thailand

* From records received from the Commonwealth Mycological Institute, Kew.

(Chandrasrikul, 1962), Trinidad (Stell, 1922), Turkey (Herb. IMI), Uganda (Herb. IMI), USA—various states (Chupp, 1930; Weber, 1932; Ramsey, 1935; Kadow and Anderson, 1940; Pound, 1946; Foster and Pinckard, 1947; Shaw and Yerkes, 1951; Borders, 1953; Kontaxis and Guerrero, 1978), Uruguay (Koch de Brotos and Boasso, 1955), USSR (Antonov, 1978), Venezuela (Herb. IMI), Vietnam (My, 1966), Yugoslavia (Sutic and Klijajić, 1954).

## C. Symptoms

Attack by *P. parasitica* on the seedlings of all common brassica hosts results initially in the development of discoloured spots on the surface of the cotyledons, which may then turn yellow and later shrivel and die. At such an early stage of growth loss of the cotyledons may be fatal. Before the seedling dies, however, sporulation of the pathogen occurs, the fine loose carpet of conidiophores and conidia developing mainly on the lower surface of the cotyledons and hypocotyl.

Later infection of hosts such as cabbage, cauliflower, Brussels sprout, turnip, swede and radish appears first as small discoloured spots and yellowing on the upper surface of the true leaves, with a sparse patchy growth of mycelium on the lower surface. While sporulation is often profuse on the mycelium on the undersurface of the leaves, it is more delicate and "wefty" than that of the powdery mildews. In slight attacks the infected areas remain discrete, as angular patches on the leaves, but under moist conditions favourable for mildew development, the infection may spread and cause the leaves to shrivel and die.

In many situations, infection and partial or complete destruction of some of the leaves may be the total expression of the disease in the field, but in certain crops, such as cauliflower and broccoli, the infection may extend to the curds both in the field (Chorin, 1946; Natti *et al.*, 1956; Davison *et al.*, 1962; Jenkins, 1964) and in store (Lund and Wyatt, 1978). In these situations a pale brown or greyish discolouration appears on the surface of the curd, and greyish or black spots and streaks develop on the stems of the immature flowering heads. By itself this discolouration inevitably leads to down-grading of cauliflower in the market, but secondary rotting of the tissues by bacteria results in destruction of the curd.

Cabbage heads also, may suffer infection in the field and in storage (Thung, 1926a; Ramsey, 1935). Extensive greyish-black discolouration spreads through the heads and even penetrates the parenchyma of the stem. As with the cauliflower curds, in addition to the damage caused by the downy mildew itself, the affected cabbage tissues are very susceptible to attack by secondary bacteria and fungi, e.g. *Erwinia*, *Alternaria* and *Rhizopus* spp.

Sporulation of *P. parasitica* may be visible on the cauliflower curds or cabbage heads in the field or in store, or will develop after incubation in humid conditions. Infection of the stems and inflorescences of radish may also occur (Butler and Jones, 1949).

In addition to infecting the parts of the plants described above, *P. parasitica* also parasitizes the "roots" of turnip (Gardner, 1920). Thus, stored turnips develop dark discoloured zones spreading down from the crown into the stelar regions of the "root". The affected tissues vary in colour from light brown to black, and though initially firm, later become predisposed to rotting by *Rhizocotonia* and soft-rotting bacteria. The "roots" of radish may also be attacked by *P. parasitica* (Ramsey *et al.*, 1954), causing a brown to black epidermal blotch or streak extending round the circumference of the "root", accompanied by slight russeting and cracking. The internal tissues are extensively explored by the fungus, and though remaining firm, are discoloured greyish-brown or black.

## III. The Pathogen

### A. Taxonomy

The earliest reference to downy mildew on crucifers is by Persoon (1796) who ascribed the cause of the disease on *Thlaspeos bursa-pastoris* (*Capsella bursa-pastoris*) to the fungus *Botrytis parasitica* Pers. In 1849, Fries (Gäumann, 1918) transferred the fungus to the genus *Peronospora* which had been established in 1837 by Corda in his description of *Peronospora ramicis* (Corda, 1837). At that time all isolates obtained from cruciferous hosts were ascribed to *P. parasitica*. From his studies, however, Gäumann (1918) considered that the various isolates obtained from different hosts should be classified as separate entities, and on this basis recognized 52 species of *Peronospora*. His conclusions were largely based upon conidial dimensions and the results of cross-inoculation tests. Subsequently, however, the value of conidial dimensions as a taxonomic criterion has been questioned since they can vary according to environmental conditions (Thung, 1926b), and Fraymouth (1956) considered that Gäumann should have classified his isolates as strains of *P. parasitica* rather than separate species within the genus.

The view expressed by Waterhouse (1973) and supported by Dickinson and Greenhalgh (1977) was that there should be an aggregate species of *Peronospora* on members of the Cruciferae. The current situation, as indicated by Yerkes and Shaw (1959), is that all isolates of downy mildew infecting members of the Cruciferae are ascribed to *P. parasitica*.

The merits of some separate speciation must not be ruled out, however, in view of the apparent differences in antheridial structure in the isolates of *Peronospora* studies on *Capsella bursa-pastoris* by Wager (1889) and on *Brassica oleracea* by McMeekin (1960).

## B. Morphology and disease cycle

The general morphology and infection cycle of *P. parasitica* is similar to that of other members of the Peronosporaceae in that the fungus possesses a coenocytic mycelium which ramifies in the intercellular spaces of the host and forms haustoria which penetrate the cells of the infected tissues. In due course sporophores (conidiophores* in *Peronospora*) emerge from the tissues of the host and bear asexual spores (conidia* in *Peronospora*), which form the principal means by which the disease is spread. In view of the obligate parasitism of the fungus, *P. parasitica* remains active only in the living plant, and survives the death of the host tissues in the form of sexually produced oospores.

In outlining in more detail the main features of the morphology of the fungus and the disease cycle it is most convenient to consider first the reproductive stages of the fungus.

## 1. Reproduction

### (a) Asexual

The process of asexual sporulation has been studied by Davison (1968) and Chou (1970). The conidiophores are produced during darkness, initially as unbranched hyphae from aggregations of intercellular mycelium beneath the host epidermis. They project either singly or in groups through the stomatal opening. Chou (1970) reported that similar hyphae could penetrate and emerge between the ordinary epidermal cells. The hyphae differentiate to form the conidiophores (200–500 $\mu$m in height) and bear primary and secondary branches, which ultimately bifurcate to form pairs of finely pointed and in-curved sterigmata, bearing single terminal conidia. These spores are spherical at first, but become broadly ellipsoidal (24–27 × 12–22 $\mu$m) and are delimited from the sterigmata by cross-walls at maturity. At 8°C, the rate of elonga-tion of the conidiophore reaches 100–200 $\mu$m h$^{-1}$, and the whole process from emergence to spore formation takes approximately 4–6 h (Davison, 1968). Felton and Walker (1946) reported that *P. parasitica* sporulates most readily at 8–16°C.

Spore release is effected mainly by hygroscopic twisting and untwisting

---

* The terms "conidiophores" and "conidia" quoted in Butler and Jones (1949) and Waterhouse (1973) have been used since spore germination in *Peronospora* spp. is always by a germ tube.

of the conidiophores in response to change in atmospheric humidity (Pinckard, 1942), though Hirst and Stedman (1963) showed that a transient period of release occurred at the onset of rainfall while the spores were still dry.

The survival of conidia of *P. parasitica* depends greatly upon the environment in which they are maintained. Thus Krober (1970) showed that while conidia stored in the open air on the leaves of kohlrabi on which they were formed survived for a maximum of about 10 days, spores maintained in air-dry soil during the winter remained viable for up to 110 days. The survival period was greatly reduced (to a maximum of 22 days) if the storage soil was moist. Marked reductions in survival were also recorded after storage in both dry and moist soils during the summer. Retention of viability was longest (up to 130 days) when the spores were stored in air-dry soil at a constant temperature of 5°C.

(b) Sexual

During sexual reproduction *P. parasitica* forms spherical oogonia and paragynous antheridia. Wager (1889), studying the fungus on *Capsella bursa-pastoris* figured the male and female sex organs arising from a common hypha, but indicated that they could be borne on separate hyphae. The antheridia were relatively short structures which became adpressed to the sides of the oogonia. By contrast, McMeekin (1960), using *Brassica oleracea*, showed that in this host, *P. parasitica* formed long tendril-like antheridia which appeared to arise on hyphae separate from those bearing the oogonia.

Observing the mode of fertilization of *P. parasitica* in *Capsella*, Wager (1900) noted that the protoplasm of the oogonium becomes differentiated into a central vacuolated ooplasm and a peripheral multinucleate granular periplasm. A receptive papilla with a very thin wall forms on the oogonium at the point of contact with the antheridium. A fertilizing tube grows from the antheridium through the receptive papilla towards an ill-defined but apparently attractive "central body" in the ooplasm, and there discharges a single "male" nucleus. Meanwhile a single "female" nucleus detaches itself from the periplasm and also migrates towards the central body. After some delay the two nuclei fuse and initiate the uninucleate oospore. During ripening of the oospore the periplasm is deposited on its wall as an exosporial layer. The mature oospore is thick-walled, yellow-brown and spherical, and measures 26–45 $\mu$m diameter.

Oospore formation is favoured by conditions which induce senescence of the host tissues, such as a deficiency of nitrogen, phosphorus or potash (McMeekin, 1960). For this reason oospores are found abundantly in necrotic or chlorotic leaves, but not in green tissues, and occur more frequently in cotyledons than true leaves since the former senesce more rapidly.

There is evidence of both homothallism and heterothallism in *P. parasitica* (De Bruyn, 1935b), and McMeekin (1960) showed that separate antheridial and oogonial strains exist.

## 2. Infection

(a) From conidia

When environmental conditions are suitable, conidia of *P. parasitica* alighting on the surface of a susceptible host form germ tubes from which appressoria develop. In hosts such as cauliflower, that have been studied closely, the appressoria in a high proportion of cases are found at the junction of the anticlinal walls of the epidermal cells (Preece *et al.*, 1967). In electron microscope studies Chou (1970) found that the contents of the conidium pass into the appressorium from which an infection hypha develops. This may occasionally enter the tissues through a stoma but more usually penetrates direct (Shiraishi *et al.*, 1975), when it breaks a hole (4–5 μm in diameter) through the cuticle which then fits closely round the infection hypha (Chou, 1970). After entering the host the hypha expands to a diameter of 7–8 μm and grows initially in the region of the middle lamella between the anticlinal walls of the epidermal cells. Such penetration between adjacent epidermal cells rather than via stomata had been noted earlier by Chu (1935). Chou further observed that the infection hypha continues its growth between the cells of the host tissues branching in all directions and varying in diameter and form according to the size and shape of the intercellular spaces.

Krober (1969) showed that single conidia could initiate infection in seedlings of kohlrabi, but that both the intensity and rate of infection increased with increasing numbers of spores in the inoculum. Commensurately more conidia were required to produce comparable responses in older plants which were less susceptible to infection.

The rates of spore germination and host penetration are markedly affected by temperature, thus Chu (1935) found that, at 15°C conidia germinated in 4–6 h, appressoria formed in 12 h, and penetration occurred in 18–24 h. Felton and Walker (1946), however, reported that on cabbage, germination of the conidia and subsequent penetration of the host took place most rapidly at 8–12°C and 16°C respectively. Jonsson (1966) found that development of the disease on winter rape was also favoured by a temperature of 8–16°C. By contrast, Chou (1970) noted that at 20–25°C infection occurred within 6 h of the arrival of the conidium on the surface of the host cotyledon.

(b) From oospores

There is little doubt that oospores constitute an important means of survival of *P. parasitica* over periods of unfavourable conditions. Chang *et al.* (1963)

obtained infection of Chinese cabbage seedlings from soil-borne oospores, and Chiu (1959) considered that oospores were the primary agents of over-wintering infection of Chinese cabbage by downy mildew. LeBeau (1945) also obtained strong circumstantial evidence that oospores could cause primary infection of cabbage seedlings. However, McMeekin (1960) using various experimental treatments, including storage for 5 months at steady temperatures between $-10°C$ and $+27°C$, and frequently alternating temperatures between $-10°C$ and $+15°C$, failed to induce germination of, or obtain plant infection from oospores.

Although Chang et al. (1963) found mycelium of P. parasitica in the seed coats, and oospores contaminating seed samples of Chinese cabbage, they reported that seedling infection from these sources was unreliable.

Whether or not oospores are an important natural source of plant infection, it is clear that invasion of the hypocotyls of brassica plants such as cabbage, cauliflower and broccoli from inoculum in the soil can occur. LeBeau (1945) demonstrated by artificial inoculation of cabbage seedlings with conidia of P. parasitica, or by growing seedlings in soil contaminated with oospores, that systemic invasion of the hypocotyls and cotyledons took place. While he also reported that the fungus did not progress beyond the cotyledonary node in cabbage, Chang et al. (1963) found that in Chinese cabbage it grew into the first pair of true leaves. In both these cases, and possibly in general, it must be assumed that further spread of the pathogen is by dissemination of conidia released from conidiophores formed on the cotyledons and hypocotyls.

(c) Haustorium formation

Haustoria form as branches of the intercellular hyphae within the host cells and vary in size and shape even in the same host species. Thus in turnip and radish roots the haustoria are initially spherical to pyriform, but later become cylindrical or clavate, and often dichotomously or trichotomously branched (Chu, 1935). In cabbage some haustoria are large irregular vesicles while others are bilobed and regular in shape. On the other hand the haustoria in cauliflower are single, globose and uniform in size (Fraymouth, 1956). Variation in shape and size of the haustoria of P. parasitica also occurs in hosts other than Brassica spp., such as Matthiola incana, Cheiranthus cheiri, Capsella bursa-pastoris, Diplotaxis muralis and Rhynchosynapis monensis (Fraymouth, 1956). Moderately high temperatures (20–24°C) favour the most rapid development of the haustoria (Felton and Walker, 1946).

Electron microscope studies by Chou (1970) have shown that during haustorium initiation, localized swelling of the host cell wall occurs in the areas in contact with the hyphae. The appearance of a microfibrillar structure in these areas of the host cell walls (possibly by enzymic removal of amorphous carbohydrates) suggests that initial weakening of the host cell wall is achieved at least in part by chemical means. Penetration by the haustorial branch occurs through a hole 1–2 $\mu$m diameter in the cell wall which may form a collar-like structure round the base of the primordial haustorium. As the haustorium enlarges, invagination of the host plasmalemma occurs and a sheath, possibly of callose forms round the intrusive fungous feeding organ (Fraymouth, 1956). The exact composition and structure of the interface between host and haustorium is at present unclear (Ehrlich and Ehrlich, 1971). The effect of haustoria on host protoplasts varies, in that the response in epidermal cells of cotyledons is much more disastrous than in mesophyll cells. In the former the cytoplasm becomes markedly disorganized by the fungal invasion, whereas in the latter the cytoplasm is merely invaginated by the haustorium and the tonoplast and plasmalemma seem to be unbroken.

## 3. Effect of the environment on disease development

The rate and severity of development of any disease depends upon the interplay of various factors on host and pathogen. Among the factors which markedly influence the development of downy mildew on brassicas are air temperature and humidity. Thus Felton and Walker (1946) found that in the pre-penetration stage, conidial germination was most rapid at 8–12°C, while penetration of the host by the infection hyphae and formation of the haustoria occurred quickest at 16°C and 20–24°C respectively. Subsequent disease development (and presumably, therefore, the growth of the fungus in the tissues) was reported by them to be most rapid at 24°C, but damage was so severe that the infected leaves were quickly destroyed and no re-infection occurred. They found that the lower temperature of 16°C favoured slower growth of both host and pathogen, resulting in less drastic damage and hence more prolific sporulation, more re-infection and consequently more profuse disease development. It is probable, therefore, that the greater prevalence of downy mildew on brassicas at 10–15°C results from the favourability of this temperature range balanced against the more rapid invasion and self-inhibiting destruction of the host tissues by the pathogen at higher temperatures. Similar temperature optima for disease development of 8–16°C were reported by Jonsson (1966) and of 8–14°C by d'Ercole (1975), though Nakov (1972) found that 15–20°C was most favourable. All the workers agreed that high humidity favoured disease development.

   There appears to be no consistent effect of fertilizers on the development of downy mildew on brassicas (Butler and Jones, 1949), though increased susceptibility in cauliflowers has been associated with potash deficiency (Quanjer, 1928). Felton and Walker (1946), however, found no direct relationship between mildew incidence and any excess or deficiency of nitrogen, phosphorus or potash, or any deficiency of sulphur.

## C. Culturing *P. parasitica*

The obligate parasitic nature of *P. parasitica* implies a sophistication of nutritional requirement which is not satisfied once the host plant has died. Concomitant with this is an inability of the fungus to grow on the artificial media used by mycologists and plant pathologists for facultative saprophytes. For this reason the commonest method of maintaining an isolate of *P. parasitica* has either been on young crucifer plants, or on detached cabbage cotyledons on moist filter paper in clear plastic boxes in an incubator at 15°C under 12 h fluorescent illumination per day (Channon and Hampson, 1968). Guttenberg and Schmoller (1958) cultured *P. parasitica* on disinfected slices of swede root, obtaining growth of aerial mycelium with conidiophores and conidia, antheridia, oogonia and oospores.

Ingram (1969b) successfully established and maintained cultures of the fungus on callus tissues derived from (a) a mature leaf of cabbage, (b) a mature root/hypocotyl of rape and (c) a seedling hypocotyl of swede. The infected calluses were incubated either at 22°C in the dark or at 15°C with 12 h fluorescent illumination per day, on an agar medium developed from that used for the culture of *Plasmodiophora brassicae* in callus tissue (Ingram, 1969a). Due to rapid senescence of the infected cabbage leaf callus, it was necessary when sub-culturing to transfer an explant of infected callus to fresh uninfected callus every 14–21 days. Greater success was obtained with calluses originating from the root/hypocotyl of rape, which not only grew faster than those from cabbage leaf, but could also be transferred directly when sub-culturing. About four weeks after inoculation of the rape callus, small nodules of new tissue developed on the infected callus, and either these nodules or the whole callus could be transferred to fresh culture medium. Conidia typical of *P. parasitica* were produced on infected callus tissue maintained in a 15°C illuminated (12 h/day) incubator, but production was much less at 22°C in the dark. Such conidia could be used to infect detached cotyledons or leaf callus of cabbage or callus of rape.

While callus tissue clearly provided a satisfactory vehicle for maintaining cultures of *P. parasitica*, it did not provide a means of testing material for resistance since its response did not necessarily reflect that of the whole plant (Ingram, 1969b).

Attempts were made with limited success, by Guttenberg and Schmoller (1958) to culture the fungus in the absence of living plant tissue. Thus they obtained visible mycelial growth in filter-sterilized swede juice, but it ceased after 3 days when a yellow discolouration appeared, suggesting that the medium was chemically unstable. Very limited mycelium developed on swede seed-glucose agar and maize decoction-glucose agar, but more success accrued with an agar medium containing 2% beer wort + 0·1% phosphate in which hyphae and conidiophores developed within and outside the agar substrate. Similar, but less marked growth was achieved in oatmeal agar and rice-starch agar.

## D. Physiological specialization

There is well documented evidence that isolates of *P. parasitica* differ in the range of cruciferous hosts that they can infect. Thus Gardner (1920) found that an isolate obtained from turnip, while able to infect seedling turnips, was unable to infect rutabaga or radish. Specialization of parasitism in *P. parasitica* may be exhibited at the generic, specific and lower taxonomic levels of the host.

At the generic level, Wang (1944) and Chang *et al.* (1964) recognized separate pathogenic varieties or *formae speciales* of the fungus on species of *Brassica* (*P. parasitica* var./f.sp. *brassicae*), *Raphanus* (var./f.sp. *raphani*) and *Capsella* (var./f.sp. *capsellae*). Gäumann (1926) also recognized three *formae speciales*: one (*P. parasitica* f.sp. *brassicae*) infected and sporulated on various *Brassica* spp. (e.g. *B. oleracea*, *B. napus*, *B. rapa* etc.), a second (f.sp. *sinapidis*) infected *Sinapis arvensis* and *S. alba*, and a third (f.sp. *raphani*) infected *Raphanus sativus* and *R. raphanistrum*. With each of these *formae speciales* some infection, but little or no sporulation, occurred on the main hosts of the other two *formae speciales* (i.e. f.sp. *brassicae* caused only sub-infections on *Sinapis* and *Raphanus* spp.). Physiological specialization at the generic level of the host has also been reported by Hiura and Kanegae (1934) on *Brassica* and *Raphanus*, by Dzhanuzakov (1962) on *Brassica*, *Raphanus* and *Sinapis*, and by Jafar (1963) on *Matthiola*.

At the specific level of the host, Wang (1944) separated his *P. parasitica* var. *brassicae* into six specialized forms which differed in their ability to infect various *Brassica* spp. (cabbage, Chinese cabbage, *B. juncea*, and swede), while Chang *et al.* (1964) divided their *P. parasitica* f.sp. *brassicae* into three sub-forms (*pekinensis*, *oleracea* and *juncea*).

Further subdivision of certain *formae speciales* must be recognized since the differing reactions of broccoli plants (U.S.D.A. Plant Introductions) to different isolates of *P. parasitica* (presumably f.sp. *brassicae*) reported by Natti *et al.* (1967) indicated the existence of separate physiologic races (1 and 2) of the fungus.

More work needs to be done on the physiological specialization of *P. parasitica* before the full extent of the threat to brassica crops of the spread of downy mildew infection from other cruciferous hosts or weeds can be assessed. Unfortunately specialization by the pathogen may not be a clear cut case of pathogenicity or non-pathogenicity to various hosts, but rather one of varying degrees of pathogenicity to them. Thus, although McMeekin

(1969) recognized specialization in the isolates she tested, she concluded that cruciferous weeds could serve as hosts for forms of *P. parasitica* found on commercial brassica crops.

## IV. Disease Control

### A. Fungicides

Although downy mildew on brassica crops is widespread, the most damaging effects are restricted to young seedlings and to particular organs on certain crops, e.g. cauliflower curds and the buttons of Brussels sprouts. For this reason the main call for fungicidal treatments has been on cabbage and Brussels sprout seedlings, and on seedlings and young plants of cauliflower, particularly those overwintering in frames, polythene tunnels or glasshouses.

During the period from the mid-1940s to the mid-1960s reliance for control of the disease rested on frequent applications of sprays or dusts of fungicides such as chloranil (Spergon), copper-based materials and zineb (Foster, 1947; Gram and Weber, 1952; Borders, 1953; Conroy, 1960; Schmidt, 1960; Ogilvie, 1969). These materials were subsequently superseded by other non-systemic fungicides, outstanding among which were captafol, daconil, dichlofluanid and propineb selected from laboratory tests on detached cotyledons (Channon and Hampson, 1968). The success of dichlofluanid was confirmed in the field when it gave excellent control of mildew on the cotyledons of cabbage and cauliflower, though less effective protection to the leaves (Channon *et al.*, 1970). Whitwell and Griffin (1967) reported that both dichlofluanid and propineb, not only reduced early mildew infection, but also increased the size and dry weight of the plants, as well as the number of plantable early summer cauliflower seedlings. There was evidence that the inferior control obtained on the true leaves with dichlofluanid was perhaps associated with inadequate coverage, since disease control was improved by the addition of a non-ionic wetter. Some phytotoxicity occurred, however, with the combination of fungicide and wetter (Channon *et al.*, 1970).

Relatively little work appears to have been done on the control of mildew on maturing plants. Davison *et al.* (1962) obtained no reduction in the infection of broccoli heads with dust or spray applications of maneb, or sprays of actidione (in three forms), chloranil, copper compound, folpet + copper combination or nickel sulphate. Gabrielson (1964) was more successful, obtaining some control with three applications of Bordeaux mixture (8-8-100) and of maneb. Clearly, however, a limiting factor with any fungicide applied to an edible crop shortly before harvest must be that the level of chemical residues is acceptable.

The advent in the mid-1970s of systemic fungicides active against phyco-mycete fungi presented the opportunity to improve on the control of *P. parasitica* on seedlings and growing brassica plants. Ryan (1977), working on plants growing both in walk-in tunnels and under cloches, showed that prothiocarb (Dynone) incorporated into the soil at 5 g m$^{-2}$ before sowing, greatly reduced the disease in overwintered cauliflower plants between December and April. Comparable effects were obtained with fosetyl-aluminium (Aliette) at an incorporation rate at 10 g m$^{-2}$. Although less success was obtained with three foliar sprays during the same period, both of the above fungicides were as effective as eight sprays of dichlofluanid.

Prothiocarb also reduced infection of radish leaves and bulbs when applied at 0·1% ai as a drench (4 litres m$^{-2}$) at 50% seedling emergence, and was much more effective than sprays of dichlofluanid, zineb, captafol or maneb which gave no disease control (Anon., 1974).

Considerable success in controlling downy mildew fungi including *P. parasitica* has also been recorded with another systemic fungicide, metalaxyl. It was first introduced in 1977 (Schwinn *et al.*, 1977) and its characteristics described by Urech *et al.* (1977). Used as a seed treatment (at 0·31 and 0·62 g ai kg$^{-1}$ seed) metalaxyl markedly reduced the development of mildew on broccoli seedlings, and when followed with one or two sprays of the same compound, significantly increased the yield of marketable heads (Paulus *et al.*, 1977). In the absence of a seed treatment a single spray of metalaxyl (0·28 kg ai ha$^{-1}$) had little effect on mildew, but two sprays markedly reduced the disease. Granular applications (0·56 and 1·12 kg ai ha$^{-1}$), prior to sowing also controlled mildew and significantly improved the early cutting yields, but not the total yields of broccoli. In cauliflower a granular application at 0·28 kg ai ha$^{-1}$ significantly improved both the early cutting and total yield of curds.

Work by the Agricultural Development and Advisory Service in England and Wales and reported by Smith (1979) confirmed the activity of metalaxyl against mildew on autumn-sown cauliflowers. A pre-sowing incorporation or a single post-sowing drench (1·5 kg ai ha$^{-1}$) or three high volume sprays (0·8 g ai litre$^{-1}$) gave much better disease control and more plantable plants than nine sprays of dichlofluanid applied during a 6–8 week period. Unfortunately, subsequent to these investigations, insensitivity to metalaxyl, on its own, was detected in *Phytophthora infestans* in Holland and Eire and it was summarily withdrawn from the market in late 1980 (T. G. Marks, personal communication, 1980). It was therefore, unavailable for practical use on its own against downy mildews after that time.

Only limited use has been made of fungicides against *P. parasitica* on other crops. There are reports of the application of proprietary compounds to false flax (*Camelina sativa*) (Zarzycka and Kloczowska, 1964; Zarzycka and Kloczowska, 1967), of Bordeaux mixture to rape (Perwaiz *et al.*, 1969), and of the systemic fungicide furalaxyl to *Matthiola incana* (Trimboli and Hampshire, 1978).

## B. Cultural methods

While fungicidal sprays provide the most effective method of controlling *P. parasitica* on brassica seedlings, measures to reduce the relative humidity around the plants by adequate aeration and avoidance of dense sowing and growth of weeds also help to reduce the disease (Conroy, 1960; Schmidt, 1960; Nakov, 1968). In addition, destruction of infected debris and crop rotation are recommended (Nakov, 1968). Avoidance of continuous cropping of rape on the same land or adjacent to fields sown to rape in the previous year is advised to reduce the chance of attack by *P. parasitica* (Downey and Bolton, 1961).

## C. Resistance

While clear evidence has been obtained on the physiological specialization of *P. parasitica* at the generic and specific levels of the host, less work has been done on the resistance of host cultivars to different populations of the pathogen.

In glasshouse tests on broccoli and cauliflower, Natti (1958) recorded high resistance to a representative isolate of *P. parasitica* in one out of a total of 75 lines of *Brassica oleracea* obtained from the Regional Plant Introduction Station, Geneva, New York. A further eight lines showed moderate resistance. In field tests, early-maturing cultivars of broccoli were generally more susceptible to mildew than late-maturing ones, though the most resistant cultivar recorded was in fact an early-maturing one (Italian Green Sprouting). In subsequent work, Natti *et al* (1967) identified resistance to two races (1 and 2) of *P. parasitica* governed by separately inherited single dominant genes, in 16 out of 230 Plant Introduction accessions. Although disease resistance was inherited independently from waxiness of the foliage it was apparent that plants with heavy foliar wax were significantly less subject to downy mildew than plants with no wax or with only a light deposit of wax.

Other sources of mildew resistance have been identified in lines of *Brassica oleracea* by Barnes (1968) and by workers in commercial seed firms in the United States, and these are being used in breeding programmes there (Prof. P. H. Williams, personal communication). In view, however, of the known heterothallism in *P. parasitica* (De Bruyn, 1935b; McMeekin, 1960) it is possible that reliance on major genes may give only temporary resistance to mildew in the host plants.

Different levels of susceptibility to downy mildew infection have also been reported in cultivars of winter oilseed rape by Jonsson (1966) and Dixon (1975). The latter showed that among the cultivars on the UK National List in 1975, Eurora, Janetski and Primor were highly resistant, while Argus, Hector, Tonus and Victor were very susceptible to infection by *P. parasitica*.

## Acknowledgements

I wish to thank Paul Holliday for providing information on the geographical distribution of *P. parasitica* from the records at the Commonwealth Mycological Institute, Kew, and Drs Mike Burge and Don Clarke of the Universities of Strathclyde and Glasgow respectively for their help with the literature on a number of occasions.

## References

Anon. (1955). 24th Annual Report NSW Department of Agriculture. Biological Branch, Division of Science Services. 37 pp.
Anon. (1957). "Annual Report. Department of Agriculture" Kenya, 1955. Vol. **2**, 237 pp.
Anon. (1962). Quarterly Report for July–September 1962 Plant Protection Committee for the South-East Asia and Pacific Region. F.A.O. Publication, Bangkok, Thailand. 19 pp.
Anon. (1968). *Rev. Appl. Mycol.* Plant host-Pathogen Index to Volumes 1–40 (1922–1961). 820 pp. Comm. Mycol. Inst., Kew, Surrey, England.
Anon. (1969). *Rep. Dep. Agric. Fiji 1968*, 29–30.
Anon. (1972a). Plant pests of importance to the Caribbean. F.A.O. Caribbean Pl. Prot. Commission. 29 pp.
Anon. (1972b). A list of plant diseases, insect pests, and weeds in Korea. Korean Soc. Pl. Prot. 424 pp.
Anon. (1974). *Pl. Path. Annual Report for 1974. ADAS Science Service, MAFF*, 180.
Antonov, Yu P. (1978). *Zashch. Rast.* **3**, 54–55.
Barnes, W. C. (1968). *Hort. Science* **3**, 79–138.
Bhatt, D. D. (1966). *J. Sci. Tri-Chandra Coll. Sci. Assoc.* **2**, 13–20.
Borders, H. I. (1953). *Pl. Dis. Reptr.* **37**, 363–364.
Butler, E. J. and Jones, S. G. (1949). "Plant Pathology". McMillan, London.
Butt, D. J. (1978). *In* "The Powdery Mildews" (D. M. Spencer, Ed.). Academic Press, London and New York.
Chandrasrikul, A. (1962). *Tech. Bull. Dep. Agric. Bangkok* **6**, 23 pp.
Chang, I. H., Shih, N. L. and Chiu, W. F. (1964). *Acta Phytopath. Sin.* **7**, 33–44.
Chang, I. H., Xu, R. F. and Chiu, W. F. (1963). *Acta Phytopath. Sin.* **6**, 153–162.
Channon, A. G. and Hampson, R. J. (1968). *Ann. Appl. Biol.* **62**, 23–33.
Channon, A. G., Hampson, R. J., Gibson, M. and Turner, M. K. (1970). *Pl. Path.* **19**, 151–155.
Chiu, W. F. (1959). *Bull. Agric. Sci.* (*China*) **9**, 314–315.
Chorin, M. (1946). *Palest. J. Bot. Rehovot Ser.* **5**, 258–259.
Chou, C. K. (1970). *Ann. Bot.* **34**, 189–204.
Chu, H. T. (1935). *Ann. Phytopath. Soc. Japan* **5**, 150–157.
Chupp, C. (1930). *Phytopathology* **20**, 307–318.
Ciferri, R. (1961). *Riv. Patol. Veg.* Pavia Ser. 3, **1**, 333–348.
Conroy, R. J. (1960). *Agric. Gaz. NSW* **71**, 462–468.

Corda, A. C. J. (1837). *Icones Fungorum Hucusque Cognitorum* 1, 20.

Da Costa, M. E. A. D. and Da Camara, E. de S. (1954). *Port. Acta Biol.* 4, 162–176.

Darpoux, H. (1945). *Annls Épiphyt.* N.S. 11, 71–103.

Davison, E. M. (1968). *Ann. Bot.* 32, 623–631.

Davison, A. D., Vaughan, E. K. and Hikida, H. R. (1962). *Pl. Dis. Reptr.* 46, 310–311.

Decarvalho, T. (1948). *Colónia Moçambique, Rep. Agric. Seccâo de Micolozia,* 1948.

De Bruyn, H. L. G. (1935a). *Tijdschr. PlZiekt.* 16, 57–64.

De Bruyn, H. L. G. (1935b). *Phytopathology* 25, 8 (Abs.).

D'Ercole, N. (1975). *Inftore Fitopatol.* 25, 21–23.

Dias, M. R. de S. and Da Camara, E. de S. (1953). *Agronomia Lusit.* 15, 17–37.

Dickinson, C. H. and Greenhalgh, J. R. (1977). *Trans. Br. Mycol. Soc.* 69, 111–116.

Dixon, G. R. (1975). *Proc. 8th Br. Insect. Fung. Conf. 1975,* 503–506.

Doidge, E., Bottomley, A., Van der Plank, J. E. and Pauer, G. D. (1953). *Bull. Sci. Dep. Agric. S. Afr.* 346, 122 pp.

Downey, R. K. and Bolton, J. L. (1961). *Publ. Dep. Agric. Can.* 1021, 19 pp.

Dzhanuzakov, A. (1962). *Bot. Zh. USSR* 47, 862–866.

Ehrlich, M. A. and Ehrlich, H. G. (1971). *Ann. Rev. Phytopath.* 9, 155–184.

Elarosi, H. and Assawah, M. W. (1959). *Alex. J. Agric. Res.* 7, 253–268.

Ershad, D. (1977). *Publ. Dept. Bot.* 10, 277 pp.

Felton, M. W. and Walker, J. C. (1946). *J. Agric. Res.* 72, 69–81.

Fernandez Roseñada, M. (1973). *Serie Agricola Academia de Ciencias de Cuba, Instituto de Investigaciones Tropicales.* No. 27, 78 pp.

Firman, I. D. (1975). *Information Document. South Pacific Commission,* No. 38, 7 + 3 pp.

Foster, H. H. (1947). *Phytopathology* 37, 712–720.

Foster, H. H. and Pinckard, J. A. (1947). *Phytopathology* 37, 896–911.

Fraymouth, J. (1956). *Trans. Br. Mycol. Soc.* 39, 79–107.

Gabrielson, R. L. (1964). *Pl. Dis. Reptr.* 48, 593–596.

Gardner, M. W. (1920). *Phytopathology* 10, 321–322.

Gäumann, E. (1918). *Beih. Bot. Zbl.* 35, 395–533.

Gäumann, E. (1923). *Beitr. KryptogFlora Schweiz.* 5, 1–360.

Gäumann, E. (1926). *Landw. Jbr. der Schweiz.* 40, 463–468.

Glaeser, G. (1970). *Pflanzenschutzberichte* 41, 49–62.

Gonzalez Fragroso, R. (1924). *Bol. R. Soc. Españ. Hist. Nat.* 24, 305–312.

Gram, E. and Rostrup, S. (1924). *Tidsskr. PlAvl* 30, 361–414.

Gram, E. and Weber, A. (1952). "Plant Diseases, in Orchard, Nursery and Garden Crops" (R. W. G. Dennis, Ed.). MacDonald and Co. Ltd., London.

Grillo, H. V. S. (1937). *Rodriguésia* 2, 39–96.

Guttenberg, H. von and Schmoller, H. (1958). *Arch. Mikrobiol.* 30, 268–279.

Hirst, J. M. and Stedman, O. J. (1963). *J. Gen. Microbiol.* 33, 335–344.

Hiura, M. and Kanegae, H. (1934). *Trans. Sapporo Nat. Hist. Soc.* 13, 125–133.

Ikata, S. (1930). *J. Pl. Prot., Tokyo* 17, 6 pp.

Ingram, D. S. (1969a). *J. Gen. Microbiol.* 55, 9–18.

Ingram, D. S. (1969b). *J. Gen. Microbiol.* 58, 391–401.

Jafar, H. (1963). *N.Z. Jl. Agric. Res.* 6, 70–82.

Jenkins, J. E. E. (1964). *Pl. Path.* 13, 46.

Johnston, A. (1963). A preliminary plant disease survey in Hong-Kong. Plant Prod. and Prot. Div. F.A.O. Rome, 32 pp.

Jones, W. (1944). *Scient. Agric.* 24, 282–284.

Jonsson, R. (1966). *Sver. Utsädesför. Tidskr.* 76, 54–62.

Kadow, K. J. and Anderson, H. W. (1940). *Bull. Ill. Agric. Exp. Stn* 469, 531–583.

Koch de Brotos, L. and Boasso, C. (1955). *Publ. Minist. Ganad. Agric., Montevideo,* 106, 65 pp.

Kontaxis, D. G. and Guerrero, P. (1978). *Pl. Dis. Reptr.* **62**, 170–171.
Krober, H. (1969). *Phytopath. Z.* **66**, 180–187.
Krober, H. (1970). *Phytopath. Z.* **69**, 64–70.
Langdon, R. F. N. (1948). *Pap. Dep. Biol. Univ. Qd* **2**, 9 pp.
Leather, R. I. (1967). *Bull. Minist. Agric. Lds Jamaica* **61** (N.S.), 92 pp.
LeBeau, F. J. (1945). *J. Agric. Res.* **71**, 453–463.
Lindquist, J. C. (1939). *Physis. B. Aires* **15**, 13–20.
Lo, T. T. (1961). *Pl. Ind. Ser. Chin.-Am. jt Comm. Rur. Reconstr.* **23**, 52 pp.
Lund, B. M. and Wyatt, G. M. (1978). *Pl. Path.* **27**, 143–144.
McGuire, J. U. and Crandall, B. S. (1967). *USDA Int. Agric. Development Service*, 157 pp.
McIntosh, A. E. S. (1951). Annual Report of the Department of Agriculture, Malaya for the year 1949, 87 pp.
McKee, R. K. (1971). *Rep. Res. Tech. Work Minist. Agric. N. Ireland 1970*, 105–113.
McMeekin, D. (1960). *Phytopathology* **50**, 93–97.
McMeekin, D. (1969). *Phytopathology* **59**, 693–696.
Moore, W. C. (1959). "British Parasitic Fungi". Cambridge University Press.
Mujica, F. and Vergara, C. (1960). *Boln téc. No. 6 Dep. Invest. agric. Chili* **6**, 60 pp.
Muller, A. S. (1950). *Pl. Dis. Reptr.* **34**, 161–164.
My, H. T. (1966). A preliminary list of plant diseases in South Vietnam. 142 pp. Saigon, Directorate of Research.
Nakov, B. (1968). *Rastit. Zasht.* **16**, 25–30.
Nakov, B. (1972). *Nauchni Trud. Vissh Selskostop. Inst. Vasil Kolarov* **21**, 109–116.
Natti, J. J. (1958). *Pl. Dis. Reptr.* **42**, 656–662.
Natti, J. J., Dickson, M. H. and Atkin, J. D. (1967). *Phytopathology* **57**, 144–147.
Natti, J. J., Hervey, G. E. R. and Sayre, C. B. (1956). *Pl. Dis. Reptr.* **40**, 118–124.
Neumann, P. (1955). *Z. Pflanzenschutz* **7**, 39–44.
Nicolas, G. and Aggéry, B. (1940). *Revue Mycol. N.S.* **5**, 14–19.
Nieuwhof, W. (1969). Cole crops. Leonard Hill, London.
Nilsson, L. (1949). *Växtskyddsnotiser, Växtskyddsanst. Stockh. 1949* **6**, 1–3.
Ocfemia, G. O. (1925). *Philipp. Agric.* **14**, 289–296.
Ogilvie, L. (1969). Diseases of vegetables. *Bulletin 123*, 110 pp. HMSO, London.
Orian, G. (1951). *Rep. Dep. Agric. Maurit. 1949*, 66–72; *for 1950*, 80–85.
Orjuela, N. J. (1965). *Boln Téc. Inst. Colomb. Agrop.* **11**, 66 pp.
Pai, C. K. (1957). *Acta Phytopath. Sin.* **3**, 137–154.
Park, M. (1932). *Ceylon Adm. Rep. Agric.* 1931, D103–D111.
Paulus, A. O., Nelson, J., Gafney, J. and Snyder, M. (1977). *Proc. 1977 Brit. Crop Prot. Conf.— Pests and Diseases*, 929–935.
Peleg, J. (1953). *FAO Pl. Prot. Bull.* **1**, 60–61.
Peregrine, W. T. H. and Siddiqi, (1972). *Phytopathological Papers*, No. 16, 51 pp. Comm. Mycol. Inst., Kew, UK.
Persoon, C. H. (1796). Observationes mycologicae seu descriptiones tam novorum quam notibilium fungorum exhibitae Part 1, 115 pp.
Perwaiz, M. S., Moghal, S. M. and Kamal, M. (1969). *W. Pakist. J. Agric. Res.* **7**, 71–75.
Pinckard, J. A. (1942). *Phytopathology* **32**, 505–511.
Porter, R. H. (1926). *Pl. Dis. Reptr.* Supplement **46**, 153–166.
Pound, G. S. (1946). *Bull. Wash. Agric. Exp. Stn* **475**, 27 pp.
Preece, T. F., Barnes, G. and Bayley, J. M. (1967). *Pl. Path.* **16**, 117–118.
Quanjer, H. M. (1928). *Tijdschr. PlZiekt.* **34**, 254–256.
Ramsey, G. B. (1935). *Phytopathology* **25**, 955–957.
Ramsey, G. B., Smith, M. A. and Wright, W. R. (1954). *Phytopathology* **44**, 384–385.
Ramsfjell, T. (1960). *Nytt Mag. Bot.* **8**, 147–178.

Rodriguez, S. H. (1972). *Folleto Misceláneo INIA* No. 23, 58 pp.
Ryan, E. W. (1977). *Proc. 1977 Br. Crop Prot. Conf.—Pests and Diseases*, 297–300.
Rýdl, R. (1968). *Uroda* **16**, 186–187.
Samuel, G. (1925). *Rep. Dep. Agric. S. Aust.* for year ending 30th June, 1924, 76–78.
Săvulescu, O. (1960). Communicärile de botanică, Bucuresti, 1957–1959, 263–267.
Schmidt, T. (1960). *Pflanzenarzt* **13**, 120–121.
Schwinn, F., Staub, T. and Urech, P. A. (1977). *Meded. Fac. Landbouww. R. Gent.* **42**, 1181–1188.
Serafim, F. J. D. and Serafim, M. C. (1968). Lista des doenças de culturas de Angola. 22 pp.
    Nova Lisboa, Instituto Investigação agronómica de Angola.
Shaw, C. G. and Yerkes, W. D. (1951). *NW Sci.* **25**, 76–82.
Shiraishi, M., Sakomoto, K., Asada, Y., Nagatani, T. and Hidaka, H. (1975). *Ann. Phytopath. Soc. Japan* **41**, 24–32.
Smith, J. M. (1979). *Proc. 1979 Br. Crop Prot. Conf.—Pests and Diseases*, 331–339.
Soonthronpoct, P. (1969). *Tech. Docum. F.A.O. Pl. Prot. Comm. S.E. Asia* **70**, 23 pp.
Stell, F. (1922). *Proc. Agric. Soc. Trin.* **22**, 779–785.
Sutić, C. and Klijajić, R. (1954). *Zašt. Bilja* **24**, 104–108.
Thind, K. S. (1942). *J. Indian Bot. Soc.* **21**, 197–215.
Thung, T. H. (1926a). *Phytopathology* **16**, 365–366.
Thung, T. H. (1926b). *Tijdschr. PlZiekt.* **32**, 161–179.
Trimboli, D. S. and Hampshire, F. (1978). *APPS Newsletter* **7**, 9–10.
Urech, P. A., Schwinn, F. J. and Staub, T. (1977). *Proc. 1977 Br. Crop Prot. Conf.—Pests and Diseases*, 623–631.
Wager, H. (1889). *Ann. Bot.* **4**, 127–146.
Wager, H. (1900). *Ann. Bot.* **14**, 263–279.
Wang, C. M. (1944). *Chin. J. Scient. Agric.* **1**, 249–257.
Waterhouse, G. M. (1973). *In* "The Fungi" Vol. IVB. (Ainsworth, Sparrow, and Sussman, Eds). Academic Press, New York and London.
Waterston, J. M. (1940). *Rep. Dep. Agric. Bermuda 1939*, 13 pp.
Weber, G. F. (1932). *Bull. Fla. Agric. Exp. Stn* No. 256.
Whitwell, J. D. and Griffin, G. W. (1967). *Proc. 4th Br Insect. Fung. Conf.*, 239–242.
Yerkes, W. D. and Shaw, C. B. (1959). *Phytopathology* **49**, 499–507.
Zarzycka, H. (1970). *Acta Mycol.* **6**, 7–19.
Zarzycka, H. and Kloczowska, T. (1964). *Biul. Inst. Ochr. Rośl. Poznań* **26**, 231–248.
Zarzycka, H. and Kloczowska, T. (1967). *Biul. Inst. Hodowl. Aklim. Rośl.* **6**, 147–159.

Chapter 16

# Downy Mildew of Cucurbits

## YIGAL COHEN

*Department of Life Sciences, Bar-Ilan University, Ramat-Gan, Israel*

## I. Introduction

Downy mildew of cucurbits occurs throughout the world on wild and cultivated Cucurbitacae, and on the genus *Cucumis* alone it has been reported
from 70 countries. It infects plants in the open field, and protected crops
under glass or plastic structures. It is widespread in tropical regions all over
the world, in some semi-arid regions such as southern USA, in the Middle
East, and in temperate regions of America, Europe, Japan, Australia and
South Africa (Palti, 1974).

Only 12 out of the 750 species of the Cucurbitaceae have been cultivated
by man (Sitterly, 1978) and of these, nine are known to be attacked by
*Pseudoperonospora cubensis* (Berk. et Curt.) Rost., the causal agent of downy
mildew in cucurbits. A full list of the host plants was recently published by
Palti and Cohen (1980). *Cucumis sativus* L., *Cucumis melo* L., *Cucurbita* spp.
and *Citrullus vulgaris* L., are the most important cultivated plant species
attacked.

The literature provides strong indications that various races of *P. cubensis* exist (Bains and Jhooty, 1976b; Palti and Cohen, 1979). As early as 1932, Doran observed that wild and cultivated curcurbits, reported to be susceptible to the pathogen elsewhere in America (Clinton, 1905), were resistant in Massachusetts. Iwata (1953a, b) followed the findings that in Japan three races of the pathogen exist (*Cucumis* type, *Cucurbita* type and *Benincasa* type) with none of them attacking *Citrullus vulgaris* and probably not *Lufa cylindrica*.

Other supporting evidence for the existence of races of *P. cubensis* are the sudden breakdown of resistance in the cucumber cultivar Palmetto in South Carolina in 1950 (Epps and Barnes, 1952), the existence of two separate biological forms of the fungus on cucumber in South Carolina, and on watermelon in Georgia (Hughes and Van Haltern, 1952), the susceptibility of Hollandian lines, which are resistant in Holland, to both Israeli (Cohen, 1976) and Japanese pathotypes (Meysing, personal communication) and the susceptible reaction of PI 142112 in Israel (Cohen, unpublished). Thomas (personal communication) noted that three races of the pathogen apparently exist in the south of Texas.

International efforts are needed to identify and code the various races of the pathogen and to determine their distribution. In the meantime the existence of different forms of the pathogen must be taken into account in quarantine and breeding work (Palti, 1974).

In this chapter emphasis is placed on host and environmental factors and the ways in which they affect disease and pathogen development. An understanding of the interaction between the host, pathogen, and environment places man in a better position to fight the disease.


## II. The Process of Infection

Infection commences when free moisture is present on a leaf on which sporangia have alighted. A successful interaction with host tissue, i.e. infection, occurs in the following sequence: zoospore are released, they encyst on stomata, after germination the germ tubes penetrate stomatal openings and this is succeeded by the production of intercellular hyphae and haustoria. Most stages of this process occur on either susceptible or resistant host plants (Cohen, 1976), as well as on many non-hosts (Iwata, 1938). In resistant cucurbits, tiny spots appear at the plant surface, indicating that fungal growth has stopped after a few haustoria have been formed. No haustoria have been observed in non-host plant tissues.

Zoospore release in water depends greatly on temperature. It does not occur under anaerobic conditions or in the presence of respiration inhibitors.

Zoospore release is similar on slide glass and leaf tissue, and is accompanied by protein synthesis, but not by DNA synthesis. Zoosporangia which have been wetted for a period shorter than is needed to ensure germination, will not release zoospores upon rewetting (Cohen et al., 1974). Storage on detached leaves at $-18°C$ for six months had no adverse effect on the viability of sporangia (Parris, 1951).

Iwata (1952a) reported a zoospore release of 99% at 6–24°C, 65% at 3°C, and 13% at 33°C whereas we found (unpublished) that zoospores were released over a much narrower range of 5–28°C. Minimal and optimal temperatures for release were relative, and dependent upon time of incubation: Thus, if incubation lasted for only 1 h the minimum temperature for zoospore release appeared to be 10°C and the optimum 20°C but in 2 h incubation the minimum was 5°C and the optimum 15°C. Incubation for as long as 6 h gave maximum and optimum temperatures of below 5°C and 10–15°C, respectively. The different upper limits for zoospore release in Japan and Israel were also reflected in infection levels and may indicate the presence of different races of the pathogen in these two countries.

Zoospores remain motile in water for 10 min to 18 h, depending upon the temperature (Iwata, 1952a). High temperature induces an immediate encystment probably because of depletion of nutrients. The optimum temperature for cyst germination is 25°C at which germination was six times higher than at 20°C. This explains why, in spite of the relatively limited zoospore release occurring at 25°C, the level of infection after only a short dew period is similar at both 20 and 25°C (Cohen, 1977). As the swimming period is long and the percentage germination is poor at low temperatures, longer dew periods and higher inoculum concentrations are needed for infection to occur at low temperatures.

Zoospores encyst singly on stomatal openings. The germ tube then penetrates into the substomatal chamber of cucumber leaves, but for unknown reasons, not in those of stems, petioles and hypocotyls (Iwata, 1938).

Light enhances zoospore settlement on stomata of cucumber probably due to some physiological effects on the host, but differences between light and darkness decrease as the period of leaf wetness increases. This enhancement is one of the reasons, among others unknown, for the great benefit of light on infection when short wet periods prevail (Cohen and Eyal, 1980).

The shortest dew period ensuring infection is 2 h. This period is shorter than the period required for haustorium formation, i.e. 4 h according to Iwata (1952b) indicating that once penetration of the germ tube has occurred, free water is no longer essential.

Infection in nature occurs under many different climatic conditions (Cohen and Rotem, 1971c) and since reports in the literature present different environmental requirements for infection (Doran, 1932; Hiura, 1929; Iwata, 1938; Van Haltern, 1933), a multifactorial experiment was conducted to examine the combined effects on infection of dew period, leaf temperature, and inoculum concentration (Cohen, 1977). Both the length of the latent period and the infection level were dependent upon the level of each factor. The minimal dew period required to establish infection gradually increased as either inoculum concentration or temperature decreased. With 2 h of dew, infection occurred with $> 100$ sporangia cm$^{-2}$ at a temperature of 20–25°C, as compared to 6 h of dew with $> 10$ sporangia cm$^{-2}$ at 10–25°C.

Minimum, optimum and maximum temperatures for infection were relative rather than absolute and dependent on the levels of the other factors. The range was wider when the other two factors were at optimal level. Optimum temperature lay between 5 and 25°C, but the maximum never exceeded 28°C.

These results explain the differences observed in disease outbreak in the field. In Israel, disease outbreak might occur from 8 to 50 days after planting, depending mainly on inoculum availability, providing that lack of dew is not a limiting factor.

## III. Symptom Production and Pathogen Development

With a heavy inoculum dose (1000 sporangia cm$^{-2}$) detection of the disease was first visualized at 3 days after inoculation, compared to 7 days with a low dose of inoculum (10 sporangia cm$^{-2}$; Cohen, 1977). Early appearance of the disease occurs if both temperature and dew period at time of inoculation are kept at favourable levels. Under field conditions in Israel, the length of the incubation period was dependent more on inoculum concentration than on prevailing temperatures. Duvdevani et al. (1946) observed that in early season 10 days elapsed between infection and symptom appearance compared with 4 days in late season. They attributed the early appearance of the disease in late season to a higher level of inoculum.

Symptom production and fungal colonization within the tissues are not necessarily corresponding processes. Whereas low temperature encourages fungal colonization with a minimal effect on symptoms, high temperature usually enhances symptom production and ends fungal development.

In laboratory experiments, an incubation temperature of 25°C was optimal for both symptom and pathogen development at an early stage of pathogenesis. At a later stage, symptom expression was maximal at 25°C whereas the pathogen developed best at 15°C. Minimum and maximum temperatures for symptom development are 10 and 35°C respectively, but sterile lesions might appear at 40°C if high inoculum doses are introduced.

Incubation at day:night temperatures of 30:20°C, and 25:20°C, favoured both symptoms and pathogen at the early stage, but 20:15°C and 25:10°C, were found to be favourable for the pathogen later on. A combination of high day temperature and low night temperature is therefore best for disease development in the field.

Usually, sporulating potential is highest in chlorotic lesions, and negligible on either greenish or necrotic lesions (Cohen and Rotem, 1969). The chlorotic state would normally develop about a week after inoculation. Environmental and biotic conditions which promote necrosis, would therefore reduce sporulation.

In growth chambers, in which dew was supplied every night, and sporangia collected in the early morning and late evening, it was found that min. night:max. day temperatures of 20:25°C or 20:30°C were associated not only with early initiation of sporulation but also with its early termination, due to rapid necrotization of lesions. Sporulation started later but lasted longer at 10:15°C and 10:20°C. Lesions remained fertile for 5, 8, 10 and 16 days, producing a total number of about 140, 190, 200, and $145 \times 10^3$ sporangia/cotyledon at 20:30, 20:25, 10:20 and 10:15°C, respectively.

In studying the effect of light intensity and photoperiod Inaba and Kajiwara (1971b) showed that disease development at 20°C, assessed as "percentage leaf area infected", increased with increasing light intensity during incubation. This was positively correlated with hyphal and haustorial development in tissue (Inaba and Kajiwara, 1972). Low light intensity reduced both the number and the size of the lesions.

Cohen and Rotem (1971a) on the other hand showed that light intensity had a minimal effect on symptoms after a 6-day incubation period. They showed that sporulating potential was directly proportional to light intensity, an effect which was dependent upon temperature. The relative increase in spore production per unit leaf area was higher at 15 and 20°C than at 25 or 30°C and long photoperiods were associated with a better development of both lesion and pathogen. At advanced stages of pathogenesis, lesions became necrotic at a photoperiod of 20 h. Under continuous illumination sporulating potential was lower than at a 20 h photoperiod.

Darkness induces lesion necrosis and hence reduces sporulation potential, especially if associated with high temperature. This is the reason for the fast necrotization occurring at high night (but not day) temperatures.

It should finally be mentioned that leaf age affects both lesion and fungal development. Disease appears earlier and proceeds slower but the sporulating potential is higher, on younger than older leaves.

## IV. Sporangium Production, Dispersal, and Survivability

Emergence of sporangiophores through stomatal openings, the first event in sporulation, does not occur until the air over the lesion is moisture-saturated. Free leaf moisture is not only not necessary for sporangium formation, but is inhibitory. Due to the higher frequency of hyphae in the spongy parenchyma most sporangia are produced on the lower leaf surface, a phenomenon more noticcable in cucumber than in cantaloupe.

Sporangial yield per unit area of affected leaf tissue in a moist chamber depends upon many factors, the most important of which are: plant species and variety, age and size of lesions, nutritional status of the host, temperature, and illumination.

Under favourable conditions, numbers of sporangia may reach $70 \times 10^3$ $cm^{-2}$ on susceptible cucumber (Cohen and Rotem, 1969) about $100 \times 10^3$ $cm^{-2}$ on cantaloupe (Cohen, unpublished) and about $4 \times 10^3$ $cm^{-2}$ on watermelon (Thomas, 1970). This last figure seems surprisingly low, especially since it was measured under epidemic field conditions. Cohen (1976) found that the yield of sporangia from the resistant cucumber Poinsett was only about 5% of that obtained from the susceptible Bet-Alfa on which yields per unit area are higher on small rather than on large lesions, probably because of a better supply of nutrients (Cohen and Eyal, 1977).

No data appear to exist on the effect of relative humidity on sporulation, but it has been clearly established that sporulation is strongly associated with dew formation (Duvdevani et al., 1946; Thomas, 1970). Rainy weather is usually less favourable for abundant sporulation than are nights with heavy dews (Van Haltern, 1933).

A minimum dew period of 6 h is needed for sporangia to be formed. In the presence of dew and with favourable temperatures, about 50–70% of the total number of sporangia produced in 24 h are formed during the first 12 h, as compared with 70–90% at unfavourable temperatures (Cohen and Rotem, 1969). Bains and Jhooty (1978) reported that on cantaloupe 95% of the sporangia are formed within 8 h, whereas Van Haltern (1933) noted that 9–10 h or longer are needed according to the temperature. Kajiwara and Iwata (1957) showed that in cucumber, sporangiophores emerge at $2–4\frac{1}{2}$ h and sporangia are fully developed at $7\frac{1}{2}–10$ h.

The optimum temperature for sporulation on cantaloupe is $18–22°C$ (Bains and Jhooty, 1979; Van Haltern, 1933). On cucumber, the optimum lies between 16 and $22°C$ according to Doran (1932), and between 15 and $19°C$ according to Hiura (1929). Cohen and Rotem (1969) found that $15°C$ is optimal in chlorotic lesions while $20°C$ is optimal in partially necrotic ones. The highest temperature at which sporangia are produced is $30°C$ (Van Haltern, 1933; Cohen and Rotem, 1969), and the lowest is $5°C$, at which $1·5 \times 10^3$ sporangia $cm^{-2}$ were formed on cucumber in a 24 h moist period.

Environmental conditions prevailing before the onset of dew have a remarkable influence on sporangial yield. Generally, any factor which enhances photosynthesis of the host will increase spore production. Thus, lengthening the light period, increasing the light intensity, improving the light spectrum, and raising the temperature will all favour subsequent sporulation in the dark (Cohen and Rotem, 1970; Inaba and Kajiwara, 1971a). In contrast, the photosynthesis inhibitor DCMU (dichloromethyldiphenylurea) reduced spore yield by 80% if given before the last period of illumination, but had no effect if supplied just prior to the dark moist period (Cohen and Rotem, 1970).

On plants sampled at midnight, sporulation occurred earlier (Kajiwara and Inaba, 1959) and was more abundant than on those sampled in the evening (Cohen, unpublished), indicating that a dry dark period of about 6 h enhances sporulation.

Cohen et al. (1971) showed that 6 h of darkness with free moisture on the leaves preceded by 6 h of darkness when the leaves were dry was as favourable for sporulation as 12 h of darkness with free water present. This is consistent with the observation of Thomas (1970) who found that 5 h of dew was minimal for abundant sporulation in the field. In later studies, Perl et al (1972) and Inaba and Kajiwara (1974a, b) elucidated that a 6 h period of either wet or dry darkness was needed for "processing" the accumulated photosynthate into compounds available for sporangial building.

Lesions behave as strong sinks for nutrients. When infected plants were exposed to $^{14}CO_2$, a higher amount of radioactivity was extracted, 1 h after exposure, from surrounding green tissue than from lesions, but lesions gradually accumulated $^{14}C$ compounds from 5 h to five days after exposure (Inaba and Kajiwara, 1974a; Perl et al., 1972). Upon placing the labelled plants in a moist chamber in darkness, there was a massive transfer of radioactivity from both diseased and healthy tissues into the developing sporangia, so that the $^{14}C$ level became 20–25 higher in the sporangia than in the lesions (Inaba and Kajiwara, 1974b; Perl et al., 1972). Before this transfer high-molecular-weight carbohydrates became hydrolyzed to hexose-like materials (Perl et al., 1972), and a large amount of labelled protein was synthesized (Inaba and Kajiwara, 1975). Carbohydrates and proteins make up about 75% of the $^{14}C$ label in sporangia produced on $^{14}CO_2$-fed leaves. Blocking the process of protein synthesis by cycloheximide inhibited sporulation (Cohen and Eyal, 1977).

Sporangium production is inhibited on lesions illuminated in the presence of free water at the terminal phase of sporaulation. Studying the inhibitory effect of light, Cohen and Eyal (1977) showed that the percentage inhibition increased as the temperature rose from 15 to 25°C, with blue light being most inhibitory.

The inhibitory effect of light was partially removed if plants were kept dry and in darkness for 2–4 h before the onset of dew (Kajiwara and Iwata, 1957; Cohen and Eyal, 1977). Light does not seem to act through the photosynthetic mechanism, as: (a) inhibition occurs under relatively low light levels (70% at 20°C with 30 $\mu E$ $m^{-2}$ $s^{-1}$); (b) red light is ineffective, and (c) DCMU does not remove the inhibition. It should be mentioned that light does not prevent sporangiophore growth through stomata even at a high light level of 165 $\mu$E m$^{-2}$ s$^{-1}$.

The inhibitory effect of light on sporulation may have some practical importance since Cohen et al. (1978) showed that inhibition could be induced by intermittent illumination (5–10 min every hour) with incandescent light. Thus, in suitable greenhouses and plastic tunnels, controlled artificial light could be used as a potent device for controlled disease.

Sporangia do not disperse from the sporulating leaf surface unless it is dry. Relative humidity (r.h.) is the major environmental factor affecting dispersal. In growth chambers (22°C, wind velocity of 1 m s$^{-1}$), peak dispersal in cucumbers was recorded at 1 h after plants had been placed in an unsaturated atmosphere, and at 2–3 h in cantaloupe. About 70% of the total number of sporangia dispersed from cucumbers during a 24 h period was dispersed within 1 h at an r.h. of 40 ± 5%, as compared to 35% at an r.h. of 80 ± 5%. Temperature and illumination had only a slight effect on dispersal (Cohen, unpublished).

In nature, peak dispersal was recorded at about 0800 in cucumbers (Israel), 0930 in cantaloupe (Georgia), and between 0900–1000 in watermelon (Florida) (see Cohen and Rotem, 1971b; Van Haltern, 1933; Schenck, 1968). Thomas (1977a) observed a peak in cantaloupe at 1300, when the dew had dried off at 0930.

Greatest dispersal of sporangia in cucumbers was recorded when disease incidence reached a level of 33% ⅔ of the leaf area was covered with chlorotic lesions) and lasted for about 2 weeks (Cohen and Rotem, 1971b). In cantaloupe, greatest dispersal was associated with a disease index of 4 out of 5 (Thomas, 1977a), and lasted for 2 weeks. In watermelon, sporangia were collected for a period of 30 days (Schenck, 1968).

Dispersed sporangia have to withstand a period of dryness until free leaf moisture is encountered. Using live trap plants Van Haltern (1933) found that sporangia remained infective after 57 h of storage under dry greenhouse conditions while Cohen and Rotem (1971b) showed that infectivity was inversely related to temperature and r.h. during storage. Twenty-two hours of storage at 30°C resulted in 10, 30 and 100% infectivity at r.h.'s of about 90, 50 and 30% respectively, compared to infectivity of 0, 0, and 50% at 35°C. Surprisingly enough, detached sporangia survived longer on slide glasses: 5 days at 17–30°C, and 16 days at 17–21°C (Kajiwara and Iwata, 1968), and after exposure to sunlight for 12 h retained their germinability irrespective of temperature within the range of 20–28°C or r.h. (38–71%).

Germinability and, hence, infectivity, were greatly reduced if sporangia were wetted for a period too short to ensure germination (Doran, 1932). A drying period of 10–15 min was enough to destroy the inner membraneous integrity of previously-wetted sporangia (Cohen et al., 1974) thus ending the metabolic process required for germination.

The adverse effect of wetting–drying–rewetting on sporangial germination was successfully used by Cohen (unpublished) to control the disease on young cucumber plants in the field. Spraying water daily at noon was efficient until the canopy became so dense that water droplets did not dry off quickly enough.

## V. Epidemiology and Crop Losses

In those areas where downy mildew has been present for many years, it is not known how the pathogen overwinters, nor is it known by what means it entered previously disease-free zones such as those in Holland (Van Vliet and Meysing, 1974) and Italy (D'Ercole, 1975). The role of oospores in initiating the disease on cultivated cucurbits has never been elucidated though oospores would seem to be the most suitable means for overwintering. They have been found in cucumber in Japan (Hiura and Kawada, 1933) and China (Chen et al., 1961), and in wild cucurbits in other countries (Bains et al., 1977; Khosla et al., 1973). The only circumstantial evidence to indicate the involvement of oospores in epidemic outbreaks came from North China (Tang, personal communication) in which the disease appeared to initiate from foci, especially in fields previously exposed to cucumbers.

In northern India and southern USA, the pathogen overwinters as active mycelium (Bains and Jhooty, 1976a; Van Haltern, 1933) on either cultivated or wild cucurbits. Having thus overwintered, air-borne inoculum then migrates from the southern USA to the Atlantic coastal states to cause infections in spring and early summer. Nusbaum (1944, 1947) was therefore able to show that monitoring the disease in the south could enable its appearance in the north to be predicted. In Israel, initial inoculum comes from unknown air-borne sources and it is possible that this involves a long distance air-borne sporangial migration since wild cucurbits growing nearby during the winter have nerver been seen to be infected. With modern agricultural techniques whereby cucurbits may be cultivated all the year round, the question of primary inoculum sources seems of little practical importance.

Sporangium concentration in the air seems to be the most important biotic factor determining disease onset (Cohen, 1977). Cohen and Rotem (1971a) showed that disease occurred at the cotyledon stage 8 days after sowing in plants growing nearby infected fields. When no such inoculum source was available disease occurrence was noticed at 7 weeks after sowing.

Dew is the most important environmental factor affecting infection and hence disease outbreak and development. Duvdevani et al. (1946) showed that prevention of dew formation largely precluded the disease. In the absence of dew, guttation drops may be used by the pathogen for infection (Duvdevani et al., 1946). In such a case leaf margins would be affected.

Sporulation is associated with dew formation not because of the need for free leaf moisture, but because of the moisture-saturated air it induces. Dew periods shorter than 6 h prevent the disease because sporulation cannot take place. Overhead sprinkling irrigation enhances disease development, slightly because of a direct effect on infection, and mainly because of its indirect effect on dew formation and hence on sporulation.

Unlike the infection process, the progress of an epidemic is influenced by temperature. In two experimental fields in Israel, sown in February and June, the disease appeared about 35 days after sowing, but the apparent infection rate ($r$) was higher in the February plot ($r = 0.27$ per unit per day) than in the June plot ($r = 0.14$ per unit per day) primarily it would seem, because of the lower temperatures that prevailed during the earlier growing season (mean day:night of 23:13°C in February, as compared to 30:21°C in June) (Cohen unpublished).

By stimulating epidemics in growth chambers, we found (Cohen and Rotem, 1971c) that lower temperatures reduced plant growth and correspondingly increased the percentage of the leaf area infected. Plants with poorer foliage would therefore suffer comparatively greater damage.

High temperatures may kill the pathogen in leaves. In the laboratory, 8 h of exposure to 40°C was fatal, while in the field hot and dry spells (35–44°C, r.h. of 15%) stopped the disease completely (Cohen, unpublished, and Van Haltern, 1933), with many tiny chlorotic, sterile lesions appearing some days afterwards. Similarly, it was found, that if plastic tunnels remained closed on sunny days, the pathogen was killed because of the great rise in temperature. It still remains obscure how, when favourable temperatures return, pathogen renewal occurs. Van Haltern attributed this to fungal mycelium which he found in stems of cantaloupes.

No quantitative data are available on losses caused by downy mildew in cucurbits, but it is widely held that this disease is a limiting factor in crop production. Our observations showed that losses are directly related to the length of the lag period between planting and the onset of the disease. An earlier outbreak, associated with a close proximity to infected fields, may cause a total loss. Unfavourable temperatures for plant growth may increase losses. Overhead sprinkling irrigation is usually associated with higher losses than those encountered where trickle irrigation is used.

## VI. Resistance and Resistant Varieties

Cucumbers have been cultivated in India for more than 3000 years and most genes for resistance can be traced back to that country (Leppik, 1970). The tolerant Indian cucumber cultivar Bangalore was the first to be used as a resistant parent in breeding against *P. cubensis* (Sitterly, 1972). Jenkins (1946) followed with the discovery of two multiple-gene highly-resistant cultivars Chinese Long and Puerto Rico 37, from which Barnes produced Palmetto and Ashley cultivars (Sitterly, 1972). Because of instability (Epps and Barnes, 1952), the Puerto Rican source was replaced by PI 197087 (Barnes and Epps, 1954) which appears to be polygenic, with the inheritance controlled by one or two major genes and one or more minor genes. Resistance was expressed by lesions which turned dark brown without passing the yellow stage. This is still the major source of resistance in the USA, from which the GY 14A, Poinsett and other varieties were produced. The only other source in use is PI 212233 from Japan used by Munger at Cornell University (Thomas, personal communication).

In spite of the widely-held notion that PI 197087 had multiple gene resistance (Barnes and Epps, 1954; McFerson *et al.*, 1978), Van Vliet and Meysing (1974, 1977) demonstrated that the resistance of Poinsett in Holland is based on one recessive gene which is either linked or identical to the gene for resistance to powdery mildew.

Poinsett (Cohen, 1976) and GY 14A (McFerson *et al.*, 1978) were found to inhibit both colonization by the pathogen and its sporulation. Lesions on Poinsett are small and circular (0·5–1 mm) and sporangial yield is negligible, but whether these two characteristics of resistance are controlled by one or more genes is still to be determined.

The genetic control of resistance in cantaloupes is also not fully understood. Ivanoff (1944) was the first to approach the problem, showing that the resistant cultivar Smith Perfect, which originated from the West Indies, had a partial dominance when crossed with susceptible Hale's Best cultivars. The Smith Perfect is in the pedigree of all material in use in the USA today (Thomas, personal communication).

Indian sources currently constitute the background material for resistance in cantaloupe in the USA. Almost all of this resistance can be traced back to PI 124111 and especially to PI 124112, collected in Calcutta in extreme eastern India. Thomas and Webb, in Texas, utilize the genetic material PI 180280 from Rajkot in extreme west India, which carries some resistance to downy mildew, and is also resistant to the watermelon virus I (Thomas, personal communication). The honeydew variety Floridew, developed in Florida, was obtained from PI 223637 from Iran (Thomas, personal communication to J. Palti).

Resistance of PI 124112 was incorporated into Georgia 47. This cultivar, crossed with Smith's Perfect in Taiwan, produced after selection, a high-quality cantaloupe, introduced into the USA as PI 321005 (Sowell and Corley, 1974; Thomas, personal communication). The cantaloupe cultivar Seminole, described by Sitterly (1972) as immune, has Georgia 47 and consequently PI 124112 in its pedigree (Sowell *et al.*, 1973).

Cohen (unpublished) examined the resistance of 19 PI entries (kindly supplied by G. Sowell, Jr.) under field conditions in Israel. With the heavy epidemic that developed, PI 124111 (Calcutta, India), PI 164323 (Madras, India) and PI 165449 (Oaxala, Mexico) exhibited a considerable amount of resistance. The lesions that developed were either yellow, uniformly small and circular (about 0·5–1 mm in diameter) with no sporulation, or were mixed with larger, angular lesions (2–5 mm) with some sporulation. PI 124111 was used to produce the cultivars Campo and Jacumba in California, both considered to have slight resistance to downy mildew (Sowell *et al.*, 1973).

Thomas (1978) examined several breeding lines in growth chambers for lesion number and colour. He found that plants which produced brown lesions had fewer lesions and those with yellow lesions had more, confirming a previous rating made in the field (Thomas, 1977b).

Further study is needed in order to elucidate the genetic control of resistance to *P. cubensis*, especially in cantaloupe. The divergent reaction of a plant material in various countries indicates that different races of the pathogen do exist.

## VII. Disease Control

Cultural control of the disease is discussed in Chapter 13 by Palti and Rotem and cultivar resistance is described in detail in the previous section; hence only chemical control will be discussed here.

Copper fungicides have long been used to protect cucurbits from downy mildew (Van Haltern, 1933; Doran, 1932) but dithiocarbamates later replaced them and are still in use today. Efforts were made to introduce the antibiotic streptomycin as a chemical control agent (Ark and Thompson, 1957), but its usage did not spread. The dithiocarbamate fungicides exert a protective activity by inhibiting zoospore release whereas streptomycin disturbs the establishment of the fungus in host tissue (Cohen and Perl, 1973). All protectants must be applied repeatedly at short intervals so as to protect new growth.

A breakthrough in the chemical control of the disease was achieved recently when systemic fungicides active against Phytomycetes were developed. Propamocarb (propyl-(3-(dimethylamino)-propyl) carbamatemonohydrochloride) and prothiocarb (S-ethyl-N-(3-dimethylaminopropyl)-thio carbamate-monohydrochloride), applied to the root system of potted cucumbers, protected them from the disease for about 25 days (Cohen, 1979). The new systemic fungicide metalaxyl (Ridomil, CGA 48988, methyl-N-(2(methoxy-acetyl)-N-2,6-xylyl-dl-alaninate has been found to be highly efficient in controlling the disease in both the greenhouse and the field, even if applied 48 h after inoculation (Reuveni and Cohen, 1979). Unfortunately, a new race of the pathogen, resistant to the chemical, has recently appeared in Israel (Reuveni *et al.*, 1980).

## VIII. Conclusions

Many gaps in the knowledge of downy mildew of cucurbits still exist and an international effort is needed to identify more clearly the races of the pathogen. The occurrence of oospores is erratic and merits closer investigation. This is particularly so where recombinations in the pathogen's populations and hence changes in its pathogenicity might well occur in India, China and Japan, where sexual reproduction occurs. Special care should be taken to avoid transfer of contaminated plant material to areas relatively free of the disease. Breeding programmes should be expanded to produce cucurbits resistant to the disease.

## References

Ark, P. A. and Thompson, J. B. (1957). *Pl. Dis. Reptr.* **41**, 452–454.
Bains, S. S. and Jhooty, J. S. (1976a). *Indian Phytopathology* **29**, 213–214.
Bains, S. S. and Jhooty, J. S. (1976b). *Indian Phytopathology* **29**, 214–215.
Bains, S. S. and Jhooty, J. S. (1978). *Indian Phytopathology* **31**, 42–46.
Bains, S. S., Sokhi, S. S. and Jhooty, J. S. (1977). *Indian J. Mycol. Pl. Pathol.* **7**, 86.
Barnes, W. C. and Epps, W. M. (1954). *Pl. Dis. Reptr.* **38**, 620.
Chen, C. P., Sung, C. C. and Ho, C. C. (1961). *Rev. Appl. Mycol.* **40**, 383.
Clinton, C. P. (1905). *Conn. State Agr. Exp. Stn.*, Ann. Rept. for 1904, **28**, 329–362.
Cohen, Y. (1976). *Phytoparasitica* **4**, 25–31.
Cohen, Y. (1977). *Can. J. Bot.* **55**, 1478–1487.
Cohen, Y. (1979). *Phytopathology* **69**, 433–436.
Cohen, Y. and Eyal, H. (1977). *Physiol. Pl. Pathol.* **10**, 93–103.
Cohen, Y. and Eyal, H. (1980). *Physiol. Pl. Pathol.* **17**, 53–62.
Cohen, Y. and Perl, M. (1973). *Phytopathology* **63**, 1172–1180.
Cohen, Y. and Rotem, J. (1969). *Israel J. Bot.* **18**, 135–140.
Cohen, Y. and Rotem, J. (1970). *Phytopathology* **60**, 1600–1604.
Cohen, Y. and Rotem, J. (1971a). *Phytopathology* **61**, 736–737.
Cohen, Y. and Rotem, J. (1971b). *Trans. Br. Mycol. Soc.* **57**, 67–74.
Cohen, Y. and Rotem, J. (1971c). *Phytopathology* **61**, 265–268.
Cohen, Y., Perl, M. and Rotem, J. (1971). *Phytopathology* **61**, 594–595.
Cohen, Y., Perl, M., Rotem, J., Eyal, H. and Cohen, J. (1974). *Can. J. Bot.* **52**, 447–450.
Cohen, Y., Levi, Y. and Eyal, H. (1978). *Can. J. Bot.* **56**, 2538–2543.
D'Ercole, N. (1975). *Informat. Fitopatol.* **25**, 11–13.
Doran, W. L. (1932). *Bull. Mass. Agr. Exp. Stn.* **283**, 22 pp.
Duvdevani, S., Reichert, I. and Palti, J. (1946). *Palest. J. Bot.* **2**, 127–151.
Epps, W. M. and Barnes, W. C. (1952). *Pl. Dis. Reptr.* **36**, 14–15.

Hiura, M. (1929). *Res. Bull. Gifu Imp. Coll. Agr.* **6**, 58 pp.

Hiura, M. and Kawada, S. (1933). *Jap. J. Bot.* **6**, 507–513.

Hughes, M. B. and Van Haltern, F. (1952). *Pl. Dis. Reptr.* **36**, 365–367.

Inaba, T. and Kajiwara, T. (1971a). *Ann. Phytopath. Soc. Japan* **37**, 220–224.

Inaba, T. and Kajiwara, T. (1971b). *Ann. Phytopath. Soc. Japan* **37**, 340–347.

Inaba, T. and Kajiwara, T. (1972). *Ann. Phytopath. Soc. Japan* **38**, 25–29.

Inaba, T. and Kajiwara, T. (1974a). *Ann. Phytopath. Soc. Japan* **40**, 30–38.

Inaba, T. and Kajiwara, T. (1974b). *Ann. Phytopath. Soc. Japan* **40**, 296–298.

Inaba, T. and Kajiwara, T. (1975). *Bull. Nat. Inst. Agric. Sci.* Series C, **29**, 65–130.

Ivanoff, S. C. (1944). *J. Hered.* **35**, 35–39.

Iwata, Y. (1938). *Ann. Phytopath. Soc. Japan* **8**, 125–144.

Iwata, Y. (1941). *Ann. Phytopath. Soc. Japan* **11**, 101–113.

Iwata, Y. (1952a). *Bull. Fac. Agric. Mie Univ. Japan* **5**, 35–41.

Iwata, Y. (1952b). *Bull. Fac. Agric. Mie Univ. Japan* **5**, 42–48.

Iwata, Y. (1953a). *Bull. Fac. Agric. Mie Univ. Japan* **6**, 32–36.

Iwata, Y. (1953b). *Bull. Fac. Agric. Mie Univ. Japan* **7**, 28–36.

Jenkins, J. M., Jr. (1946). *J. Hered.* **37**, 267–271.

Kajiwara, T. and Iwata, Y. (1957). *Ann. Phytopath. Soc. Japan* **22**, 201–203.

Kajiwara, T. and Iwata, Y. (1959). *Ann. Phytopath. Soc. Japan* **24**, 109–113.

Kajiwara, T. and Iwata, Y. (1968). *Ann. Phytopath. Soc. Japan* **34**, 85–91.

Koshla, H. K., Dave, G. S. and Nema, K. G. (1973). *JNKVV Res. J.* **7**, 175–177.

Leppik, E. E. (1970). *Ann. Rev. Phytopath.* **8**, 323–341.

McFerson, J. R., Pike, L. M. and Frederiksen, R. A. (1978). *Phytopath. News* **12**, 183 (Abstr.).

Nusbaum, C. I. (1944). *Pl. Dis. Reptr.* **28**, 82–85.

Nusbaum, C. I. (1947). *Pl. Dis. Reptr.* **32**, 44–48.

Palti, J. (1974). *Phytoparasitica* **2**, 109–115.

Palti, J. and Cohen, Y. (1980). *Phytoparasitica* **8**, 109–147.

Parriss, G. K. (1951). *Pl. Dis. Reptr.* **35**, 52–53.

Perl, M., Cohen, Y. and Rotem, J. (1972). *Physiol. Pl. Path.* **2**, 113–122.

Reuveni, M. and Cohen, Y. (1979). *Phytoparasitica* **7** (Abstr., 51–52).

Reuveni, M., Eyal, H. and Cohen, Y. (1980). *Pl. Dis. Reptr.* **64**, 1108–1109.

Schenck, N. C. (1968). *Phytopathology* **58**, 91–94.

Sitterly, W. R. (1972). *Ann. Rev. Phytopath.* **10**, 471–490.

Sitterly, W. R. (1978). *In* "The Powdery Mildews" (D. M. Spencer, Ed.), 359–379. Academic Press, London and New York.

Sowell, G., Jr. and Corley, W. L. (1974). *Pl. Dis. Reptr.* **58**, 900–902.

Sowell, G., Jr., Braverman, S. W., Diety, S. M., Clark, R. L., Jarvis, J. L. and Pestro, G. R. (1973). *Reg. Pl. Introd. Stn. Exp. GA.* 16 pp.

Thomas, C. E. (1970). *Pl. Dis. Reptr.* **54**, 108–111.

Thomas, C. E. (1977a). *Phytopathology* **67**, 1368–1369.

Thomas, C. E. (1977b). *Pl. Dis. Reptr.* **61**, 375–377.

Thomas, C. E. (1978). *Pl. Dis. Reptr.* **62**, 221–222.

Van Haltern, F. (1933). *Georgia Exp. Stn. Bull.* **175**.

Van Vliet, G. J. A. and Meysing, W. D. (1974). *Euphytica* **23**, 251–255.

Van Vliet, G. J. A. and Meysing, W. D. (1977). *Euphytica* **26**, 793–796.

Chapter 17

# Downy Mildew of Forage Legumes

## D. L. STUTEVILLE

*Department of Plant Pathology, Kansas State University, Manhattan, Kansas 66506, USA*

## I. Introduction

Downy mildew occurs widely on many forage legumes in cool moist conditions. It is rarely the major disease of any crop but takes its toll, particularly of seedling stands. Thyr *et al.* (1978) found that autumn-sown alfalfa (lucerne) stands weakened by downy mildew were more likely than healthy stands to be killed in the following winter. In North and South America the disease is widespread on alfalfa but occurs rarely on sweetclover and clovers. In

Europe, however, it is more common on clovers. It is widespread on alfalfa in the USSR (Assaul, 1973). Downy mildew of vetch is sometimes severe in Britain (Sampson and Western, 1954) and is one of the more widespread and serious diseases of winter vetch in the Poltava region of Ukrainian USSR (Turukina and Temnokhud, 1977).

Worldwide, the economic losses from mildew in forage legumes are undoubtedly greatest on alfalfa and since most research has therefore been on that crop, this chapter will deal primarily with downy mildew of alfalfa.

## II. Host Range and History

*Peronospora trifoliorum* de Bary, the primary cause of downy mildew in the major forage legume genera, except *Vicia*, was first reported by de Bary in 1863 on *Medicago sativa* L., *Trifolium alpestre* L., *T. incarnatum* L., and *T. medium* L. In 1884, Smith reported it in England on *M. sativa*, *T. incarnatum* and *T. pratense* L.; Fischer (1892) reported additional hosts including *Coronilla varia* L., *Lotus corniculatus* L., *L. uliginosus* Schk., *Medicago falcata* L., *M. lupulina* L., *Melilotus alba* Desr., *M. officinalis* (L.) Lam., *Trifolium agrarium* L., *T. arvense* L., *T. minus* Koch, *T. repens* L., *T. rubens* L., *T. spadiceum*, and *T. striatum* L. Chilton *et al.* (1943) included an extensive synonomy and reference list for *Peronospora* on species of *Medicago*, *Melilotus*, and *Trifolium*. A current host range and geographical distribution of *P. trifoliorum*, is available in CMI Map No. 343.

Downy mildew of *Vicia* species is caused by *P. viciae* (Berk.) de Bary (de Bary, 1863). It was first described by Berkeley in England in 1846 and is widespread on vetches. Johnson *et al.* (1956) reported *P. viciae* on *Medicago sativa* and *Trifolium incarnatum* in the USA.

## III. Morphology and Taxonomy

*P. trifoliorum* fruiting structures vary widely in size and Table 1 illustrates the variation on alfalfa. Sampson and Western (1954) gave the range in size of structures on some *Trifolium* species recorded by Gäumann (1923). Gäumann's 1923 monograph of the genus *Peronospora* separated *P. trifoliorum* into several species, primarily on biometrical and minor morphological differences in the apices of the conidiophores. However, his changes have not been widely accepted.

Gäumann (1923) assigned the downy mildew fungus that attacked *Medicago* species to *P. aestivalis* and observed that it occasionally produced

TABLE 1

Size of fruiting structures of *Peronospora trifoliorum* de Bary
from *Medicago sativa* as reported by different investigators

| | Size (μm) | | |
|---|---|---|---|
| Investigator | Conidia | Conidiophores | Oospores |
| Fischer (1892) | 16–19 × 20–22 | 360–460 | 24–30 |
| Gäumann (1923)[a] | 9–27 × 16–37 | 200–500 | 20–30 |
| Melhus and Patel (1929) | 14·5–25·5 × 15·5–30·5 | 110–680 | 18–37 |
| Saccardo (1888) | 15–19 × 19–26 | — | 25–34 |

[a] Gäumann classified the form on *M. sativa* as *P. aestivalis.*

wrinkled epispores. However, descriptions of *P. trifoliorum* include only smooth-walled epispores (Fischer, 1892; Saccardo, 1888). This point deserves further attention, as I have never observed smooth-walled epispores in alfalfa.

*P. viciae* oospores are characterized by broad reticulate markings on the epispore wall (de Bary, 1863; Fischer, 1892). Gäumann (1923) also subdivided this species and reported that oospores were not produced in *Vicia sativa* but were 29–46 μm in diameter in other *Vicia* species. Gäumann's measurements of the form from *V. sativa* gave conidiophores as 400–800 μm and conidia as 19–37 × 12–31 μm whereas Fischer's (1892) measurements of *P. viciae* were, conidiophores 300–700 μm, and conidia 16–20 × 21–27 μm.

## IV. Biology

### A. Conidium germination and penetration

*P. trifoliorum* conidia germinated from 4 to 29°C with the optimum at 18°C (Patel, 1926; Melhus and Patel, 1929). They required a water film and did not germinate on a dry slide at 100% relative humidity. Conidium germination was enhanced by light. Detached conidia in water exposed to 5400 lux of fluorescent lighting for 0 min, 10 min, 1 h, or 24 h and then placed in darkness for the remainder of a 24-h period germinated 43, 72, 69, and 85% (LSD ($p > 0·05$) = 6), respectively (Fried and Stuteville, 1977). Germination percentages did not differ significantly among treatments under continuous fluorescent lighting at intensities of 540–10 800 lux. Waite (1971) reported that the germination percentage of *P. trifoliorum* conidia collected in the field was highest at 1100 h.

As with all *Peronospora* species, *P. trifoliorum* conidia germinate by producing a germ tube. Cohen (1951) showed that temperature affects the method of germ-tube penetration by some Peronosporales and this could apply to the penetration of alfalfa by *P. trifoliorum*. Rockett (1970) observed that the germ tubes formed appressoria and penetrated the epidermis directly at 20°C but Waite and Cannon (1974), using 5 and 15°C, observed only stomatal penetration.

## B. Infection and disease development

After penetrating susceptible leaves, the fungus moves intercellularly though haustoria, which are usually formed by obligate parasites, are not always produced by *P. trifoliorum* in alfalfa. Fraymouth (1956) found a few haustoria in small disintegrating areas of infected alfalfa leaves collected in May and June but none in August collections.

Ultrastructural studies of nonhaustorial *P. trifoliorum* mycelia revealed finger-like ingrowths of the fungal wall that extended up to 1 $\mu$m into the fungal cytoplasm (Martin and Stuteville, 1975). The fungus plasma membrane conformed to the contours of the wall, thus the ingrowths greatly increased its surface area. The increased absorptive area provided by this larger plasma membrane may permit the fungus to function without haustoria.

Disease development is usually reflected by symptoms, with the most resistant plants expressing none. Symptoms range from a few white pin-point dots, with no sporulation, on inoculated leaves of plants with high levels of resistance, to heavy sporulation on severely stunted and chlorotic, systemically-infected shoots of the most susceptible plants.

Systemic invasion of alfalfa by *P. trifoliorum* is apparently affected by both genetic and environmental factors. Jones and Torrie (1946) noted that only a proportion of susceptible plants were capable of being invaded systemically and Hodgden (1978), studying environmental effects on oospore production in alfalfa, noted that *P. trifoliorum* became systemic only in seedling $S_1$ progeny from one to two clones used. Systemic infections were most frequent at 12°C (the lowest temperature tested) with a 24-h photoperiod of 10 000 lux (the highest intensity used) and generally decreased as temperature increased and photoperiod and/or light intensity decreased.

## C. Conidium production

Conidium production by *P. trifoliorum* requires a dark period (Rockett and Stuteville, 1969) and relative humidity of at least 97% (Fried and Stuteville, 1977), but does not occur in free water on leaves. Fried and Stuteville (1977) studying conidium development and discharge found that immature conidia were spherical and became egg-shaped when morphologically mature. At 20°C, conidia were morphologically mature after a 10-h dark period but their germinability was highest at 12 h and declined after 16 h. A short exposure to light stopped conidium production for 2–3 h, then it was resumed and progressed normally. Conidia formed simultaneously at the apices of all branches on each conidiophore so they were the same age and size. However, conidiophores at various stages of development often protruded from the same stoma.

Conidia were released in water or were discharged into the air by the twisting of conidiophores during drying (Fried and Stuteville, 1977). The decrease in moisture tension for this form of mechanical discharge caused conidia to shrivel, and apparently reduced their germinability.

## D. Oospore production

Oospores of *P. trifoliorum* were produced sparingly in the field in late autumn (Patel, 1926) and in dead or dying alfalfa leaves (Melhus and Patel, 1929). Rockett (1970) observed oospores 7 days after inoculation in alfalfa seedlings kept at 20°C. *P. trifoliorum* is homothallic as shown by Hodgden (1978) who inoculated alfalfa seedlings with a monoconidial isolate of *P. trifoliorum* and observed them for oospore production over three weeks at 12, 16, 20, and 24°C in photoperiods of 8, 16, and 24 h at fluorescent light intensities of 6500 and 10 000 lux. Oospores were produced within one week in all treatments except at 12°C and 6500 lux. At three weeks oospores were produced in all treatments and averaged from 38 to 11 081 per seedling. Production peaked at 16°C with a 24-h photoperiod of 10 000 lux and tended to decrease as photoperiod and/or light intensity decreased. Higher light intensity also favoured early oospore production though there were fewer oospores when conidia were also produced.

## E. Perennation and spread

*P. trifoliorum* probably overwinters as hyphae and as oospores in the crown and/or crown buds. Jones and Torrie (1946), after observing conidium production in early spring on systemically infected crown shoots that had survived the winter, concluded that systemic infection in shoots is important in the overwintering of the fungus, though Patel (1926) failed to show that mycelium could survive the winter in alfalfa plants.

It has long been assumed that *P. trifoliorum* overwinters in the form of oospores, but to my knowledge their germination was not reported until 1968 by Faizieva. She noted that oospores from alfalfa leaves stored in cotton bags on and 1–10 cm under the soil surface in both the field and in the greenhouse all germinated within 7 days. The length of storage period and the methods used to initiate germination were not made clear. Hodgden (1978) used several procedures but was unable to germinate *P. trifoliorum* oospores.

Conidia are too short-lived to overwinter. Their viability declined rapidly with desiccation (Fried and Stuteville, 1977) and they did not survive 10 days when frozen in water (Melhus and Patel, 1929).

Conidia are the chief means of secondary infection and dispersal for short distances but are probably too fragile to be wind-borne over long distances. Oospores, either in hay or in the debris accompanying seed, are the most likely means of long-distance dispersal. Both Dickson (1947) and Eriksson (1930) suggested that *P. trifoliorum* might be transmitted by infected alfalfa seed but to my knowledge that has not been demonstrated.

## V. Pathogenic Specialization

*P. trifoliorum* encompasses many host-specific forms. Gäumann (1923) inoculated plants of several of 38 legume species with conidia from 12 hosts and got cross-infection in only one; conidia from *Lotus corniculatus* infected *L. uliginosus*. Melhus and Patel (1929), after inoculating 28 leguminous hosts with *P. trifoliorum* conidia from alfalfa, reported infection on all of the inoculated alfalfa plants, scanty infection on 8% of the inoculated *Medicago lupulina* plants, and no infection on any of the other 26 hosts.

By using alfalfa clones and monoconidial isolates, Stuteville (1973) demonstrated pathogenic specialization of *P. trifoliorum* in alfalfa and showed that the ranking of cultivars for mildew resistance depended on the mildew isolate used for inoculation.

## VI. Nature of Host Resistance

Little is known about the nature of the resistance of alfalfa to downy mildew. Martin (unpublished) observed that *P. trifoliorum* germ tubes penetrated resistant alfalfa leaves, but that further development ceased as the host responded hypersensitively. Susceptible leaves become resistant with age.

Saponins in alfalfa may influence downy mildew resistance. Pedersen *et al.* (1976) selected high and low saponin populations in six alfalfa cultivars and studied the relationship between the saponin index and many agronomic and pest resistance traits. The saponin index and downy mildew resistance were positively and significantly correlated ($r = 0.901$) among the high saponin populations. The parent cultivars showed a similar but less striking trend. Berkenkamp *et al.* (1978) also observed significantly less downy mildew in Pedersen's high than in his low saponin Ranger selections.

Stuteville and Sorensen (1980) tested the resistance of Pedersen's selected material and the parent cultivars to three monoconidial isolates of *P. trifoliorum* and compared the percentage of resistant plants in each entry with the saponin index assigned by Pedersen *et al.* (1976). Selecting for saponin content significantly affected mildew resistance in all six cultivars. For Ranger and Vernal, and their selections, mildew resistance was positively associated with saponin index. For the cultivars, DuPuits, Ladak, Lahontan, and Uinta, mildew resistance of both the high and low selections was lower or not significantly different from that of the parent. Except for Ladak and Lahontan, mildew resistance to each isolate was greater for the high-saponin than for the low-saponin selection. However, the correlation coefficient between the saponin index and downy mildew resistance was low primarily because a few entries with high saponin indices had little if any resistance to one or two of the fungus isolates. Entries with the lowest saponin indices generally had low levels of mildew resistance.

The involvement of isoflavonoid phytoalexins produced by *Medicago* species (Ingham, 1979) in resistance to *P. trifoliorum* has not been examined.

## VII. Genetics of Host Reaction

Jones and Smith (1953) suggested that susceptibility to downy mildew in alfalfa was inherited as a dominant character. However, the decrease in the proportion of susceptible plants from 89 to 6.5% after one backcross and one selfed generation during the development of the cultivar Caliverde (Stanford, 1952) suggests some degree of dominance for resistance. Pedersen and Barnes (1965) indicated that mildew resistance was conditioned by one tetrasomically inherited gene with incomplete dominance. The discrepancies might, in part, stem from strains of the fungus to which resistance is inherited differently. Additional research under isolated conditions with known strains of the fungus is needed.

## VIII. Breeding for Downy Mildew Resistance in Alfalfa

Genetic resistance is the most feasible means of controlling downy mildew of alfalfa and it is available in a diversity of alfalfa germplasm. Barnes *et al.*, (1977) listed nine germplasm sources introduced into the US between 1850 and 1947 with the percentage that each comprise of US cultivars. These sources probably represent most of the material currently used in alfalfa breeding in the world. In descending order of their winter-hardiness they are *Medicago falcata*, Ladak, *M. varia*, Turkistan, Flemish, Chilean, Peruvian, Indian, and African.

*Medicago falcata*, a source of extreme winter-hardiness in cultivars developed for northern US and Canada has a high frequency of mildew-resistant plants.

The cultivar Ladak, introduced from northern India, was ranked eighth among 21 strains or cultivars examined for mildew resistance in Wisconsin (Smith, 1948) but was ranked low in Canada (Berkenkamp *et al.*, 1978; Harding, 1972).

The cultivar Grimm, of 100% *M. varia* parentage, showed high downy mildew resistance under natural infection in Canada (Harding, 1972) and in Wisconsin (Smith, 1948), and was ranked intermediate among cultivars in Nevada (Thyr *et al.*, 1978).

The cultivar Lahontan, derived entirely from Turkistan germplasm rated intermediate for mildew resistance (Hanson *et al.*, 1964; Thyr *et al.*, 1978).

The Flemish cultivars are highly resistant to downy mildew (Nittler *et al.*, 1964). DuPuits, developed in northern France, was introduced into the US in 1947 and was followed by other Flemish cultivars including Alfa, Cardinal, and Socheville. Flemish germplasm has been an important source of mildew resistance in many US cultivars developed after 1963 (Barnes *et al.*, 1977).

Caliverde, released in 1951 with downy mildew resistance derived from Chilean germplasm, was the first US cultivar developed with mildew resistance as a major objective (Stanford, 1952). A modified backcross breeding method with greenhouse screening was used in its production.

Mesa-Sirsa, derived by clonal selection in Sirsa No. 9 of Indian origin, was highly resistant to downy mildew (Schonhorst *et al.*, 1966, 1968).

Sonora, a 13-clone synthetic derived from African germplasm had intermediate resistance to mildew (Schonhorst *et al.*, 1966).

Although these various examples illustrate downy mildew resistance in specific germplasm sources, most cultivars have parentage from two or more of those sources (Barnes *et al.*, 1977). Most, if not all, alfalfa cultivars have at least a low percentage of plants with mildew resistance that can be increased. No cultivar has all plants resistant to downy mildew but rather has proportions of plants with various levels of resistance. In order to characterize levels of resistance, workers often classify individual plants arbitrarily. Thus, Nittler *et al.* (1964) rated plants with least disease as 1 and those with most severe symptoms as 5 whilst Pedersen and Barnes (1965) used a scale of 1 to 4. Their class 1 plants had no infection to a trace; class 2 plants, 2–5% of leaves infected; class 3 plants, 6–25% of leaves infected; and class 4 plants, general infection over the entire plant.

Alfalfa genetic behaviour and breeding procedures appropriate for developing downy mildew-resistant cultivars are described by Busbice *et al.* (1972). To achieve rapid selection for increased resistance requires a severe screening test best accomplished by artificially created epidemics in the laboratory. Natural epidemics are unpredictable in most areas and are seldom adequately severe to identify all suceptible plants. For example, during a severe natural epidemic in Minnesota, most of 36 cultivars evaluated had 55% or more mildew-free plants, and more than 90% of Saranac plants were mildew-free (Kehr *et al.*, 1972). However, under seedling screening procedures (Section IX), approximately 16% of Saranac plants remained symptomless and some cultivars had fewer than 1% symptomless plants (Stuteville, unpublished).

## IX. Procedures for Screening Alfalfa Seedlings for Resistance

### A. Inoculum preparation

Conidia from infected alfalfa plants collected in the field are usually contaminated with bacteria and other fungi and germinate poorly, so it is preferable to produce conidia for inoculum on seedlings in the laboratory and use field collections only for initial inoculum. Conidia from several sources should be included in screening for cultivar resistance because pathogenic specialization occurs in *P. trifoliorum*. To inhibit most of the bacteria and fungi which are antagonistic to *P. trifoliorum* and also pathogenic to alfalfa, place the collected leaves in a jar with water containing 50 $\mu$g nystatin and 10 $\mu$g tetracycline per ml (Stuteville, 1977). Shake the jar to dislodge the conidia, strain out the leaves and spray the treated spore suspension onto flats of alfalfa seedlings. As tetracycline may cause some chlorosis in alfalfa, it is best to use it only on seedlings to produce stock inoculum for screening purposes.

When only old diseased material is available with conidia of questionable viability, or when mildewed leaves are received in the mail, it may be advisable, after washing off existing conidia, to attempt to produce a new crop on those leaves. This is often possible if the leaves are placed on moistened filter-paper in a petri dish or plastic bag and exposed to light for a few hours. If the leaves are then kept in darkness for 12–16 h, they will often produce conidia.

To prepare inoculum produced on seedlings, uncover and excise the infected seedlings 12–16 h after placing them in the dark to induce sporulation, then immediately place them into a jar containing chlorine-free water (Fried and Stuteville, 1975). Close and shake the jar to dislodge the conidia and pour the spore suspension through a tea strainer to remove the plants. The inoculum should be adjusted to contain at least 25 000 (preferably 100 000) viable conidia per ml.

## B. Conidium storage

*P. trifoliorum* conidia frozen in water remained viable for less than 10 days (Melhus and Patel, 1929) but when newly produced conidia on seedlings harvested and immediately placed at $-20°C$, were tested for variability 11% germinated 24 weeks later (Martin, 1971). Liquid nitrogen has been found to provide the best means of long term storage. The methods developed by Bromfield and Schmitt (1967) to store *P. tabacina* conidia in liquid nitrogen were equally effective for storing *P. trifoliorum* conidia (Martin, 1971). Stuteville (unpublished) found that *P. trifoliorum* conidia kept for 8 years in liquid nitrogen were still infective.

## C. Inoculation procedures and post-inoculation care

Plants are inoculated by spraying the freshly prepared spore suspension onto the seedlings until a drop forms between the cotyledons. To prevent plants from drying off during the infection period, they are kept in a closed box. Conidium germination is improved by light (Fried and Stuteville, 1977) but sufficient exposure to light is obtained during inoculum preparation. At the end of the incubation period so as to induce sporulation, which requires darkness and nearly 100% relative humidity (Fried and Stuteville, 1977), the trays of seedlings are kept in a closed box where the moist sand growth medium will provide sufficient humidity.

The following schedule has been established for screening plants in growth chambers at 20°C and 5000–10 000 lux of continuous fluorescent lighting, except for designated dark periods. Specific days are given for added clarity and to present a schedule that requires little weekend attention.

Day 1—Friday
Plant seeds 1·3 cm deep in rows in flats of fine sand. (Seed dormant types and small seeds a day earlier.)

Days 3–5
To insure even emergence, sprinkle water on flats twice daily to settle the sand around the seedlings.

Day 5—Tuesday, a.m.
Inoculate, cover plants, and turn off lights.

Day 6—Wednesday, a.m.
Uncover plants, turn on lights, and rogue plants emerged since inoculation.

Days 7–10               Continue roguing newly emerged seedlings.

Day 11—Monday, p.m.    Cover plants again and turn off lights 12–16 h (overnight) to induce sporulation. Do not irrigate plants just before covering as the fungus will not sporulate in free water.

Day 12—Tuesday        Rogue all plants with symptoms. Occasional plants that develop symptoms later should be removed.

As a standard to compare tests, the evaluation is based on the percentage of mildew-free plants compared with Saranac plants, included as a resistant control. Saranac is used because to date at least 15% of its plants are resistant to any *P. trifoliorum* isolate we have.

Using these screening procedures with two cycles of recurrent phenotypic selection in a population from the cultivar Arc, Sorensen *et al.* (1978) increased the frequency of downy mildew-free plants from 7% to 43% (235% that of Saranac). More recently the 3% frequency of plants in an alfalfa population that were resistant to one monoconidial isolate of *P. trifoliorum* was increased to 65% by one cycle of recurrent phenotypic selection (Stuteville, unpublished).

## References

Anon. (1978). CMI Distribution Maps of Plant Diseases, Map No. 343.

Assaul, B. D. (1973). *Zasch. Rast.* **10**, 24.

Barnes, D. K., Bingham, E. T., Murphy, R. P., Hunt, O. J., Beard, D. F., Skrdla, W. H. and Teuber, L. R. (1977). *USDA ARS Tech. Bull.* **1571**.

Berkeley, M. J. (1846). *J. R. Hort. Soc.* **1**, 31.

Berkenkamp, B., Folkins, L. P. and Meeres, J. (1978). *Can. J. Pl. Sci.* **58**, 893–894.

Bromfield, K. R. and Schmitt, C. G. (1967). *Phytopathology* **57**, 1133.

Busbice, T. H., Hill, R. R., Jr. and Carnahan, H. L. (1972). *In* "Alfalfa Science and Technology" (C. H. Hanson, Ed.) 283–318. Amer. Soc. Agron., Madison, Wisconsin.

Chilton, S. J. P., Henson, L. and Johnson, H. W. (1943). *USDA Misc. Bull.* 499.

Cohen, M. (1951). Ph.D. thesis. University of California, Berkeley.

de Bary, A. (1863). *Annls Sci. Nat. Bot.* **4** Ser. 20, 117–118.

Dickson, J. G. (1947). "Diseases of Field Crops". McGraw-Hill, New York.

Eriksson, J. (1930). "Fungous Diseases of Plants". Baillière, Tindall and Cox, London.

Faizieva, F. (1968). *Uzbek. Biol. Zh.* **12** (5) 13–15.

Fischer, A. (1892). *Rabenhorst's Kryptogamen-Flora Phycomycetes* **4**, 454–458.

Fraymouth, J. (1956). *Trans. Br. Mycol. Soc.* **39**, 79–107.

Fried, P. M. and Stuteville, D. L. (1975). *Phytopathology* **65**, 929–930.

Fried, P. M. and Stuteville, D. L. (1977). *Phytopathology* **67**, 890–894.

Gäumann, E. (1923). *Beitr. KryptogFlora Schweiz* **5**, 174–222.

Hanson, E. W., Hanson, C. H., Frosheiser, F. I., Sorensen, E. L., Sherwood, R. T., Graham, J. H., Elling, L. J., Smith, Dale and Davis, R. L. (1964). *Crop Sci.* **4**, 273–276.

Harding, H. (1972). *Can. Pl. Dis. Surv.* **52**, 149–150.

Hodgden, L. D. (1978). M.Sc. thesis, Kansas State University, Manhattan.

Ingham, J. L. (1979). *Biochem. Syst. Ecol.* **7**, 29–34.

Johnson, H. W., Kilpatrick, R. A. and Bain, D. C. (1956). *Phytopathology* **46**, 16.

Jones, F. R. and Smith, O. F. (1953). *In* "Plant Diseases the USDA Yearbook of Agriculture" 228–237. Supt. Documents, Washington, D.C.

Jones, F. R. and Torrie, J. H. (1946). *Phytopathology* **36**, 1057–1059.

Kehr, W. R., Frosheiser, F. I., Wilcoxson, R. D. and Barnes, D. K. (1972). *In* "Alfalfa Science and Technology" (C. H. Hanson, Ed.) 335–354. Amer. Soc. Agron., Madison, Wisconsin.

Martin, T. J. (1971). M.Sc. thesis, Kansas State University, Manhattan.

Martin, T. J. and Stuteville, D. L. (1975). *Phytopathology* **65**, 638–639.

Melhus, I. E. and Patel, M. K. (1929). *Proc. Iowa Acad. Sci.* **36**, 113–119.

Nittler, L. W., McKee, G. W. and Newcomer, J. L. (1964). Principles and methods of testing alfalfa seed for varietal purity. *Bull. 807. NY. Agr. Exp. Stn.* Geneva, N.Y.

Patel, M. K. (1926). *Phytopathology* **16**, 72.

Pedersen, M. W. and Barnes, D. K. (1965). *Crop Sci.* **5**, 4–5.

Pedersen, M. W., Barnes, D. K., Sorensen, E. L. *et al.* (1973). *Crop Sci.* **16**, 193–199.

Rockett, T. R. (1970). M.Sc. thesis, Kansas State University, Manhattan.

Rockett, T. R. and Stuteville, D. L. (1969). *Phytopathology* **59**, 1046.

Saccardo, P. A. (1888). *Sylloge Fungorum* **7**, 245–253.

Sampson, K. and Western, J. H. (1954). "Diseases of British Grasses and Herbage Legumes". Cambridge Univ. Press, Cambridge.

Schonhorst, M. H., Nielson, M. W., Keener, P. D., Thompson, R. K., Lieberman, F. V. and Woodrow, A. W. (1966). *Progve Agric. Ariz.* **18** (2), 22–23.

Schonhorst, M. H., Nielson, M. W., Thompson, R. K., Lieberman, F. V., Keener, P. D. and Nigh, E. L., Jr. (1968). *Crop Sci.* **8**, 396.

Smith, D. (1948). *J. Am. Soc. Agron.* **40**, 189–190.

Smith, W. G. (1884). "Diseases of Field and Garden Crops". MacMillan and Co., London.

Sorensen, E. L., Stuteville, D. L. and Horber, E. (1978). *Crop Sci.* **18**, 918.

Stanford, E. H. (1952). *Sixth Intern. Grassland Cong. Proc.* 1585–1590, Pennsylvania State College.

Stuteville, D. L. (1973). *Second Int. Cong. Pl. Path.* Abst. No. 0715. Minneapolis, Minnesota.

Stuteville, D. L. (1977). *Proc. Am. Phytopath. Soc.* **4**, 167–168.

Stuteville, D. L. and Sorensen, E. L. (1980). *Rep. 27th National Alfalfa Improvement Conf.* USDA ARM NC-19.

Thyr, B. D., Hartman, B. J., Hunt, O. J. and McCormick, J. A. (1978). *Pl. Dis. Reptr.* **62**, 338–339.

Turukina, N. F. and Temnokhud, M. P. (1977). *Zakhÿst Roslÿn* **24**, 77–80.

Waite, S. B. (1971). *Utah Sci.* **32**, 98–99.

Waite, S. B. and Cannon, O. S. (1974). *Phytopath. Z.* **79**, 368–369.

Chapter 18

# Downy Mildews of Graminaceous Crops

## R. G. KENNETH

*Department of Plant Pathology and Microbiology,
The Hebrew University of Jerusalem, Rehovot, Israel*

## I. Introduction

The downy mildew diseases of graminaceous crops are so problematical and have become of such economic importance that six international meetings devoted to them have been held in the last decade; the proceedings of most have been or are being published (*Indian Phytopathology* **23**(2), 1970; *Tropical Agricultural Research Centre, Tokyo*, Ser. 8, 1975; *Kasertsart Journal* (Thailand), **10** (2), 1977; *Proc. Consultants' Group Meetings on Downy Mildew and Ergot of Pearl Millet*, International Crops Research Institute for the Semi-Arid Tropics (ICRISAT), Hyderabad, 1–3 Oct. 1975; Proc. Workshop "Sorghum Downy Mildew of Corn and Sorghum", *Rev. Fac. Agron.*,

TABLE 1

The downy mildews of Gramineae and their known hosts

| Fungus | Common name of disease[a] | Presence of resting spores | Systemic infection | Local lesions | Hosts | W. hemisphere | E. hemisphere |
|---|---|---|---|---|---|---|---|
| *Sclerospora* (Schroet.) de Bary *S. graminicola* (Sacc.) Schroet. | Green ear of pearl millet, Graminicola Downy Mildew | ++ | + | Seldom | Maize, Pearl millet[b], Setaria spp.[b] *Panicum* spp., *Euchlaena*, *Saccharum*[c], *Sorghum*[c], *Echinochloa* | + | + |
| *Peronosclerospora* (Ito) Shirai & Hara | | | | | | | |
| *P. dichanthiicola* (Thirum. & Naras.) C. G. Shaw | ------ | - | + | - | *Dichanthium* | - | + |
| *P. heteropogoni* Siradhana, Dange, Rathore and Singh | ------ | + | + | - | Maize[b], *Heteropogon*, *Euchlaena* | - | + |
| *P. maydis* (Racib.) C. G. Shaw | Java Downy Mildew | - | + | - | Maize[b], *Euchlaena*[c], *Tripsacum*[c] | - | + |
| *P. miscanthi* (T. Miyake apud Sacc.) C. G. Shaw | Leaf-splitting Downy Mildew | ++ | + | - | Maize, *Miscanthus*, *Saccharum* spp., *Sorghum plumosum* | - | + |
| *P. noblei* (Weston) C. G. Shaw | ------ | + | + | - | *Sorghum plumosum* (*Andropogon australis*), *Sorghum leiocladum*? | - | + |
| *P. philippinensis* (Weston) C. G. Shaw | Philippine Downy Mildew | +? | + | - | Maize[b], *Saccharum*, *Avena*[c], *Euchlaena*, *Sorghum*, *Tripsacum*, *Miscanthus*[c] | - | + |
| *P. sacchari* (T. Miyake in Ito) Shirai and Hara | Sugarcane Downy Mildew | + | + | - | Maize[b], *Saccharum*[b] spp., *Andropogon* spp.[c], *Bothriochloa*[c], *Euchlaena*, *Miscanthus* (*Eulalia*), *Schizachyrium sp.*[c], *Sorghum* spp., *Tripsacum* | - | + |
| *P. sorghi* (Weston & Uppal) C. G. Shaw | Sorghum Downy Mildew | ++ | + | + | Maize[b], *Sorghum* spp.[b], *Andropogon*[c], *Euchlaena*, *Panicum trypheron* | + | + |
| *P. spontanea* (Weston) C. G. Shaw | Spontaneum Downy Mildew | +? | + | - | Maize[b], *Saccharum* spp., *Euchlaena*, *Miscanthus*, *Sorghum* | - | + |
| *P. westonii* (Srin., Naras. & | ------ | + | + | - | *Iseilema* | - | + |

| Fungus | Disease | | | | Principal hosts | | |
|---|---|---|---|---|---|---|---|
| *S. cryophila* W. Jones | | ---- | ++ | — | *Dactylis; Apluda, Dichanthium, Digitaria, Heteropogon* | + | + |
| *S. farlowii* (Griffiths) Kenneth | | ---- | ++ | — | *Chloris, Cynodon* | + | — |
| *S. lolii* Kenneth | | ---- | ++ | — | *Lolium* | — | + |
| *S. macrospora* (Sacc.) Thirum.. Shaw & Naras. | Crazy top of maize, etc.., Yellow wilt of rice | ---- | ++ | + | Maize[b], rice[b], wheat[b] and *c* 140 other species, including *Eleusine*[b], Barley, Oats, *Echinochloa, Eragrostis, Iseilema, Miscanthus, Paspalum, Pennisetum, Saccharum, Setaria, Sorghum* | + | — |
| *S. rayssiae* var. *rayssiae* Kenneth, Koltin & Wahl | Blotch Downy Mildew of barley | ---- | ++ | — | Barleys[b] | — | + |
| *S. rayssiae* var. *zeae* Payak & Renfro | Brown Stripe Downy Mildew of maize | ---- | ++ | — | Maize[b], *Digitaria* | — | + |
| *Basidiophora* Roze & Cornu | | | | | | | |
| *B. butleri* (Weston) Thirum. & Naras. | | ---- | + | + | *Eragrostis* spp. | + | — |
| *Bremia* Regel | | | | | | | |
| *B. graminicola* Naoumov | | ---- | — | — | *Arthraxon* | — | + |
| *B. graminicola* var. *indica* Patel | | ---- | — | — | *Arthraxon* | — | + |
| *Plasmopara* Schroet. | | | | | | | |
| *P. oplismeni* Viennot-Bourgin | | ---- | — | — | *Oplismenus* | — | + |
| *P. penniseti* Kenneth & Kranz | | ---- | — | — | Pearl millet | — | + |
| Unknown affinity (no asexual state known) | | | | | | | |
| "*Sclerospora*" *iseilematis* Thirum. & Naras. | | ---- | + | + | *Iseilema* | — | + |
| "*Sclerospora*" *northi* Weston | | ---- | ++ | + | *Erianthus, Saccharum* | — | + |
| "*Sclerospora*" *secalina* Naoumov | | ---- | + | — | Rye | + | + |

[a] Renfro, 1970
[b] Disease of economic significance
[c] By inoculation
++ Frequently found, + Occasionally found, — Not found

*Maracay*, Venezuela (in press); *Proc. Conf. Graminaceous Downy Mildews, Bellagio*, Italy (in press). Reviews have been written on "The global status of maize downy mildews" (Frederiksen and Renfro, 1977), on "Sorghum downy mildew, a disease of maize and sorghum" (Frederiksen *et al.*, 1973), on "Downy mildew and ergot of pearl millet" (Nene and Singh, 1975), and a book was published summarizing the results of one Indian institution's research on some of these diseases (Safeeulla, 1976).

## II. General Taxonomy

There are 20 species of downy mildews known to infect cereals and other grasses; another three are described under *Sclerospora*, but their taxonomic affinities are actually uncertain, for their asexual state is unknown (Table 1). They belong to six genera, more than are known to attack other plant families: *Sclerospora* (1 sp.), *Peronosclerospora* (10 spp.), *Sclerophthora* (5 spp.), *Plasmopara* (2 spp.), *Bremia* (1 sp.) and *Basidiophora* (1 sp.). The first three genera are known only on Gramineae; some species of these genera cause spectacular losses in yield in such crops as maize, sorghum, pearl millet, sugarcane, finger millet (*Eleusine coracana*) and foxtail millet (*Setaria italica*).

## III. Economic Effects

Heaviest losses are incurred in those diseases which are characterized by systemic infection and symptoms, the usual type encountered with *Sclerospora graminicola* and the various species of *Peronosclerospora*. Plants stricken in this manner may produce little or no seed or, as in "Green ear" of pearl millet, heads may be partially or entirely converted into a mass of leaf-like structures. Apart from the chlorosis that systemically infected plants suffer, there may be stunting, reducing the value of forage. Severe reduction in yield may also occur in local lesion infection of maize by *Sclerophthora rayssiae* var. *zeae* on maize and *P. sorghi* on forage sorghum.

Heavy losses have been recorded in the Philippines, Taiwan, Indonesia, Thailand, India, West Africa, Venezuela and other tropical countries from various downy mildew diseases on the above-named crops. Particularly high losses are caused by the tropical species of *Peronosclerospora* and *Sclerophthora rayssiae* var. *zeae* on maize, by *P. sorghi* on sorghum and maize and by *S. graminicola* on pearl millet. Since Old World *P. sorghi* was first found in the New World in 1961 in Texas (Reyes *et al.*, 1964), sorghum downy mildew has spread over much of the Western Hemisphere. In Venezuela the disease was first noticed only in 1973, on both sorghum and maize, but by 1975 it was so serious that the epiphytotic was declared a national emergency (Frederiksen and Renfro, 1977). In Taiwan, sugarcane downy mildew (*P. sacchari*) had become a limiting factor in maize production, and it was considered that "it will remain so until some desirable resistant varieties are developed" (Sun, 1970). Six years later, through a number of measures, including breeding of resistant maize and sugarcane varieties, "the disease has been reduced to such an insignificant low level that it is considered practically controlled" (Sun *et al.*, 1976). This must be considered the most successful project undertaken to date in controlling a graminaceous downy mildew. *P. philippinensis* causes yield losses of 40–60% in the Philippines and disease incidence may commonly be as high as 80–100% (Exconde, 1970). Java downy mildew (*P. maydis*) is reported to have caused infection levels ranging from 20–40% in maize crops in Indonesia (Exconde, 1970).

The importance of maize cultivation in the tropics has risen tremendously in recent years. In Indonesia, for example, the area under maize has reached 3·7 million ha (Triharso *et al.*, 1976). India is the fifth largest maize growing country in the world by area but the loss in grain production in three states alone, attributable to sorghum downy mildew, was put at over 100 000 tons (Payak, 1975a), and grain yield reductions due to the widespread brown stripe downy mildew (*S. rayssiae* var. *zeae*) varied from 20–90% in some fields (Payak, 1975a). In Nepal, maize is the second most important crop, covering half a million ha but losses are heavy from *P. philippinensis* (Moin Shah, 1976). At least 1·5 million tons of maize are now produced yearly in Nigeria, and a downy mildew which attacks maize but not sorghum has lately been discovered in one region (Fajemisin, 1980).

The growth of maize cultivation in Thailand has been spectacular; in 1950 there were 36 000 ha averaging 0·8 tons ha$^{-1}$, while in 1974 it had reached 1 240 000 ha averaging 2·2 tons ha$^{-1}$, of which 90% was for export (Syamananda, 1976). Purported sorghum downy mildew was noticed there only in 1968, but has since spread throughout the country, causing losses as high as 100% in some fields (Syamananda, 1976).

Although sorghum suffers economic losses only from sorghum downy mildew, such losses may be serious. In the United States, an estimated $2·5 million loss was incurred in sorghum and maize in 1969, with loss in grain yield in individual fields reaching as high as 70% (Frederiksen *et al.*, 1969). Broomcorn (sorghum) acreage there dropped to less than a fifth in ten years because of its great susceptibility (Frederiksen *et al.*, 1969). Introduced to Israel about 1963 (Kenneth, 1966), this disease was so severe on sudangrass and sudangrass × *S. bicolor* hybrid forage sorghum, that they are scarcely grown there any longer. Although less than one maize plant per 10 000 was found stricken by sorghum downy mildew in an Israeli locality where sudangrass was heavily infected, within 10 years, sweet corn fields with as much as 50% infection could be found (Kenneth, 1975). Sorghum ranks second in area and third in production among field crops in India, with more than 18 million ha under cultivation; sorghum downy mildew is considered to be "among the most important diseases attacking that crop in various parts of India" (Safeeulla, 1976). Over 10 million ha are grown in South America, and losses there from newly introduced sorghum downy mildew are great, particularly in Venezuela (Malaguti, 1980).

Pearl millet "is probably the most important" of the millets, being cultivated on about 65 million ha, most extensively in India and some African countries (Nene and Singh, 1975). "Green ear" disease of pearl millet, induced by *S. graminicola*, is probably the most serious disease of this crop. Nene and Singh (1975) list loss estimates of 6% in East China, 45% near Allahabad, North India, 60% in Mozambique and 30% in high yielding hybrids in India. King and Webster (1970) estimated a 10% loss in Nigeria. In Senegal, pearl millet, the main cereal, is grown on 600 000 ha and "Green ear" was one of the two most economically important diseases (Girard, 1975). In one locality in Israel, it caused an almost complete loss of a 70 ha crop grown for fodder (Kenneth, 1966).

## IV. Taxonomy of the Exclusively Graminicolous Downy Mildews

The first described species, *Sclerospora graminicola* (Sacc.) Schroet. was referred to by Saccardo as *Protomyces graminicola*, based on oospores alone; Schroeter, on the basis of the newly-discovered imperfect state, made the combination *Peronospora graminicola* (Sacc.) Schroet. in 1879, and proposed *Sclerospora* as a subcategory within that genus (Shaw, 1974). De Bary in 1881 referred to Schroeter's *Sclerospora* as a genus, thereby typifying it, as *Sclerospora* (Schroet.) de Bary (Shaw, 1974). Many new species have since been described under that generic heading. Almost all bore that very noticeable hallmark of the genus (as it was then constituted)—a white downy growth, often from both sides of a leaf, comprising the long bloated trunks and branches of sporangiophores or conidiophores that exit by way of stomata (Fig. 1). These fructifications collapse immediately after sporangia or conidia are actively discharged from the tips of pointed sterigmata and then are often not recognizable. Some species form resting spores (oospores) within the host tissues. The very name *Sclerospora* was derived from the thick-walled oospore with its dark-walled exosporium and adherent oogonial envelope.

Three great changes were eventually made in the taxonomic treatment of this genus.

(1) For decades after *S. graminicola* had first been named, it was common

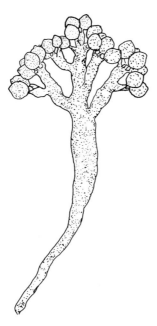

Fig. 1. Sporangiophore and sporangia of *Sclerospora graminicola* on *Setaria viridis* from W. H. Weston (1924) *J. Agric. Res.* **27** (10).

for researchers to use that nomen for downy mildews on sorghum. Kulkarni in 1913 (Waterhouse, 1964) noticed that the fungus on sorghum in India which superficially resembled *S. graminicola* formed conidia, rather than zoospore-producing sporangia. He renamed it *S. graminicola* var. *Andropogonis-sorghi*. The fungus was later accorded species rank, as *S. sorghi* Weston & Uppal (Weston and Uppal, 1932) on that basis, as well as on difference in the mode of branching of fructifications, length of sterigmata and host range. Instances of downy mildews on sorghum erroneously attributed in the literature to *S. graminicola* ever since 1932, have not been conducive to the understanding of these fungi and their diseases.

(2) All other species producing bloated asexual fructifications formed conidia rather than sporangia of the type species of the genus. Ito (1913), in recognizing the significance of this difference, erected the subgenus *Peronosclerospora* to accommodate the conidial species. Shirai and Hara (1927) raised it to generic rank, but transferred only one, sugarcane downy mildew, to *Peronosclerospora*, a step which was totally overlooked (Shaw and Waterhouse, 1980). Shaw (1978) argued successfully for the use of *Peronosclerospora* for conidial species, and transferred them all to this genus, leaving *Sclerospora* with but one species.

(3) Thirumalachar *et al.* (1953) examined material of what was known then

as *Sclerospora macrospora* Sacc., a world-wide pathogen of such crops as maize (in which it causes the disease "crazy top"), rice and wheat and about 140 other grasses (Safeeulla, 1976). They, as well as Peglion (1930), Peyronel (1929) and Tanaka (1940), observed that the sparse, barely noticeable imperfect state consisted of zoospore-producing sporangia as in *S. graminicola*, but which were borne on ephemeral hyphoid sporangiophores exiting through stomata. The imperfect state of *S. macrospora* resembled that of a *Phytophthora*, and Tanaka (1940) had called this species *Phytophthora macrospora* (Sacc.) Ito & Tanaka. Thirumalachar *et al.* (1953), realizing the affinity of the asexual states of *S. macrospora* and *Phytophthora*, nevertheless felt that the sexual state of the former more closely resembled that of *Sclerospora*, with oospore confluent with the wall of the oogonial envelope and the resulting two walls of the dark resting spore being thicker than might be expected of a *Phytophthora*. They consequently erected the genus *Sclerophthora*. Waterhouse (1963) questioned the validity of that genus, considering its members to belong better to *Phytophthora*. Payak and Renfro (1967) showed that the sporangiophore, at least in *S. rayssiae* var. *zeae*, was determinate, in that a sporangium forming at its apex prevented further growth, with other sporangia appearing in basipetal sequence below the apex. Shaw (1978) took the view that not only is *Sclerophthora* distinct from *Phytophthora* on the basis of asexual state differences, but that it should be placed in the Peronosporaceae and not the Pythiaceae. He bolstered his argument by citing the typically obligate parasitism of *Sclerophthora* species, and also differences in oogenesis between the two genera. Since 1953, four other species of this genus have been named (Jones, 1955; Kenneth, 1963, 1979; Kenneth *et al.*, 1964).

Delineation of species in *Peronosclerospora* and in *Sclerophthora* is sometimes fraught with uncertainty. The principal characters employed for the separation of species are conidial (or sporangial) dimensions and shape, conidiophore (or sporangiophore) size, degree and kind of branching, length of sterigmata, and the shape, colour, and morphology of the resting spores (including oospore and oogonial envelope). Host range and symptoms are also often used to advantage in identification.

Within the *Peronosclerospora*, there are species bearing globose or ovoid conidia e.g. *P. sorghi*, *P. maydis*, *P. heteropogoni*, *P. dichanthiicola*, *P. westonii* and *P. noblei* and others with more elongate conidia e.g. *P. philippinensis*, *P. sacchari*, *P. spontanea* and *P. miscanthi*. Some of the species within each group might possibly be conspecific with others; it has been shown that in some species, such as *P. sacchari*, conidial dimensions can change according to the temperature at which they form (Leu, 1973), or, as in *P. philippinensis*, according to host (Exconde *et al.*, 1968). Kimigafukuro (1979) demonstrated that the mean length and width of conidia of a *P. maydis* isolate was not affected by different temperature regimes during their formation, whereas they were longer and wider with increased temperature for purported *P. sorghi* from Thailand and longer and narrower for *P. philippinensis*. Schmitt *et al.* (1979) found that isolates of *P. sorghi, P. philippinensis* and *P. sacchari* from different countries could be differentiated from each other on the basis of the distribution ranges of length of conidia, formed under identical conditions on the same maize hybrid. *P. philippinensis* could also be distinguished from *P. sacchari* by its different length/width ratio. The Texas strain of *P. sorghi* differed from that of purported *P. sorghi* from Thailand, in conidial length and in dimensions of sterigmata. In Thailand, ten categories of conidial shapes, from globose to cylindrical, have been found in the field on maize, often with two or more kinds on the same leaf (Pupipat, 1976). Inoculation of a maize cultivar with conidia from a maize plant that had itself been inoculated with a single conidium, produced conidia ranging from globose to dumb-bell shaped, thus displaying astounding pleomorphism (Pupipat, 1976); the experimenter himself, though, suggested further experimentation "to confirm this conclusion". Renfro (personal communication) pointed out the possible value of the number of nuclei/conidium, found by various workers for some species, in distinguishing between those species. Strains of *P. sorghi* from South India had 16–34 nuclei/conidium (average 22) whereas those same purportedly of the species from Thailand had 8–18 (average 10) and 6–12 (average 8). Although Exconde (1977) had found "the number of nuclei of all species varied with conidial size", the Thai conidia were at best only slightly larger than those from South India.

Some species of graminaceous downy mildews have been grown on callus for long periods and two of them, *S. macrospora* and *S. graminicola*, were grown in axenic culture on known semi-synthetic media (Tokura, 1975). Bhat (in Safeeulla, 1976) found that *S. graminicola* could grow on the callus of a non-host, *Eleusine*, and that the average size of sporangia in callus culture was greater than that of sporangia formed on the host.

## V. Host Range

Most named species of graminaceous downy mildews infect more than one genus of grasses, and many grass genera have more than one downy mildew pathogen attacking them (Table 1), sometimes in the same geographical location. There are apparently very clear natural limits to the ability of a downy mildew genus to attack certain grasses. It has lately been pointed out (J. M. Daly, personal communication) that true downy mildews of Gramineae (apparently not including *Sclerophthora*) may possibly be restricted to $C_4$ syndrome plants.

*Sclerophthora macrospora* has the largest known host range, attacking over

140 species in many genera (Safeeulla, 1976) scattered throughout almost all grass assemblages and tribes; individual isolates, however, have a much more circumscribed range, and the most noticeable host in most regions, maize, may not be a host in other regions (Safeeulla, 1976; Roth, 1967). The other species of *Sclerophthora* are known to attack festucoid grasses such as barley, ryegrass and orchardgrass (Jones, 1955; Kenneth, 1963; Kenneth *et al.*, 1964), chloridoids such as *Chloris*, and bermudagrass (Kenneth, 1979). Panicoids such as maize, *Digitaria*, etc. (Payak and Renfro, 1967). The single species of *Basidiophora* attacks *Eragrostis* species, a chloridoid grass (Thirumalachar and Whitehead, 1952); that of *Bremia* infects *Arthraxon*, a panicoid grass, and the two species of *Plasmopara* infect pearl millet (Kenneth and Kranz, 1973) and *Oplismenus* (Viennot-Bourgin, 1959), both in the tribe Paniceae of the panicoid grasses.

Of particular interest is the relatively narrow host range of *Sclerospora* and of *Peronosclerospora*. Both are essentially restricted to panicoid grasses (Kenneth, 1979). *Sclerospora* (*S. graminicola*) attacks grasses in tribe Paniceae, only occasionally infecting maize of tribe Maydeae, whereas *Peronosclerospora* species attack those in tribes Andropogoneae and Maydeae. A few anomalous reports of infections outside of these limits may be ascribed mostly to error in host or pathogen identification. Others, such as *S. graminicola* infection on sorghum and sugarcane (Melhus *et al.*, 1928), and *P. philippinensis* on oats (Exconde *et al.*, 1968) resulted from artificial inoculations that might have been carried out under conditions particularly conducive to infection, such as internal stem inoculation. Bonde (personal communication) could not induce infection of oats with two Philippine isolates of *P. philippinensis*.

Maize is infected by at least ten downy mildew species (Table 1)—two of *Sclerophthora*, one of *Sclerospora*, seven of *Peronosclerospora* (*P. miscanthi* only by inoculation). Other crop plants, such as sorghum and sugarcane, and some wild grasses, such as *Saccharum spontaneum*, *Miscanthus* and *Iseilema* are hosts of more than two downy mildews.

## VI. Geographical and Climatic Ranges

The graminaceous downy mildews are typically Old World in origin. Only *Sclerophthora macrospora*, *S. cryophila*, *S. farlowii* and *Sclerospora graminicola* were known to occur in the Western Hemisphere until lately, when *P. sorghi* was introduced. *S. farlowii* (on *Chloris* and *Cynodon* in southwestern United States) is the single species known only from the New World. The first record of *P. sorghi* in the Western Hemisphere was in Texas in 1961 (Reyes *et al.*, 1964).

The majority of graminaceous downy mildews are strictly tropical or sub-tropical, but others are not: (1) *S. macrospora*, found in much of the USA, Japan, China, USSR, Australia, Germany and other countries with temperate climates (Safeeulla, 1976), attacks cool-weather crops like wheat as well as hot-weather ones like maize. Ullstrup (1970) considered "crazy-top" of maize, caused by this pathogen, not to be very prevalent in tropical areas, but felt that further critical observation was needed. In tropical southern India, however, "Green ear" disease of finger millet or ragi (*Eleusine coracana*), caused by this pathogen, is rampant and damaging to this very important crop, though incidentally, it did not infect maize there (Safeeulla, 1976). (2) *S. cryophila* infects orchardgrass in the cold early spring of British Columbia, Canada (Jones, 1955) whereas in tropical India and Thailand, other isolates considered to belong to this species attack warm-weather grass species (Srinivasan and Thirumalachar, 1962). (3) *S. rayssiae* var. *rayssiae* heavily attacks barleys in Israel during the coldest months of the winter (Kenneth *et al.*, 1964), whereas *S. rayssiae* var. *zeae* is very destructive to maize in tropical regions of India, Pakistan, Nepal and Thailand (Pupipat, 1975). (4) *Bremia graminicola* was found on *Arthraxon* in eastern Siberia near the Manchurian border (Naoumov, 1913), but *B. graminicola* var. *indica* attacks *Arthraxon* in tropical India (Patel, 1949). (5) *Sclerospora graminicola*, with probably the second widest geographical range was first described on German material and is found primarily on *Setaria* species in relatively cool countries, such as the USA, Canada, China, Japan and southern and central Europe (Safeeulla, 1976; Kenneth, 1977) and on pearl millet in Israel and South Africa. The species, however, has proven most destructive, on pearl millet in particular, in such tropical regions as southern and northern India (Safeeulla, 1976), West Africa (Girard, 1975), and in Tanzania, Malawi and Mozambique (Kenneth, 1976). It causes a severe disease on everglade millet in southern Florida (Weston and Weber, 1928).

Although the Peronosclerosporas are restricted to the tropics more than are the other downy mildews, there are two exceptions, *P. noblei* and *P. sorghi*. *P. noblei* is known, on wild *Sorghum plumosum*, only from temperate New South Wales (Weston, 1929b, 1942). *P. sorghi* was first found in India and is very common there (Payak, 1975b; Frederiksen and Renfro, 1977). There is some controversy over what constitutes this species. There are differences based mostly on whether it is able to infect sorghum and maize, and usually produces resting spores in both ("sorghum pathotype"), or maize without resting spores ("maize pathotype"). "Sorghum pathotype" is found in the south and west of India; it has a very wide world distribution, having been recorded in South Africa (Van der Westhuizen, 1977), in much of tropical eastern and central Africa, Egypt and Israel (Kenneth, 1976). In Nigeria and a few West African countries south of the Sahara, the disease has been damaging to sorghum, with resting spores produced, but *P. sorghi* was not noticed there on maize until a very few years ago (Kenneth, 1976) and the fungus might normally be restricted to sorghum. Fajemisin (1980) stated that so-called sorghum downy mildew newly-found on maize in southern Nigeria could not infect sorghum varieties nor form resting spores in maize; the disease found earlier in northern Nigeria formed resting spores in maize but it was not known if it could infect sorghum.

A few African records are misleading, or in error: *P. maydis* (as *S. maydis*) on maize in the Belgian Congo (Steyaert, 1937) is apparently *P. sorghi* (Kenneth, 1976). Castellani (1949) corrected an earlier report of his own of *P. sorghi* on maize and pearl millet in Eritrea. He considered the fungus on pearl millet to be *S. graminicola* and that on maize to be *Sclerospora* sp. Ciferri (1949) reported maize and "bultoc" (*Pennisetum typhoides* × *P. spicatum*) to be attacked by *P. maydis* in Somaliland, but later admitted that further comparative material was needed to obtain the true identity. Van der Westhuizen (1977) re-examined field and herbarium material in South Africa, and concluded that some previous workers had erred in the identification of downy mildew species there. He stated that *P. sorghi* was the pathogen on sorghum and maize, with resting spores only in the former, *S. graminicola* on pearl millet and *Setaria* sp. and apparently *S. macrospora* on sugarcane.

*P. sorghi* ("sorghum pathotype") has spread widely on sorghum and maize in North, Central and South American countries (Frederiksen, 1976) since it was first noticed in Texas in 1961. It has managed to survive as far north as Indiana, in the "corn belt" (Warren *et al.*, 1974). It is particularly destructive, though, in tropical regions such as Venezuela (Malaguti, 1976). The fungus in Argentina, in contrast to that elsewhere on the continent, was rarely observed on maize (Frezzi, 1970; Frederiksen, 1976). The original "maize pathotype" of *P. sorghi*, from Rajasthan, attacks the wild grass *Heteropogon contortus*, and resting spores form in that host (Dange *et al.*, 1974a; Payak, 1975a). Payak considered this pathotype to be present also in Thailand. *H. contortus* could not be infected by "sorghum pathotype" isolates from southern India, Texas and Israel, but "maize pathotype" isolates from Thailand, as well as *P. philippinensis* and *P. sacchari* isolates, also were not infective to that species (Bonde, personal communication). The Rajasthan fungus appears to differ from downy mildew fungi of other regions in the verrucosity of its resting spores (Shaw, 1976), and thus deserves the species rank assigned to it under the name *P. heteropogoni* (Siradhana *et al.*, 1980).

Most of the so-called "conidial downy mildews" belonging to *Peronosclerospora* are very destructive to maize. *P. sacchari* had been a problem in the sugarcane crop of Queensland, Australia (Leece, 1941) and later in Taiwan in maize and to a lesser extent in sugarcane.

## VII. Symptoms and Disease Development

The symptoms of various graminaceous downy mildew diseases differ greatly from each other. One can divide them into local lesion diseases, systemic diseases and those with both types (Table 1).

All species of *Sclerophthora* except *S. macrospora*, induce only local lesions and cause no noticeable deformation in the hosts. Most infection is by way of zoospores. These are released by short-lived sporangia, formed during dewy nights over discrete chlorotic streaks and linear blotches on leaves, causing the latter to have a fine whitish cast during the early hours of the morning. In *S. rayssiae* var. *zeae* the down is pronounced enough to be seen even in late afternoon (Payak *et al.*, 1970), and the sporangiophores and sporangia were experimentally shown (Singh, J. P. *et al.*, 1970) to be capable of forming even in daytime, if free moisture was present on leaves. With *S. cryophila* and *S. farlowii*, lesions can form on leaf sheaths and spikelets. Lesions turn brownish in most of these diseases, as prodigious numbers of dark resting spores form in the mesophyll of leaves between vascular bundles. Lesions become necrotic, whether or not resting spores appear though infected leaves do not tend to become shredded. Resting spores eventually reach the soil, where they can live for at least 3 years under certain undefined conditions (Singh, J. P. *et al.*, 1970). Some infections, particularly initial infection at the beginning of the season, must be caused by resting spores, though resting spore germination in this genus has been actually seen only in *S. macrospora* (Peglion, 1930; Semeniuk and Mankin, 1964; Safeeulla, 1976). It is very likely that the oospores of the other species behave in a similar manner in the soil, and produce short wide sporangiophores which form germ-sporangia releasing infective zoospores. It is very doubtful if these resting spores are capable of direct infection. Waterlogged conditions, or at least very wet fields about the time of infection, have been mentioned as being favourable for almost all *Sclerophthora* diseases examined. Payak *et al.* (1970), found *S. rayssiae* var. *rayssiae* not to be thus restricted, and it is just possible that the tremendous number of sporangia produced in other fields and carried by wind could suffice to cause leaf infection in non-waterlogged fields. Nevertheless, Payak *et al.* pointed out that maximum disease ratings occur in those maize-growing areas of India with heaviest rainfall. Singh, J. P. *et al* (1970), obtained heaviest infection from resting spore inoculum of *S. rayssiae* var. *zeae* when it was placed on the soil surface or incorporated in the upper 3·75 cm of soil. This should have allowed zoospores to reach the lower leaves by water splash and to initiate the first local lesions. Semeniuk and Mankin (1964) showed the presence of viable zoospore inoculum of *S. macrospora* in spring in ditch water alongside cereal fields in South Dakota, obtained from infected grasses, and these infected the cereals with the systemic disease. It is probable that such inoculum is responsible for much of the initial local lesion infections in the other *Sclerophthora* diseases. The age of the plant may be a factor in infection by some of these pathogens. Thus, in brown stripe downy mildew, infection dropped from 87% for 10-day-old inoculated plants to only 9% for 60-day-old plants.

S. *macrospora* has been known since 1890, when it was found on *Alopecurus* in Australia. It causes systemic infection, with symptoms like "crazy top", vivipary, proliferation of floral and vegetative parts into leafy structures (phyllody), excessive tillering and stunting; leaves become narrow, strap-like and often have a warty surface (Ullstrup, 1970). The failure of leaves to unfurl often leads to distortion of the upper parts of the plant. There may be mild chlorotic striping in infected plants. Ear shanks in maize may be greatly elongated. In "Green ear" disease of finger millet, caused by the fungus, the earhead may be converted into a brush-like mass (Safeeulla, 1976), but the symptoms differ from host to host. Although widespread and eye-catching, the disease on maize is generally of minor importance and only localized areas of fields may be affected (Ullstrup, 1970); on finger millet is causes heavy losses (Safeeulla, 1976). The asexual state is generally very difficult to see, but sporulation in finger millet may be profuse (Safeeulla, 1976).

The multinucleate sporangia are larger than those of the other downy mildews, being 60–100 × 43–64 μm and produce very many more mono-nucleate reniform biflagellate zoospores than do other *Sclerophthora* spp., 24–48 according to Safeeulla (1976) or as many as 117 (Ullstrup, 1970). The multinucleate resting spores are also larger than in other downy mildews, being 42–74 μm in diameter (Ullstrup, 1970). They may be produced in large numbers in various tissues, even including the pericarp of finger millet seed (Safeeulla, 1976). In leaves they are affixed to the vascular bundles, unlike other graminaceous downy mildews (Thirumalachar *et al.*, 1953). Ullstrup considered that infection of maize can take place over a fairly wide range of temperature and Safeeulla (1976) found that resting spore germination occurred at 10–30°C, with an optimum of 22–25°C. Cross-inoculations showed that in Karnataka the disease could pass from finger millet to sorghum and pearl millet, but not to maize, rice, wheat or sugarcane. Akai (1959) found that zoospores showed a chemotactic reaction towards germinating rice seeds.

No systemic infection has been found in *Oplismenus* stricken by *Plasmopara oplismeni* in Guinea (Viennot-Bourgin, 1959), pearl millet by *Plasmopara penniseti* in Ethiopia (Kenneth and Kranz, 1973), and *Arthraxon* by *Bremia graminicola*; neither have resting spores been found in local lesions produced on the leaves of hosts. In *Eragrostis* species systemically stricken by *Basidiophora butleri* in Malawi (Weston, 1933) and India (Thirumalachar and Whitehead, 1952), resting spores are produced in leaves, which then become shredded lengthwise.

*Sclerospora graminicola* and all species of *Peronosclerospora* have a number of characters in common, apart from bloated ephemeral fructifications and panicoid hosts. They all induce systemic infection in the plant and only a few cause local lesions as well. They all are apparently incapable of systemically infecting plants after the plants have reached a certain age (new tillers or regrowth of cut shoots of old plants, however, may be susceptible). In all, the most prominent systemic symptom is chlorosis of leaves. In the few that form oospores (Table 1) they do so only in systemically-infected plants.

The first noticeable symptom of systemic infection is chlorosis, which may appear on the very first or second seedling leaf or may be much delayed. In order for systemic infection to occur, hyphae must penetrate to the growing point at the base of the shoot just above ground. There it colonizes all embryonic leaves and continues to grow into those developing leaves which have not emerged from the whorl. Once young leaves have elongated within the whorl, they are no longer juvenile enough for the hyphae to invade systemically, except for the area near the base of the blade. Thus, the first systemically infected leaf is likely to show chlorosis only at the proximal part of the blade—"half-leaf symptoms" (Weston, 1923a), and each succeeding leaf displays more chlorosis, until leaves emerge completely chlorotic or striped. When resting spores in the soil are responsible for infection, the first seedling leaf is never chlorotic, since the infective hyphae are slow to reach the growing point. Chlorotic symptoms may be sometimes considerably delayed, though shortening of internodes might occur at an early stage. If conidia infect the coleoptile shortly after germination, the first seedling leaf may be chlorotic and such plants often die (Kenneth and Shahor, 1973); if the disease is transmitted by hyphae within seed, the first leaf may be chlorotic (Exconde, 1977). Late conidial inoculation may result in the first chlorotic symptoms appearing only on upper leaves, but inoculation of plants four weeks old almost always fails, perhaps because the growing point has then risen within the true stem and is inaccessible to hyphae.

The degree of chlorosis (light green, yellow, white, mottled) depends largely upon the species of downy mildew, the host species and even the cultivar. Likewise there may be a solid continuous chlorosis at the base of the blade with jagged fan-like extensions either spreading or narrowing towards the apex, or an open yellow stripe system or a closed system of stripes the length of the blade, or a very fine wavy etching of white within a green area. *P. sacchari*, for instance, tends to give a streaked or striped appearance to maize and sugarcane leaves. The type of chlorosis may vary according to stage of development of the plant, e.g. striping caused by *P. sorghi* may appear in upper leaves of sorghum and maize.

On dewy nights, heavy sporulation may take place over the pallid areas (but not on leaves in which almost no chlorophyll remains), and may appear on lower or sometimes both sides of leaves. Leaves may be white from down, and Weston (1923a) estimated the number of conidia of *P. philippinensis* produced on a single maize plant in one night at 6 billion. Sporulation may start at midnight or a few hours later, depending on time of dewfall, and may last a few hours. Although in nature, sporulation is normally nocturnal and this had been considered as its inherent periodicity (Weston, 1923a), others (Renfro *et al.*, 1980; Chang and Wu, 1970; Safeeulla and Thirumalachar, 1956; Siradhana *et al.*, 1976; Schmitt and Freytag, 1974) could obtain sporulation at will by manipulation, by collecting infected leaves at certain daylight hours, after some hours of bright sunlight and then holding them at r.h. 100% in darkness.

Conidia of *P. sorghi, P. spontanea* and *P. miscanthi* sometimes germinate while still attached to sterigmata and they are then not forcibly discharged as are non-germinated conidia. Conidia and sporangia germinate quickly in water or on wet leaves and have a life span of but a few hours although if held at abnormally low temperatures they may survive much longer, e.g. *S. graminicola* sporangia in 10% dimethyl sulphoxide at 6°C survived for 5 days (Safeeulla, 1976).

Systemic infection by conidia is often preceded by penetration of the leaf through stomata and ramification of hyphae intercellularly into and through the leaf sheath and eventually to the shoot apex (Dalmacio and Exconde, 1969). The pathway in the leaf blade may often be marked by a local lesion (Dalmacio and Exconde, 1969; Cohen and Sherman, 1977).

In the "sorghum pathotype" of *P. sorghi*, resting spores may appear in sorghum in great numbers within streaks and stripes, particularly of upper leaves, with a concomitant reduction in sporulation over that area. The stripes turn brownish due to the dark resting spores within, and the leaves become shredded but in some maize cultivars resting spores appear in smaller numbers, often in pallid stripes on ear bracts and leaves, but leaves do not shred into strips. Among other species, *S. graminicola* induces leaf shredding in *Setaria* species, but not in pearl millet (though resting spores form in both), *P. sacchari* in sugarcane but not in maize (resting spores form only in the former except in India) and *P. miscanthi* in *Miscanthus*. There is shredding in leaves of *Heteropogon contortus*, in which spores of *P. heteropogoni* are formed.

As a rule, plants displaying very early systemic symptoms of various graminaceous downy mildews die prematurely or are stunted, and seldom flower if they survive. If systemic infection occurs later, grain production is prevented or reduced. In maize, some cultivars produce fewer ears when infected and others produce many small empty ears, whereas proliferation of heads of sorghum and pearl millet and phyllodied tassels are not uncommon. Although early systemic symptoms caused by *S. graminicola* in pearl millet almost invariably lead to a dwarfed plant devoid of earheads, symptomless maturing pearl millet plants with normal ears can still produce worthless "green ears" out of axillary buds. In semi-arid Rajasthan, sporulation in infected pearl millet is hardly known (Chaudhuri, 1962), yet the "green ear" phase of the disease is particularly damaging there (Suryanarayana, 1961a). Kenneth (1966) observed in Israel a field of pearl millet growing out of season, in which the mature crop displayed only a trace of downy mildew; continuing irrigation encouraged the appearance of new earheads, when 15% of these previously "healthy" plants displayed "Green ear". As there had been no subsequent sporangial inoculum, Kenneth considered that the disease had remained latent and symptomless from early infection. Safeeulla (1976) doubted that sporangia play a role in infection under Karnataka conditions and suggested that resting spores were all-important, but Safeeulla (1979) has now claimed that zoospores in soil may bring about infection through roots. Systemic infection of pearl millet has been effected with zoospores by artificial inoculation (Safeeulla *et al.*, 1963; Girard, 1975) and Girard (1975) attributed elongated local leaf lesions in pearl millet in Senegal to possible sporangial infection. R. J. Williams (personal communication) demonstrated conclusively at ICRISAT that sporangia can be important in the epidemiology of this disease, and developed a screening technique for a breeding programme based on sporangial inoculum.

*P. sorghi* conidia are capable of inducing severe local lesions in any and all leaves of sorghum and to a lesser extent in maize. In forage sorghum the lesions are rectilinear, pallid at first and turn purplish; before becoming necrotic they support profuse sporulation, which, as inoculum, can cause new local lesions or induce systemic infection in nearby young plants or in new regrowth of older ones.

In sugarcane stricken with downy mildew, plants derived from infected seed-pieces are usually severely stunted, whereas when healthy plants are infected by conidial inoculum they produce elongated stalks with short striped leaves (Edgerton, 1959).

Only in *S. macrospora* have resting spores actually been seen to germinate in large numbers. Since Hiura (1930) succeeded in germinating those of *S. graminicola*, *in vitro*, many claims have been made that this phenomenon has been seen in *S. graminicola* and a number of other species, but always in small numbers. Oospores of *Peronosclerospora* spp. apparently germinate by wide coenocytic germ tubes, and germination appears to be stimulated by the presence of young growing roots (Pratt, 1978).

# VIII. Control

## A. Chemical control

Attempts have been made to control most of these mildews by (1) protectant foliage sprays, (2) soil treatment, or (3) seed dressing, either to inhibit resting spore germination or to prevent the inoculation of emerging seedlings, and the growing plant.

(1) The greatest problem in controlling these diseases by foliage sprays concerns the question of economic feasibility. Since it has been found that several sprayings are necessary, this becomes a problem in field crops which have relatively low value/area sown. Exconde (1975, 1976) summarized most of the work done throughout the world on chemical control of maize downy mildews. In the Philippines, eight sprays of fentin hydroxide + maneb on a susceptible variety gave an excellent return, particularly in the wet season, and more than paid for itself when used on a more resistant variety, UPCA var. 2. He and co-workers found that the use of resistant varieties generally gave higher additional income than did four or eight sprayings of fentin hydroxide + maneb. Four sprayings of a mineral oil (F-1243) alone or together with maneb resulted in an increase in economic return over untreated controls or maneb alone (Exconde, 1976).

(2) In Texas, potassium azide incorporated into oospore-infested soil effectively reduced sorghum downy mildew incidence in greenhouse and field trials (Frederiksen et al., 1973). Singh, J. P. et al. (1970) obtained good control of brown stripe downy mildew of maize by soil drenching with bleaching powder or with carboxin. As carboxin is considered effective only on Basidio-mycetes, the drop in downy mildew disease index and rise in yield might possibly reflect indirect effects of the material on the plant itself, but carboxin (and oxycarboxin) surprisingly inhibited sporangium germination to a fair extent. Dange et al. (1974b) obtained no reduction in downy mildew on maize in Rajasthan, caused by P. heteropogoni, when using carboxin or oxycarboxin as a seed-dressing.

(3) Seed treatment for maize seed was first initiated by Weston (1923b) who suggested an alcohol dip followed by concentrated sulphuric acid for 5–10 minutes for the elimination of Peronosclerospora spp. resting spores.

Schultz and Dalmacio (1971) found that a combination of seed treatment and four foliar applications, using 60% chloroneb, controlled Philippine downy mildew of maize for 28 days whereas systemic fungicides like benomyl and thiabendazole had no effect on the disease. Lately, metalaxyl (Ciba-Geigy), a new systemic fungicide affecting some Phycomycetes, has been evaluated as a seed-dressing by a number of workers, with excellent results against graminaceous downy mildews. Exconde and Molina (1978), for instance, in testing four maize varieties in field plots against *P. philippinensis*, and using the 25% w.p. as a slurry at 2 g ai kg$^{-1}$ seed for seed treatment, found no downy mildew 26 days after emergence, whereas controls showed 79–100% infection. A slight temporary depressive effect on seed germination and seedling growth was found when 4 or 6 g ai kg$^{-1}$ seed was used. Singh and Williams (1979) obtained complete recovery of infected pearl millet after spraying with metalaxyl (as Ridomil at 500 ppm) during the first 21 days after planting non-treated seed.

The great advantage in using a systemic fungicide by seed or furrow application for most graminaceous downy mildews is that one treatment may be effective. The host plant becomes almost immune from infection by *Sclerospora* and *Peronosclerospora* from about 4 weeks after emergence, and as a rule even 3-week-old plants are more resistant than are younger ones. Even in brown stripe downy mildew of maize, a local lesion disease, plants become much more resistant with age (Singh, J. P. *et al.*, 1970). Thus, with seed treatment giving near immunity for the first few weeks, the plants will have gained age resistance by the time immunity wears off. Exceptions to the rule regarding age resistance are (1) local lesions in sorghum, induced by *P. sorghi* at all ages, and the susceptibility of young regrowth such as occurs in cut forage sorghum; (2) new susceptible nodular tillers appearing on pearl millet until near-maturity. Singh and Williams (1978b) considered that pearl millet seed-treatment with metalaxyl would be expected to give even better control in a farmer's field situation than was effected in plots sited alongside non-treated controls which would provide sporangial inoculum. It is not yet known if isolates of these downy mildews will eventually develop resistance to the new systemics.


## B. Nutritional control

There are conflicting reports on the effect of host nutrition on the downy mildews of Gramineae. Triharso *et al.* (1976) mentioned that in Java an overdose of nitrogen in soil increased susceptibility to Java downy mildew whereas high potassium reduced susceptibility; Frederiksen *et al.* (1973) showed graphically that high available soil nitrogen favoured sorghum downy mildew incidence in maize; and Yamada and Aday (1977) stated that only nitrogen in soil, regardless of phosphorus and potassium levels, increased the susceptibility of susceptible cultivars of maize to *P. philippinensis*. Yet, Singh, R. *et al.* (1970) stated that the application of nitrogen, phosphorus and potassium lessened the degree of infection by *S. rayssiae* var. *zeae* and *P. sacchari* while Tantera (1975) found that no combination of nitrogen, phosphorus and potassium had any effect on the level of infection in Java downy mildew. Moin Shah (1976) found in Nepal that the addition of 180 kg ha$^{-1}$ and 240 kg ha$^{-1}$ of nitrogen resulted in much lower severity of *P. philippinensis* on maize (21% and 26%) than did no addition (52%) or 60 kg ha$^{-1}$ (53%).

Apparently zinc deficiency predisposes maize plants to sugarcane downy mildew and brown stripe downy mildew so that a foliar application of zinc reduced disease incidence significantly (Singh, R. *et al.*, 1970). Frederiksen *et al.* (1973) showed that zinc (as $ZnSO_4$) in the seedbed reduced sorghum downy mildew incidence in maize.

Deshmukh *et al.* (1978) found that phosphorus decreased both the incidence and the intensity of pearl millet downy mildew, as did high rates of nitrogen application, but potassium had no effect on the disease.

## C. Cultural and manipulative control

Ullstrup (1970) stressed adequate soil drainage for prevention of "crazy top" and this is likely to hold true for other *Sclerophthora* diseases, bearing in mind that both sporangia and oospores produce zoospores. According to Nene and Singh (1975), *S. graminicola* on pearl millet had been found to be serious in low-lying poorly drained areas in India but there is a general belief in India that high soil moisture at sowing time and germination does not encourage infection because of faster seedling growth and consequent escape from infection. Girard (1975) found that disease incidence in Senegal does not necessarily increase with higher rainfall.

In the *Peronosclerospora* diseases, there does not appear to always be a clear correlation between soil moisture and disease incidence. Balasubramanian (1974) found that sorghum downy mildew on sorghum in Karnataka was suppressed by a high available soil moisture of 74–79% for the 16 days after sowing. Sorghum downy mildew on sorghum and maize in South Africa is severe only on heavy clay soils where sowing must be delayed until the soil has absorbed enough moisture after the winter's drought, at which time high temperatures prevail, allowing infection by oospores (Van der Westhuizen, 1977); however, where early rains allow early sowing on cold heavy soils, there is virtually no downy mildew apparently eliminating both soil texture and waterlogging *per se* as factors favouring the disease. Pratt and Janke (1978), though, found evidence suggesting that fewer oospores are required to initiate sorghum downy mildew in sorghum on sandy soils than in clay soils in Texas.

Resting spores of these downy mildews have been found to be infective for as long as ten years (Nene and Singh, 1975) though there are numerous conflicting reports of samples behaving differently in infection experiments after being held for various periods under different conditions. Lack of infection in these cases might signify dormancy rather than loss of viability. Newly formed resting spores of *S. graminicola* (Girard, 1975) and *S. sorghi* (Kenneth, 1970) can be infective. Consistently high resting-spore germination *in vitro* has not yet been achieved, so that actual resting spore response in the field to various treatments is often uncertain. Tetrazolium chloride has been used lately to verify the viability of resting spores (Shetty *et al.*, 1980).

Time of sowing is important in preventing disease when resting spores are the primary or only effective inoculum, and sometimes where there are no resting spores. Kenneth (1970) could not obtain sorghum downy mildew infection in infested soil with a constant temperature of 20°C or below, whereas good infection occurred at 22–32°C; susceptible grain sorghum varieties sown in the cool early spring in Israel escape infection, as does early-sown maize. In Rajasthan, maize sown early, with the start of the monsoon in late June, escapes infection by *P. heteropogoni* (previously considered to be *P. sorghi*) there, it would seem not to be connected with soil temperature, but with the absence at that time of the assumed source of initial inoculum, infected wild *Heteropogon contortus* (Dange, 1976).

When the plants in a field are of one age, rather than at different stages of growth, systemic downy mildew incidence may be greatly reduced (Tantera, 1975; Cohen and Sherman, 1977; Kenneth and Shahor, 1973). A field in which strips are sown successively on different days would allow young susceptible plants to become infected by any older plants which would provide inoculum; the last-sown strip or plot in the field would often be a complete loss. Neighbouring fields of the same or other host crop constitute a hazard for late-sown ones, particularly if the prevailing night winds favour transfer of viable conidia. In Taiwan and Queensland, sugarcane downy mildew is restricted to those areas where both maize and sugarcane are extensively grown. In both countries, regulations prohibit the growing of maize near sugarcane fields and in Taiwan delaying the sowing of maize until sugarcane has been harvested is also practised, considerably reducing the damage from the disease. In Queensland the disease has been no problem for many years, and in Taiwan, these steps, along with new resistant varieties developed for both crops, have reduced downy mildew to the state that maize and sugarcane may now be intercropped in the same field (Sun *et al.*, 1976). Roguing of young infected plants, particularly on the earliest crop of a season, has been effective and feasible for reducing inoculum where manpower is not a problem. Roguing is effective, as conidial infection appears to be restricted essentially to the near vicinity of diseased crops; Sun (1970) mentions less than half a mile from its source as the maximum that viable *P. sacchari* could reach, but heavy infection for these species occurs only metres away (Tantera, 1975; Kenneth and Shahor, 1975; Cohen and Sherman, 1977; Mikoshiba *et al.*, 1978).

The eradication of wild collateral hosts near maize and sorghum fields has been recommended. Kans (*Saccharum spontaneum*) harbours *P. philip-pinensis* in its underground parts (Suryanarayana, 1961b). Wild, ubiquitous Johnsongrass (*Sorghum halepense*) fortunately constitutes no threat in Israel, as few lines of this perennial are susceptible to sorghum downy mildew, but false Johnsongrass (*S. verticilliflorum* and *S. arundinaceum*) in Venezuela, ubiquitous and very susceptible, has presented a serious problem (Malaguti, 1980); usually there is no sorghum downy mildew, or a very low incidence, in those areas where false Johnsongrass does not occur.

Elimination of resting spore inoculum would drastically reduce disease incidence caused by those downy mildews which produce them. In Texas, deep ploughing as opposed to shallow gave a fourfold reduction in sorghum downy mildew infection where the disease had been prevalent the previous year (Frederiksen and Renfro, 1977). Bacteria (Frederiksen and Ullstrup, 1975) and fungi have been found to destroy resting spores of these and other downy mildews. Among the fungi, the one-celled chytrids such as *Rhizophydium* (Melhus, 1914) and *Phlyctochytrium* (Kenneth and Shahor, 1975; Dogma, 1975) appear to be common. *Phlyctochytrium* can be grown in shake culture, survives drying and destroys resting spores of *P. sorghi* in natural soil (Kenneth, unpublished) though much soil moisture would be needed to promote its activity. *Fusarium semitectum* was shown to attack resting spores (Raghavendra and Pavgi, 1976).

Some downy mildews can be spread by resting spores in leaf debris in seed lots or on seed itself, and it is likely that sorghum downy mildew reached the New World in that manner. The role of hyphae of downy mildews in seed pericarp or embryo on transmission of various downy mildews is hotly contested. Strong evidence of such internal coenocytic hyphae has been shown by many workers, but Williams (1980) takes issue with most claims of seed transmission by internal mycelium, on the grounds of faulty technique. With very few exceptions, claims for transmission have been only for seeds that are fresh and still have a very high water content, and which came from systemically infected plants. Drying of such seed almost invariably prevented disease transmission but Shetty *et al.* (1980) claim proof that pearl millet downy mildew can be transmitted by internal mycelium in fully-dried old seed, with symptoms appearing as late as 54 days after sowing. If indeed, there is some transmission of *P. maydis* from undried seed with internal mycelium, as claimed, it would help to explain the source of some initial infection in Java by this pathogen. Since it produces no resting spores and has no host other than maize, self-sown wet seed in left-over nubbins from infected plants might possibly allow carry-over of the disease, whence it could spread to sown maize.

## D. Genetic control

It is generally considered that host resistance is "the most efficient, effective and economical means of controlling downy mildew diseases" (Frederiksen and Renfro, 1977), and great efforts, some very successful, have been made in that direction during the past dozen years. Only at the beginning of these breeding programmes, was basic knowledge obtained regarding physiological specialization in the various pathogens, the mode of inheritance in the crop plants, and efficient methods for inoculation.

Pathogenic variability within morphologically delineated downy mildew species is sometimes pronounced, so that sugarcane downy mildew (*P. sacchari*), common on maize in part of northern India, is unknown there on sugarcane; *P. philippinensis*, known to attack sugarcane, does not do so in Thailand; and *P. sorghi* of sorghum apparently does not affect maize in West Africa.

Resistance to these diseases is at present generally measured as the percentage of plants infected referred to as the "incidence". The downy mildew rating scale for pearl millet in use today is calculated from incidence as well as severity ratings on stricken plants (Deshmukh *et al.*, 1978). Immunity of cultivars is apparently unknown for most of these diseases. The sorghum variety QL-3, tested for three years at numerous international sorghum downy mildew nurseries organized by ICRISAT in India, Venezuela and Botswana, remained completely free of the disease in the presence of heavily infected varieties, and might possibly be immune (Dange and Williams, 1978). Disease incidence, even in resistant varieties, rises with the inoculum load. Thus, inoculum pressure may be critical in selecting resistant materials, for too low an amount will result in escapes and too high may cause all plants to become infected.

It appears that certain geographical regions have more strongly pathogenic isolates than do others e.g. West Africa for pearl millet downy mildew and the Philippines for any downy mildew attacking maize. Evidence for the existence of physiological races (isolates inducing different patterns of incidence in a list of differential varieties or cultivars) has been meagre; (1) *P. philippinensis* on maize in the Philippines (Titatarn and Exconde, 1974; Exconde, 1977). (2) *P. sorghi* in Texas, where two isolates behaved differentially on sorghum lines (Craig and Frederiksen, personal communication); they considered these isolates to be pathotypes, in contrast to the maize and sorghum "pathotypes" of Payak which they designated as *formae speciales*, in line with the classification system of Robinson (1969). (3) *S. graminicola* on pearl millet lines tested in Mali and Nigeria (Girard, 1975); Rasheed *et al.*, (1978) in Karnataka, found two such isolates which not only differed morphologically but also in various chemotaxonomical criteria such as isozymes; the results from the many international pearl millet downy mildew nurseries organized by ICRISAT for three years in West Africa and India have also considerably strengthened the evidence for the existence of physiological races (Singh and Williams, 1978a).

Resistance of pearl millet to *S. graminicola* has been found, with the most resistant sources originating in northern Nigeria and southern Niger. Resistant hybrids and varieties have been developed in India from these materials from 1975 to 1979, some of which broke down under high sporangial inoculum, and some which do not withstand normal inoculum in the field in West Africa (Williams, personal communication). Safeeulla (1979) asserted that resistance is a quantitative character, and that epistatic dominance and additive effects, in that order, are important in inheritance of resistance.

Resistance of sorghums to downy mildew is common, and in an early screening in Nigeria of the World Sorghum Collection (Futrell and Webster, 1966), 14% were resistant, in most classification groups, 67% among South African entries, mainly in the Caffrorum group. Inheritance studies in Texas (Frederiksen *et al.*, 1973) indicated that resistance is partially dominant or dominant, depending on the genetic background, and that more than one gene is involved. Pathogen-resistant grain sorghum hybrids were released in the USA in 1972 and agronomically acceptable resistant forage-sorghums became available in 1980.

Mochizuki (1975) summed up the mode of inheritance of resistance of maize to *Peronosclerospora* spp. as gleaned from investigations by various workers. He showed that resistance is controlled (1) by dominant gene(s) when materials are inbred lines (e.g. in Taiwan), with a few factors controlling inheritance, and with evidence of dominant susceptible gene and gene interaction; (2) by a polygenic system when materials are open-pollinated varieties (as in most Asian countries). De Leon (1980) stated that later studies, in Thailand, Philippines, India and Texas, confirmed that resistance is controlled by polygenic systems and is additive.

Of paramount importance is the realization, mostly through comparisons among many varieties sown at multi-locational nurseries (International Corn Downy Mildew Nurseries and others) in various countries, that genes for resistance to one *Peronosclerospora* also confer resistance to others (De Leon, 1980; Renfro *et al.*, 1980). This realization has simplified and accelerated cooperative breeding programmes.

Resistant native maize varieties have been found in Southeast Asia, but have low yield potential. Maize breeding programmes in Asia, mostly for polygenic resistance, are directed towards the construction of superior composites or synthetics, yielding incomplete resistance which would allow the pathogen to maintain itself but keep the level of survival low enough so as not to cause serious economic losses (Jinahyon, 1975). Philippine downy mildew resistant sources are used exclusively in breeding programmes in south and southeast Asia, and a high yielding Thai composite, Suwan I, based on Philippine downy mildew resistance, was released to Thai growers in 1975.

## Acknowledgements

This chapter is dedicated to Dr B. L. Renfro, Plant Pathologist of the Rockefeller Foundation, and to that Organization, for having urged and helped sustain the revival of interest in all aspects of the graminaceous downy mildew problem, as a truly international endeavour.

## References

Akai, S. (1959). *Pl. Prot. Sect. Bur. Agric. Admin.*, Min. Agric. and Forestry, Japan **17**, 1–18.

Balasubramanian, K. A. (1974). *Pl. Soil* **41**, 233–241.

Castellani, E. (1949). *Manip. I., Nuovo Giorn. Bot. It.*, n.s. **49**.

Chang, Y. O. and Wu, T. H. (1970). *Rep. Corn. Res. Center* (Taiwan) **8**, 1–10.

Chaudhuri, H. (1962). *Phytopathology* **22**, 241–246.

Ciferri, R. (1949). *Notiz. Malatt. Piante* **1**, 9.

Cohen, Y. and Sherman, Y. (1977). *Phytopathology* **67**, 515–521.

Dalmacio, S. C. and Exconde, O. R. (1969). *Philipp. Agric.* **53**, 35–52.

Dange, S. R. S. (1976). *Kasertsart J.* **10**, 121–127.

Dange, S. R. S. and Williams, R. J. (1978). *Progress Rep. ICRISAT SPDM* **7901**, 10 pp.

Dange, S. R. S., Jain, K. L., Rathore, R. S. and Siradhana, B. S. (1974a). *Pl. Dis. Reptr.* **58**, 285–286.

Dange, S. R. S., Jain, K. L., Rathore, R. S. and Siradhana, B. S. (1974b). *Rajasthan J. Pesticides* **1**, 46–48.

Deshmukh, S. S., Mayee, C. D. and Kulkarni, B. S. (1978). *Phytopathology* **68**, 1350–1353.

De Leon, C. (1980). *Proc. Intl. Conf. Gramin. Downy Mildews*, Bellagio (in press).

Dogma, I. J., Jr. (1975). *Kalikasan, Philipp. J. Biol.* **4**, 69–105.

Edgerton, C. W. (1959). "Sugarcane and its Diseases". 2nd edn. Louisiana State Univ. Press, Baton Rouge.

Exconde, O. R. (1970). *Indian Phytopathol.* **23**, 275–284.

Exconde, O. R. (1975). *Trop. Agric. Res. Ser., Tokyo* **8**, 157–163.

Exconde, O. R. (1976). *Kasertsart J.* **10**, 94–100.

Exconde, O. R. (1977). *Proc. Sympos. on Downy Mildews of Gramineae, Maracay, Venez.*, Aug. 1977 (in press).

Exconde, O. R. and Molina, A. B., Jr. (1978). *Philipp. J. Crop Sci.* **3**, 60–64.

Exconde, O. R., Elec, J. V. and Advincula, B. A. (1968). *Philipp. Agric.* **52**, 175–188.

Fajemisin, J. M. (1980). *Proc. Intl. Conf. on Gramin. Downy Mildews, Bellagio* (in press).

Frederiksen, R. A. (1976). *Kasertsart, J.* **10**, 164–167.

Frederiksen, R. A. and Renfro, B. L. (1977). *Ann. Rev. Phytopathol.* **15**, 249–275.

Frederiksen, R. A. and Ullstrup, A. (1975). *Trop. Agric. Res. Ser., Tokyo* **8**, 39–43.

Frederiksen, R. A., Amador, J., Jones, B. L. and Reyes, L. (1969). *Pl. Dis. Reptr.* **53**, 995–998.

Frederiksen, R. A., Bockholt, A. J., Clark, L. E., Cosper, J. W. *et al.* (1973). "Sorghum Downy Mildew, a Disease of Maize and Sorghum". *Texas A and M Univ. Res. Monograph* **2**, College Stn., Texas.

Frezzi, M. J. (1970). *Idia* **274**, 16–24.

Futrell, M. C. and Webster, O. J. (1966). *Pl. Dis. Reptr.* **50**, 641–644.

Girard, J. C. (1975). *Proc. Consultants' Gp. Meetings on Downy Mildew and Ergot of Pearl Millet*, ICRISAT, Hyderabad (R. J. Williams, Ed.).

Hiura, M. (1930). *Science, n.s.* **72**, 95.

Ito, S. (1913). *Bot. Mag. Tokyo* **27**, 218.

Jinahyon, S. (1975). *Trop. Agric. Res. Ser. Tokyo* **8**, 221–230.

Jones, W. (1955). *Can. J. Bot.* **33**, 350–354.

Kenneth, R. (1963). *Israel J. Bot.* **12**, 136–139.

Kenneth, R. (1966). *Scripta Acad. Hierosolymitana* **18**, 142–172.

Kenneth, R. (1970). *Indian Phytopathol.* **23**, 371–377.

Kenneth, R. (1975). *Trop. Agric. Res. Ser. Tokyo* **8**, 35–38.

Kenneth, R. (1976). *Kasertsart J.* **10**, 148–159.

Kenneth, R. (1977). *In* "Diseases, Pests and Weeds in Tropical Crops". (J. Kranz, H. Schmutterer and W. Koch, Eds). 96–99, Paul Parey, Berlin.

Kenneth, R. (1979). *Phytoparasitica* **7**, 50.

Kenneth, R. and Kranz, J. (1973). *Trans. Brit. Mycol. Soc.* **60**, 590–593.

Kenneth, R. and Shahor, G. (1973). *Phytoparasitica* **1**, 13–21.

Kenneth, R. and Shahor, G. (1975). *Trop. Agric. Res. Ser. Tokyo* **8**, 125–127.

Kenneth, R., Koltin, Y. and Wahl, I. (1964). *Bull Torrey Bot. Club* **91**, 185–193.

Kimigafukuro, T. (1979). *JARQ.* **13**, 76–77.

King, S. B. and Webster, O. J. (1970). *Indian Phytopathol.* **23**, 342–349.

Leece, C. W. (1941). "Downy Mildew Disease of Sugarcane and other Grasses". *Bull. Sugar Expt. Stn., Queensland Tech. Comm.* **5**, 111–135.

Leu, L. S. (1973). *Bull. Plant Prot.* (Taiwan) **15**, 106–115.

Malaguti, G. (1976). *Kasertsart J.* **10**, 160–163.

Malaguti, G. (1980). *Proc. Intl. Conf. Gramin. Downy Mildews, Bellagio* (in press).

Melhus, I. E. (1914). *Phytopathology* **4**, 55–62.

Melhus, I. E., Van Haltern, F. H. and Bliss, D. E. (1928). *Iowa Agric. Exp. Stn. Res. Bull.* **111**, 297–338.

Mikoshiba, H., Sudjadi, M. and Soediarto, A. (1978). *JARQ* **11**, 186–189.

Moin Shah, S. (1976). *Kasertsart J.* **10**, 137–142.

Mochizuki, N. (1975). *Trop. Agric. Res. Ser. Tokyo* **8**, 179–193.

Naoumov, N. (1913). *Bull. Soc. Mycol. France* **29**, 273–278.

Nene, Y. L. and Singh, S. D. (1975). A comprehensive review of downy mildew and ergot of pearl millet. *Proc. Consultants' Gp. Meetings on Downy Mildew and Ergot of Pearl Millet*, Hyderabad. (R. J. Williams, Ed.), 15–53.

Patel, M. K. (1949). *Indian Phytopathol.* **1**, 104–106.

Payak, M. M. (1975a). *Trop. Agric. Res. Ser. Tokyo* **8**, 81–91.

Payak, M. M. (1975b). *Trop. Agric. Res. Ser. Tokyo* **8**, 13–20.

Payak, M. M. and Renfro, B. L. (1967). *Phytopathology* **57**, 394–397.

Payak, M. M., Renfro, B. L. and Lal, S. (1970). *Indian Phytopathol.* **23**, 183–193.

Peglion, K. (1930). *Boll. R. Staz. Patol. Veg. Roma, n.s.* **10**, 153–164.

Peyronel, B. (1929). *Boll. R. Staz. Patol. Veg. Roma, n.s.* **9**, 353–357.

Pratt, R. G. (1978). *Phytopathology* **68**, 1606–1613.

Pratt, R. G. and Janke, G. D. (1978). *Phytopathology* **68**, 1600–1605.

Pupipat, U. (1975). *Trop. Agric. Res. Ser. Tokyo* **8**, 63–80.

Pupipat, U. (1976). *Kasertsart J.* **10**, 106–110.

Raghavendra Rao, N. and Pavgi, M. (1976). *Can. J. Bot.* **54**, 220–223.

Rasheed, A., Shetty, H. S. and Safeeulla, K. M. (1978). *Third Intl. Congr. Pl. Pathol. Munich, Abstr.* **123**.

Renfro, B. L. (1970). *Indian Phytopathol.* **23**, 177–179.

Renfro, B. L., Pupipat, U., Singburaudom Choonhawangse, K. *et al.* (1980). *Proc. Intl. Conf. on Gramin. Downy Mildews, Bellagio* (in press).
Reyes, L., Rosenow, D. T., Berry, R. W. and Futrell, M. C. (1964). *Pl. Dis. Reptr.* **48**, 249–253.
Robinson, R. A. (1969). *Rev. Appl. Mycol.* **48**, 593–606.
Roth, G. (1967). *Z. Pflanzenkr.-Pflanzenschutz* **74**, 83–100.
Safeeulla, K. M. (1976). "Biology and Control of the Downy Mildews of Pearl Millet, Sorghum and Finger Millet". 304 pp. Wesley Press, Mysore, India.
Safeeulla, K. M. (1979). *Newsletter, Intl. Working Gp. on Gramin. DMs* **1**, 3 and 6.
Safeeulla, K. M. and Thirumalachar, M. J. (1956). *Phytopath. Z.* **26**, 41–48.
Safeeulla, K. M., Shaw, C. G. and Thirumalachar, M. J. (1963). *Pl. Dis. Reptr.* **47**, 679–681.
Schmitt, C. G. and Freytag, R. E. (1974). *Pl. Dis. Reptr.* **58**, 825–829.
Schmitt, C. G., Woods, J. M., Shaw, C. G. and Stansbury, E. (1979). *Pl. Dis. Reptr.* **63**, 621–625.
Schultz, O. E. and Dalmacio, S. C. (1971). *Proc. 2nd Ann. Conf. on Corn, Sorghum, Mango and Peanut.* UPCA, College, Laguna, Philippines, 41–47.
Semeniuk, G. and Mankin, C. J. (1964). *Phytopathology* **54**, 409–416.
Shaw, C. G. (1974). *Trop. Agric. Res. Ser. Tokyo* **8**, 47–55.
Shaw, C. G. (1976). *Kasertsart J.* (Thailand) **10**, 85–88.
Shaw, C. G. (1978). *Mycologia* **70**, 594–604.
Shaw, C. G. and Waterhouse, G. M. (1980). *Mycologia* **72**, 425–426.
Shetty, H. S., Mathur, S. B. and Neergaard, P. (1980). *Trans. Br. Mycol. Soc.* **74**, 127–134.
Shirai, M. and Hara, K. (1927). "A list of Japanese Fungi hitherto known" 3rd edn. Shizuoka, Japan.
Singh, J. P., Renfro, B. L. and Payak, M. M. (1970). *Indian Phytopathology* **23**, 194–208.
Singh, R., Chaube, H. S., Singh, H., Asnani, V. and Singh, R. (1970). *Indian Phytopathology* **23**, 209–215.
Singh, S. D. and Williams, R. J. (1978a). *Progress Rep., PMPDM 7901, ICRISAT* 17 pp.
Singh, S. D. and Williams, R. J. (1978b). *Progress Rep., PMPDM 7906, ICRISAT* 28 pp.
Singh, S. D. and Williams, R. J. (1979). *Newsletter, Intl. Working Gp. on Gramin. DMs, ICRISAT* **1**, 6.
Siradhana, B. S., Dange, S. R., Rathore, R. S. and Jain, K. L. (1976). *Pl. Dis. Reptr.* **60**, 603–605.
Siradhana, B. S., Dange, S. R., Rathore, R. S. and Singh, S. D. (1980). *Curr. Sci.* **49**, 316–317.
Srinivasan, M. C. and Thirumalachar, M. J. (1962). *Bull. Torrey Bot. Club* **89**, 91–96.
Steyaert, R. L. (1937). *Inst. Nat. l'Etude Agron. Congo Belge, Ser. Scient.* **13**, 1–16.
Sun, M. H. (1970). *Indian Phytopathology* **23**, 262–269.
Sun, M. H., Chang, S. C. and Tseng, C. M. (1976). *Kasertsart J.* **10**, 89–93.
Suryanarayana, D. (1961a). *Indian Farming* **11**, 11.
Suryanarayana, D. (1961b). *Note, Second All-India Conf. of Millet Workers*, Kanpur, 1–6.
Syamananda, Riksh (1976). *Kasertsart J.* **10**, 79–80.
Tanaka, J. (1940). *Ann. Phytopath. Soc. Japan* **10**, 127–138.
Tantera, D. M. (1975). *Trop. Agric. Res. Ser. Tokyo* **8**, 165–175.
Thirumalachar, M. J. and Whitehead, M. D. (1952). *Amer. J. Bot.* **39**, 416–418.
Thirumalachar, M. J., Shaw, C. G. and Narasimhan, M. J. (1953). *Bull. Torrey Bot. Club* **80**, 299–307.
Titatarn, S. and Exconde, O. R. (1974). *Philipp. Agric. J.* **58**, 90–104.
Tokura, R. (1975). *Trop. Agric. Res. Ser. Tokyo* **8**, 57–60.
Triharso, Martoredjo, T. and Kudiarti, L. (1976). *Kasertsart J.* **10**, 101–105.
Ullstrup, A. (1970). *Indian Phytopathology* **23**, 250–262.
Van der Westhuizen, G. C. A. (1977). *Phytophylactica* **9**, 83–89.
Viennot-Bourgin, G. (1959). *Bull. Soc. Mycol. France* **75**, 33–37.

Warren, H. L., Scott, D. H. and Nicholson, R. L. (1974). *Pl. Dis. Reptr.* **58**, 430–432.
Waterhouse, Grace M. (1963). "Key to the species of *Phytophthora* de Bary". *C.M.I. Mycol. Paper* **92**, 22 pp. C.M.I., Kew.
Waterhouse, Grace M. (1964). "The Genus *Sclerospora*". *C.M.I. Misc. Public.* **17**, 30 pp. C.M.I., Kew.
Weston, W. H. (1923a). *J. Agric. Res.* **23**, 239–278.
Weston, W. H. (1923b). *J. Agric. Res.* **24**, 853–860.
Weston, W. H. (1929). *Phytopathology* **19**, 1107–1115.
Weston, W. H. (1933). *Phytopathology* **23**, 587–595.
Weston, W. H. (1942). *Phytopathology* **32**, 206–213.
Weston, W. H. and Uppal, B. N. (1932). *Phytopathology* **22**, 573–586.
Weston, W. H. and Weber, G. F. (1928). *J. Agric. Res.* **36**, 935–963.
Williams, R. J. (1980). *Proc. Intl. Conf. on Gramin. Downy Mildews, Bellagio* (in press).
Yamada, M. and Aday, B. (1977). *Ann. Phytopath. Soc. Japan* **43**, 291–293.

Chapter 19

# Downy Mildew of the Hop

## D. J. ROYLE[1] AND H. TH. KREMHELLER[2]

[1] *Department of Hop Research, Wye College (University of London), near Ashford, Kent, England*
[2] *Bayerische Landesanstalt für Bodenkultur und Pflanzenbau, München-Freising, Federal Republic of Germany*

## I. Introduction

Hop downy mildew (*Pseudoperonospora humuli* (Miy. & Tak.) Wilson) is one of the most serious diseases of the cultivated hop (Humulus lupulus L.). It occurs in all of the many hop-growing countries of the Northern Hemisphere, but is confined to Argentina in southern regions, not having spread to Australia, New Zealand or South Africa.

The disease was first reported in 1905 in Japan, on both cultivated hops in the north island and hops growing wild in the south island, almost 400 km away (Miyabe and Takahashi, 1906). The pathogen was then ascribed to *Peronoplasmopara humuli* (Miy. & Tak.). In 1909 downy mildew was noted on wild hops in Wisconsin, USA (Davis, 1910) and the pathogen was then renamed by Wilson (1914) *Pseudoperonospora*, which antedates *Peronoplasmopara* (Waterhouse, 1973). Nothing more seems to have been heard of the disease until 1920 when it was recorded for the first time in Europe as tiny spots on leaves of some Italian wild hops in an experimental garden at Wye College, England (Salmon and Ware, 1923). It was not seen in 1921, a dry year, but reappeared in 1922 in the same locality. By 1924 it had become widespread in commercial hop gardens throughout south-east England and was reported on wild hops in the county of Devon, about 300 km from the nearest cultivations. Downy mildew also appeared for the first time in Germany in 1923 (Lang, 1925) and, between 1923–1926, in Belgium, Czechoslovakia, France, Poland, Yugoslavia and the USSR (Salmon and Ware, 1927; Arens, 1929). Within a few years the disease had spread pandemically throughout Europe and North America where it had started to attack cultivated crops in 1928 (Hoerner, 1932). Crop losses were devastating. In Germany, for example, the scale of damage was quoted in national economic terms and, in 1926, crops valued at more than 30 million marks were lost (Zattler, 1927). In the USA downy mildew regularly caused crop losses in excess of 30% (Magie, 1942) and compounded problems due to powdery mildew (*Sphaerotheca humuli* (DC.) Burr.). It eventually forced a major shift in cultivation of the susceptible Cluster cultivar from the high rainfall areas of the east to the relatively arid regions of Washington and Idaho (Skotland, 1961).

In Europe hops are grown in most countries but are confined to localities which are mostly widely separated. It is remarkable, therefore, that downy mildew spread so rapidly and so destructively. There are claims that before its discovery on cultivated hops the pathogen may have existed in several countries in forms so avirulent as to pass unnoticed and that new aggressive strains then arose. However, it is also possible that the disease was introduced, at least into Europe, since hop planting material was internationally and freely exchanged in the early part of this century (Salmon and Ware, 1923, 1925).

Recognition of the serious threat that the disease presented to hop growing prompted immediate research in several countries which continued unabated until about 1940. Elements of the summer disease cycle were studied, and the dependence of outbreaks on wet weather was quickly recognized in the early work of Salmon and Ware (e.g. 1925), Blattny (1927), Zattler (1927, 1931), Millasseau (1929) and Hoerner (1932, 1939). Control measures based on cultural operations and routine application of protective sprays of

Bordeaux and Burgundy mixtures were advocated which, except in the worst years, succeeded in reducing losses from downy mildew to less severe proportions.

Research interest then appears to have dwindled and, at least in England and Germany as attention was diverted to new problems caused by aggressive attacks of wilt (*Verticillium albo-atrum* Reinke and Berth.). Nevertheless, downy mildew continued to take a periodic toll of crops everywhere and in England became worse in the 1950s partly because the downy mildew-tolerant, but wilt-sensitive Fuggle cultivar declined in popularity in favour of Whitbread's Golding Variety which is very susceptible to downy mildew in the rootstock. There was then once again active inquiry into the disease and, around 1960, investigators began to explore new opportunities for control based upon a better understanding of its life-history and epidemiology, and upon incorporation of resistance into newly bred commercial cultivars. In recent years much effort has gone into developing warning schemes aimed at reducing or re-scheduling fungicide applications for greater

Fig. 1.   Primary basal spike of cv. Northern Brewer in April. A young healthy shoot is to the left ( × 0·5).

efficiency and economy of control. This chapter describes the main characteristics of the disease and pathogen, and considers prospects for control in
the future in relation to measures currently adopted. Some emphasis is placed
on circumstances in England and in the German Federal Republic (GFR).

## II. Symptoms and Damage on Annual Growth

Like many other downy mildews, the hop disease is most conspicuous on
the above-ground portions of the plant where it causes both localized disease
on leaves, flowers and cones (the harvested crop), and systemic disease of
shoots. Infection of shoots results in so-called "spikes" which are its most
characteristic feature.

The disease first becomes apparent in March or April when a small proportion of the many shoots which arise from a perennial rootstock appear sickly
and stunted due to the presence systemically of mycelium of *P. humuli* (Fig. 1).
Growth of these "primary basal spikes" is restricted and if left on a plant
they eventually die. They typically have short, fat internodes and yellowish,

Fig. 2.  Angular leaf spots on leaves of field plants (× 0·5).

downcurled and brittle leaves which develop on their lower surface dark brown or black masses of zoosporangia, the initial inoculum in the seasonal chain of disease.

Because primary spikes are close to the ground and often sheltered within a canopy of healthy basal shoots, dispersal of primary inoculum is short-range and secondary infections begin on neighbouring leaves and shoots. On leaves, irregular chlorotic patches appear which sporulate on the lower surfaces and turn necrotic. These lesions may be coalescent but are localized by the veins and are thus termed "angular leaf spots" (Fig. 2). Infection of healthy basal shoots is via the stipules of the growing tips and leads to "secondary basal spikes". These are often confused with primaries but can be distinguished by their normally elongated lower internodes with at least one pair of healthy leaves at the base.

In favourable weather, and if not controlled, the pathogen multiplies rapidly within the abundant basal foliage. Early in May, selected shoots

Fig. 3. A young terminal spike which has ceased to climb and is falling away from the string. A healthy, climbing shoot is to the right (×0·5).

(usually four or eight per plant) are hand-trained on to strings supported by an overhead wirework which in England is 4–5 m but elsewhere 6–7 m above ground. Surplus basal growth is then removed by hand, or with defoliant chemicals a little later, but often not before some disease has spread on to the rapidly climbing shoots ("bines"). Angular leaf-spots and spikes continue to appear during the season at any height above ground, and dispersal of the resulting inoculum becomes more widespread, although never apparently to great distances even downwind from a source (Kremheller, 1979). Stunted lateral shoots ("lateral spikes") arise in response to infection of axillary buds and shoots from June onwards, and "terminal spikes" can develop at the tips of main bines which then suddenly cease growth and fall away from the strings (Fig. 3). Apart from their contribution to the inoculum pool, both these spike forms of the disease can seriously impair potential yield. Infection of lateral shoots prevents formation of the flowers ("burr") from which cones develop. Disease of the main shoots removes apical dominance and stimulates healthy axillary buds or shoots from lower down the bine to elongate. These may compensate for the main shoots if trained in their place early in the season, but main shoots lost late cannot adequately be replaced.

Outbreaks of terminal spike are particularly common 2–3 weeks after training when the length of bine of 1·5–2 m, and it is probable that mycelia from basal shoot infections can be transported within the climbing shoot and colonize the tip much later (Ware, 1926, 1929). However, the relationship between shoot extension, which in the hop can reach 25 cm in a day, and development of the spike symptom is not clearly understood.

Direct loss in weight of the crop and in its content of alpha-acid (the principal constituent of hop resin which gives beer its bitterness), results from infection of the burr in July and of young, ripening or mature cones from

Fig. 4. Cone disease: the bracteoles are more severely diseased than the bracts giving a variegated appearance ($\times$ 1).

July to harvest in September. Diseased burr hardens, turns brown and falls off so that cones are not formed. Later attacks on the cones turn the bracts and bracteoles reddish-brown or, as often happens, only a proportion is discoloured giving a brown and green variegation (Fig. 4). The pathogen sporulates profusely within the cone's favourable microclimate and if unchecked can multiply quickly through several generations and rapidly ruin a crop at any stage of development.

## III. Sources of Infection and Rootstock Disease

It has long been thought that primary basal spikes could arise either from germinating oospores of *P. humuli* which in spring might produce sporangia and zoospores to infect the shoots as they emerge through the soil, or from the spread into developing shoots of a mycelium from a perennially diseased rootstock. Alternate hosts have never been regarded as likely sources, though there are reports that the pathogen can cause limited infection of certain species within the Urticales, to which the family containing hop (Cannabinaceae) is closely related (Salmon and Ware, 1928; Hoerner, 1941; Glazewska, 1971d).

Oospores are often produced in abundance in autumn in diseased leaves, shoots and especially cones (Figs 5 and 6), and they become distributed over the ground of hop gardens in plant residues. Because they are abundant and have occasionally been reported to germinate in the laboratory (Arens, 1929; Bressman and Nicholls, 1933; Magie, 1942; Mori, 1962a, 1966), it has been generally assumed, especially on the continent (Zattler, 1951; Petrlik and Stys, 1978) and in N. America (Jones, 1932; Skotland, 1959), that oospores are at least partly responsible for the seasonal production of primary spikes. However, their importance seems to have been exaggerated. In England and Poland repeated attempts to induce germination in artificial conditions have failed and they have never seriously been considered to play a role in overwintering. Elsewhere, with the possible exception of Japan, germination in laboratory or field conditions has not been demonstrated and evidence that they can cause infection of field hops appears to be entirely circumstantial.

Mycelium of *P. humuli* in diseased rootstocks was discovered by Ware (1926) and in dormant rootstock buds by Salmon and Ware (1927). Much later, Skotland (1961) produced field evidence to suggest that in the arid regions of western Washington, where rootstock disease is the major concern (despite primary spikes being produced relatively infrequently), the pathogen overwinters as mycelium in diseased rootstocks. Field surveys and experiments carried out by Coley-Smith (1962) then showed clearly that in England primary spikes arise from buds infected by an overwintering mycelium in

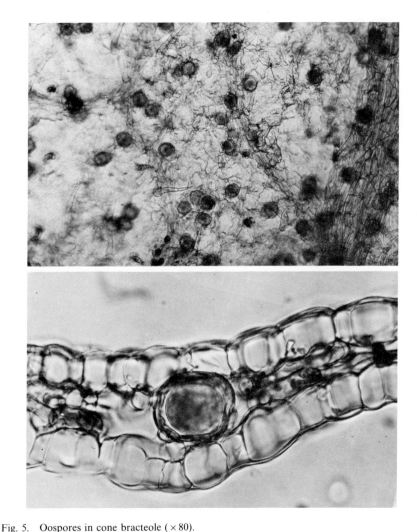

Fig. 5.   Oospores in cone bracteole ( × 80).
Fig. 6.   Oospore in transverse section of cone bracteole ( × 400). (Figs 5, 6 and 8 reproduced here with permission of Dr J. R. Coley-Smith.)


hop rootstocks. Many diseased buds were observed to rot during the winter whilst others co-exist with healthy buds at the same nodes, by no means all of them becoming primary spikes in spring. The rootstock origin of seasonal disease is now accepted in most other countries (Mori, 1962b, 1966; Romanko, 1965; Glazewska, 1971c; Kremheller, 1979).

Rootstocks can become infected in several ways. Direct infection of roots by zoospores washing down through the soil occurs but its importance is

Fig. 7. Pairs of healthy (left), moderately diseased (centre) and well rotted "strap cuts" in early March (× 0·5).

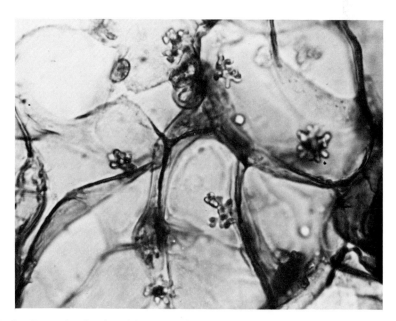

Fig. 8. Haustoria of *P. humuli* in cortex of rootstock (× 400).

difficult to assess. Isolated root infections have been encountered on field plants but it is not known if they occur frequently. Growth of the fungus downwards into the crown from stems is, however, common and can arise in spring following secondary infection of basal shoots or at any time during the growing season from zoospores infecting bine bases, probably through lenticels or wounds (Mori, 1962a, 1969; Coley-Smith, 1965). The relative importance of these two methods of infection appears to vary between countries.

Apart from its importance in the disease cycle, rootstock disease is also associated with death of plants, feeble shoot growth and a considerable loss in yield. In some cultivars diseased rootstocks usually die and are replaced with healthy ones which take three years to give maximum yield. In others the disease has a debilitating effect and rootstocks survive in a weakened form; yields can then be reduced by up to 28% (Coley-Smith, 1964). Disease in rootstocks can conveniently be detected by inspection of the previous year's bine bases ("strap cuts") when they are cut away in winter (Fig. 7). This pruning operation, however, is only carried out when gardens are cultivated and the crown earthed up in July. In England, nowadays, herbicides have replaced tillage for weed control, a practice which is spreading to other countries. Bine bases then shrivel naturally and the disease status of root-stocks is apparent only if primary spikes arise in spring. The fungus can persist in diseased rootstocks for at least four years. The presence of inter-cellular mycelium with characteristic haustoria in sectioned material (Fig. 8) confirms a field diagnosis (Coley-Smith, 1964).


## IV. Epidemiology


### A. The infection phase

Infection in hop downy mildew is wholly dependent upon liquid water which is required for each constituent process—sporangium germination, zoospore motility and encystment, and the emergence, growth and entry of zoospore germ tubes into the host tissues. The pathogen is well adapted to take advantage of brief spells of wetness but during infection is nevertheless very vulnerable to the effects of adverse conditions. Shoot infection requires a longer minimum wet period (3 h) and proceeds over a more restricted tem-perature range (8–23°C) than leaf infection (1·5 h, 5–29°C). Appreciable infection of leaves and shoots proceeds only when wetness lasts 4–8 h and moderate temperatures occur (Magie, 1942; Royle, 1970).

Sporangia germinate only indirectly, by zoospores, which are the infective units. In this respect *P. humuli* is primitive among downy mildew fungi. Each

sporangium discharges 4–8 zoospores and to do so requires wetness lasting
1 h at 20–22°C to 10 h at 2°C (Arens, 1929; Zattler, 1931; Acimovic, 1962;
Petrlik and Stys, 1966). The proportion germinating within a population of
fresh sporangia is high (>90%) but germination declines with age and may
also be reduced slightly by illumination (Glazewska, 1971b; Petrlik and Stys,
1973).

Pathogen entry into hop tissues occurs almost entirely through stomata
which, in leaves, are confined to the lower surfaces. The amount of infection
is therefore dependent upon the frequency with which zoospores encounter
stomata and then settle, encyst and penetrate by their germ tubes. Royle
and Thomas (1971a, b) showed that zoospore distribution on a leaf varies
with the state of stomatal opening and that 4–5 times more disease results
after infection in the light, when most stomata are open, than in the dark,
when they are mostly closed. In light the majority of zoospores quickly settle
and encyst singly on stomata, producing germ tubes which directly enter
between the guard cells (Figs 9–13), but in darkness zoospores remain motile
for much longer and then gradually settle randomly over the leaf surface
resulting in relatively few (<25%) encountering stomata (Figs 14 and 15).

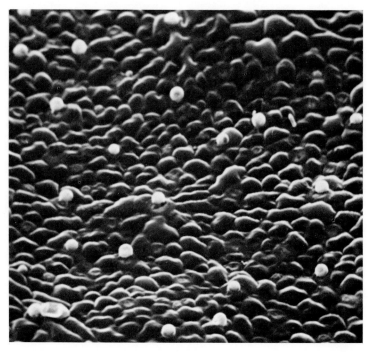

Fig. 9.   Typical zoospore distribution pattern over a lower hop leaf surface where each has
settled singly on a stoma (× 300).

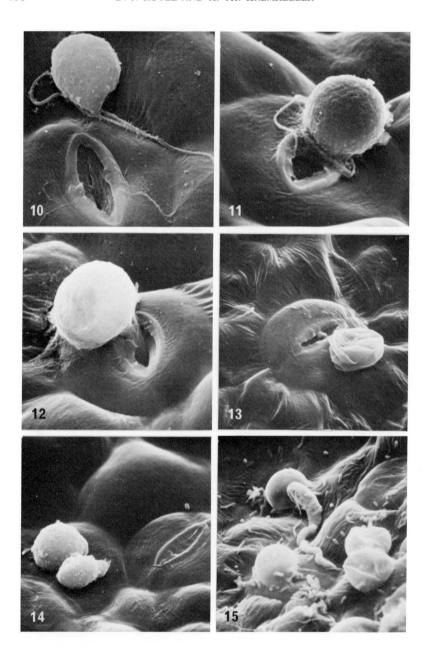

Zoospore behaviour in relation to stomata of hop leaves in the light (Figs 10–13), and in the dark (Figs 14 and 15).

Patterns of zoospore response to stomata in *P. humuli* show interesting contrasts to those in *Plasmopara viticola* on grapevine leaves (Royle and Thomas, 1973). Zoospores of *P. viticola* aggregate on stomata in groups as in response to diffusion gradients; larger numbers (up to 30) gather on fewer stomata in darkness than in light. By contrast, the great majority of *P. humuli* zoospores in an inoculum drop settle singly on stomata in light by a trapping rather than a taxis mechanism. This has been attributed to two stimuli, one of which is chemical and dependent upon photosynthesis which causes zoospores both to select stomata and to settle and germinate quickly. The other is purely physical and has no influence on the speed of the zoospore processes but causes them to favour only the topographically prominent open stomata. *P. humuli* zoospores have the remarkable property of settling on open but not closed "stomata" of perspex leaf replicas.

There is some evidence that infection in darkness can be as high as that in light when provision of a high inoculum dose and a prolonged wet period ensures that randomly-settling zoospores occupy most ports of entry to a leaf (Kremheller, 1979). Nevertheless, the infection level is conditioned by light or dark over a wide range of natural conditions and this is consistent with strong field evidence that outbreaks of downy mildew follow infection periods brought about by daytime rain, never by dew (Royle, 1970, 1973; Kremheller, 1979), (see Section IV.D).

## B. Colonization

Colonization is characterized by the incubation period, ending in symptoms, and the latent period, ending in sporulation. It is this phase of the disease cycle in which the pathogen is within the host and therefore protected from the direct effects of external conditions.

---

Fig. 10. Zoospore approaching a stoma, 2 min after inoculation ($\times$ 2900).

Fig. 11. Settling on a stoma, 2 min after inoculation ($\times$ 3300).

Fig. 12. Zoospore germ tube entering a stoma, 30 min after inoculation ($\times$ 3500). Note the web-like attachments of the cyst to the leaf surface and the ruptured cuticular membrane over the stoma.

Fig. 13. Evacuated zoospore cyst, 6 h after inoculation ($\times$ 1700).

Fig. 14. Zoospores settled randomly, apart from a closed stoma, 30 min after inoculation ($\times$ 1600).

Fig. 15. Germ tube growing over the leaf surface and failing to infect, 6 h after inoculation ($\times$ 1400).

Signs of leaf disease usually precede the beginning of sporulation but symptoms and spores may arise simultaneously when conditions persistently encourage sporulation. Systemic colonization of shoots is also associated with a shorter incubation than latent period, but spikes may never yield sporangia or may take a long time to do so when the weather is dry or when they develop in hop cultivars having some resistance.

The rate of colonization is primarily governed by temperature; there is no evidence that wetness or humidity directly affect the process. Despite confusion in the literature over whether incubation or latent periods have been measured in experiments using constant environments, results of a number of investigations indicate that in susceptible cultivars leaf symptoms develop in 3–10 days according to temperature over a wide range (7–28°C), whilst spikes require 7–22 days over a more limited range (9–20°C). There has been little examination of the effects of unsteady temperatures in natural conditions, but occasional observations (Acimovic, 1962; Royle, 1970; Glazewska, 1971a) indicate that incubation periods correspond when average outdoor temperatures match those in constant environments. However, more study of this aspect is needed.

## C. Inoculum production and dispersal

Sporulation is a diurnally rhythmic process and, like infection, occurs only via stomata. Sporangia borne on each dichotomously-branched sporangiophore develop and mature together during the course of one night, though several sporangiophores usually emerge through a single stoma over several days. Yarwood (1937) followed sporulation on leaves of artificially-inoculated plants in the field in California. He observed that sporangiophores had emerged by midnight, their differentiation was complete and small sporangia had formed by 0300 h, sporangia were full-sized at 0600 h and were mature and being liberated at 0900 h.

Of a number of external factors which interact in sporulation, vapour pressure is the most important. Profuse sporulation occurs only at 95–100% r.h. and although it is unlikely that significant spore production occurs below 90% (Royle, 1968), a little has occasionally been recorded at relative humidities between 40–70% (Arens, 1929; Petrlik and Stys, 1967; Dolinar, 1975). Wet surfaces do not seem to impede sporulation (Sonoda and Ogawa, 1970) but they are unnecessary and, since they encourage sporangia to germinate at night when only slight infection can result, are probably a disadvantage to the pathogen.

Except in the spring, when low temperatures can delay sporulation, temperature is usually favourable throughout the growing season. The optimum range is about 16–20°C. If low (c. 6°C) or high (c. 30°C) temperatures persist for more than 4 h during a night then sporulation is appreciably reduced (Glazewska, 1972).

The role of light is unclear but sporulation seems to be favoured by a natural alternation of light and dark conditions. It is reduced in continuous light and dark or when light interrupts a dark period (Yarwood, 1937; Glazewska, 1972).

Sporangium liberation may involve hygroscopic twisting of sporangiophores in response to declining vapour pressure, as described for *Peronospora tabacina* by Pinckard (1942). However, it has been studied only indirectly in *P. humuli* by sampling the air of hop gardens with volumetric traps (Glazewska, 1967, 1974; Royle, 1968; Mostade, 1971; Kremheller, 1979). Mass release has been claimed to occur at 5–10 day intervals (Glazewska, 1967) which corresponds to the generation time, and day-to-day variation has otherwise been loosely related to the favourability of weather for sporulation and to the amount of disease (Glazewska, 1974). Attempts to associate spore release quantitatively with meteorological conditions occurring previously have revealed none which consistently may be responsible (Royle and Thomas, 1975).

Diurnal patterns of release in *P. humuli* show features similar to those reported for many other downy mildew fungi. In dry weather, and in direct response to the declining r.h. of the air (Royle, 1968; Sonoda and Ogawa, 1972; Kremheller, 1979), sporangia are released during the morning in increasing hourly amounts until 0900–1000 h. Thereafter, the concentration falls gradually as the supply dwindles with none detected at night. Since conditions (high r.h., low temperature) which preserve the life of sporangia do not usually prevail in dry weather during a growing season, a large proportion of the sporangia which are daily deposited dry on hop leaves probably perish unless taken up in dew or rain within a short time of release. Comprehensive data on survival of sporangia are lacking, but Sonoda and Ogawa (1972) found that in the hot, arid climate of California only 1% lived longer than 14 h and that they survived best on lower rather than upper leaf surfaces, and when skies were overcast rather than clear. In the less extreme conditions of the southern GFR, 50% died within 1·5 days, though a few survived 8 days (Kremheller, 1979). Both systemically and in leaf lesions the fungus can often survive appreciable desiccation of diseased tissue and sporulate afresh upon return of more favourable conditions.

The concentration of airborne sporangia often increases at the start of rain due to the shock action of raindrops on the upper, nonsporing leaf surfaces (Hirst and Stedman, 1963). These transient increases often follow dry-air peaks and appear to be most pronounced in response to heavy rain and in daytime when r.h. at the start is low. Continued rain washes sporangia from the air and further release is then prevented by a corresponding rise in r.h. A succession of release peaks during a day can occur when intermittent rain alternates with changing humidity conditions. Rain is responsible for significant inoculum dispersal, by its mechanical action and by splash. At the same time it provides the principle condition for infection and so infectivity of rain-dispersed sporangia is likely to be high. However, the importance of rain dispersal to disease outbreaks is only surmised, mainly from experiments with living plant traps (Royle, 1973; Kremheller, 1979).

## D. Factors affecting disease in the field

It has long been known that wet weather favours hop downy mildew attacks. Salmon and Ware (1933), for instance, recognized early that severe attacks were associated with above-average monthly rainfall and that dry periods checked the disease. Acimovic (personal communication) reports a close relationship between disease levels over 25 years in Yugoslavia and the annual rainfall and average r.h. Similar general relationships, in which the importance of frequent, intermittent rain seems to have been emphasized, have frequently been documented over the years. More precise analyses, carried out during the last 12 years, have been aimed at defining and quantifying the main biotic and abiotic variables which account for seasonal disease variation in field hops as a prelude to developing methods for predicting attacks so that control might be rationalized.

Since wetness is the key determinant of disease progress there has understandably been an emphasis on the nature of infection periods. The development in time of disease on susceptible, unsprayed hops has been recorded and related to weather factors in several background investigations (Glazewska, 1968; Royle, 1970; Kremheller, 1979). Each has indicated that the relationship of disease development to host growth and to inoculum and weather factors is complex and so conclusions have inevitably been of a broad nature and largely inferred, not proven. Royle (1970) simplified interpretation by using data from controlled environment experiments to relate disease fluctuations on leaves and shoots to occasions when conditions of temperature/wetness duration were suitable for either high or low levels of infection (= major and minor infection periods). Disease increased when there were major or minor infection periods but severe outbreaks appeared to arise after rain had contributed to a major infection period, never when dew alone was responsible for plant wetness. Subsequent observations on field plants (Kremheller, 1979) and experiments embracing a wide range of weather over several years (Royle, 1973; Dolinar, 1975; Kremheller, 1979) in which groups of healthy, susceptible pot plants were exposed in hop gardens to "trap" infection in successive 48 h periods, confirmed that dew fails to encourage substantial infection except when there are very high inoculum levels. Dew is thought to be inefficient for three reasons (i) dew starts to form in the dark when zoospores do not respond to stomata, (ii) inoculum which is splash-dispersed by rain augments the component deposited dry after daily release, and (iii) longevity of dry-deposited inoculum is reduced during the interval between release and wetting by dew, especially in the dry weather commonly associated with dewy nights. Rain-initiated wet periods in daylight are probably always required for epidemics to develop. Even in arid regions dew seems to allow the disease only to perpetuate; severe attacks probably need daytime sprinkler irrigation.

Multiple regression analysis of data from plant trap experiments shows that the degree of infection is most highly correlated with variables expressing

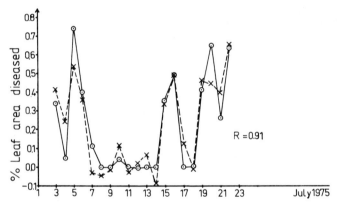

Fig. 16.  Observed levels (o–o) of leaf disease on potted hop plants exposed in a hop garden for 48-h periods in relation to levels predicted (x–x) from an equation: $Y = -0.275 + 0.044 \, RWD + 0.012T$, where $RWD$ = rain wetness duration from 0400–1700 h, $T$ = hours with temperatures between 15–22°C. Data from GFR.

wetness. Interestingly, in both England and the GFR, rain-wetness is consistently the single most important meteorological variable and, together with the airborne spore catch, explains $>70\%$, sometimes $90\%$, of the variation in infection experienced. Consequently, multiple regression equations incorporating several combinations of significant variables estimate observed levels of infection with a high level of accuracy (Fig. 16) and are now used as the basis of prediction schemes currently being evaluated (see Section VI.D). Recently, an alternative empirical analysis of plant trap data has suggested improvements for prediction and given better biological explanations of the relationships exposed (Waggoner, Norvell and Royle, 1980). Formalization in a mechanistic, computer-based model, of the effects of biotic and abiotic factors on events of the infection process leading to leaf lesions is also currently being attempted. This may provide a possible alternative approach to prediction and should identify relationships we do not yet understand (Royle, Hau and Kranz, unpublished).

## V. Aspects of Resistance

Although most of the older cultivars of commercial interest have been susceptible to downy mildew, resistance has nevertheless been recognized and utilized in selective breeding programmes since the 1920s. Yet, little is known of the nature and inheritance of resistance mechanisms and the development of satisfactory levels of resistance in new cultivars has been further complicated by the different relative levels of susceptibility between plant parts of a given cultivar. For example, in the GFR the cultivar Hallertauer Mittelfrüher has greater resistance in the rootstock than Brewer's Gold yet both are equally susceptible to cone disease. The cones of Northern Brewer in England are more resistant than those of Bullion but both cultivars are equally sensitive to the disease in shoots and leaves.

The nature of resistance and susceptibility in rootstocks is undefined. In some cultivars, such as Whitbread's Golding Variety and Bullion in England and Clusters in the USA, downy mildew causes widespread death of rootstocks and those that survive yield very few primary spikes. In others, such as Northern Brewer, rootstocks can be perennially diseased but they rarely rot completely and usually yield many spikes. Rootstocks of still other cultivars remain resistant despite repeated exposure to the pathogen.

A little more is known about the interactions within leaves differing in resistance to localized infection by *P. humuli* (Royle, 1967). Resistance is here expressed, after pathogen entry, as an incomplete hypersensitivity. Within 12–16 h of inoculation of the most resistant genotypes mesophyll cells around the entry sites die. However, the fungus continues to grow feebly, to cause further host cell death and, in continuously favourable conditions, it may sporulate slightly. Moderately resistant genotypes, which represent the resistant cultivars now in commerce, exhibit less extreme hypersensitivity and though the fungus may not sporulate in natural conditions it can do so quite profusely in the laboratory. Susceptible cultivars allow vigorous hyphal growth within leaves followed by dense sporulation; cell necrosis occurs but only late in colonization. These and other observations over the years of the effects of environment upon resistance expression, suggest that resistance is essentially quantitative and controlled polygenically rather than by major genes.

Apart from possible differences in strains of *P. humuli* between continents, there is no evidence that the pathogen exists in physiologically specialized forms. With the possible exception of some breakdown of resistance in cones of the cultivars Fuggle and Saaz many years ago (Salmon and Ware, 1931, 1932), increased virulence in response to selection pressure imposed by large plantings of resistant cultivars has not been experienced. The scope for new races is possibly limited because the products of sexual fusion, oospores, do not seem to play a role in the disease cycle.

## VI. Control

### A. Cultural and sanitation practices

Mechanical removal of diseased tissues and attention to the health of planting material have traditionally been important components of downy mildew control. Several of the procedures may not always be carried out nowadays because they prove to be inconvenient in busy farm work programmes. Increased mechanization and changes in cultivation practices in hop gardens over the last 20 years have not encouraged downy mildew as they have powdery mildew. As a result, cultural measures for downy mildew control are increasingly regarded with less importance.

Hand removal of basal spikes, the primary sources, and lower leaves has traditionally been carried out, often punctiliously, at weekly or more frequent intervals early in the season. In many countries it is still a routine practice but in England it receives less attention these days. Superior fungicides, resistant cultivars, and defoliant chemicals to remove excess basal growth with its residual inoculum sources all help to compensate. Neither defoliants nor herbicides are yet used in most countries and under normal tillage rootstock crowns are pruned late in winter. Deep pruning is sometimes done and may reduce numbers of basal spikes by up to 10%.

### B. Resistant cultivars

Of the hop cultivars grown in England during the early epidemics in the 1920s all were susceptible except Fuggle which was highly resistant, particularly in the cones (Salmon and Ware, 1930). Subsequently, the resistance of Fuggle became unstable and the cultivar suffered some severe attacks both in England and the USA. Nevertheless, it appears to have retained a useful degree of resistance to this day and in the USA it is still the most resistant to rootstock infection of all the older cultivars. It has been used there, in normal diploid or tetraploid form, as the basis of resistance in breeding since 1957 and several new commercial cultivars have emerged, containing the essential rootstock resistance and requiring little, if any, chemical means to control downy mildew. In Yugoslavia also, the most widely grown traditional cultivar, Savinia Golding, which is thought to be synonymous with Fuggle, is considerably resistant although it requires up to five annual sprays of protectant fungicides.

Resistance breeding in Western Europe has largely been based on several German lines, derived from wild hops early in the GFR breeding programme. These lines contain probably the highest levels of resistance yet encountered and have led to a number of resistant cultivars in the GFR, some of which have not succeeded due either to unsatisfactory yields or to susceptibility to wilt. However, a new introduction, Perle, possesses more favourable attributes and allows a substantial reduction in the number of annual fungicide sprays.

Combination of the German sources of resistance with high alpha-acid qualities of English Northern Brewer at Wye College has resulted in the two resistant cultivars Wye Northdown and Wye Challenger which were intro-duced in the early 1970s and are now widely planted. In seasons favourable to downy mildew, Northdown usually develops a little disease which can be controlled with 2–3 sprays whilst Challenger is so highly resistant that in large plantings it only rarely requires fungicides.

## C. Fungicides

Control of downy mildew in susceptible cultivars relies heavily on the use of protectant fungicides. Nowadays, growers have become accustomed to a comprehensive, routine set of applications aimed at a continuous cover of chemical as an insurance against the risk of infection at any time. In countries with only moderate rainfall, like England and Yugoslavia, some 8–12 sprays and dusts may be applied each season whilst in the GFR, where the disease threatens more frequently, 12–16 or more are needed. In practice a continuous cover is difficult to maintain on fast-growing hop tissues and occasional lapses in a regular spray frequency have in the past resulted in bad disease attacks. The regular, arbitrary sequence of applications has little other biological rationale but nevertheless has evolved of necessity without benefit of knowledge of the conditions under which the disease develops and spreads.

Until the late 1950s fungicide applications in England were not usually begun until the bines had reached the top wire, in June; little encouragement was given to use fungicides earlier in the season, mainly because of lack of evidence as to their value. However, re-examination by Coley-Smith (1965, 1966) of the practice of applying dusts to the crowns and young shoots in spring confirmed growers' suspicions at the time of the value of this operation, not only to reduce secondary spread from already diseased plants, but also to delay rootstock infection in new plantings. As a result, so-called "hill-dusting" with a number of fungicides such as Bordeaux, zineb and copper oxychloride, became firmly recommended as an adjunct to routine manual removal of spikes and grubbing of severely diseased rootstocks.

Improvements in the degree of control achieved by early-season dusts and greater ease by which control could be achieved then came about with the systemic antibiotic, streptomycin. Following a favourable report of its use in the USA (Hoerner and Maier, 1957), but conflicting results in some other countries, Coley-Smith (1966) showed that two early-season sprays to severely diseased plantings significantly reduced secondary infection without the need for laboriously hand-removing spikes. Streptomycin was then used as a routine by many growers in England, as in the USA, and there can be little doubt that its sustained use over 15 years or so greatly facilitated later control and caused downy mildew to decline in importance, a trend enforced by the introduction since 1971 of resistant cultivars. Pesticide regulations in the GFR prevented its adoption, and in England it has recently been withdrawn. If they did not choose to use the expensive streptomycin, most English growers came to prefer low volume sprays, rather than dusts, of organic and copper fungicides for use on young hops. In the GFR fentin acetate (Brestan 60) has become the standard treatment in the early-season period.

Fungicides based on copper, as Bordeaux and Burgundy mixtures, have been the mainstay of control since early times and until organics were intro-duced in the mid-1950s were the only fungicides available. Since then English growers have increasingly preferred dithiocarbamates, such as zineb and propineb, mainly because they encourage more luxuriant foliage and less brittle cones which pick more easily in modern machine harvesters. As a result copper usage had declined to only 20% of the total market by 1977 (Umpelby and Sly, 1978). There is no doubt that this decline contributed significantly to deteriorating control in the 1960s of powdery mildew, largely because copper renders hop foliage more resistant to this disease (Royle, 1978). In contrast, copper fungicides are still popular in the GFR and retain two-thirds of total usage. They are used there extensively in the post-flowering period partly because the harder cones are preferred and also because there are restrictions on use of dithiocarbamates within a month of harvest due to toxic residues, a problem which is currently attracting concern in England. Introduction of resistant cultivars in England has reduced by 50% total fungicides used against downy mildew in 5 years (1973–1977).

In the last 2 years hop downy mildew control has begun to become profoundly influenced by the advent of new groups of highly efficient, Phycomycete-specific systemic fungicides, especially metalaxyl, based on acyl alanine, and also the aluminium ethyl phosphites. A single application of metalaxyl on crowns bearing small shoots in April practically restores primary spikes to normal shoots and completely suppresses spore production. It offers protection to healthy shoots for up to 8–12 weeks when the inoculum density is not excessively high. Metalaxyl has been used commercially now for only two years and so, although the long-term effects of its continued use on the downy mildew status of hop gardens are not known, it seems reasonable to suppose that sources of downy mildew could dwindle to such an extent that the disease no longer poses a threat to the health of the crop. In such a situation concern is naturally felt of the dangers of tolerance to the fungicide developing within the pathogen population. Indiscriminate use is therefore to be avoided and recommendations in England are now being reviewed. In the GFR metalaxyl is applied in mixture with copper as a precaution against pathogen tolerance. Investigations are proceeding in England on the levels of metalaxyl in hop tissues in relation to its biological effectiveness.

### D. In prospect—directed control

The desire to utilize conventional protectants and, for that matter, new highly competent fungicides like metalaxyl and aluminium ethyl phosphites, in rational strategies of disease control has led to the development of pre-diction (prognosis) schemes for discriminate scheduling of fungicide appli-cations. In the GFR, the aim of such an approach is to reduce the large number of sprays used each season, on economic and environmental grounds, whilst in England, as in Yugoslavia and probably Czechoslovakia also, the objective is primarily to deploy the existing number of applications more effectively, reduction being a welcome bonus if seasonal conditions permit.

These aims are achieved using several formulae which summarize numeri-cally the relationships between some measure of disease activity and key environmental factors and which can be used to warn of favourable, or unfavourable conditions, for disease attacks. Pejml (1969) in Czechoslovakia proposed the first formula which integrates various meteorological para-meters (daily temperature, r.h., rainfall, and length of dry periods) for calculating an index expressing the degree to which weather is conducive to disease development. Royle (1973, 1975), Dolinar (1978) and Kremheller (1979) have independently derived multiple regression equations from analysis of their plant trap and field experiments described in section IV.D. Each equation uses rain-wetness variables, and some also include an inoculum variable to give an estimate of the expected degree of infection. The equations have been validated on commercial field crops.

Of course, none of the formulae is able to forecast infection. Each can only inform of when and how much infection is likely to have occurred. The prediction scheme developed at Wye College (Royle, 1979) couples this information with the time delay given by the incubation period to indicate when to coincide application of a protective fungicide with the expected appearance of symptoms and fresh spores. This gives a grower 5–7 days notice of a required spray treatment. At the present time no eradicant fungicides are recommended in the UK but where they have been examined experimentally clear possibilities emerge for effective control action shortly after infection is predicted. Since it is impractical to monitor airborne spore levels on hop farms, ways are being sought to utilize observations of disease within crops, as is now done in the Czech scheme (Pejml *et al.*, 1978), to give the desirable biological dimension. For the time being an equation composed only of abiotic variables (rain amount, hours r.h. > 90%) and with a temperature threshold and a dew variable as constraints, is being used to guide spray tactics with or without the benefit of disease information.

The German system (Kremheller, 1979), which is based on a regression equation incorporating airborne spore concentration and rain-wetness duration as a product variable, operates more stringently. Because disease attacks occur more frequently and severely than in England, it is dangerous to adopt the retrospectively directed control measures used in the Wye system. Instead, protectant fungicides must be applied when the level of airborne sporangia exceeds a threshold value (different for leaves and cones) in advance of a possible infection period. Alternatively, systemic fungicides may be used to eradicate infection, as long as spraying can be accomplished within 48 h of an assessed infection period, i.e. on small areas of hops only. The main difficulty with the German and Wye methods is obtaining appropriate local records of spore concentration in the air. If no simple method is forthcoming in the GFR, development of a warning service may depend first upon establishing relationships of aerial spore levels at selected representative sites throughout the hop region with those on individual farms in each vicinity.

In addition to the problem of acquiring information on inoculum, various other difficulties caused by the time-dependence of a directed spray strategy need to be resolved. For example, growers may find it difficult to respond to a spray warning if the weather is bad or if other farm operations tie up labour and equipment; the time needed to spray a large area of hops may be too long to take full advantage of a disease warning; and timed downy mildew sprays will not necessarily coincide with routine sprays against other pests and diseases. However, the benefits of directed control for dealing economically and efficiently with downy mildew in future are considerable. The stated aims of the approach seem already to have been satisfied. In the Hallertau region of GFR in 1979 downy mildew prevention required about 16 routine protective sprays but only five protective or four systemic fungicide applications under directed control. Similarly, on commercial trial sites in England healthy crops were picked in 1979 after reducing and re-scheduling sprays applied to susceptible cultivars. Possibilities exist for integrating the directed control approach with resistance in some commercial cultivars which need only 2–3 sprays each year.

418    D. J. ROYLE AND H. TH. KREMHELLER

# References

Acimovic, M. (1962). *Contemp. Agric.* (Yugoslavia) **5**, 373–391.
Arens, K. (1929). *Phytopath. Z.* **1**, 169–193.
Blattny, C. (1927). *Sb. vyzk. Ust. zemed. RCS* **27a**, 3–274.
Bressman, E. N. and Nicholls, R. A. (1933). *Phytopathology* **23**, 485.
Coley-Smith, J. R. (1962). *Ann. Appl. Biol.* **50**, 235–243.
Coley-Smith, J. R. (1964). *Ann. Appl. Biol.* **53**, 129–132.
Coley-Smith, J. R. (1965). *Ann. Appl. Biol.* **56**, 381–388.
Coley-Smith, J. R. (1966). *Ann. Appl. Biol.* **57**, 183–191.
Davis, J. J. (1910). *Science* **31**, 752.
Dolinar, M. (1975). *Bull. Inst. Hop Res. Zalec, Yugoslavia* **3**, 1–9.
Dolinar, M. (1976). *4th Yugoslav Hop Symposium Zalec, Yugoslavia.* 139–150.
Glazewska, Z. (1967). *Pamietnik Pulawski* **28**, 117–132.
Glazewska, Z. (1968). *Pamietnik Pulawski* **31**, 263–278.
Glazewska, Z. (1971a). *Pamietnik Pulawski* **43**, 181–188.
Glazewska, Z. (1971b). *Pamietnik Pulawski* **43**, 189–199.
Glazewska, Z. (1971c). *Pamietnik Pulawski* **49**, 175–190.
Glazewska, Z. (1971d). *Pamietnik Pulawski* **49**, 191–204.
Glazewska, Z. (1972). *Roczn. Naukro ln.* **2**, 39–52.
Glazewska, Z. (1974). *Roczn. Naukro ln.* **4**, 139–158.
Hoerner, G. R. (1932). *Bull. Oregon St. Agric. Coll. Extn. Serv.* **440**, 1–11.
Hoerner, G. R. (1939). *Pl. Dis. Reptr.* **23**, 361–366.
Hoerner, G. R. (1941). *J. Agric. Res.* **61**, 331–334.
Horner, C. E. and Maier, C. R. (1957). *Phytopathology* **47**, 525.
Hirst, J. M. and Stedman, O. J. (1963). *J. Gen. Microbiol.* **33**, 335–344.
Jones, W. (1932). *J. Inst. Brew.* **38**, 194–196.
Kremheller, H. Th. (1979). *Dissertation, Tech. Univ. München.*
Lang, W. (1925). *NachrBl. dt. PflSchutzdienst, Berlin* **5**, 63.
Magie, R. O. (1942). *Tech. Bull. N.Y. State Agric. Exp. Stn.* **267**, 1–48.
Millasseau, J. (1929). *Annls. Epiphyt.* **14**, 177–198.
Miyabe, K. and Takahashi, Y. (1906). *Trans. Sapporo Nat. Hist. Soc.* **1**, 150–157.
Mori, Y. (1962a). *Bull. Brew. Sci.* **7**, 1–5.
Mori, Y. (1962b). *Bull. Brew. Sci.* **7**, 7–12.
Mori, Y. (1966). *Ann. Phytopath. Soc. Japan* **32**, 275–284.
Mori, Y. (1969). *Bull. Brew. Sci.* **15**, 5–10.
Mostade, J. M. (1971). *Parasitica* **27**, 64–69.
Pejml, K. (1969). *Met. Zpr.* **22**, 18–22.
Pejml, K., Petrlik, Z. and Stys, Z. (1978). *Sb. Uvtiz Ochr. Ros.* **14**, 41–46.
Petrlik, Z. and Stys, Z. (1966). *Ceska Mykol.* **20**, 105–110.
Petrlik, Z. and Stys, Z. (1967). *Ceska Mykol.* **21**, 242–246.
Petrlik, Z. and Stys, Z. (1973). *Ceska Mykol.* **27**, 112–117.
Petrlik, Z. and Stys, Z. (1978). *Chmelarstvi* **51**, 52–53.
Pinckard, J. A. (1942). *Phytopathology* **32**, 505–511.
Romanko, R. R. (1965). *Pl. Dis. Reptr.* **49**, 247–250.
Royle, D. J. (1967). *Rep. Dep. Hop Res. Wye Coll.* 1966, 49–56.
Royle, D. J. (1968). *Rep. Dep. Hop Res. Wye Coll.* 1967, 22–26.
Royle, D. J. (1970). *Ann. Appl. Biol.* **66**, 281–291.
Royle, D. J. (1973). *Ann. Appl. Biol.* **73**, 19–30.
Royle, D. J. (1975). *Proc. Sem. d'Ét. Agric. Hyg. des Plantes, Gembloux*, 381–394.

Royle, D. J. (1978). *In* "The Powdery Mildews" (D. M. Spencer, Ed.) 381–409. Academic Press., London and New York.
Royle, D. J. (1979). *Rep. Dep. Hop Res. Wye Coll.* 1978, 49–56.
Royle, D. J. and Thomas, G. E. (1975). *Trans. Br. Mycol. Soc.* **58**, 79–89.
Royle, D. J. and Thomas, G. G. (1971a). *Physiol. Pl. Path.* **1**, 329–343.
Royle, D. J. and Thomas, G. G. (1971b). *Physiol. Pl. Path.* **1**, 345–349.
Royle, D. J. and Thomas, G. G. (1973). *Physiol. Pl. Path.* **3**, 405–417.
Salmon, E. S. and Ware, W. M. (1925). *J. Ministr. Agric. Fish.* **31**, 1–13.
Salmon, E. S. and Ware, W. M. (1927). *J. Ministr. Agric. Fish.* **33**, 1108–1121.
Salmon, E. S. and Ware, W. M. (1928). *Ann. Appl. Biol.* **15**, 352–370.
Salmon, E. S. and Ware, W. M. (1930). *J. Inst. Brew.* **36**, 63–66.
Salmon, E. S. and Ware, W. M. (1931). *J. Inst. Brew.* **37**, 24–31.
Salmon, E. S. and Ware, W. M. (1932). *J. Inst. Brew.* **38**, 37–44.
Salmon, E. S. and Ware, W. M. (1933). *Jl. S.-e agric. Coll. Wye* **32**, 108–119.
Salmon, E. S. and Wormald, H. (1923). *J. Ministr. Agric. Fish.* **30**, 1–5.
Skotland, C. B. (1959). *Mod. Brew. Age* **59**, 47–48.
Skotland, C. B. (1961). *Phytopathology* **51**, 241–244.
Sonoda, R. M. and Ogawa, J. M. (1970). *Mycologia* **62**, 1067–1069.
Sonoda, R. M. and Ogawa, J. M. (1972). *Hilgardia* **41**, 457–474.
Umpelby, R. A. and Sly, R. M. A. (1978). *Pesticide Usage: Survey Report* 16. Hops. Ministry of Agriculture, Fisheries and Food.
Waggoner, P. E., Norvell, W. A. and Royle, D. J. (1980). *Phytopathology* **70**, 59–64.
Ware, W. M. (1926). *Trans. Br. Mycol. Soc.* **11**, 91–107.
Ware, W. M. (1929). *Ann. Bot.* **43**, 683–710.
Waterhouse, G. M. (1973). *In* "The Fungi" Vol. IVB. (G. C. Ainsworth, F. K. Sparrow and A. S. Sussman, Eds.) 165–183. Academic Press, London and New York.
Wilson, G. W. (1914). *Mycologia* **6**, 194.
Yarwood, C. E. (1937). *J. Agric. Res.* **54**, 365–373.
Zattler, F. (1927). *Arb. Bayer Landesanst. Pflbau* **5**.
Zattler, F. (1931). *Phytopath. Z.* **3**, 281–302.
Zattler, F. (1951). Bericht über die Versuchs- u Forschunfstätigkeit auf dem Hopfenversuchsgut Hüll, 1926–51. Jubiläumsfestschrift der Deutschen Gesellschaft für Hopfenforschung e.V.

Chapter 20

# Downy Mildew Diseases Caused by the Genus *Bremia* Regel

I. R. CRUTE[1] AND G. R. DIXON[2]

[1]*National Vegetable Research Station, Wellesbourne, Warwick, UK*
[2]*School of Agriculture, Aberdeen, UK*

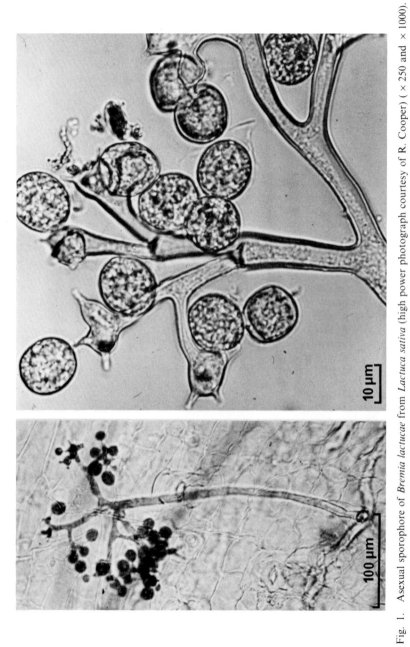

Fig. 1. Asexual sporophore of *Bremia lactucae* from *Lactuca sativa* (high power photograph courtesy of R. Cooper) ( × 250 and × 1000).

# I. Diagnostic Characteristics of the Genus *Bremia* Regel

The genus *Bremia* Regel is distinguished morphologically from other genera of the Peronosporaceae by its asexual sporophores which are slender structures, dichotomously branched with disc-like tips bordered by sterigmata (Fig. 1).

*Bremia* primarily parasitizes members of the family Compositae although one species assigned to this genus is found in the graminicolous genus *Arthraxon* Beauv.

# II. Taxonomy and Nomenclature of *Bremia* Species

Following the taxonomic principles of Yerkes and Shaw (1959) for the Peronosporaceae, two species of *Bremia* are recognized which parasitize different host families.

## A. *Bremia graminicola* Naoumov

*Bremia graminicola* Naoumov is the only species of this genus recorded on a non-composite host. Only the asexual stage has been described and there are four records of its occurrence. Naoumov (1913) noted its presence on the grass *Arthraxon ciliaris* Beauv. in Austria and Russia and there are further reports from China (Miyake, 1914) and Japan (Togashi, 1926). Patel (1948) found *Bremia* on *Arthraxon lancifolius* Hoch. in India and since he considered that it was not morphologically identical to the fungus described previously he named it *Bremia graminicola* Naoumov var. *indica* Patel. Table 1 summarizes the reports and morphology of *Bremia* from *Arthraxon* species.

## B. *Bremia lactucae* Regel

All other records of *Bremia* have been of fungi collected from hosts in the family Compositae and can be considered to belong to a single species, *Bremia lactucae* Regel. There was much synonymy in early reports (Table 2). As Shaw (1949) indicates, the binomial *Bremia lactucae* was used by Regel (1843) to name the asexual stage while oospores were first recorded by Tulasne (1854) and described by Caspary (1855) under the name *Peronospora ganglioniformis*. Despite this, *Bremia lactucae* is the binomial which has been retained since in the Peronosporaceae, it is the asexual stage which is diagnostic.

TABLE 1

Morphology of *Bremia* from *Arthraxon* spp.

| *Bremia* species | Author | Host species | Conidiosporangium morphology | | | | Sporophore morphology | | | |
|---|---|---|---|---|---|---|---|---|---|---|
| | | | Length (μm) | | Breadth (μm) | | Length (μm) | | Breadth (μm) | |
| | | | Range | Mean | Range | Mean | Range | Mean | Range | Mean |
| *B. graminicola* | Naoumov (1913) | *A. ciliaris* | — | 12<br>Round | — | 12 | — | 600 | 9–10 | — |
| *B. graminicola* | Miyake (1914) | *A. ciliaris* | 14–17 | —<br>Oval | 11–13 | — | — | 400 | 10 | — |
| *B. graminicola* | Togashi (1926) | *A. ciliaris* | 11–15 | —<br>Papillate | 11–15 | — | 460–760 | — | 7·5–10·0 | — |
| *B. graminicola* var. *indica* | Patel (1948) | *A. lancifolius* | 11–18 | 14·5<br>Ovate | 9–13 | 10·6 | 330–840 | — | 8·0–10·5 | — |

TABLE 2

Early synonyms of *Bremia lactucae* Regel

| Binomial | Author | Host species | Stage described |
|---|---|---|---|
| *Bremia lactucae* | Regel (1843) | *Lactuca angustana* Bisch (= *L. sativa* L.) | Asexual |
| *Botrytis stellata* | Desmazieres (1846) | *Lactuca sativa* | Asexual |
| *Botrytis ganglioniformis* | Berkley (1846) | *Lactuca sativa* | Asexual |
| *Botrytis parasitica* var. *lactucae* | Berkley (unknown) | *Lactuca sativa* | Asexual |
| *Botrytis germinata* | Unger (1847) | *Lapsana communis* L. | Asexual |
| *Botrytis lactucae* | Unger (1847) | *Lactuca sativa* | Asexual |
| *Botrytis gangliformis* | Berkley and Broome (1851) | *Lactuca sativa* | Asexual |
| *Botrytis sonchicola* | Von Schlechtendal (1852) | *Sonchus oleraceus* L. | Asexual |
| *Peronospora ganglioniformis* | Tulasne (1854) | *Lactuca sativa* | Sexual and asexual |
| *Actinobotrytis tulasnei* | Hoffman (1856) | *Senecio vulgaris* | Asexual |
| *Peronospora ganglioniformis* | de Bary (1863) | *Lactuca sativa* | Sexual and asexual |
| *Peronospora stellata* | Delacroix (1867) | *Lactuca sativa* | Asexual |
| *Peronospora senecionis* | Fuckel (1869) | *Senecio vulgaris* | Asexual |

Some workers have erected additional species to *Bremia lactucae* to accommodate fungal collections from different genera of Compositae and with slight morphological differences from the type species. Details of these and their comparative morphology are given in Table 3. With the possible exception of collections from *Hemistepta carthamoides* Kuntze (= *Saussurea affinis* Spreng.) named by Sawada (1919) as *Bremia saussurea* Saw. and accepted by Ling and Tai (1945) there appears to be little justification for recognizing more than one taxon on morphological grounds. This argument is supported by evidence that age and environmental factors affect the size and shape of sporophores and conidiosporangia of downy mildew (Schweizer, 1919; Ling and Tai, 1945).

On the basis of host range and morphology studies, Ling and Tai (1945) recognized and named four *formae speciales* of *Bremia lactucae* (f. *chinensis* on *Lactuca chinensis* Makino; f. *sonchicola* (Schlecht.) on *Sonchus oleraceus* L.; f. *taraxaci* (Ito and Tokunaga) on *Taraxicum mongolicum* Hand. -Mzt. and f. *ovata* (Saw.) on *Crepis japonicus* Benth.) while Milovtzova (1937) had previously named a collection of the fungus from *Carthamus tinctorius* L. as *B. lactucae* f. *carthami*. Host range and specificity of *Bremia lactucae* are discussed in Section III while the implications and dangers of categorizing

TABLE 3

Comparative morphology of *Bremia* collections which have been considered to constitute distinct taxa

| *Bremia* species | Authors | Hosts | Conidiosporangium morphology Length (μm) Range | Mean | Breadth (μm) Range | Mean | Sporophore morphology Length (μm) Range | Mean | Breadth (μm) Range | Mean | Pro-portion unbranched | Oospore morphology Diameter (μm) Range | Mean |
|---|---|---|---|---|---|---|---|---|---|---|---|---|---|
| *B. centaurea* Syd. | Oescu and Radulescu (1933) | *Centaurea* spp. | 16–28 | — | 14–22 | — | — | — | — | — | — | — | — |
| *B. elliptica* Sawada | Savulescu (1962) | | 16–22 | 18·75 | 15–22 | 17·2 | 360–500 | — | 6–9 | — | $\frac{2}{3}-\frac{3}{4}$ | — | — |
| | Ito and Tokunaga (1935) | *Lactuca* spp. | 15–24 | — | 12–20 | — | 240–570 | — | 8–16 | — | $\frac{1}{2}-\frac{2}{3}$ | — | — |
| *B. lactucae* Regel | Fischer (1892) | Many | — | 17·0 | — | 15·0 | 240–400 | — | 8–10 | — | $\frac{2}{3}$ | 26–34 | — |
| | Oescu and Radulescu (1932) | | 15–26 | — | 14–19·5 | — | — | — | — | — | — | — | — |
| | Oescu and Radulescu (1933) | | 15–24 | — | 12–21·6 | — | — | — | — | — | — | — | — |
| | Ling and Tai (1945) | | 10–24 | 17·55 | 10–23 | 16·3 | 275–610 | — | 8–15 | — | — | — | — |
| | Savulescu (1962) | | 15–22 | 17·75 | 14–18 | 16·2 | 225–510 | — | 7–12 | — | $\frac{1}{2}-\frac{2}{3}$ | 27–35 | — |
| *B. lactucae* f. sp. carthami Milovtzova | Milovtzova (1937) | *Carthamus tinctorius* | | | | | | | | | | | |
| *B. lactucae* f. sp. chinensis Ling and Tai | Ling and Tai (1945) | *Lactuca chinensis* | 10–22 | 15·3 | 8–20 | 14·1 | 310–510 | — | 10–13 | — | — | — | — |
| *B. lactucae* f. sp. ovata Ling and Tai | Ling and Tai (1945) | *Crepis japonicus* | 10–22 | 15·6 | 7–19 | 11·9 | 850–1050 | — | 7–11 | — | — | — | — |

| Species | Author | Host | | | | | | | | | |
|---|---|---|---|---|---|---|---|---|---|---|---|
| *B. lactucae* f. sp. *sonchicola* Ling and Tai | Ling and Tai (1945) | *Sonchus oleraceus* | 13–24 | 18·5 | 9–22 | 16·1 | 332–534 | 11–13 | — | — | — |
| *B. lactucae* f. sp. *taraxaci* Ling and Tai | Ling and Tai (1945) | *Taraxicum mongolicum* | 13–24 | 18·5 | 11–21 | 16·5 | 428–545 | 10–13 | — | — | — |
| *B. lampsanae* Syd. | Savulescu (1962) | *Lampsana communis* | 15–20 | 17·2 | 13–17 | 15·1 | 380–540 | 7–10 | $\frac{2}{3}$–$\frac{3}{4}$ | — | — |
| *B. microspora* Sawada | Ito and Tokunaga (1935) | *Lactuca debilis* | 9–19 | — | 8–15 | — | 270–600 | 6–16 | $\frac{1}{2}$–$\frac{2}{3}$ | — | — |
| *B. ovata* Sawada | Tanaka (1919) | *Crepis japonicus* | 14–18 | — | 10–13 | — | — | 3–8 | — | — | — |
| *B. picridis* Ito and Tokunaga | Ito and Tokunaga (1935) | *Picris hieracoides* | 12–22 | 16·9 | 15–18 | 14·5 | 300–850 | 7–15 | $\frac{1}{3}$–$\frac{3}{4}$ | — | — |
| *B. saussurea* Sawada | Tanaka (1919) | *Hemistepta carthamoides* | 24–57 | — | 18–28 | — | 270–1021 | 4–10 | — | — | — |
| *B. sonchi* Sawada | Ling and Tai (1945) | *Sonchus* spp. | 23–37 | 30·4 | 22–33 | 27·8 | 206–650 | 8–11 | — | — | — |
| | Tanaka (1919) | | 17–24 | — | 13–21 | — | 230–560 | 4–9 | $\frac{1}{2}$ | — | — |
| | Oescu and Radulescu (1933) | | 16–24 | — | 17–23 | — | — | — | — | — | — |
| *B. sonchi* Sawada | Savulescu (1962) | *Taraxacum* spp. | 18–24 | 20·5 | 17–22 | 19·2 | 260–480 | 8–12 | $\frac{1}{2}$–$\frac{2}{3}$ | 30–38 | — |
| *B. taraxaci* Ito and Tokunaga | Ito and Tokunaga (1935) | | 16–24 | — | 15–22 | — | 200–720 | 7–19 | $\frac{1}{3}$–$\frac{3}{4}$ | — | — |
| *B. tulasnei* (Hoffm.) Syd. | Sydow (1923) | *Senecio* spp. | 17–19 | — | 14–16 | — | — | — | — | — | — |
| | Savulescu (1962) | | 15–19 | 17·2 | 13–17 | 15·1 | 400–480 | 9–12 | $\frac{1}{2}$–$\frac{2}{3}$ | — | — |

pathotypes of Peronosporaceous species taxonomically are reviewed in Chapter 11 (Crute, 1981).

## III. Host Range and Specificity

Over two hundred species of Compositae from 36 genera have been recorded as hosts for *Bremia*. The genera affected are principally assigned to the tribes Cichoriae and Cynareae with fewer host genera in the tribes Arctotideae, Helenieae, Astereae, Senecioneae and Inuleae. Listed below are genera containing species recorded as hosts of *Bremia*: *Agoseris* Rafin., *Andryala* L., *Arctium* L. (syn. *Lappa* Scop.), *Carduus* L., *Carlina* L., *Carthamus* L., *Centaurea* L., *Cicerbita* Wallr. (syn. *Lactuca* L. and *Mulgedium* Cass.), *Cichorium* L., *Cirsium* Mill. (syn. *Barkhausia* Moench.), *Crepis* L. (syn. *Lagoseris* Hoff. and Link), *Cynara* L., *Dendroseris* D. Don., *Dimorphotheca* Moench., *Erechtites* Rafin., *Gaillardia* Fougeroux, *Helichrysum* Mill., *Hemistepta* Bunge (syn. *Saussurea* D.C.), *Hieraceum* L., *Inula* L., *Krigia* Schreb., *Lactuca* L., *Lapsana* L. (syn. *Lampsana* Mill.), *Launaea* Cass., *Leontodon* L., *Mycelis* Cass. (syn. *Lactuca* L.), *Onopordum* L., *Parthenium* L., *Prenanthes* L. (syn. *Nabalus* Cass.), *Picris* L. (syn. *Helminthia* Juss.), *Senecio* L. (syn. *Jacobaea* Mill and *Cineraria* L.), *Solidago* L., *Sonchus* L., *Taraxacum* Weber, *Tragopogon* L., *Venidium* Less.

Despite its extensive host range (Marlatt, 1974), collections of this fungus from one host species are usually specific to that species or a few others from the same genus (Schweizer, 1919; Ling and Tai, 1945; Wild, 1947; Powlesland, 1954; Dzhanuzakov, 1962; Savulescu, 1962). Crute and Davis (1976) recorded that isolates from lettuce (*Lactuca sativa*) were compatible only with collections of other *Lactuca* species from the same sub-section of the genus (i.e. *L. serriola*, *L. saligna* and *L. virosa*).

The fungus from lettuce also exists as pathotypes which vary in their ability to parasitize different *L. sativa* genotypes. This variation has been described by categorizing the pathogen as physiological races (Jagger, 1926; Jagger and Chandler, 1933; Schultz and Roder, 1938; Jagger and Whitaker, 1940; Ogilvie, 1944, 1946; Channon *et al.*, 1965; Sleeth and Leeper, 1966; Rodenburg, 1966; Tjallingii and Rodenburg, 1967, 1969; Zink and Duffus, 1969, 1975; Channon and Smith, 1970; Channon and Higginson, 1971; Ventura *et al.*, 1971; Netzer, 1971, 1973; Sequeira and Raffray, 1971; Jones and Leeper, 1971; Zink, 1973; Dixon and Doodson, 1973; Blok and Eenink, 1974; Eenink, 1974; Globerson *et al.*, 1974; Boulidard and Bannerot, 1975; Zinkernagel, 1975a, b; Crute and Dickinson, 1976; Crute and Johnson, 1976a; Crute and Davis, 1976; Dixon, 1976; Blok and Van Bakel, 1976; Cruger, 1976; Blok and Van der Schaff-van Waadenoyen Kernekamp, 1977a, b; Wellving and Crute, 1978; Lebeda, 1979a, b).

Crute and Johnson (1976a) interpreted data on race–cultivar interactions in terms of a gene-for-gene type relationship (Person, 1959). At present, eleven race specific resistance factors (R factors) are recognized in cultivated lettuce genotypes and most are inherited as single dominant genes (Johnson *et al.*, 1977, 1978; Norwood and Crute, 1980). Twenty-two different combinations of these factors are recognized in lettuce cultivars, and lists of R-factors published at intervals (Crute and Johnson, 1976b; Johnson, 1978; Crute, 1979a, b). With eleven specific resistance factors matched by eleven comparable factors for virulence in the pathogen, it is theoretically possible to identify $2^{11}$ ($= 2048$) different pathotypes of *B. lactucae*. Additionally other sources of seedling resistance have been identified in wild *Lactuca* species related to *L. sativa* some of which may be race specific (Crute and Norwood, unpublished data).

## IV. Economic Importance

The economic importance of a pathogen may be assessed either directly or indirectly. Surveys of pathogen incidence and severity are needed to make direct assessments of economic losses, such studies have largely been confined to crops where easily determined factors such as seed weight can be correlated with pathogen incidence. For high value crops, such as lettuce, grown by numerous relatively small producers indirect techniques are often used which provide valuable pointers to the significance of particular pathogens. The interpretation of such information, however, requires an intimate knowledge of such crops.

### A. Lettuce

The significance of *B. lactucae* as a lettuce pathogen may be gauged from the extensive efforts made to control it. Some growers are prepared to apply fungicides every four days throughout a crop's life; while plant breeders have worked for 50 years in an attempt to produce resistant cultivars. Breeding work has been most active in the Netherlands, UK and USA indicating the importance of the crop to the agricultural economy of these countries and their need for technology to attempt to control this pathogen. Caution must, however, be exercised in assessing pathogen importance by plant breeding effort. The inception of variety (cultivar) registration in Europe for which distinctness, uniformity and stability are primary considerations, has distorted the aims of some plant breeders. There has been a trend towards using pathogen resistance simply as markers to achieve cultivar novelty in comparison with the products of competing breeders.

Purchasers (either wholesale or retail) require lettuce which is free from damage and since the product is consumed in the fresh state, even minute blemishes will diminish its value. The first important reference to *B. lactucae* as a cause of loss in lettuce came from California, USA; Milbrath (1923), indicated that *B. lactucae* caused damage not only while plants were in the field but after harvest, the quality continued to deteriorate during transit. This was still the position forty years later (Zink and Welch, 1962). More recently, losses in transit from California to New York, USA, due to *B. lactucae* were estimated at 0·3% (Ceponis, 1970). Lettuce is grown as an extensive monocrop in California, consequently dramatic losses could be expected; in 1923 Milbraith reported an area of 8000 ha producing the single cultivar New York. Forty-six years later Zink and Duffus (1969) reported an area of 10 000 ha planted with the cultivar Calmar. In the intervening years *B. lactucae* caused serious losses but Calmar remained resistant for about 10 years. Eventually this cultivar became badly infected with consequent heavy losses. Elsewhere in the USA Sequeira and Raffray (1971) considered *B. lactucae* as the most significant pathogen of the lettuce crop which was surpassed in cash value only by potatoes and tomatoes. Macpherson (1932) described losses of up to 50% in unheated overwintered glasshouse crops in the UK, describing downy mildew as a "troublesome and costly disease". Fletcher (1976) estimated annual losses in the UK protected lettuce crop to be 1–3%. In outdoor crops in the UK losses are largely confined to late summer and autumn crops but losses could be of a similar level, although the value of field grown lettuce is subject to large fluctuations due to market forces. The farmgate value of the total UK lettuce trade was estimated to be £40–50 million (Anon., 1980) indicating possible annual losses of £400 000. To all loss figures the costs of attempting to prevent infection by chemical, husbandry and resistance breeding techniques must be added.

In Israel, an important world supplier of fresh vegetables, particularly salads, *B. lactucae* is of significance on winter lettuce crops (Netzer, 1973). In the Netherlands Verhoeff (1960) reported that lettuce yield losses of 40% were not uncommon. The overall figure for losses may now be higher since in the Westland area there are 5000 ha of glass with lettuce as a major crop.

Other countries where *B. lactucae* has been cited as an important pathogen of lettuce include Australia (Anon., 1945; Blackford, 1944; Geard, 1961), Austria (Schmidt, 1957), China (Ling and Tai, 1945; Tao, 1965), Czechoslovakia (Lebeda, 1978), France (Dyke, 1977), Germany (Schreiber, 1944), Sweden (Wellving and Crute, 1978), Turkey (Bremner *et al.*, 1947), USSR (Kanchaveli *et al.*, 1957; Vlasova and Komarovara, 1977).

## B. Other hosts

There are few reports of *B. lactucae* causing losses on hosts of economic

TABLE 4

Reports of *B. lactucae* on hosts other than lettuce where chemical control has been advised

| Country | Authority | Host |
|---|---|---|
| Austria | Baudys (1935) | *Cichorium endivia* L. |
| | | *C. intybus* L. |
| | | *Cynara scolymus* L. |
| France | Moreau and Moreau (1962) | *Cynara scolymus* L. |
| Germany | Wittmann (1972) | *Helichrysum* sp. |
| Italy | Ferraris (1936) | *Senecio* sp. |
| Israel | Moreau (1961) | *Cynara scolymus* L. |
| Morocco | Anon. (1937) | *Cynara scolymus* L. |
| Switzerland | Anon. (1949) | *Cynara cardunculus* L. |
| Tunisia | Valdeyron (1958) | *Cynara scolymus* L. |
| United Kingdom | Moore (1948) | *Cynara scolymus* L. |
| Yugoslavia | Martinovic (1960) | *Helichrysum* sp. |

value other than lettuce. Such hosts comprise a range of ornamental species as well as safflower, chicory, endive and cardoon. Reports which indicate damage as being sufficient to warrant chemical control are summarized in Table 4. This suggests that on these hosts *B. lactucae* is of localized and sporadic significance only.

## V. Geographical Distribution

The incidence of *B. lactucae* on chicory (*Cichorium intybus*), endive (*C. endiva*), lettuce (*L. sativa*), *Sonchus* spp. and other hosts is illustrated in Fig. 2 (Anon., 1969).

## VI. Disease Symptoms

### A. Localized

Lettuce plants at all stages of growth are susceptible to *B. lactucae*. Following penetration and mycelical development (see Sections VIII, IX), the appearance of sporophores from the stomata may be the first visible symptom. *En masse* they are readily visible to the naked eye as white discrete projections bearing spores. Individually the sporophores and asexual spores are hyaline. Usually the fructifications are confined to the under-surface of mature leaves

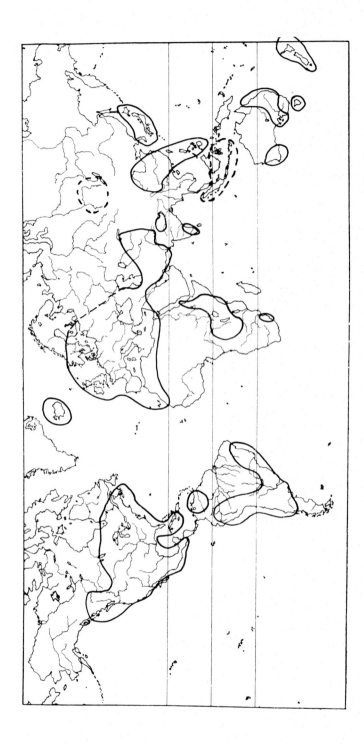

Fig. 2. World distribution map of *Bremia lactuca* Regel. (Reproduced by permission of Commonwealth Mycological Institute, Kew, UK).

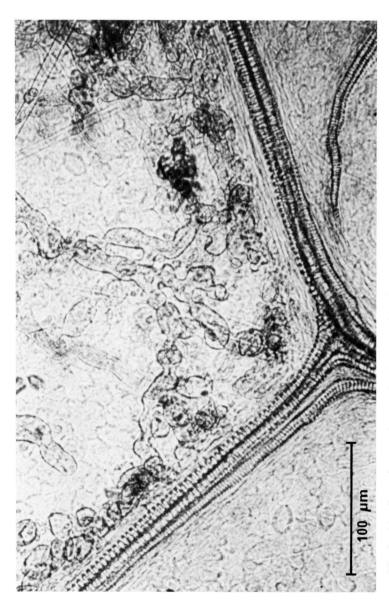

100 μm

Fig. 3. Mycelium of *Bremia lactucae* in a lettuce leaf restricted by the veins to produce an angular lesion (×350).

but they may occur on the upper-surface and will commonly completely cover the cotyledons and primary leaves. The leaf may show light green or yellow lesions either before or after sporulation depending on whether environmental conditions have favoured early sporophore production. Lesions are very variable in size and are often angular, being delimited by the larger leaf veins (Fig. 3). On older leaves the chlorotic lesions may turn necrotic or translucent becoming brittle especially near leaf margins. Vlasova and Komarovara (1977) categorized the localized symptoms of lettuce downy mildew as follows: (1) typical lesions with sporulation on lower surface, (2) diffuse lesions on both surfaces, (3) large lesions, (4) small angular necrotic lesions and (5) abundant sporulation on both leaf surfaces. Only infrequently are adult plants killed by *B. lactucae* but seedlings may wither and die as a result of invasion. Where an infected plant produces a flower stalk, the bracts also become infected, and seed production and seed quality are reduced substantially where seed plants are infected with downy mildew (Janyska, 1957).

A severe early infection by *B. lactucae* may delay maturity and produce lighter crops of inferior quality (Wild, 1947; Smieton and Brown, 1940). Rodenburg (1966) recorded that when seeds were inoculated with *B. lactucae* spores just prior to germination, the emergent radicle was discoloured brown regardless of whether the cultivar was resistant or susceptible.

Symptoms of *Bremia* infection on *Arthraxon* spp. are seen as chlorotic lesions with white to grey sporulation in the lower surface. The lesions are either delimited by the veins or coalesce over the entire leaf and severely infected leaves wilt and die (Miyake, 1914; Naoumov, 1913; Togashi, 1926; Patel, 1948).

## B. Systemic

Following initial field observations, Marlatt *et al.* (1962) reported that lettuce seedlings could become systemically infected by *B. lactucae* following artificial inoculation. Symptoms of systemic infection in a mature lettuce crop were described by Phillips and Lipton (1974) who showed that the fungus could cause a black-brown discolouration of the stem tissues and leaf bases near the shoot tip in a mature head. Systemically infected plants were slightly smaller than healthy plants, no mycelium was found in the roots and only susceptible cultivars showed symptoms. Recently, systemic infection has been recorded in glasshouse-grown lettuce in the Netherlands (I. Blok, personal communication). Mycelium and haustoria of *B. lactucae* were found in the stem pith of *Sonchus oleraceus* and stem cortex of *Senecio vulgaris* (Fraymouth, 1956).

## C. Physiological

A few changes in host physiology have been shown to occur as a result of infection by *B. lactucae*. Sempio (1942, 1950) reported that *B. lactucae* disturbed the glycolysis:respiration ratio which in healthy lettuce leaves was 1:1 but in infected plants became 0·32:0·70. Possibly related to this phenomenon, Yarwood (1953) found that the average temperature of lettuce leaves infected by *B. lactucae* was 1·59°C above ambient compared with 0·63°C above ambient for healthy ones. It is possible that this difference could allow downy mildew infections to be detected by infra-red photography or related techniques.

While investigating host/pathogen physiology, Sempio (1938, 1959) reported that infected lettuce plants could be kept in the dark for up to six days after inoculation without permanent damage. Prolonged darkness during the stage of fungal development when sporulation was expected (seven to nine days after inoculation) however caused leaves to become flaccid, rolled and to die. This may demonstrate that the pathogen draws more heavily on host nutrients, especially carbohydrate during this period. In related studies, using leaf pieces floated on various sugar solutions, Sempio (1942) concluded that the fungus did not use xylose but readily used glucose, lactose and sucrose and to a lesser extent fructose (levulose) and galactose.

## VII. Spore Germination

### A. Conidiosporangial germination

Asexual spores of *B. lactucae* germinate directly by means of a germ tube which emerges from the distal end. Germination requires free water and occurs over a wide temperature range. Schultz (1937) reported germination between 1 and 19°C with an optimum at 10°C and the same optimum was recorded by Powlesland (1954) although she found that germ tube extension was most rapid at 15°C. Verhoeff (1960) demonstrated that spores would germinate between −3°C and 31°C with an optimum at 4–10°C, but the temperature during spore production affected this optimum. Sargent and Payne (1974) and Sargent (1976) reported that germination occurred between 0°C and 21°C and was completely inhibited at 28°C. *Bremia lactucae* spores contain a water soluble auto-inhibitor which must be removed before germination can start (Mason, 1973).

Studies of the processes occurring during germination have been made by Sargent and Payne (1974); Andrews (1975); Sargent (1976) and Duddridge and Sargent (1978). The major storage material in the spore is lipid although exogenously applied glucose is utilized during the germination process. High temperatures may inhibit germination by affecting lipid metabolism. During the early stages of germination, the endoplasmic reticulum proliferates and cytoplasmic vesicles, probably containing essential enzymes and cell wall precursors accumulate at the papilla from which a germ tube emerges. Ribosomes assemble into polysomes, indicating the activation of protein synthesis and concomitantly lipid bodies migrate to the centre possibly providing energy for germination. Germ tube extension follows enzymatic softening and rupture of the papilla wall.

Nuclear division has not been demonstrated during germination and each spore contains up to twenty-two nuclei. Under ideal conditions the process commences within one hour of inhibitor removal, and *in vitro* germ tube extension continues until nutrient reserves are depleted and a long germ tube almost devoid of cytoplasm results (Crute and Dickinson, 1976) (Fig. 4). Germination readily takes place in distilled water and although there have been reports of enhanced germination *in vivo* (Schultz, 1937), it has been consistently stated that exogenously applied nutrients exert no influence on the process (Powlesland, 1954; Mason, 1973). Acidic conditions (pH 4·0) reduce germination (Wild, 1947).

There are two reports of the conidiosporangia of *B. lactucae* producing zoospores (Milbrath, 1923; Arens, 1937). These were apparently developed from sporangia formed under cool conditions, with eight zoospores produced by each sporangium and they formed germ tubes after several hours motility. If the motile bodies observed were *B. lactucae* zoospores rather than other organisms often seen in spore suspensions prepared from infected lettuce leaves, this suggests that some collections of the fungus may sporadically germinate by this means.

### B. Oospore germination

In common with several other members of the Peronosporales, oospores of *B. lactucae* cannot be germinated regularly. Morgan (1978) observed oospore germination taking place by means of a germ tube, a process stimulated by, but not dependent upon, the presence of germinating lettuce seed. No other workers have reported this phenomenon.

## VIII. Penetration and Early Development

*Bremia lactucae* occasionally enters the host plant via the stomata (Schultz, 1937; Cohen, 1952; Viranyi and Henstra, 1976) but more often by direct

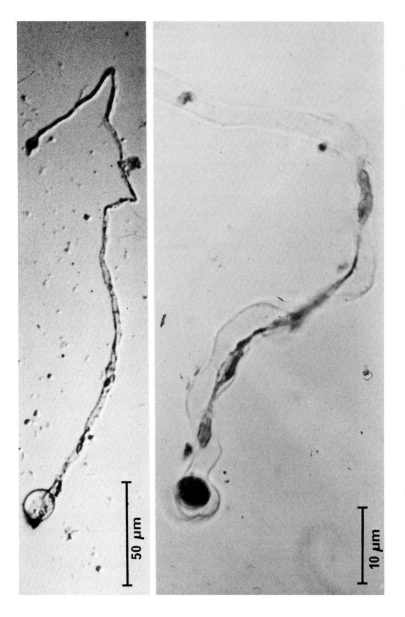

Fig. 4.  Long germ tube, devoid of cytoplasm, produced by *Bremia lactucae* following germination *in vitro* (Top ×550, Bottom ×2000).

cuticular penetration within three hours of spore deposition under optimal conditions (10–22°C) (Verhoeff, 1960). Several workers suggest that direct penetration occurs through the anticlinal cell wall of the epidermis (Arens, 1937; Iwata, 1943; Marlatt et al., 1962; Crute and Dickinson, 1976), others have demonstrated invasion via the periclinal walls (Verhoeff, 1960; Sargent et al., 1973). This discrepancy probably results from the fact that an appressorium usually forms at the junction of two epidermal cells hence electron and fluorescence microscopy will reveal that penetration occurs through the periclinal wall, but this is less obvious in bright-field light microscopy. Raffray and Sequeira (1971) reported that the process is inhibited by light.

After spore germination, cytoplasm migrates along the germ tube which ceases growth at appressorium formation. Penetration is effected enzymatically (Sargent et al., 1973; Duddridge and Sargent, 1978) and the contents of the germ tube and appressorium, flow via an infection peg, into the epidermal cell forming a spherical structure, just smaller than the spore, termed the primary vesicle (Marlatt et al., 1962; Sargent et al., 1973; Ingram et al., 1976a). A callose plug seals off the point of entry. At this stage there is no nuclear division but the fungus is capable of utilizing both leucine and glucose (Andrews, 1975). Following primary vesicle formation, an elongated secondary vesicle expands within the same epidermal cell (Fig. 5). Both vesicles invaginate but do not penetrate the host plasmalemna. A ring of callose surrounds the neck between primary and secondary vesicles. Nuclear division commences as the secondary vesicle expands eventually filling most of the host cell.

## IX. Mycelial Growth, Haustoria and Systemic Development

After penetration of the first epidermal cell the fungus produces intercellular hyphae which branch and ramify throughout the host tissue (Fig. 5). Haustoria are produced at intervals from each intercellular hypha, which penetrate and develop within adjacent host cells (Fig. 6). Intercellular hyphae vary in diameter dependent upon host tissue morphology, being thinner in areas where cells are densely packed (Verhoeff, 1960). The haustoria are "flask-shaped" (Fraymouth, 1956), size being related to host cell volume, and several may form in each cell. Host cell penetration is effected enzymatically and the haustoria invaginate the host plasmalemma. A callose deposit surrounds the neck of the haustorium and may develop more extensively over the surface. Detailed haustorial fine structure and the events associated with host cell response to invasion are reported by Ingram et al. (1976a) and by Sargent (see Chapter 10).

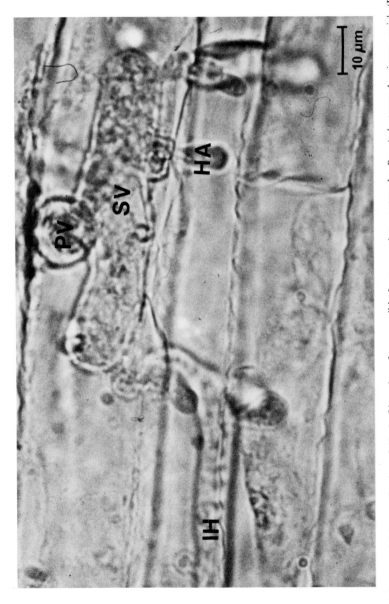

Fig. 5.  Early stages (48 hours after inoculation) of invasion of a susceptible *Lactuca sativa* genotype by *Bremia lactucae*. A primary vesicle (PV), secondary vesicle (SV), intercellular hypha (IH) and haustoria (HA) are present ( × 1200).

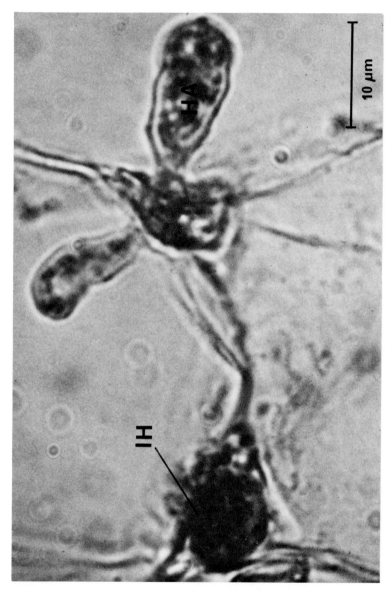

Fig. 6. Cross-section of haustoria (HA) and intercellular hyphae (IH) of *Bremia lactucae* in lettuce parenchyma tissue ( × 3000).

Normally *B. lactucae* produces only localized infections which are physically restricted by the larger leaf veins, since sclerenchyma sheathing the larger vascular bundles prevents fungal growth (Fig. 3). Systemic infection, however, of both seedlings (Marlatt *et al.* 1962, 1963, 1966; Dickinson and Crute, 1974; Crute and Dickinson, 1976) and mature plants (Phillips and Lipton, 1974) has been recorded (Section VI.B). In the case of seedlings, mycelium may grow from the cotyledons, into the hypocotyl and the main and lateral roots. Susceptibility to systemic infection is dependent upon age (Marlatt *et al.*, 1966; Dickinson and Crute, 1974) with root and stem tissues becoming resistant to invasion after a few weeks of development. Rate of mycelial growth also appears to be affected by physiological age and the nutrition of seedlings (Dickinson and Crute, 1974).

# X. Sporulation

## A. Asexual

Asexual sporulation occurs 5–14 days after inoculation, depending on environmental conditions and inoculum concentration which determine how long the pathogen takes to make sufficient growth (Verhoeff, 1960; Dickinson and Crute, 1974). Sporulation occurs over a wide range of temperatures. According to Schultz (1937) the range was 1–19°C with an optimum at 15–17°C; Powlesland (1954) reported a range of 4–20°C with an optimum at 6–11°C while comparable figures from Verhoeff (1960) were 5–24°C with an optimum of 20°C. The process requires 80–100% relative humidity (Schultz, 1937; Powlesland, 1954) although Verhoeff (1960) also reported a requirement for free water.

Sporulation of *B. lactucae* is a nocturnal process being greatly inhibited, by light. Yarwood (1937) reported a lower level of sporulation on infected leaves incubated for eleven hours at a light intensity of 800 foot-candles (*c.* 8700 lux) compared to leaves incubated in the dark. Verhoeff (1960) found that sporulation usually started after a dark period of 6–7 h and was retarded by 2–3 h in a light intensity of 10 000 lux but was not completely inhibited in continous light. Verhoeff (1960) also found that a dry dark period followed by a wet dark period was required and that short light intervals during the dry dark period retarded sporulation. The longer plants were in the dark, prior to the wet period, the more rapidly sporulation developed. In a further study of this phenomenon, Raffray and Sequeira (1971) recorded complete inhibition of sporulation for 24 h at a light intensity of 2000 foot-candles (*c.* 21 700 lux). At least a six hour dark period was needed for maximum sporulation. Green light (wavelength 526 nm) was most inhibitory.

## B. Sexual

Oospores of *B. lactucae* were first described by Caspary (1855) and reference
is made to their occurrence by de Bary (1863), Smith (1884), Cook (1906),
Schweizer (1919), Melhus (1921), Lavrov (1933) and Baudys (1935). Most
reports emphasized, however, that they were rarely present in lettuce.
Jörstad (1964) recorded oospores in cultivated lettuce and the following wild
hosts: *Lactuca scariola, Lampsana communis, Leotodon autumnalis, Senecio
vulgaris, Sonchus arvensis* and *S. asper*. Although most accounts of this disease
tend to assume that oospores are important in perennation (for example,
Walker, 1952) there was little evidence of this until Humphreys- Jones (1971)
demonstrated that oospores of *B. lactucae* were formed in stem tissues of
diseased plants and following their incorporation into compost used them
to infect lettuce seedlings. Oospores have been shown to develop in artificially
inoculated cotyledons and in leaves of mature lettuce plants infected naturally
by *B. lactucae* (Ingram *et al.*, 1975). These workers also established that
certain isolates of the pathogen were more readily able than others to produce
oospores. The widespread occurrence of oospores in lettuce debris has also
been demonstrated by Fletcher (1976) and Sanders (1975).

Sexual organs form from the intercellular hyphae of *B. lactucae*; gametangia
arising from swollen hyphal tips containing up to six nuclei. Oospore
ontogeny was studied by Tommerup *et al.* (1974) and is summarized in
Table 5. The fine structure of oogonia, antheridia, oospheres and oospores
is described by Sargent *et al.*, (1977). It is postulated that meiosis takes place
in the antheridia and oogonia (Ingram *et al.*, 1976b). This agrees with the

TABLE 5

Ontogeny of *Bremia lactucae* oospores

| Time after inoculation (days) | Event |
|---|---|
| 3–4 | Oogonial and antheridial initials develop. |
| 4–5 | Attachment of oogonium and antheridium. |
| 5–7 | Enlargment of oogonium. |
| 8 | Differentiation of the oosphere and periplasm, formation of a fertilization pore. |
| 9 | Formation of a thick wall around the oosphere. |
| 10 | Development of a fertilization tube and fertilization. |
| 11 | Nuclear fusion and degeneration of the fertilization tube. |
| 12–15 | Maturation of the oospore and degeneration of the periplasm and antheridium. |

For further details see Tommerup *et al.* (1974).

proposition that the vegetative state of Oomycete fungi is diploid, a suggestion recently supported by Michelmore and Ingram (1980). Their work has established a relationship between frequency of sexual reproduction by *B. lactucae* and concentration of asexual sporangia used to infect lettuce cotyledons. Sterols, growth regulators, inorganic ions, soluble carbohydrate levels, light intensity, temperature and host phenotype neither markedly stimulated nor markedly inhibited oospore formation. There was an indication that factors which favoured the growth of the host enhanced sexual reproduction while factors resulting in plant stress depressed the process. Oospore production, however, led to accelerated host senescence which may explain why Fletcher (1976) found that tissues containing oospores tended to be senescent. Sexual reproduction was not however stimulated by host tissue senescence and was inversely related to asexual reproduction (Michelmore, 1979; Michelmore and Ingram, 1980; Michelmore, 1981). Many isolates of *B. lactucae* were shown to be heterothallic and oospore production occurred only when these isolates were cultured in certain combinations (Michelmore and Ingram, 1980). A survey of 39 isolates demonstrated the existence of two compatibility types, designated B1 and B2. Establishment of both compatibility types in the same zone of host tissue was the primary determinant of sexual reproduction. Nonetheless other isolates of *B. lactucae* produced oospores regularly when cultured alone suggesting there are both homo and heterothallic types. Analysis of homothallic isolates by producing single spore isolates indicated that stable heterothallic derivatives of both B1 and B2 compatibility types segregated continually from a heterokaryon (Michelmore, 1979, 1981) (see Chapter 9 in this book).

## XI. Epidemiology

### A. Source of primary inoculum

#### 1. Seed

Transmission of *B. lactucae* by seed is unlikely, although in a series of experiments using 23 775 seedlings, raised from seeds produced on infected plants, Wild (1947) reported that 0·025% of them became infected.

#### 2. Alternative host species

Despite the wide host range of the species, *B. lactucae* isolates are highly host specific (Section III). Hence, the fungus from one group of related species does not pose a threat to another. Cultivated lettuce is attacked by *B. lactucae* from the related wild species *L. saligna*, *L. virosa* and *L. serriola* (Ogilvie, 1946; Powlesland, 1954; Schweizer, 1919; Wild, 1947) but there is no conclusive evidence of the importance of these species in disease epidemiology.

### 3. Soil

Mycelia of *B. lactucae* from lettuce debris may initiate infections in subsequent crops (Brien *et al.*, 1957). Since, however, lettuce debris decays rapidly and the fungal hyphae depend on host tissue for survival, this mode of perennation can be of only minor importance (Wild, 1947; Verhoeff, 1960). It is now realized that oospores may be an important primary source of inoculum (Section X.B). Debris bearing oospores, incorporated into soil provides a perennation mechanism and a means whereby infection of subsequent host plants might develop. Whether oospores infect directly via the roots or are splash dispersed from the soil surface onto leaves has not been established.

### 4. Asexual propagules

Asexual spores of *B. lactucae* are important in the build-up of disease epidemics. They may also initiate new infections when they arrive from outside the crop. These spores are reported to remain viable for several days following release (Angell and Hill, 1931; Powlesland, 1954) but detailed studies of the effect of environment on viability are lacking.

There is a general assumption that the asexual spores are disseminated by wind and/or water splash but the relative importance of these factors is unknown. Similarly, the distance over which spores may be carried has not been examined (Verhoeff, 1960). Spore release occurs as relative humidity falls so that spore concentrations in the atmosphere over a lettuce crop reach a peak at 1000–1200 h (Fletcher, 1976; Burchill *et al.*, 1980).

### B. Conditions for epidemic development

For the initiation of infection and for propagule production, high humidity and/or free water are essential but a detailed quantitative analysis of the environmental conditions necessary for downy mildew epidemics in lettuce is lacking.

Consideration of the environmental conditions required for pathogen development indicates that *B. lactucae* poses a threat primarily to crops produced in cool, temperate and Mediterranean regions. Such environmental requirements also give a seasonal nature to disease development so that crops grown in hotter, drier months tend to be unaffected (Schnathorst, 1962). In

TABLE 6

Percentage of total *Bremia lactucae* samples obtained at NIAB[a] during each month of 1976 and 1977

|  | 1976 | 1977 |  | 1976 | 1977 |
|---|---|---|---|---|---|
| January | — | 2 | July | 0 | 6 |
| February | — | 0 | August | 3 | 10 |
| March | 0 | 6 | September | 12 | 19 |
| April | 4 | 5 | October | 34 | 19 |
| May | 2 | 7 | November | 21 | 17 |
| June | 2 | 3 | December | 20 | 6 |

— = No samples obtained.
[a] Anon. (1977; 1978) National Institute of Agricultural Botany, Cambridge, UK.

the UK, seasonal occurrence of lettuce downy mildew can be gauged from Table 6 which shows the number of samples of diseased lettuce received at the National Institute of Agricultural Botany in Cambridge during 1976–1977. An autumnal peak is clear from these figures. Also, from these data, the comparative frequency of *B. lactucae* in different crop systems was determined (Table 7), and this indicates that protected crops are more liable to infection.

TABLE 7

Number and percentage of *Bremia lactucae* samples obtained at NIAB[a] from protected and field crops in 1976 and 1977

|  | Number | | Percentage | |
|---|---|---|---|---|
|  | 1976 | 1977 | 1976 | 1977 |
| Protected | 74 | 118 | 68 | 63 |
| Field | 25 | 70 | 32 | 37 |
| Total | 109 | 188 | 100 | 100 |

[a] Anon. (1977, 1978) National Institute of Agricultural Botany, Cambridge, UK.

Changes in husbandry, such as the production of summer lettuce under some form of protection (for example, polyethylene, hessian or plastic mesh) to produce soil free leaves militates against crops being free from *B. lactucae* infection. These systems produce high inoculum levels in localized areas and in the summer in cool temperate areas there will normally be periods suited to the establishment of infection and rapid epidemic development.

## XII. Development on Resistant Plants

Spores of *B. lactucae* will probably germinate normally on the surface of any plant provided the physical environment is suitable (Arens, 1937; Crute and Dickinson, 1976; Viranyi and Henstra, 1976; Maclean and Tommerup, 1979) although Viranyi and Henstra (1976) recorded that on resistant lettuce cultivars, incompatible isolates of *B. lactucae* produced longer germ tubes which grew irregularly.

The events which determine the specificity of *B. lactucae* on some non-hosts and on lettuce cultivars carrying specific resistance factors primarily occur after penetration although some of the reports are conflicting. Arens (1937) stated that *B. lactucae* from lettuce could only penetrate this species and *Senecio cenentis* D.C. while three other composites (all non-hosts to *Bremia*) and four non-composite species were not penetrated. Verhoeff (1960) stated that the fungus penetrated resistant lettuce cultivars but not four composite species known to be common hosts of other pathotypes of the fungus. Crute and Dickinson (1976) however, observed penetration of ten non-host Compositae by *B. lactucae* from lettuce plus a crucifer, *Brassica oleracea* L., and Maclean and Tommerup (1979) recorded that the fungus could penetrate two non-host Compositae and three non-composites; but wheat (*Triticum aestivum* L.) was not penetrated.

The histology and symptoms associated with the interactions between *B. lactucae* from lettuce, and from both non-hosts and lettuce cultivars carrying different specific resistance factors, have been studied recently. Maclean *et al.* (1974) recorded that the action of a race specific factor in lettuce, now-identified as *R8* (Crute and Johnson, 1976a) was associated with rapid and restricted host cell death following invasion by an incompatible isolate of the fungus. Further, Maclean and Tommerup (1979) demonstrated how the detail and rate of response depended on the host resistance factor(s) controlling the association. Thus the presence of *R3* and *R8* allowed only restricted fungal development with rapid host cell necrosis and no increase in parasite nuclei while the action of *R7* was associated with a slower host response allowing more extensive fungal development, some nuclear division within the first 24 h of growth and eventually limited sporulation. These observations accord with those of Crute and Dickinson (1976), Viranyi and Blok (1976) and Crute and Norwood (1978a).

Detailed differences of response in the same species carrying different specific resistance factors is to be expected but variation of response between non-host species and between collections of the same non-host have been recorded (Crute and Dickinson, 1976; Crute and Davis, 1976; Maclean and Tommerup, 1979). Similarly, it has been shown that the details of response due to single factors can be modified by genetic background and environment (Crute and Norwood, 1978a).

## XIII. Disease Control

### A. Chemical control

Early attempts at control used copper-based fungicides such as Bordeaux mixture (Erwin, 1920; Kendrick, 1929; Wild, 1947) which were applied in the early stages of propagation. Sulphur was also found to be effective (Erwin, 1920; Ogilvie, 1944) although this was disputed by others (Kendrick, 1929; Crogan, *et al.*, 1955; Verhoeff, 1960).

More recently dithiocarbamates have been used extensively to control *B. lactucae* and of these the most efficacious was zineb (Haasis and Ellis, 1950; Powlesland and Brown, 1954; Cox, 1955, 1956, 1957; Verhoeff, 1960; Channon and Webb, 1967) though a range of related compounds exerted some control. Standard methods of application have been dusting or spraying at two-weekly intervals at a rate of approximately $1.5$ kg ai ha$^{-1}$ although Haasis and Ellis (1950) found soil drench applications of zineb gave good control, presumably because of vapour activity.

Conventional surface protectant chemicals like dithiocarbamates operate by inhibiting spore germination, however, the degree of control achieved by zineb particularly in the field lettuce crop was often unsatisfactory (Channon and Webb, 1967). This is possibly due to the difficulty of achieving complete cover of the upper and lower leaf surfaces, the appearance of unprotected leaves between applications and a lack of persistence after rain and irrigation. For these reasons and because of undesirable residue levels alternatives to the dithiocarbamates were sought.

Several experimental compounds with claimed systemic activity against phycomycetous pathogens have recently been introduced. These have been evaluated in laboratory, glasshouse and field tests for activity against *B. lactucae* (Crute *et al.*, 1977; Crute and Gregory, 1976; Fletcher, 1976; Crute and Norwood, 1977, 1978b; Paulus *et al.*, 1977; Smith *et al.*, 1977; Griffin and Griffin, 1977; Williams *et al.*, 1977; Bertrand *et al.*, 1977; Schwinn *et al.*, 1977; Crute, 1978, Crute and Jagger, 1979; and Anon, 1979). Of these chemicals, the metallic ethyl phosphites (sodium and aluminium ethyl· phosphite), the carbamic acid esters (prothiocarb and propamocarb), and the acylalanines (furalaxyl and metalaxyl) possess systemic activity against *B. lactucae*. Most active are the acylalanines which provide effective control under field conditions. Metalaxyl (Ridomil, 25 w.p.; Ciba Geigy Ltd) is cleared for use on lettuce crops in several countries. This fungicide provided excellent control as a high volume foliar spray (two-weekly foliar spray, 0·2 kg ai ha$^{-1}$ in 1000 litres of water), a soil-drench applied after drilling (2 kg ai ha$^{-1}$ in 25 000 litres) or incorporated into peat blocks in which plants are raised prior to transplanting (50 g ai m$^{-3}$ peat in 5 litres of water approximately 0·007 gm ai per block). Metalaxyl also exerts a curative effect in that complete control was achieved in laboratory experiments with applications applied up to three days after inoculation and in the field when applications were delayed until the disease first appeared (Crute, unpublished data). Recently, however, formulations containing only metalaxyl as the fungicidally active ingredient have been withdrawn from commerce by Ciba-Geigy. This decision was made for all crops in an attempt to avoid the possibility of metalaxyl insensitive pathogen strains becoming widespread and threatening effective control. Crute and Jagger (1979) found that isolates of *B. lactucae* did vary in sensitivity to the chemical at the lowest limit of its activity but so far no metalaxyl insensitive strains of *B. lactucae* which could affect the effective commercial use of the chemical have been reported. In the future, a formulated mixture of metalaxyl together with another fungicide may become commercially available for use on lettuce.

None of these systemic fungicides apparently exert their effect on spore germination since the concentration of them required to inhibit spore germination is greater than that required to achieve complete control *in vivo*. *Bremia lactucae* penetrates treated seedlings but further fungal development is restricted. Invaded host cells become necrotic and the association between fungus and fungicide-treated host is indistinguishable from the hypersensitive response following penetration of resistant cultivars by incompatible spores (Crute *et al.*, 1977; Crute, 1978; Crute and Jagger, 1979) (Section XII).

## B. Cultural control

Crop management is an essential means of controlling *B. lactucae*. Under protected culture (glass, polythene sheets, hessian and plastic mesh) prevention of free water formation and retention on leaf surfaces at all stages of crop production is an essential requirement for control. Where crops are heated this is easier to attain than in unheated conditions. Most plants are now raised from pelleted seed sown in peat blocks and these produce regular and vigorous growth. Such techniques obviate the problems of overcrowding which used to beset seedling production in Dutch light structures and frames. It has been considered important, however, to provide only limited nitrogen supplies in the peat compost. This prevents over-rapid etiolated growth, which is thought to be more susceptible to *B. lactucae*. The blocks are placed in the growing position before the roots emerge through their sides to prevent any check to growth which is also claimed to render the seedlings more liable to *B. lactucae* attack. The area to be cropped is irrigated to field capacity before the blocks are stood out and by this method rooting takes place into the border soil and the developing plant is held above soil level thereby encouraging air movement and reducing relative humidity beneath the leaves. Many crops do not require further irrigation until 10–14 days before cutting. This prevents irrigation water from standing on the foliage during the major growth phase and helps prevent infection.

Temperature manipulation can be used to lower humidity. In the Westland area of The Netherlands, which probably contains the world's largest concentration of protected lettuce crops, the "quiet growing system" has evolved to combat *B. lactucae*. Growers use 12°C day and 5°C night temperatures, the latter starting at 1500 h. Ventilators remain open until the temperature has fallen to the required night time value. This has a two-fold effect, there is a positive "pull down" of temperature which with open ventilators reduces humidity and also stops growth, otherwise the leaves contain thin walled palisade cells which are thought to be highly susceptible to *B. lactucae* invasion. If conditions outside the glasshouse are such as to prevent humidity being dispelled, then temperatures are raised to 20–25 C thereby drying out the atmosphere.

Since protected crops are grown in rapid succession, up to 7–9 crops per annum on the same land, good management involves the removal of all crop debris between crops. This is particularly important in view of recent findings concerning the potential role of oospores (Sections X.B and XI.A3). A recent trend is to produce crops on sheets of light reflective polyethylene with holes through which roots can develop from the propagation peat blocks into the border soil. At the end of each crop this sheet is rolled up with all the plant residues and incinerated. The only factor which militates against this technique is that the plastic tends to encourage high humidity levels in the microclimate beneath the leaf surfaces. This can be overcome by temperature manipulation and the use of extractor fans fitted to the ends of production houses. Nutrient film production techniques either on concrete or using gutter piping should similarly permit the hygienic removal of all crop debris but will also create a moist microclimate near the plants.

Where crops are grown in border soil, sterilization by steam has been advocated as a means of reducing crop losses (Anon., 1946; Sprau and Minckwitz, 1949). This technique is now too expensive due to its energy requirements, but chemical sterilization with materials such as dazomet, metham sodium or methyl isothiocyanate mixtures are useful alternatives. Soil fumigation with methyl bromide is not attractive as a means of controlling lettuce downy mildew because of the facility with which the lettuce plant appears to accumulate bromide residues from the soil, resulting in undesirably high bromide levels in the edible tissues (Kempton, 1979).

For field produced crops similar principles of crop management apply. Increasingly, plants for these crops are produced in peat blocks permitting irrigation to be applied before planting and then again near crop maturity. When irrigation must be applied this should be done as early in the day as possible permitting free moisture to evaporate from the foliage before nightfall. Plant spacings should be as wide as is economically feasible to promote air movement between plants. Pre-emergence, pre-planting and crop phase herbicides should be used to prevent weed growth which will increase humidity within the crop. Only sufficient nutrients should be applied in order to ensure unhindered crop growth. At the end of each crop residues should be ploughed deeply into the top soil. Successional field crops must be rotated thereby giving time for residues to decay before a further lettuce crop is grown on that land. Plantings or drillings should be kept away from existing crops particularly if these are already carrying the pathogen. In addition, late season field crops should be sited away from protected crops.

## C. Resistant cultivars

### 1. Genetics of host-pathogen interaction and breeding for resistance

Control of *B. lactucae* by the use of resistant cultivars has been attempted for more than 50 years but with only limited success. All the resistance so far utilized is race specific and governed by major genes.

The first lettuce types bred for downy mildew resistance were Imperial strains produced in the USA; these incorporated resistance from the French cultivar Romaine Blonde lente à Monter now identified as possessing the resistance factor *R7* (Jagger, 1924, 1926; Jagger and Chandler, 1933; Jagger *et al.*, 1941; Crute and Johnson, 1976a). Subsequently, this factor was found in some Great Lakes types (Bohn and Whitaker, 1951). When this resistance became ineffective, a *Lactuca serriola* accession, PI 91532 (shown by Zink (1973) to have been also quoted, incorrectly by others as PI 104854) was used to confer resistance (Jagger and Whitaker, 1940; Whitaker *et al.*, 1958; Thompson and Ryder, 1961; Leeper *et al.*, 1963; Welch *et al.*, 1964). This resistance factor was termed *R8* and remained effective in California for 13 years where the cultivar Calmer was free from mildew between 1960 and 1973. This is one of the few successes in breeding downy mildew resistant cultivars (Crute and Johnson, 1976c). During the early stages of the breeding programme which involved *R8*, a further source of resistance was identified in the cultivar Bourguignonne Grosse Blonde d'Hiver (Whitaker *et al.*, 1958). This source now said to carry *R9* has not been used in further cultivar development. Material from the Californian breeding programme carrying *R7*, *R8* plus *R6* was used at the National Vegetable Research Station, UK, to develop the cultivars Avoncrisp and Avondefiance (Watts and George, 1961). In contrast to the USA, *R8* was rendered ineffective in Europe, by pathogen isolates before it had been used extensively in commercial cultivars. Further resistance from *Lactuca serriola* (PI 167150) (*R5*) was used in Texas to breed the cos cultivar Valmaine (Anon., 1963; Zink, 1973).

In Europe, lettuce breeding concentrated primarily on glasshouse genotypes. Early Dutch material carried factors *R2* and *R4* from cv. Meikoningin (≡ May Queen) (Schultz and Roder, 1938) and *R3* from cv. Proeftuins Black-pool (Rodenburg and Huyskes, 1962). In the UK, the field cultivar Mildura carrying *R3* was bred from cv. Passion Blanche à Graine Noire identified as resistant by I. C. Jagger (Dawson, 1976). The cos cultivar Sucrine was identified as resistant to some fungal isolates by Channon and Smith (1970) and Tjallingii and Rodenburg (1967, 1969) and is now said to carry *R10*. This factor has been transferred to certain glasshouse genotypes bred for an upright habit. Recently breeding glasshouse and field cultivars in Europe has concentrated on combining resistance from the USA programmes with that present in European cultivars and cultivars with combinations of the R-factors *R2, 3, 4, 6, 7* and *8* have been produced. In Israel, factor *R3* from cv. Mildura was used to produce a locally adapted cos type (Netzer and Globerson, 1976) and in the USA, European sources of resistance (*R2, 3* and *4*) are being evaluated in breeding programmes. More recent develop-ments in resistance breeding involve a Hilde × *L. serriola* cross released as *F4* material from Instituut voor Tuinbougewassen, Wageningen, The Nether-lands. This new factor (*R11*) is already in some cultivars.

All resistance factors with the exception of *R7* are inherited as single dominant genes (Jagger, 1926; Jagger and Whitaker, 1940; Ventura *et al.*, 1971; Sequeira and Raffray, 1971; Zink, 1973; Globerson *et al.*, 1974; Bannerot and Bouledard, 1976; Johnson *et al.*, 1977, 1978; Norwood and Crute, 1980). Under some circumstances *R7* has been noted as a comple-mentary pair of independent dominant genes and at other times as a single dominant gene. This could be due to incomplete expression (Crute and Norwood, 1978a) causing a proportion of heterozygotes to be recorded as susceptible or as a result of one gene of the pair being effective alone against some isolates (Johnson *et al.*, 1977, 1978). Resistance factors of known genetic identity have been designated *Dm* genes following Zink and Duffus (1973). Norwood and Crute (1980) studied linkage between some *Dm* genes and suggest that *Dm2, 3* and *6* constitute a tight linkage group, while *Dm11* and *R7* and *Dm8* and *Dm9* may be tightly linked or allelic. The independent inheritance suggested by other genetic studies indicates that genes of specific resistance to *B. lactucae* are located on at least three chromosomes.

Major genes have been ineffective in achieving lasting control because the appropriate specific virulence factors (v-factors), in the pathogen population which render them ineffective, have been common and well distributed. Mention has been made (Section III) of the identification of physiological races of *B. lactucae*. It is now considered more appropriate to consider the question of variation for specific virulence in *B. lactucae* in terms of the frequency of occurrence and distribution of particular v-factors and v-factors combinations. In Europe, specific virulence combinations, rendering in-effective all the currently available combinations of *R1–10* are common

(Crute and Davis, 1976; Blok and Van Bakel, 1976; Blok and Van der Schaaf-Van Waadenoyen Kernekamp, 1977a, b; Wellving and Crute, 1978; Dixon and Wright, 1978; Dixon, 1978; Crute and Norwood, 1978b, 1979; Lebeda *et al.*, 1980). In Sweden and UK a relatively new R-factor, *R11* is infrequently overcome and may provide control. By examining data from surveys by phenotype analysis (Wolfe *et al.*, 1976) the usefulness of different *R*-genes and *R*-gene combinations can be evaluated.

Clearly, at least in the UK, consistent control cannot be expected from combinations of the factors *R1–10* and novel sources of resistance are required.

The origins and emergence of races of *B. lactucae* have been the subject of some debate. Early workers assumed that one race simply mutated to give rise to another (Jagger and Whitaker, 1940; Channon *et al.*, 1965). Mutation must play a role in creating new virulence genes, but it is likely that "new" races emerge as a result of selection for components of the pathogen population which were previously at a level too low to be detected in surveys. Crute and Dickinson (1976) presented evidence that a single collection of *B. lactucae* was heterogenous for specific virulence and this was confirmed by Crute and Norwood (1980). If, as seems probable, *B. lactucae* is diploid and outbreeding occurs (Tommerup *et al.*, 1974; Michelmore and Ingram, 1980), then heterozygosity can be expected. If specific virulence determinants are recessive as indicated by Michelmore (personal communication) alleles for virulence will be masked within a non-virulent genotype. Were mitotic recombination to occur, however, it would be possible for "virulent" nuclei to develop within the coenocytic mycelium and be transferred to the next generation of condiosporangia. If these propagules are exposed to a resistant host then those with the necessary virulence would appear as a "new" race. Crute and Norwood (1980) suggested that the sexual process is the most likely means whereby new specific virulence combinations are produced and that heterozygosity and somatic recombination could account for the race changes which they and others have studied (Channon *et al.*, 1965; Dixon and Doodson, 1973; Boulidard and Bannerot, 1975).

If heterokaryosis occurs in *B. lactucae*, then there are further complications when considering racial evolution. Infections result from a multinucleate coenocytic spore hence the presence of a dominant allele at a locus determining specific virulence in only one of the several nuclei could determine the host reaction to that spore. Alternatively, critical proportions of nuclei may determine which host reaction follows penetration. When appropriate isolates were grown together and inoculated on to indicator lettuce genotypes, Crute (unpublished) found no evidence of novel virulence combinations resulting from vegetative hybridity over several generations of culture.

Crute and Johnson (1976b, c) have discussed the strategies which might be employed in utilizing host resistance to achieve more successful control.

As lettuce is a quality crop, sold by appearance, useful resistance must be expressed at a high level. With the realization that R-factors *1–10* are of little value regardless of their mode of use, breeders are seeking alternative sources of resistance. Numerous novel sources, not attributable to factors *R1–11* and expressed at the seedling stage, have been identified in certain wild *Lactuca* species capable of crossing with *L. sativa* (Eenink, 1974; Crute and Davis, 1976; Netzer *et al.*, 1976; Crute and Norwood, 1977, 1978b, 1979; Johnson and Norwood, 1978, 1979). Post-penetrational host cell necrosis (hypersensitivity) appears to be associated with this resistance (Crute and Davis, 1976; Norwood and Crute, unpublished) and some novel sources already appear to be race specific. Resistance factors are required, for which the comparable pathogen virulence is infrequently encountered in the pathogen population. Once such resistance has been identified, the question of how best to use it must be considered. This could involve: (1) exploiting disruptive selection, either by cultivar rotation, or by growing cultivars of mixed resistance genotype (2) introducing resistant cultivars sequentially as they become ineffective or (3) introducing a combination of new resistance factors into a single cultivar before any are exposed individually to the pathogen population. At present, little information exists to determine which strategy is most likely to succeed and convenience rather than scientific considerations may ultimately be the criterion.

Some lettuce genotypes which were recorded as susceptible following artificial inoculation show significantly less disease as adult plants under conditions of a natural field epidemic (Dixon *et al.*, 1973; Crute and Norwood, 1977, 1978b, 1979). Such field resistance may withstand shifts in the virulence characteristics of the pathogen population. Although the effectiveness of this resistance is dependent upon inoculum pressure the result is an overall reduction in the amount of disease within the crop. Field resistant genotypes develop fewer infected leaves and smaller lesions when compared to a standard susceptible cultivar (Crute and Norwood, unpublished).

Additionally, investigations suggest that the field resistant cultivar Iceberg can be protected completely with lower levels of fungicide (four applications of 0·0025% ai metalaxyl) than are required to provide control on a standard susceptible cultivar (Hilde) (four applications of 0·02% ai metalaxyl) (Crute and Gordon, 1980). This raises the possibility of integrated use of genetic and chemical controls providing economic and lasting control which neither can achieve alone.

## 2. *Evaluation of cultivar reaction*

Studies on the reaction of lettuce cultivars to different isolates of *B. lactucae* have produced a voluminous literature (see Section III). This work was directed towards determining the resistance of cultivars, the identification of pathotypes (races) and the development of new resistant cultivars. To facilitate these studies separate and distinct host cultivar differential series were erected in Germany, Israel, the Netherlands, Sweden, UK and USA. Attempts to rationalize this situation began with the formation of the International *Bremia* Survey (the results of whose studies were privately circulated by Instituut voor Plantenziektenkundig Onderzoek (IPO), Wageningen, the Netherlands). Subsequently Crute and Johnson (1976a) developed their gene-for-gene model to describe the specific interaction between host cultivars and pathogen isolates. As a result, a further standardized series of differential cultivars has been proposed representing all 11 R-factors which can be used to characterize the virulence of *B. lactucae* populations with respect to R-factors currently deployed commercially (Temperate Downy Mildews Newsletter No. 1, Dixon, 1979).

Lettuce cultivars are tested for specific resistance to *B. lactucae* using seedlings, detached cotyledons or detached primary leaves. The methods employed in Europe are documented by Channon et al. (1965), Rodenburg (1966), Kapooria and Tjallingii (1969), Dixon and Doodson (1970, 1971, 1973), Huyskes (1971) and Crute and Dickinson (1976). Usually after incubation, cultivars are either completely susceptible when all seedlings bear profuse sporulation (recorded +) or are totally resistant and when there are no symptoms (recorded −). Occasionally, intermediate reactions occur where either sparse sporulation or a lower incidence of profuse sporulation is observed. The latter response results from isolate heterogeneity with respect to the ability to overcome a particular R-factor, with spores capable of compatible infection present at low concentrations (Crute and Dickinson, 1976). The former response results from an incomplete expression of resistance characteristics by certain R-factors (Section XII). Keys to assess such seedling reaction were proposed by Dixon and Doodson (1971) where levels of sporophore production were evaluated as a percentage of leaf area covered. A 0–3 scale and a 0–4 scale based on sporophore cover for seedlings and leaf discs respectively was used by Crute and Dickinson (1976), Crute and Norwood (1978a), while extent and degree of cell death (hypersensitive necrosis) was assessed by Crute and Davis (1976).

Evaluation of resistance in mature plants has received scant attention. Techniques for such work were described by Dixon et al. (1973) and Dixon (1976). Artificial epidemics of *B. lactucae* may be induced by growing test plants in polyethylene structures and maintaining high humidity with overhead irrigation with artificially infected spreader cultivars. Natural epidemics may be studied in areas where climatic conditions are conducive to frequent infection such as coastal areas subject to sea mists, e.g. Tamar Valley, UK. Assessment keys are available for evaluating infection on single mature leaves and whole plants (Dixon and Doodson, 1971; Dixon et al., 1973). These two assessment methods produced comparable results under conditions of artificial inoculation and in natural epidemics. At least two assessment dates were required since on individual cultivars there were differences in the leaf area infected at separate dates which may result from the cycle of host growth and pathogen sporulation varying in length between particular cultivars.

## References

Andrews, J. H. (1975). *Can. J. Bot.* **53**, 1103–1115.
Angell, H. R. and Hill, A. V. (1931). *J. Aust. Çouncil Sci. Indust. Res.* **4**, 178–181.
Anon. (1937). Abst. *Rev. Appl. Mycol.* **17**, 216–217 (1938).
Anon. (1945). *Agric. Gaz. New South Wales* **56**, 251–254.
Anon. (1946). *Annu. Agric. Suisse* **47**, 741–842.
Anon. (1949). *Annu. Agric. Suisse* **50**, 767–904.
Anon. (1963). *Texas Agric. Expt. Stn. Leaflet* No. 610, 1–3.
Anon. (1969). Distribution Maps of Plant Disease—*Bremia lactucae* Regel, No. 86. Comm. Mycol. Inst. Kew, UK.
Anon. (1977). The distribution of specific virulences in lettuce downy mildew (*Bremia lactucae*) populations in 1976. Cyclostyled report by Ministry of Agriculture, Fisheries and Food, London, 25 pp.
Anon. (1978). The distribution of specific virulences in lettuce downy mildew (*Bremia lactucae*) populations in 1977. Cyclostyled report by Ministry of Agriculture, Fisheries and Food, London, 26 pp.
Anon. (1979). *The Grower* **91**(8), 48.
Anon. (1980). *The Grower* **93**(8), 6.
Arens, K. (1937). *Botanica Univ. Sao Paulo* **1**, 39–54.
Bannerot, H. and Boulidard, L. (1976). *Proceedings Eucarpia Meeting on Leafy Vegetables, Wageningen*, 86–87.
Baudys, E. (1935). Abst. *Rev. Appl. Mycol.* **15**, 420 (1936).
Berkeley, M. J. (1846). *J. Hort. Soc. Lond.* **1**, 9–35.
Berkeley, M. J. and Broome, C. E. (1851). *Ann. Nat. Hist. Ser. 2* **7**, 101.
Bertrand, A., Ducret, J., De Bourge, D. C. and Horrière, D. (1977). *Phytiatrie-Phytopharmacie* **26**, 3–17.
Blackford, F. W. (1944). *Qd. Agric. J.* **54**, 221–223.
Blok, I. and Van Bakel, J. J. M. (1976). *Groenten en Fruit* **31**, 172.
Blok, I. and Eenink, A. H. (1974). *Zaadbelangen* **28**, 138–140.
Blok, I. and Van der Schaff-van Waadenoyen Kernekamp, K. (1977a). *Zaadbelangen* **31**, 57–60.
Blok, I. and Van der Schaff-van Waadenoyen Kernekamp, K. (1977b). *Groenten en Fruit* **32**, 1477.
Bohn, G. W. and Whitaker, T. W. (1951). *USDA Dept. of Agriculture Circular* 881.
Boulidard, L. and Bannerot, H. (1975). *Rapport d'Activite Stn. Genetique d'Amelioration des Plantes, Versailles* 1972–1974, M1–18.
Bremer, H., Ismen, H., Karel, G., Ozkan, H. and Ozkan, M. (1947). *Rev. Fac. Sci. Univ. Istanbul, Ser. B.* **13**, 122–172.
Brien, R. M., Dye, D. W., Fry, P. R., Harrison, R. A., Jacks, H. and Newhook, F. J. (1957). *Inform. Serv. Dept. Sci. Industr. Res. New Zealand* **14**, 1–38.
Burchill, R. T., Crute, I. R. and Gordon, P. L. (1980). *Rep. Natn. Veg. Res. Stn. for 1979*, 75.
Caspary, R. (1855). *Monatsber. Preuss. Akad. Wissensch. Z. Berlin* (1855), 308–331.
Ceponis, M. J. (1970). *Pl. Dis. Reptr.* **54**, 964–966.
Channon, A. G. and Higginson, Y. (1971). *Ann. Appl. Biol.* **68**, 185–192.

Channon, A. G. and Smith, Y. (1970). *Hort. Res.* **10**, 14–19.
Channon, A. G. and Webb, M. J. W. (1967). *Ann. Appl. Biol.* **59**, 355–362.
Channon, A. G., Webb, M. J. W. and Watts, L. E. (1965). *Ann. Appl. Biol.* **56**, 389–397.
Cohen, M. (1952). *Phytopathology* **42**, 512–513.
Cook, M. C. (1906). *In* "Fungoid Pests of Cultivated Plants". Spottiswoode, London.
Cox, R. S. (1955). *Pl. Dis. Reptr.* **39**, 421–423.
Cox, R. S. (1956). *Phytopathology* **46**, 10.
Cox, R. S. (1957). *Pl. Dis. Reptr.* **41**, 455–459.
Crogan, R. G., Snijder, W. C. and Bardin, R. (1955). *Agric. Expt. Stn. Calif. Circ.* **448**, 14–15.
Cruger, G. (1976). *Gemüse* **12**, 350–351.
Crute, I. R. (1978). *PANS* **24**, 519–520.
Crute, I. R. (1979a). *ARC Res. Rev.* **5**, 9–12.
Crute, I. R. (1979b). *The Grower* **92** (Veg. 1979 Supplement), 15–17.
Crute, I. R. (1981). *In* "The Downy Mildews" (D. M. Spencer, Ed.). Academic Press, New York and London.
Crute, I. R. and Davis, A. A. (1977). *Trans. Br. Mycol. Soc.* **69**, 405–410.
Crute, I. R. and Dickinson, C. H. (1976). *Ann. Appl. Biol.* **82**, 433–450.
Crute, I. R. and Gordon, P. L. (1980). *Rep. Natn. Veg. Res. Stn. 1979* **75**.
Crute, I. R. and Gregory, A. (1976). *Rep. Natn. Veg. Res. Stn. 1975*, 101–104.
Crute, I. R. and Jagger, B. M. (1979). *Rep. Natn. Veg. Res. Stn. 1978*, 78–79.
Crute, I. R. and Johnson, A. G. (1976a). *Ann. Appl. Biol.* **83**, 125–137.
Crute, I. R. and Johnson, A. G. (1976b). *Proceedings Eucarpia Meeting on Leafy Vegetables, Wageningen*, 88–94.
Crute, I. R. and Johnson, A. G. (1976c). *Ann. Appl. Biol.* **84**, 287–290.
Crute, I. R. and Norwood, J. M. (1977). *Rep. Natn. Veg. Res. Stn. 1976*, 99–102.
Crute, I. R. and Norwood, J. M. (1978a). *Ann. Appl. Biol.* **89**, 467–474.
Crute, I. R. and Norwood, J. M. (1978b). *Rep. Natn. Veg. Res. Stn. 1977*, 101–103.
Crute, I. R. and Norwood, J. M. (1979). *Rep. Natn. Veg. Res. Stn. 1978* **77**.
Crute, I. R. and Norwood, J. M. (1980). *Ann. Appl. Biol.* **94**, 275–278.
Crute, I. R., Wolfman, S. A. and Davis, A. A. (1977). *Ann. Appl. Biol.* **85**, 147–152.
Dawson, P. R. (1976). *Ann. Appl. Biol.* **84**, 282–283.
De Bary, A. (1863). *Ann. Sci. Nat. Bot.* (Ser 4), **20**, 5–148.
Delacroix. (1867). As quoted by A. Fischer (1892).
Desmazieres, J. B. H. J. (1846). *Ann. Sci. Nat. Ser. VI* **3**, 65.
Dickinson, C. H. and Crute, I. R. (1974). *Ann. Appl. Biol.* **76**, 49–61.
Dixon, G. R. (1976). *Ann. Appl. Biol.* **84**, 283–287.
Dixon, G. R. (1978). *In* "Plant Disease Epidemiology" (P. R. Scott and A. Bainbridge, Eds) 71–78. Blackwell, London.
Dixon, G. R. (1979). Temperate Downy Mildews Newsletter No. 1. *Federation of British Plant Pathologists News* **3**, 33–34.
Dixon, G. R. and Doodson, J. K. (1970). *J. Natn. Inst. Agric. Bot.* **12**, 124–129.
Dixon, G. R. and Doodson, J. K. (1971). *J. Natn. Inst. Agric. Bot.* **12**, 299–307.
Dixon, G. R. and Doodson, J. K. (1973). *Hort. Res.* **13**, 89–95.
Dixon, G. R., Tonkin, M. H. and Doodson, J. K. (1973). *Ann. Appl. Biol.* **74**, 307–313.
Dixon, G. R. and Wright, I. R. (1978). *Ann. Appl. Biol.* **88**, 287–294.
Duddridge, J. and Sargent, J. A. (1978). *Phys. Pl. Path.* **12**, 289–296.
Dyke, J. (1977). *The Grower* **88**, 441–444.
Dzhanuzakov, A. (1962). *Botan. Zhur. SSSR* **47**, 862–866.
Eenink, A. H. (1974). *Euphytica* **23**, 411–416.
Erwin, A. T. (1920). *Proc. Amer. Soc. Hort. Sci.* **17**, 161–168.
Ferraris, T. (1936). *Riv. Agric. Roma* **32**, 26–27.

Fischer, E. (1892). *Rabenhorst's Krypt-Flor Deutchl.* **4**, 439–442.
Fletcher, J. T. (1976). *Ann. Appl. Biol.* **84**, 294–298.
Fraymouth, J. (1956). *Trans. Br. Mycol. Soc.* **39**, 79–107.
Fuckel, L. (1869). *Jb. Nassau. Ver. Naturk.* **13**, 69.
Geard, I. D. (1961). *Tasmania, J. Agric.* **32**, 369–377.
Globerson, D., Netzer, D. and Tjallingii, F. (1974). *Euphytica* **23**, 54–60.
Griffin, M. J. and Griffin, G. W. (1977). *Proc. Brit. Crop. Prot. Conf. 1977*, 301–308.
Haasis, F. A. and Ellis, D. E. (1950). *Pl. Dis. Reptr.* **34**, 310–311.
Hoffmann, H. (1856). *Bot. Zeit.* **14**, 154.
Humphreys-Jones, D. R. (1971). *Pl. Soil* **35**, 187–188.
Huyskes, J. A. (1971). *Euphytica* **22**, 235–238.
Ingram, D. S., Tommerup, I. C. and Dixon, G. R. (1975). *Trans. Br. Mycol. Soc.* **64**, 149–153.
Ingram, D. S., Sargent, J. A. and Tommerup, I. C. (1976a). *In* "Biological Aspects of Plant Parasite Relationships" (J. Friend and D. R. Threlfall, Eds) 43–78. Academic Press, London and New York.
Ingram, D. S., Tommerup, I. C. and Searle, L. M. (1976b). *Ann. Appl. Biol.* **84**, 299–303.
Ito, S. and Tokunaga, Y. (1935). *Trans. Sapporo Nat. Hist. Soc.* **14**, 11–33.
Iwata, Y. (1943). *Ann. Phytopath. Soc. Japan* **12**, 97–108.
Jagger, I. C. (1924). *Phytopathology* **14**, 122.
Jagger, I. C. (1926). *Ann. Rep. Cheshunt Expt. Res. Stn* **12**, 35.
Jagger, I. C. and Chandler, N. (1933). *Phytopathology* **23**, 18–19.
Jagger, I. C. and Whitaker, T. W. (1940). *Phytopathology* **30**, 427–433.
Jagger, I. C., Whitaker, T. W. and Uselman, J. J. (1941). *USDA Dept. Agric. Circular* No. 596.
Janyska, A. (1957). *Ceskoslov. Akad. Zemedel. ved. Sborn Rostlinna Vyroba* **3**, 47–56.
Johnson, A. G. (1978). *Hort. Ind.* (May) **13**, 15–16.
Johnson, A. G., Crute, I. R. and Gordon, P. L. (1977). *Ann. Appl. Biol.* **86**, 87–103.
Johnson, A. G., Laxton, S. A., Crute, I. R., Gordon, P. L. and Norwood, J. M. (1978). *Ann. Appl. Biol.* **89**, 257–264.
Johnson, A. G. and Norwood, J. M. (1978). *Rep. Natn. Veg. Res. Stn. 1977*, 27–28.
Johnson, A. G. and Norwood, J. M. (1979). *Rep. Natn. Veg. Res. Stn. 1978*, 61–62.
Jones, B. L. and Leeper, P. W. (1971). *Pl. Dis. Reptr.* **55**, 794–796.
Jörstad, I. (1964). *Mytt. Mag. F. Bot.* **11**, 47–82.
Kanchaveli, L. A., Natsvlishvili, A. A. and Gvritishvili, M. N. (1957). *Trud. Inst. Zashch. Rast. Acad. Sci. Georgia* **12**, 181–194.
Kapooria, R. G. and Tjallingii, F. (1969). *Neth. J. Plant Path.* **75**, 224–226.
Kempton, R. J. (1979). *Ann. Rep. Glasshouse Crops Res. Inst.* 1978, 79–80.
Kendrick, J. B. (1929). *Phytopathology* **19**, 1143–1144.
Lavrov, M. N. (1933). *Rev. Appl. Mycol.* Abst. **12**, 306–307.
Lebeda, A. (1978). *Zahradnictvo* **2**, 83–85/124.
Lebeda, A. (1979a). *Phytopath. Z*, 94, 208–217.
Lebeda, A. (1979b). *Z. PflKrankh. PflPath. PflSchutz.* **86**, 729–734.
Lebeda, A., Crute, I. R., Blok, I. and Norwood, J. M. (1980). *Z. PflZücht.* (in press).
Leeper, P. W., Whitaker, T. W. and Bohn, G. W. (1963). *Am. Veg. Grow.* **11** (9), 7–16.
Ling, L. and Tai, M. C. (1945). *Trans. Br. Mycol. Soc.* **28**, 16–25.
Maclean, D. J. and Tommerup, I. C. (1979). *Phys. Pl. Path.* **14**, 291–312.
MacLean, D. J., Sargent, J. A., Tommerup, I. C. and Ingram, D. S. (1974). *Nature* (Lond.) **249**, 186–187.
MacPherson, N. J. (1932). *J. Min. Agric.* (London) **38**, 998–1003.
Marlatt, R. B. (1974). *Florida Agric. Exp. Stn. Tech. Bull.* **764**, 1–25.
Marlatt, R. B., Lewis, R. W. and McKittrick, R. T. (1962). *Phytopathology* **52**, 888–890.

Marlatt, R. B., Lewitt, R. W. and McKittrick, R. T. (1963). *Pl. Dis. Reptr.* **47**, 126–128.
Marlatt, R. B., McKittrick, R. T. and Lewis, R. W. (1966). *Phytopathology* **56**, 856–857.
Martinovic, M. (1960). *Rev. Appl. Mycol.* Abstr. **41**, 36 (1962).
Mason, P. A. (1973). "Studies on the biology of *Bremia lactucae* Regel". PhD Thesis, University of Cambridge, UK.
Melhus, I. E. (1921). *Phytopathology* **11**, 54.
Michelmore, R. W. (1979). "A study of sexual reproduction by *Bremia lactucae* Regel". PhD Thesis, University of Cambridge, UK.
Michelmore, R. W. (1981). *In* "The Downy Mildews" (D. M. Spencer, Ed.). Academic Press, New York and London.
Michelmore, R. W. and Ingram, D. S. (1980). *Trans. Br. Mycol. Soc.* **75**, 47–56.
Milbrath, D. G. (1923). *J. Agric. Res.* **23**, 989–994.
Milovtzova, M. O. (1937). *Abst. Rev. Appl. Mycol.* **17**, 838 (1938).
Miyake, I. (1914). *Bot. Mag. Tokyo.* **28**, 36–56.
Moore, W. C. (1948). Bull. 139, Ministry of Agriculture and Fisheries, London, 41.
Moreau, M. (1961). *Bull. Res. Council Israel, Sect. D.* **10**, 219–222.
Moreau, C. and Moreau, M. (1962). *Sciences* **19–20**, 42–53.
Morgan, W. M. (1978). *Trans. Br. Mycol. Soc.* **71**, 337–340.
Naoumov, N. (1913). *Bull. Mycol. Francaise* **29**, 273–278.
Netzer, D. (1971). *Israel J. Agric. Res.* **21**, 143–144.
Netzer, D. (1973). *Trans. Br. Mycol. Soc.* **61**, 375–378.
Netzer, D. and Globerson, D. (1976). *Phytoparasitica* **4**, 210.
Netzer, D., Globerson, D. and Sacks, J. (1976). *Hort. Sci.* **11**, 612–613.
Norwood, J. M. and Crute, I. R. (1980). *Ann. Appl. Biol.* **94**, 127–135.
Oescu, C. and Radulescu, T. (1932). *Bull. Sect. Sci. Acad. Roum.* **15**, 181–191.
Oescu, C. and Radulescu, T. (1933). *As quoted by O. Savulescu* (1962).
Ogilvie, L. (1944). *Rep. Agric. Hort. Res. Stn. Bristol 1943*, 90–94.
Ogilvie, L. (1946). *Rep. Agric. Hort. Res. Stn. Bristol 1945*, 147–150.
Patel, M. K. (1948). *Indian Phytopath.* **1**, 104–106.
Paulus, A. O., Nelson, J., Gafney, J. and Snyder, M. (1977). *Proceedings of the 1977 British Crop Protection Conference* 929–935.
Person, C. C. (1959). *Can. J. Bot.* **37**, 1101–1130.
Phillips, D. J. and Lipton, W. J. (1974). *Pl. Dis. Reptr.* **58**, 118–119.
Powlesland, R. (1954). *Trans. Br. Mycol. Soc.* **37**, 362–371.
Powlesland, R. and Brown, W. (1954). *Ann. Appl. Biol.* **41**, 461–469.
Raffray, J. B. and Sequeira, L. (1971). *Can. J. Bot.* **49**, 237–239.
Regel, E. (1843). *Bot. Zeit.* **1**, 665–667.
Rodenburg, C. M. (1966). *Euphytica* **15**, 141–148.
Rodenburg, C. M. and Huyskes, J. A. (1962). *The Grower* **57**, 300–301.
Sanders, P. F. (1975). "The development of oospores of *Bremia lactucae* Regel". MSc Thesis, University of Exeter.
Sargent, J. A. (1976). *Ann. Appl. Biol.* **84**, 290–294.
Sargent, J. A. and Payne, H. L. (1974). *Trans. Br. Mycol. Soc.* **63**, 509–518.
Sargent, J. A., Ingram, D. S. and Tommerup, I. C. (1977). *Proc. Roy. Soc. Lond.* **B198**, 129–138.
Sargent, J. A., Tommerup, I. C. and Ingram, D. S. (1973). *Phys. Pl. Path.* **3**, 231–239.
Savulescu, O. (1962). *Rev. Biol.* (*Roumania*) **7**, 43–62.
Sawada, K. (1919). *Agric. Expt. Stn. Gov. Formosa. Spec. Bull. No. 19.*
Schmidt, T. (1957). *Pflanzenarzt.* **10**, 62–63.
Schnathorst, W. C. (1962). *Phytopathology* **52**, 41–46.
Schreiber, F. (1944). *Kuhn-Arch.* **60**, 462–465.
Schultz, H. (1937). *Phytopath. Z.* **10**, 490–503.

Schultz, H. and Roder, K. (1938). *Züchter* **10**, 185–194.
Schweizer, J. (1919). *Verh. Thuring Naturf. Ges.* **23**, 17–61.
Schwinn, F. J., Staub, T. and Urech, P. A. (1977). *Med. Fac. Lanbouww. Rijksuniv. Gent* **42**, 1181–1188.
Sempio, C. (1938). *Riv. Patol. Veg.* **28**, 393–397.
Sempio, C. (1942). *Ann. Fac. Agr. Univ. Perugia for 1942,* 129–143.
Sempio, C. (1950). *Phytopathology* **40**, 799–819.
Sempio, C. (1959). *In* "Plant Pathology—An Advanced Treatise" (J. G. Horsfall and A. E. Dimond, Eds). Vol. I, 278–312. Academic Press, New York and London.
Sequeira, L. and Raffray, J. B. (1971). *Phytopathology* **61**, 578–579.
Shaw, C. G. (1949). *Mycologia* **41**, 326.
Sleeth, B. and Leeper, P. W. (1966). *Pl. Dis. Reptr.* **50**, 460.
Smieton, M. J. and Brown, W. (1940). *Ann. Appl. Biol.* **27**, 489–501.
Smith, W. G. (1884). *Gardeners Chronicle* **21**, 418.
Smith, J. M., Cartwright, J. H. M. and Smith, E. M. (1977). *Proc. Brit. Crop Prot. Conf. 1977,* 633–640.
Sprau, F. and Minckwitz, A. V. (1949). *Pflanzenschutz.* **1**, 66–69.
Sydow, H. (1923). *Annls. Mycol.* **21**, 168–169.
Tanaka, T. (1919). *Mycologia* **11**, 84–85.
Tao, C. F. (1965). *Acta Phytophylac. Sin.* **4**, 15–20.
Thompson, R. C. and Ryder, E. J. (1961). *USDA Tech. Bull.* No. **1244**.
Tjallingii, F. and Rodenburg, C. M. (1967). *Zaadbelangen* **21**, 104–105.
Tjallingii, F. and Rodenburg, C. M. (1969). *Zaadbelangen* **23**, 436–438.
Togashii, K. (1926). *Bull. Imper. Coll. Agric. and Forestry, Morioka, Japan* **9**, 17–29.
Tommerup, I. C., Ingram, D. S. and Sargent, J. A. (1974). *Trans. Br. Mycol. Soc.* **62**, 145–150.
Tulasne, L. R. (1854). *C. r. Acad. Sci.* **38**, 1101–1104.
Unger, F. J. A. N. (1847). *Bot. Zeit.* **5**, 313–316.
Valdeyron, G. (1955). *Rev. Appl. Mycol.* Abst. **35**, 816 (1956).
Ventura, J., Netzer, D. and Globerson, D. (1971). *J. Amer. Soc. Hort. Sci.* **96**, 103–104.
Verhoeff, K. (1960). *Tijdschr. Pl. Ziekt.* **66**, 133–203.
Viranyi, F. and Blok, I. (1976). *Neth. J. Pl. Path.* **82**, 251–254.
Viranyi, F. and Henestra, S. (1976). *Acta Phytopathol. Acad. Sci. Hung.* **11**, 173–182.
Vlasova, E. A. and Komarovara, A. (1977). *Trudy Prikl. Bot., Genet. Selek.* **61**, 130–142.
Von Schlechtendal, D. F. L. (1852). *Bot. Zeit.* **10**, 620.
Walker, J. C. (1952). *In* "Diseases of Vegetable Crops". McGraw-Hill, New York.
Watts, L. E. and George, R. A. T. (1961). *Rep. Natn. Veg. Res. Stn. 1960* **14**.
Welch, J. E., Grogan, R. G., Zink, F. W., Kihara, G. M. and Kimble, K. A. (1964). *Calif. Agric.* **19**, 3–4.
Wellving, A. and Crute, I. R. (1978). *Ann. Appl. Biol.* **89**, 251–256.
Whitaker, T. W., Bohn, G. W., Welch, J. E. and Grogan, R. G. (1958). *Proc. Amer. Soc. Hort. Sci.* **72**, 410–416.
Wild, H. (1947). *Trans. Br. Mycol. Soc.* **31**, 112–125.
Williams, D. J., Beach, B. G. W., Horrier, D. and Marechal, G. (1977). *Proceedings of the 1977 British Crop Protection Conference*, 565–573.
Wittman, W. (1972). *Rev. Appl. Mycol.* Abst. **51**, 4075.
Wolfe, M. S., Barrett, J. A., Shattock, R. C., Shaw, D. S. and Whitbread, R. (1976). *Ann. Appl. Biol.* **82**, 369–374.
Yarwood, C. E. (1937). *J. Agric. Res.* **54**, 365–373.
Yarwood, C. E. (1953). *Phytopathology* **43**, 675–681.
Yerkes, W. D. and Shaw, C. G. (1959). *Phytopathology* **49**, 499–507.
Zink, F. W. (1973). *J. Amer. Soc. Hort. Sci.* **98**, 293–296.

Zink, F. W. and Duffus, J. E. (1969). *J. Amer. Soc. Hort. Sci.* **94**, 404–407.
Zink, F. W. and Duffus, J. E. (1973). *J. Amer. Soc. Hort. Sci.* **98**, 49–50.
Zink, F. W. and Duffus, J. E. (1975). *Phytopathology* **65**, 243–245.
Zink, F. W. and Welch, J. E. (1962). *Pl. Dis. Reptr.* **46**, 719–721.
Zinkernagel, V. (1975a). *Rheinische Monatsschift Gemüse Obst. Schnittblumen* **63**, 408–410.
Zinkernegel, V. (1975b). *NachrBl.dt PflSchutzdienst., Stuttg.* **27**, 185–188.

Chapter 21

# Downy Mildew of Onion

### F. VIRÁNYI

*Research Institute for Plant Protection, Budapest, Hungary*

## I. Distribution and Importance

Downy mildew of onion, caused by the fungus *Peronospora destructor* (Berk.) Casp. ex Berk., is practically world-wide in distribution. In 1841, Berkeley made the first record of this disease on a species of *Allium*, probably *A. cepa*, in England (Yarwood, 1943). Since then it has been studied intensively in many countries, e.g. Ireland (Murphy and McKay, 1926, 1932; McKay, 1957), USA (Cook, 1932; Yarwood, 1943), the Netherlands (van Doorn, 1959), Yugoslavia (Jovićević, 1964), Poland (Rondomański, 1967), and Hungary (Virányi, 1974b, c, 1975).

The pathogen attacks various kinds of onion, but is especially destructive to the common onion (*A. cepa*). According to Cook (1932), the actual reduction in yield is sometimes as high as 60–75% of the crop. If leaf damage is severe, bulb development is markedly retarded and as a result, a large number of "bottle-necked" onion bulbs develop (Rondomański, 1967). Such bulbs do not become closed at the neck and are vulnerable to bacterial neck rot during storage.

As a consequence of the collapse of the seed-stalks, the disease also causes serious damage to the seed crop. In addition, seeds obtained from infected plants germinate very poorly, and seedlings grown from such seeds develop much more slowly than those from healthy seeds (Virányi, 1974a).

## II. Symptoms

Depending on the mode of infection, the disease appears on onion leaves growing from systemically infected bulbs or as local lesions on leaves or seed-stalks, originating from secondary infections by wind-borne conidia.

Systemically infected plants are dwarfed and the leaves tend to be distorted and pale-green. They frequently curve downward beyond the normal angle of the healthy leaves. In humid weather, the fungus sporulates on the surface

Fig. 1.   Leaf damage caused by *Peronospora destructor* on onion plants grown from sets

of the leaves, to give a greyish-violet, fine, downy growth characteristic of this disease.

Local lesions, resulting from secondary infection, first appear as pale spots, oval to elongated in shape, and variable in size. When moist conditions prevail, the pale-green areas become covered with conidiophores and conidia, giving the spots a greyish-violet colour. The invaded leaves shrivel and collapse, and finally die (Fig. 1). Secondary symptoms on seed-stalks are quite similar, except that the lesions are more elongated, and the mildew layer often has a zonate appearance (Fig. 2).

Severe downy mildew infections on leaves and seed-stalks are very often accompanied by facultative parasites, such as *Stemphylium botryosum*, or

Fig. 2.    Typical appearance of downy mildew on an onion seed-stalk showing fungus sporulation

*Macrosporium* sp., both causing a dark-coloured appearance of the affected spots.

Under certain conditions no external symptoms of the disease prior to sporulation may be visible. This was first noticed by Yarwood (1943), but he did not attempt to explain this phenomenon. Rondomański (1967) described the same feature in Poland and stated, that, as a rule, secondary symptoms on the leaves begin with fungus sporulation, and that the light-green discolouration of the infected tissues develops later. Rondomański's statement may be true under Polish conditions, but not in general. Based on numerous field observations made by the author, it seems that the appearance of the secondary symptoms, either discolouration, or sporulation, mainly depends on the weather situation prevailing during the incubation period (Virányi, 1975). According to the data shown in Fig. 3, the most important factor determining symptom appearance is air humidity.

## III. Causal Organism

*Peronospora destructor* is a typical downy mildew fungus belonging to the family Peronosporaceae. The conidiophores, emerging from the stomata, are

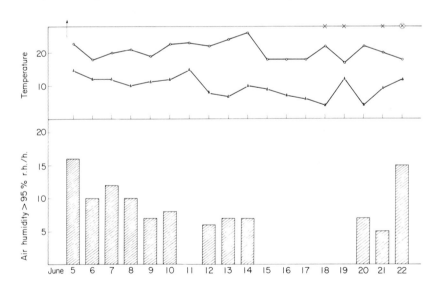

Fig. 3.    One infection-cycle of *Peronospora destructor* associated with the weather situation at Júlia Major, near Budapest in 1973. Primary sporulation ↑; light-green discolouration (secondary symptom) ×; sporulation (secondary symptom) ⊗; maximum temperature O——O; minimum temperature △——△

non-septate, 122–820 μm in length, and 7–18 μm in diameter at the base (Yarwood, 1943). Branching is monopodial with two sterigmata at the tips bearing pyriform to broad fusiform conidia (Fig. 4). Conidia are unusually large in size, measuring 17–34 by 29–82 μm, thin-walled, slightly papillate at the proximal end, and germinate by one or two germ tubes. The nonseptate mycelium is intercellular with simple or forked haustoria within the cells (Fraymouth, 1956). Oospores are more or less spherical in shape, possessing a smooth, uniform, thick wall. The mean diameter of the oospores is 28–35 μm (McKay, 1939).

Fig. 4. Epidermal strip of an infected onion leaf with a conidiophore and conidia emerging out of a stoma (× 175)

## IV. Host Range

*P. destructor* appears to be confined to the genus *Allium*. Besides the common onion (*A. cepa*) a number of other *Allium* species, both cultivated, and living wild, are recorded as host plants of this fungus. Murphy and McKay

(1926) reported that *P. destructor* obtained from *A. cepa* readily attacks *A. ascalonicum* and *A. cepa* var. *bulbelliferum*, but under the same conditions *A. porrum*, *A. fistulosum*, *A. sativum*, *A. scorodoprasum*, and *A. schoenoprasum* were free from the disease. Cook (1932), however, obtained infection only on *A. cepa* and *A. schoenoprasum*. A fairly complete list of the host plants of *P. destructor* was given by Yarwood (1943), the data presented, however, mainly originated from earlier records rather than from his own observations.

In order to clear up the contradictions existing in this field, an extensive study was carried out by Rondomański (1967), which included greenhouse and field observations. He succeeded in detecting some new hosts among wild *Allium* species but, in contrast, he proposed to remove *A. porrum* and *A. sativum* from the list of onion downy mildew hosts.

Based on these results together with a recent report by Furst (1976), the following cultivated, as well as wild *Allium* species are considered to be host plants of *P. destructor*: *A. albidum*, *A. alataviense*, *A. altaicum*, *A. ascalonicum*, *A. cepa*, *A. fistulosum*, *A. flavum*, *A. galanthum*, *A. obliquum*, *A. ochroleucum*, *A. oschanini*, *A. pskemense*, *A. schoenoprasum*, and *A. tuberosum*.

## V. Overwintering and Sources of Primary Infection

The term "overwintering" is not quite correct because of the differences between various locations in climate and manner of onion cultivation, but this word is generally used to define a period during which the pathogen survives from one growing season to another.

Although there are several ways, theoretically, for *P. destructor* to survive, it is generally accepted that overwintering mostly takes place in systemically infected onion bulbs. This fact was first demonstrated by Murphy (1921) in Ireland, and further evidence was given by Cook (1932) and Yarwood (1943). The data reported by several workers, however, differ greatly in regard to the level of infection of such bulbs. The two extremes are represented by Murphy and McKay (1926) in Ireland with up to 100% of infection, and by Yarwood (1943) in California with 0·004%. Almost equally low infection rates have been found in other countries, the percentage of diseased bulbs varying between 0·004 and 1·3% (van Doorn, 1959; Rondomański, 1967; Virányi, 1974b). This contradiction can be explained by differences in the climatic conditions of the various countries, as was suggested by Rondomański (1967). He affirmed that the percentage of systemic infection depends to a large extent on disease development, as well as on the length of the growing season, after infection, in the previous year.

In any case, systemically infected onion bulbs, occurring even in relatively low numbers, have to be considered as an important, if not the most important source of primary infection at the beginning of the growing season.

The survival of the fungus on other *Allium* species is difficult to estimate, but it seems likely that, at least in some European countries, *P. destructor* can survive on systemically infected *A. fistulosum* (McKay, 1957; Rondomański, 1967; Virányi, 1974b). If this is so, then these plants could supply conidia as primary inoculum much earlier than would systemically infected onion bulbs.

Although the formation of oospores of *P. destructor* seems to be a sporadical process, the long viability of these spores supports their possible involvement in the overwintering of the fungus. McKay (1957) revealed that oospores, exposed to circumstances existing in the field over 25 years, were still capable of germinating to a certain extent. Successful germinations of the oospores of various ages were also obtained by Berry and Davis (1957), Takahashi *et al.* (1958), and Jovićević (1964). On the other hand, oospores, when developed, failed to germinate in experiments conducted by Rondomański (1967), and by the author (Virányi, unpublished data).

Nevertheless, as yet, no one has been able to induce the disease process under controlled conditions by means of oospores.

Several investigations have demonstrated the occurrence of *P. destructor* mycelium in the flower parts of onion plants, but no conclusive evidence of seed transmission has yet been presented (Cook, 1932; Yarwood, 1943; Rondomański, 1967; Virányi, 1974a).

## VI. Epidemiology

The epidemiological factors with most influence on the incidence and severity of downy mildew disease are: source of inoculum, moisture conditions, temperature, and the size of the onion field. Any attempt, therefore, to investigate the epidemic spread of onion downy mildew demands knowledge of both fungal development and the environment.

### A. Role of environmental factors

According to Yarwood (1943), conditions influencing sporulation can be divided into those preceding and those during sporulation. In earlier work he found that sporulation occurred on those plants placed in a darkened humidity chamber in the afternoon, but not on those placed in the chamber earlier in the day. This indicates that some process taking place during daylight gives rise to a natural precursor of sporulation (Yarwood, 1937). Evidence for this diurnal cycle was given later by van Doorn (1959), who suggested that the rate of sporulation probably depends on the rate of assimilation of nutrients during the day. In fact Talieva (1966) found a close correlation between the sugar content of the infected plants and the sporulation process of the pathogen.

It seems likely that during the light period the host tissues must build up a reserve of nutrients, necessary for sporulation. My own observations, which indicate that sporulation prevents further production of conidiophores and conidia during the next 24 h, also support this hypothesis (Virányi, 1974c).

Van Doorn (1959) indicated that sporulation is strongly inhibited when high temperatures of 25–30°C occur during the preceding day. Furthermore, Rondomański (1967) found that both low and high temperature delayed sporulation and caused certain morphological changes in the reproductive organs. The most luxuriant growth of conidiophores and conidia was found at temperatures of about 10–13°C, which is generally considered optimum for the sporulation of this fungus.

Though light and temperature are evidently important regulating factors, the most critical, is humidity. Cook (1932) noticed conidia being produced when the leaves were wet with dew. The importance of leaf wetness was also underlined by Murphy and McKay (1932), and McKay (1939). According to van Doorn (1959), a relative humidity of more than 96% appears to be necessary for sporulation, and the presence of free water on the leaf strongly stimulates this process. Similar observations were made by Virányi (1974c) who showed that *P. destructor* requires at least 95% relative humidity for the formation of condiophores and conidia.

Once the thin-walled conidia have developed, they retain their viability for a few days depending on the temperature and moisture conditions. Conidia attached to the conidiophores remain viable much longer, than those which are detached, the latter being more sensitive to direct sunlight (Rondomański, 1967). Sunlight itself is not lethal to conidia if they do not become dried or too warm (Newhall, 1938). This may explain the survival of inoculum in the field during warm and sunny days.

As with other downy mildews, germination of the conida of *P. destructor* takes place only in free water. Under laboratory conditions maxium germination was found in twice distilled water containing 0·001 mol glucose (Rondomański, 1967; Virányi, 1974c).

The optimum temperature for germination of conidia is about 10–13°C, but it can occur at temperatures between 1 and 28°C (Cook, 1932; Yarwood, 1943; Jovićević, 1964; van Doorn, 1959). In general, germination begins in about 2–2·5 h from the start of favourable conditions and the majority of conidia germinate within 4 h.

For infection to occur *P. destructor* requires a prolonged period of wetness on the plant surface subsequent to conidium germination, and an optimum temperature of about 13°C, though penetration can occur between 3·5 and

25°C (van Doorn, 1959). Berry (1959) reported that 3 h are usually sufficient for the initiation of infection, which takes place when germ tubes enter via the stomata of leaves and seed-stalks.

A relatively high temperature of 25–30°C may lengthen the incubation period up to 18–20 days from the 10–17 considered an average length of time from infection to symptom expression (van Doorn, 1959).

## B. Infection cycle and disease spread

It has been shown by several workers that sporulation and infection usually take place overnight, while dissemination of the newly formed conidia occurs during day-time, mostly in the morning hours (Yarwood, 1943; van Doorn, 1959; Rondomański, 1967). Furthermore, Yarwood (1943) indicated that, as a rule, two humid nights are necessary for the whole infection process. On the other hand, Rondomański (1967) found that infection sometimes occurred immediately after sporulation, either at night, or during the subsequent morning hours.

P. destructor, like other downy mildews, shows a periodic daily dispersal of conidia beginning early in the morning, and reaching a peak between 1000 and 1200 h. This regularity was found to be associated with alterations in humidity of the surrounding air (Rondomański, 1967).

Since most downy mildews are characterized as anemophilous fungi, it is not surprising that the downy mildew of onion has a remarkable wind-borne nature, as was first established by Newhall (1938). He could detect viable conidia over diseased onion fields as high as 1500 feet (about 450 m). The direct influence of wind on the spread of the disease was observed by van Doorn (1959), and by Virányi (1975). Starting from an infection source, containing one or more systemically infected onion plants, the area of diseased plants increased and assumed an elongated shape, according to the prevailing wind direction (Virányi, 1975).

By comparing the environmental requirements of the pathogen and the corresponding meterological factors, it may be possible to determine the "critical periods" for infection. Van Doorn (1959) suggested that weather favourable for infection includes an 11 h period with at least 95% air humidity followed by a period of 6 h during which air humidity is not less than 80%. A similar observation was made when recording the duration of air humidity in the field (Virányi, 1975). Nevertheless, neither the estimation of the "critical periods", nor the bioclimatological model recently elaborated by de Weille (1975) will give a complete explanation of the infection process of P. destructor.

It is interesting to note that Yarwood (1939) registered a lower overnight temperature on onion leaves than in the surrounding atmosphere. He concluded that this difference in temperature can be responsible for the condensation of moisture on the leaves to give favourable conditions for infection. These various observations suggest that air humidity as well as other factors such as duration of leaf wetness and temperature, are of great importance in promoting the infection process. Fig. 3 illustrates one infection-cycle of *P. destructor* in relation to the prevailing weather situation. Because of the lack of leaf-wetness recording, successive infections could be related only to relative air humidity and temperature (Virányi, 1975), but it can be seen that sporulation and infection took place only when both of these factors coincided with the optimum for the pathogen.

## VII. Host–Parasite Relations

*P. destructor* can cause infections of two different kinds. The primary injury of leaves and seed-stalks always comes from local infections by means of conidia. The appearance of disease spots resulting from local infections indicates that the pathogen usually colonizes the host tissues to only a limited extent. Some investigations, however, revealed that *P. destructor* is able to grow within the leaves, either downwards or in an opposite direction (Murphy and McKay, 1926; Berry, 1959). This systemic infection provides a means for the longer term survival of the pathogen but the process involved as a local infection becomes systemic is not yet clear.

Changes in the metabolism of infected onion plants has not yet been extensively studied. From recent reports it is known that nucleic acid synthesis increases following infection, and later, as pathogenesis progresses, the nucleic acid content decreases indicating the destruction of host cells prior to sporulation (Ogolevets and Talieva, 1967). In addition, a significant increase in IAA content was observed in diseased onion leaves (Talieva, 1966).

There are only a few data available concerning the resistance of onions to downy mildew. It appears that the common onion (*Allium cepa*) is highly susceptible to *P. destructor*. On the other hand, lack of infection has often been found in onion crops for 2–3 weeks (Murphy and McKay, 1926; Virányi, unpublished data).

Recently, Talieva (1976) and Furst (1976) indicated some correlation between anatomical features of several *Allium* species and their degree of resistance to downy mildew. Berry (1959), however, found no such correlations though he noticed that once the pathogen had entered the resistant plant, the extent of mycelial growth was less than in any of the susceptible hosts studied. This he suggested indicates a physiological, rather than an anatomical basis of plant resistance.

## VIII. Control

Various practical measures, to preclude sources of infection and to prevent spread of disease, are of great importance in controlling onion downy mildew. For example, elimination of infected plant-remains, heat treatment of bulbs, or eradication of diseased volunteer plants, are highly recommended (Murphy and McKay, 1926; Yarwood, 1943; Cs.Kovács, 1964).

Fungicides, containing zineb or other dithiocarbamate compounds with the addition of a surfactant, are generally considered most effective as protectants against downy mildew if applied at the correct time before infection has occurred.

In order to determine when control measures should be applied Rondomański (1967) has worked out a method for predicting the intensity of primary infection, which is based on the estimation of the amount of mycelium overwintering in the bulbs. In addition, the timing of chemical control can also be indicated by the use of artifically inoculated bulbs planted outside early in spring. It is time to take precautions when sporulation occurs in these bulbs (Rondomański, 1963; Szakál, 1974).

Sprinkler irrigation of onion crops, commonly used in some countries has been found to be an important factor in the promotion of downy mildew epidemics (Palti et al., 1972). This should therefore be used with caution when other conditions are favourable for the disease.

The most obvious means of prevention of the disease would be the cultivation of resistant onions. Despite Berry's (1959) efforts, no one has yet been able to produce such cultivars with the other necessary characteristics to make them useful in practice.

### References

Berry, S. Z. (1959). *Phytopathology* **49**, 486–496.
Berry, S. Z. and Davis, G. N. (1957). *Pl. Dis. Reptr.* **41**, 3–6.
Cook. H. T. (1932). *NY Agric. Exp. Stn. Ithaca. Mem.* **143**, 1–40.
Cs.Kovács, L. (1964). *Délalföldi Mg. Kis. Int. Közl.* **5**, 119–130.
Fraymouth, J. (1956). *Trans. Br. Mycol. Soc.* **39**, 79–107.
Furst, G. G. (1976). *In* "Physiology and Immunity of Cultivated Plants" (N. V. Tsitsin Ed.), 51–63, Nauka, Moscow.
Jovićević, B. M. (1964). *Zast. Bilja* **78**, 117–172.
McKay, R. (1939). *J. R. Hort. Soc.* **64**, 272–285.
McKay, R. (1957). *Scient. Proc. R. Dubl. Soc.* **27**, 295–307.
Murphy, P. A. (1921). *Nature* **108**, 304.
Murphy, P. A. and McKay, R. (1926). *Scient. Proc. R. Dubl. Soc.* **18**, 237–261.
Murphy, P. A. and McKay, R. (1932). *J. Dep. Lds Agric. Dubl.* **31**, 60–76.

Newhall, A. G. (1938). *Phytopathology* **28**, 257–269.
Ogolevets, Ya. G. and Talieva, M. N. (1967). *Byull. Glavn. Bot. Sada, Moskva* **67**, 73–76.
Palti, J., Brosh, S., Stettiner, M. and Zilkha, M. (1972). *Phytopathol. Mediterr.* **11**, 30–36.
Rondomański, W. (1963). *Ochr. Rośl.* **4**, 1–4.
Rondomański, W. (1967). Final Techn. Rep. 1962–67., *Res. Inst. Veg. Crops, Skierniewice*, 59 pp.
Szakál, M. (1974). *Növényvédelem* **10**, 320–321.
Takahashi, M., Tanaka, Y. and Oishi, C. (1958). *Ann. Phytopath. Soc. Japan* **23**, 117–120.
Talieva, M. N. (1966). *Byull. Glavn. Bot. Sada, Moskva* **62**, 64–72.
Talieva, M. N. (1976). *In* "Physiology and Immunity of Cultivated Plants" (N. V. Tsitsin Ed.) 36–50, Nauka, Moscow.
Van Doorn, A. M. (1959). *Tijdschr. Plziekt. Meded.* **210**, 1–63.
Virányi, F. (1974a). *Növényvédelem* **10**, 205–209.
Virányi, F. (1974b). *Acta Phytopathol. Acad. Sci. Hung.* **9**, 311–314.
Virányi, F. (1974c). *Acta Phytopathol. Acad. Sci. Hung.* **9**, 315–318.
Virányi, F. (1975). *Acta Phytopathol. Acad. Sci. Hung.* **10**, 321–328.
Weille, G. A. de (1975). *Meded. Verh. K. Ned. Met. Inst.* **97**, 1–83.
Yarwood, C. E. (1937). *J. Agric. Res.* **54**, 365–373.
Yarwood, C. E. (1939). *Phytopathology* **29**, 933–945.
Yarwood, C. E. (1943). *Hilgardia* **14**, 595–691.

Chapter 22

# Downy Mildews of Ornamentals

## B. E. J. WHEELER

*Imperial College Field Station, Silwood Park, Berkshire, England*

## I. Introduction

Downy mildew fungi have been reported on a wide range of ornamentals, especially those in home gardens, but only in a few instances have the diseases they cause been sufficiently severe and widespread to warrant much investigation. Generally the diseases have been of most concern in a few ornamentals grown as crops either outdoors (e.g. anemones), or under glass (e.g. roses) or in nurseries, where many seedlings (e.g. antirrhinums) are raised in close proximity, often under humid conditions which favour these fungi. Even in these situations disease outbreaks tend to be sporadic, a feature itself not conducive to sustained research. As a result the literature on downy mildews of ornamentals is diffuse, consisting mainly of reports of particular fungi or popular articles on some diseases of interest to growers; few papers contain experimental data.

This review has three aims: to summarize present knowledge on those few downy mildews which occur or have occurred fairly frequently on ornamental crops, to present an annotated list of those of minor importance and to indicate others that have been recorded occasionally.

## II. Downy Mildews of Plant Pathological Interest

### A. Anemone

Two downy mildews are reported on cultivated anemones. One, caused by *Plasmopara pygmaea*, seems to be rare. It was first recorded in Britain on de Caen anemones near Liskeard, Cornwall, in September 1937 (Gregory, 1950), where, within four days, it caused blackening of the foliage in a 0·4 ha planting. Oospores were found in the discoloured tissues and conidiophores emerged from stomata at the periphery of the blackened zone. There appear to have been no reports since then of this fungus as a destructive pathogen of anemone crops.

The other downy mildew, caused by a *Peronospora* sp., is more common. The first British record in 1935 from south-west England was also by Gregory (1950). In this collection only isolated conidiophores were found and there were no visible lesions, but in later collections, from a different site, affected leaves had marginal, blackish lesions and tended to curl upwards bringing the conidiophores which emerged from stomata on the lower surface into an erect position. No oospores were found and the conidia (sporangia) which measured 28–39 × 23–30 μm germinated by zoospores. Gregory referred the fungus provisionally to *Peronospora ficariae*.

From 1950 onwards, the fungus became widespread in south-west England (Moore, 1959). It was then found in Jersey (Phillips, 1958) and the Netherlands (Boerema and Silvar, 1959). The Dutch workers considered the disease to be identical to that described in England but cast doubts on the identity of the *Peronospora* sp. They found many brown, thick-walled oospores in the leaf stalks, also occasionally in the flower stalks which they suggested, remained infective for 3–4 years.

The fungus continued to spread south and was next reported on anemone crops in the Var and Alpes–Maritimes departments of southern France by Tramier (1960). This and subsequent papers (Tramier, 1963, 1965) provide the most detailed account of the disease and its pathogen. The parasitized leaves are typically erect and at first a dull greyish-yellow but the tissues later become necrotic and blackish. When the disease is severe leaf development is also much reduced, the flower stalks remain short, the flowers sometimes abort and the plant has a stunted, dome-shaped appearance. The mycelium of the fungus may be found throughout the plant. It consists of intercellular, non-septate hyphae, 4–6 μm diameter with conspicuous digitate haustoria in parenchymatous cells. The conidiophores, which emerge from stomata, are 200–250 μm long and bear ovoid conidia 24·6 × 19·8 μm and oospores, 27–44 μm diameter, form abundantly in the dying leaves and flowers. Tramier (1960) considered the spore measurements to be sufficiently different from those of *Peronospora ficariae* to distinguish the fungus as a new species, confined to *Anemone coronaria* and *A. globosa*, which he described as *Peronospora anemones* (Tramier, 1963).

Infection may be initiated either by oospores via the roots or by conidia on shoots. The former is common where the crop is planted too often and is particularly damaging so that rotation is advised, sometimes to allow as long as 7 years between anemone crops (Anon., 1971). In the south of France high incidence of mildew is associated with rain in August and September. Maximum conidial production requires 11 h with 100% r.h. at 12–18°C and germination to penetration about 8 h of leaf surface wetness. Tramier (1963) found that conidial germination was optimal at 11–13°C and that leaves were penetrated directly without the formation of zoospores (cf. Gregory, 1950). The conidia are disseminated particularly at flower picking and also, presumably, in rain splash though this has not been systematically examined, but winds seem not to affect the spread of conidia.

Good control of the disease has been obtained with soil sterilants such as metham-sodium (Tramier, 1965) but in many areas this is too costly (Anon., 1971), and attention must be given to crop hygiene and rotation. The disease meanwhile, continues to concern anemone growers, the latest report being of severe outbreaks in Italy (Garibaldi and Gullino, 1972).

## B. Antirrhinum

Downy mildew of antirrhinum is something of an enigma. The causal fungus, *Peronospora antirrhini*, was described by Schroeter as early as 1874 on the wild *Antirrhinum orontium* from Germany but it was not recorded as a pathogen of cultivated antirrhinums until 1936. For a few years it devastated nursery stock in several localities and spread to many countries but by the late 1950s it had ceased to be of major concern to growers. It remains today as a disease of relatively minor importance but with sufficient history of damage in commercial nurseries to command respect.

This history started in Carlow, Ireland, with an outbreak which destroyed

50 000 seedlings (Murphy, 1937). In the following year (1937), the disease
was found in Sussex, England (Green, 1937, 1938) and by 1952 over 50
outbreaks had been recorded in 25 counties throughout England and Wales
(Moore and Moore, 1952). During that period the disease had also been
reported from Scotland (Wallace, 1948), various European countries, e.g.
Denmark (Weber, 1943), Sweden (Gréen, 1943), Norway (Jørstad, 1946),
from Canada (Conners and Savile, 1950), several states of the USA
(Yarwood, 1947; Jeffers, 1952) and from Australia (Anon., 1941; Simmonds,
1951) and New Zealand (Brien, 1946). In many instances the outbreaks were
severe. Yarwood (1947) comments that during 1940–1942 the disease was so
generally destructive in southern California that it was almost impossible to
obtain healthy seedlings. In many instances, too, the outbreaks were sporadic
and their origins obscure. For example, *P. antirrhini* was first observed as
a parasite of cultivated antirrhinum in Norway near Oslo in 1935 but did
not reappear until 1944 and then in the same locality (Jørstad, 1946). The
first record in Sweden was in January 1943 from the Weibullsholm Plant
Breeding Institute in an area where the wild host, *A. orontium* did not occur
(Gréen, 1943). Speculation that the fungus is seed-borne has not been substantiated by the limited experiments (Yarwood, 1947; Moore and Moore,
1952) but contamination of seed with oospore-containing plant debris cannot
be ruled out.

The infections on antirrhinums are of two kinds, systemic and local. The
former are more important, and result in a downward curling of the leaves
and a marked reduction in growth which results in a rosette of small leaves
at the tips of shortened shoots and gives the affected seedling a stunted
appearance. In most instances, the tips die and affected plants then produce
many secondary shoots from the base. The local, non-systemic infections
commonly appear on leaves as round, pale lesions but are rarely destructive.

The intercellular mycelium within the tissues gives rise to intracellular haustoria, characteristically with 4–8 finger-like branches, and to
dichotomously-branched conidiophores, 350–700 μm long, which emerge
from stomata and produce at their tips ovid conidia, 14–17 × 21–29 μm,
that *en masse* are mealy white to yellowish-brown. The conidia germinate
by a germ tube, optimally at 13°C in water films. Sporulation, too, is favoured
by relatively low temperatures and high humidities. In experiments by
Yarwood (1947), sporulation was abundant on systemically-infected leaves
kept in humid conditions at 13°C and moderate at 19°C; there was no
sporulation at or below 7°C or at 22°C and above. Oospores (30–38 μm
diameter), form abundantly in the petioles, stems and roots of infected plants
and are the means by which the fungus perennates (McKay, 1949).

In glasshouses some control of the disease has been achieved by reducing
humidity either through forced ventilation (Yarwood, 1947) or by raising
the temperature (Pettersson, 1954) and by protective spraying with copper

or dithiocarbamate fungicides (Green, 1938; Weber, 1943; Jeffers, 1952; Anon., 1953) and by sterilization of soil and equipment to destroy the oospores (McKay, 1949).

## C. Rose

The history of rose downy mildew and information on the disease and its pathogen to 1951 have been admirably summarized by Baker (1953). The disease was first reported in 1862 on roses in a cool glasshouse in England by M. J. Berkeley who described the causal fungus as *Peronospora sparsa* by which name it is still known. Subsequently the disease was reported in most European countries, Canada and the USA and it was regarded mainly as a disease of roses in the northern hemisphere. As Baker (1953) points out all records (until 1953) were from north of the Tropic of Cancer except one from Sao Paulo, Brazil. Since then the fungus has expanded its distribution. It was found in Mauritius in 1965 (Anon., 1967), in Israel in 1968 (Brosh, 1970) and in Australia (N.S.W.) in 1971 (Johns, 1972) where it has since become widespread (Bertus, 1977).

The fungus infects young leaves and stems at the shoot apex and also peduncles, calyces and petals. Frequently, though not invariably, infected leaves and stems develop purplish to black spots, a feature which has merited the name, black mildew, for the disease. Infected leaves often become yellow and drop prematurely, and infected shoots are sometimes killed. Under humid conditions, conidiophores *c.* 350 $\mu$m long bearing sub-elliptical conidia, $17-22 \times 14-18$ $\mu$m, are produced abundantly through stomata on the lower leaf surface, though in drier conditions they are often sparse (hence, *sparsa*) and thus easily overlooked. Conidia can be produced over long periods from an infected leaf, a feature which assists the development of epidemics. However, these rarely develop outdoors; the disease occurs mainly in glasshouses and even here outbreaks tend to occur sporadically. Disease development is favoured by humidities of 90–100% and relatively low temperatures. The conidia germinate best at *c.* 18°C and require only 4 h in water for this but they do not germinate at all below 4°C or above 27°C. Thus some control of the disease can be achieved in glasshouses by temporarily raising the temperature to 27°C and ventilating to reduce humidity (MacLean and Baker, 1951; Jaude, 1959; Brosh, 1970; Gill, 1977).

The formation of oospores seems to be variable but even where they do not form, there is some evidence that the fungus perennates as dormant mycelium in stems (Pickel, 1939; Stahl, 1973) so that control measures should include the destruction of infected leaves and stems. The severity of the disease can be reduced by spraying with fungicides of which, hitherto, the most popular have been various formulations containing copper and/or dithiocarbamates (Baker, 1953; Baresi, 1957; Taylor, 1963; Marziano *et al.*, 1973; Aleksandrova, 1976; Bertus, 1977). Varieties differ in their susceptibility to *P. sparsa* (Baker, 1953; Jaude, 1959; Rumberg, 1974), so some selection is advantageous where downy mildew is a problem or at least the confinement of susceptible varieties to drier areas or heated glasshouses (Brosh, 1970).

## III. Downy Mildews of Minor Plant Pathological Interest

The downy mildews of anemone, antirrhinum and rose have attracted attention because of their impact on the commercial production of these ornamentals. There are other downy mildews about which relatively little is known simply because in economic terms they are unimportant though intrinsically they are no less interesting as diseases. The following section attempts to summarize the more noteworthy of these and includes only those downy mildews for which there is some information other than a description of the causal fungus.

### A. *Alyssum*

The above remarks apply particularly to the downy mildew of *Alyssum saxatile* caused by *Peronospora galligena* because this fungus, unlike most other species, causes small blisters or gall-like growths on the leaves of its hosts. It was first described by Blumer (1938a) from collections in 1937 in the Berne district of Switzerland on *A. saxatile* and its varieties *citrinum* and *compactum*. He gives the following measurements; profusely branched conidiophores 400–800 μm long, ending in curved prongs, 10–22 μm long, and ellipsoid or globose conidia, 16–19 × 14–17 μm. Attempts to infect other *Alyssum* spp. and several crucifers failed as did later experiments by Ullrich and Schöber (1968). Although designating it a new species, Blumer (*loc. cit.*) considered it to be a highly specialized form of the *Peronospora parasitica* group. It was subsequently reported in Germany in 1938 and in England in 1946 (Moore, 1949). A similar type of disease, recorded in west Switzerland, involving malformations of the shoots and inflorescences of *A. saxatile* is also caused by a *Peronospora* sp. but its relationships with *P. galligena* have not been completely resolved. The symptoms are not typically those of *P. galligena* and the host range appears wider; in inoculations with the fungus Mayor (1963) was able to infect 18 out of 38 species and subspecies of *Alyssum*. However, specimens of *A. saxatile* with distorted inflorescences collected in Bohemia, Czechoslovakia, are attributed to *P. galligena* by Skalický (1953). Inoculations to *A. murale* and *A. montanum* were not successful, whereas Mayor (*loc. cit.*) obtained infection of cotyledons and young shoots of the latter species.

## B. *Chrysanthemum*

*Peronospora radii* was isolated from dark, pin-head lesions on the outer florets of the American spray variety Yellow Top in Sussex, England by Wilcox (1961). The fungus was also found on the varieties White Top, Igloo, Iceberg and Shasta in nearby beds. This appears to be the first record on cultivated chrysanthemums. The conidia appeared dark violet *en masse* and were ovate, 22–36 × 18–25 μm. Some oospores (mean diameter *c*. 30 μm) were found in affected florets but there was no leaf infection. Limited inoculations to other varieties suggested that not all were susceptible. The same fungus was also found on *C. cinerariifolium* by Mikusković (1968) in Yugoslavia who gives the following measurements: conidiophores 280–630 μm (average 423 μm); conidia, ellipsoid-ovid, 26–33 × 15–20 μm.

## C. *Cineraria*

Two downy mildews have been reported, *Plasmopara halstedii* and *Bremia lactucae*. Gill (1933) describes an outbreak of *P. halstedii* on *Senecio cruentus* in a glasshouse at Long Island, USA causing leaf spots up to 3 cm in diameter, white on the upper surface and slightly brown on the lower leaf surface. He gives the following measurements: conidiophores 247–640 μm long, conidia 15–30 × 13–23 μm. These are smaller than those given for this fungus by Wilson (1907) but forms of *P. halstedii* attack a wide range of composites, including many grown as ornamentals, and their inter-relationships are complex (Leppik, 1966). Like *P. halstedii*, forms of *B. lactucae* attack ornamentals within several genera of the Compositae such as *Centaurea*, *Gaillardia* (Blumer, 1938b; Viennot-Bourgin, 1954) and *Helichrysum* (Moore, 1959; Wittman, 1972). On cinerarias it causes yellowish to red, irregular leaf spots followed by withering of the leaves and defoliation. Ferraris (1936) recommends keeping plants in warm but not overdamp conditions and spraying with copper fungicides. Wittman (1972) reports that the fungus was controlled on *Helichrysum bracteatum* and other species by spraying early with copper formulations or with zineb, maneb, ziram, captan or daconil.

## D. *Clarkia*

*Clarkia elegans* is attacked by *Peronospora arthuri* (Lewis, 1937; Moore,

1959), sometimes severely so, if the host is growing under adverse conditions, as reported by Neergaard (1943) for various localities in Denmark in 1942. In this outbreak several varieties were affected but adjacent plantings of *C. pulchella* and its variety, *rubra* were not, nor was the related *Godetia hybrida*. The first infections were found on the lower leaves as white masses of conidiophores on the abaxial surfaces which later became grey with a faint purplish tinge and eventually brownish-grey. On the adaxial surfaces there were pale yellow areas associated with this sporulation. Neergaard (1943) suspected that the fungus was seed-borne and recommended a 30 min soak in 0·5% upsulun. Gustavsson (1959) casts some doubt on the identity of this peronospora on *Clarkia*.

### E. *Helleborus*

Downy mildew of the Christmas rose (*H. niger*) is caused by *Peronospora pulveracea*. The fungus induces precocious flowering, discolouration of the petals, irregular curling of the leaves and general stunting of the plant, and it sometimes seriously affects glasshouse crops of this ornamental (von Arx and Noordam, 1951). An early histological examination of infected tissues by Klebahn (1925) indicates several features of interest which deserve further investigation. For example, he did not observe haustoria associated with the intercellular mycelium in the mesophyll as was also noted in downy mildew of alfalfa (*P. trifoliorum* see Chapter 17) and in the leaf petioles and buds from rhizomes Klebahn (1925) found intracellular hyphae which sometimes penetrated adjacent cell walls.

### F. *Matthiola*

Downy mildew is common on the cultivated stock (*Matthiola incana*) especially at the seedling stage. The first symptoms are light green patches on the upper leaf surface with corresponding areas on the lower surface covered with the white down of the massed conidiophores and conidia. Infected leaves turn yellow, then become necrotic and often fall prematurely. The causal fungus is considered to be a form of *Peronospora parasitica*, restricted to *Matthiola*, and normally given that name but which Gaumann (1918) split off as a distinct species, *P. matthiolae*. The most recent information on the biology of the fungus is given by Jafar (1963) from New Zealand following the first outbreak there in 1958. His inoculations to a range of cruciferous plants indicated that only *M. incana* and *M. bicornis* were susceptible. None of the cultivated stocks were resistant though some supported less sporulation than others. Infection and sporulation were optimal at 15·5–21°C and under these conditions lesions developed in 5–6 days. The conidia germinated best at 15·5°C, rather poorly at 10°C or 21°C and not at all at 4·5 and 27°C. No oospores were found. Some control of this disease has been achieved with 0·5% buisol (Anon., 1951), by spraying weekly with dithiocarbamates (Jafar, 1963) or every 3 to 4 days with a copper oxychloride-zineb mixture (Bertus, 1968).

### G. *Mecanopsis* (and *Papaver*)

*Peronospora arborescens* has been reported on several species of *Mecanopsis* (Cotton, 1929; Alcock, 1933; Moore, 1959) and *Papaver* particularly the opium poppy, *P. somniferum* (Yossifovitch, 1929; Pape, 1934; Behr, 1956; Kothari and Prasad, 1971). Black spots or blotches on the upper surfaces of leaves are associated with white-grey or faintly mauve masses of conidiophores emerging from the lower leaf surface. Ornamental garden poppies are sometimes killed at the cotyledon stage whilst older plants become stunted with twisted or swollen flower pedicels and flower buds which fail to open. The fungus is presumed to overwinter as oospores in soil and possibly is distributed as oospores on seed. Spraying seedbeds with copper fungicides has been recommended (Beaumont, 1953).

### H. *Mimulus*

A downy mildew fungus on *M. guttatus* from Oregon, USA was first described by Shaw (1951) as *Peronospora jacksonii*. The same fungus was subsequently found affecting *Mimulus* cuttings at Wisley, England by Mence (1971) and this appears to be the first European record though it was quickly followed by another from Germany (Doppelbaur and Doppelbaur, 1972). Mence (1971) found Orchid to be the most susceptible cultivar and she was able to control the disease by spraying with zineb and improving the ventilation.

## IV. Other Downy Mildews of Ornamentals

Many other downy mildew fungi have been recorded on ornamentals mainly by mycologists interested in the genera to which they belong. The pathology of these remains uninvestigated. Some that are most commonly encountered are listed in Table 1 and in publications by Pape (1955), Moore (1959), Forsberg (1963), Wager (1970), Stahl and Umgelter (1976) and Pirone (1978).

TABLE 1

Downy mildews reported occasionally on ornamentals

| Fungus | Principal host genera | Notes |
|---|---|---|
| *Basidiophora entospora* Rose and Cornu. *Ann. Sci. Nat.* (Ser. 5) **11**, 84, 1869. | *Aster, Callistephus, Erigeron* | Sometimes troublesome in N. America, see Pirone (1978); also *Rep. Fla. agric. Exp. Stn.* for year ending 30th June 1967, p. 327. |
| *Bremiella megasperma* (Berlese) Wilson, G. W. *Mycologia* **6**, 195, 1914. | *Viola* | Apparently more common on *Viola* in USA than *Peronospora violae*, see Constantinescu, O. (1979) *Trans. Br. Mycol. Soc.* **72**, 510–515. |
| *Peronospora corollae* Transch., *Hedwigia* **34**, 214, 1895. | *Campanula* | see Moore (1959), p. 233, Gustavsson (1959), p. 187. |
| *Peronospora dianthi* de Bary, *Ann. Sci. Nat.* 4 Ser. 20, **114**, 1863. | *Dianthus* | Another species, *P. dianthicola* Barthelet has been described (see Gustavsson, 1959, p. 60) but this may be the same as *P. dianthi* de Bary (G. M. Waterhouse, personal communication). |
| *Peronospora gei* Sydow *In* Gäumann (1923), *Beitrage zu einer Monographie der Gattung Peronospora Corda*, p. 291. | *Geum* | see Dennis, R. W. G. and Wakefield, E. M. (1946) *Trans. Br. Mycol. Soc.* **29**, 153; also Gustavsson (1959), p. 125. |
| *Peronospora grisea* Unger, *Bot. Ztg* **5**, 315, 1847. | *Veronica* | see Moore (1959), p. 236, Gustavsson (1959), p. 187. |
| *Peronospora leptoclada* Sacc., *Syll.* **7**, 250. | *Helianthemum* | see Moore, W. C. and Moore, F. J. (1950), *Trans. Br. Mycol. Soc.* **32**, 274; also Gustavsson (1959), p. 160. |
| *Peronospora mesembryanthemi* Verwoerd, *Annale Univ. Stellenbosch* **2A**, 13–20, 124. | *Mesembryanthemum* | see Bradley-Jones, J. and Dickens, J. S. W. (1967) *Pl. Path.* **16**, 192; also Gustavsson (1959), p. 35. |

| Fungus | Principal host genera | Notes |
|---|---|---|
| *Peronospora myosotidis* de Bary, *Ann. Sci. Nat.* **4**, Ser. 20, 112, 1863. | *Myosotis* | see Moore (1959), p. 236; Gustavsson (1959), p. 173. |
| *Peronospora oerteliana* Kühn, *Hedwigia* **23**, 173, 1884. | *Primula* | see Moore (1959), p. 236. |
| *Peronospora parasitica* (Fr.) Tul., *C.R. Acad. Sci., Paris* **38**, 1103, 1854. | Various forms on *Arabis, Cheiranthus, Draba, Hesperis* and other Cruciferae | Sometimes severe (e.g. Weise, (1927). *Gartenwelt*, **31**, 486). Gaumann (1918) recognizes some forms as distinct species. |
| *Peronospora viciae* (Berk.) Casp., *Monatsber. Kgl. Preuss Akad.* **308**, 1855. | *Lathyrus* | see Moore (1959), p. 238. |
| *Peronospora violae* de Bary, *Ann. Sci. Nat.* **4**, Ser. 20, 125, 1863. | *Viola* | see Moore (1959), p. 238; Gustavsson (1959), 161–166; also Hammarlund, C. (1933), *Arkiv. Bot.* **25**, A2, 1–126. |

## V. Discussion

An examination of the plant pathological literature reveals that downy mildews on ornamentals attract attention only infrequently. For example, in the 57 volumes of the *Review of Plant Pathology* (formerly *Review of Applied Mycology*) published between 1922 and 1978 there are references to *Peronospora sparsa* and rose downy mildew only in 28 (Vols 4, 6–8, 11, 13, 14, 17, 19, 23, 26, 30–34, 37, 39, 43, 44, 46, 48, 50, 52, 53, 55–57) and these total only some 40 entries, mostly reports of occurrence. This suggests that either the disease occurs sporadically or that serious outbreaks are rather rare. As Baker (1953) points out, the former presupposes that the fungus has an effective means of perennation. It might be thought that oospores have this function but for many downy mildews on ornamentals there is little experimental evidence either to defend or support this hypothesis.

That these diseases are rarely destructive suggests they have rather special requirements for infection and this, indeed, appears to be so. Most attack young tissue, often only seedlings grown in crowded conditions which give the high humidities favouring sporulation or water films on leaves necessary for conidial germination and penetration of the tissues. In the sense that their parasitism is so limited, these fungi, on ornamentals at least, are not unlike the damping-off fungi e.g. *Pythium*. They differ in their biotrophic habit and in their apparent (though seldom proven) host specialization. Nevertheless, their co-existence with their hosts is ephemeral.

Despite the obvious difficulties of working with biotrophs that cause disease infrequently on hosts with a changing population of cultivars, the parasitism of these fungi presents many challenging problems for the plant pathologist.

## References

Alcock, N. L. (1933). *New Flora Silva* **5**, 279–282.
Aleksandrova, I. (1976). *Rastit. Zasht.* **24**, 10–11.
Anon. (1941). *Agric. Gaz. N.S.W.* **52**, 538.
Anon. (1951). *Tidsskr. PlAvl.* **55**, 70.
Anon. (1953). *Tasm. J. Agric.* **24**, 366–369.
Anon. (1967). *Rep. Dep. Agric. Maurit. 1965*, 91–97.
Anon. (1971). *Advis. Leafl. Minist. Agric. Fish* no. 353, 7 pp.
Arx, J. A. von and Noordam, D. (1951). *Tijdschr. Plziekt.* **57**, 32–35.
Baker, K. F. (1953). *Pl. Dis. Reptr.* **37**, 331–339.
Baresi, F. (1957). *Notiz. Mal. Piante* **40–41** (n.s. 19–20), 145.
Beaumont, A. (1953). *Gardeners' Chronicle Ser. 3*, **134**, 71.
Behr, L. (1956). *Phytopath. Z.* **27**, 287–334.
Berkeley, M. J. (1862). *Gardeners' Chronicle* **22**, 308.
Bertus, A. L. (1968). *Agric. Gaz. N.S.W.* **79**, 178–179.
Bertus, A. L. (1977). *Agric. Gaz. N.S.W.* **88**, 8–9.
Blumer, S. (1938a). *Mitt. Naturf. Ges. Bern. 1937*, 17–25.
Blumer, S. (1938b). *Ber. Schweiz. Bot. Ges.* **48**, 239–252.
Boerema, G. H. and Silver, C. N. (1959). *Versl. Meded. Plziektenk. Dienst Wageningen* **134**, 155–157.
Brien, R. M. (1946). *N.Z. Jl. Sci. Technol.* A **28**, 221–224.
Brosh, S. (1970). *Hassadeh* **50**, 434–436.
Conners, I. L. and Savile, D. B. O. (1950). *Rep. Can. Pl. Dis. Surv. 1949.*
Cotton, A. D. (1929). *Gardeners' Chronicle* **85**, 143–144.
Doppelbaur, H. and Doppelbaur, Hanna (1972). *Ber. Bayer. Bot. Ges.* **43**, 145–148.
Ferraris, T. (1936). *Riv. Agric.* **32**, 26–27.
Forsberg, J. L. (1963). "Diseases of Ornamental Plants". Special Publication no. 33. University of Illinois.
Garibaldi, A. and Gullino, G. (1972). *Inftore fitopatol.* **22**, 9–11.
Gäumann, E. (1918). *Beitr. Bot. Centrabl.* **35**, 395–533.
Gill, D. L. (1933). *Mycologia* **25**, 446–447.
Gill, D. L. (1977). *Pl. Dis. Reptr.* **61**, 230–231.
Green, D. E. (1937). *Gardeners' Chronicle* **102**, 27–28.
Green, D. E. (1938). *J. R. Hort. Soc.* **63**, 159–165.
Green, S. (1943). *Agric. Hort. Genet.* **1**, 97–98.
Gregory, P. H. (1950). *Trans. Br. Mycol. Soc.* **32**, 241–245.

Gustavsson, A. (1959). *Op. Bot. Soc. Bot. Lund.* **3**(1), 271 pp.
Jafar, H. (1963). *N.Z. J. Agric. Res.* **6**, 70–82.
Jaude, Clotilde (1959). *An. Soc. Cient. Argent.* **168**, 52–59.
Jeffers, W. F. (1952). *Pl. Dis. Reptr.* **36**, 211.
Johns, T. H. (1972). *Rep. Dep. Agric. N.S.W. 1971*, 200–210.
Jørstad, I. (1946). *Norg. GartForen Tidsskr.* **36**, 497–498.
Klebahn, H. (1925). *Z. PflKrankh. PflPath. PflSchutz* **35**, 15–22.
Kothari, K. L. and Prasad, N. (1971). *Indian Phytopathology* **23**, 674–688.
Leppik, E. E. (1966). *Bull. Pl. Prot. F.A.O.* **14**, 72–76.
Lewis, Esther A. (1937). *Phytopathology* **27**, 951–953,
MacLean, N. A. and Baker, L. F. (1951). *Bull. Roses Inc.* no. 160, 5–6.
Marziano, F., Calarese, S. and Stefanis, D. (1973). *Annali Fac. Sci. Agr. Univ. Napoli* **7**, 42–56.
Mayor, E. (1963). *Phytopath. Z.* **48**, 322–328.
McKay, R. (1949). *Gardeners' Chronicle* **126**, 28.
Mence, J. M. (1971). *J. R. Hort. Soc.* **96**, 393–394.
Mikušković, M. (1968). *Zast. Bilja* **19**, 197–200.
Moore, W. C. (1949). *Trans. Br. Mycol. Soc.* **32**, 95–97.
Moore, W. C. (1959). "British Parasitic Fungi". Cambridge University Press, Cambridge.
Moore, W. C. and Moore, F. J. (1952). *Pl. Path.* **1**, 135–136.
Murphy, P. A. (1937). *Int. Bull. Pl. Prot.* **11**, 176.
Neergaard, P. (1943). *Gartnertidende 1943* **8**, 95–98.
Pape, H. (1934). *Gartenwelt* **37**, 289–290.
Pape, H. (1955). "Krankheiten und Schadlinge der Zierpflanzen und ihre Bekampfung". Paul Parey, Berlin.
Pettersson, S. (1954). *Växtskyddsnotiser 1954*, 27–29.
Phillips, D. H. (1958). *Rep. Sts. Exp. Stn. Jersey 1957*, pp. 24–33.
Pickel, B. (1939). *Biológico* **5**, 192–194.
Pirone, P. P. (1978). "Diseases and pests of ornamental plants". Wiley, New York.
Rumberg, V. (1974). *In* "Boleznevstoichivost' Rastenii". (A. Semenova, Ed.) 68–120. Tallin Botanical Gardens, USSR.
Schroeter, J. (1874). *Hedwigia* **13**, 183.
Shaw, C. G. (1951). *Mycologia* **43**, 448–449.
Simmonds, J. H. (1951). *Rep. Dep. Agric. Stk. Qd. 1950–51* p. 48.
Skalický, V. (1953). *Česká Mykol.* **7**, 133–136.
Stahl, M. (1973). *NachBl. dt. PflSchutzdienst., Stuttg.* **25**, 161–162.
Stahl, M. and Umgelter, H. (1976). "Pflanzenschutz im Zierpflanzenbau" Eugen Ulmer, Stuttgart.
Taylor, J. C. (1963). *Sci. Hort.* **16**, 31–34.
Tramier, R. (1960). *C.r. Hebd. Séanc. Acad. Agric. Fr.* **46**, 622–624.
Tramier, R. (1963). *Annls Épiphyt.* **14**, 311–323.
Tramier, R. (1965). *Phytiat. Phytopharm.* **14**, 49–56.
Ullrich, J. and Schöber, B. (1968). *Jber. Biol. BundAnst. Land-u. Forstw. Braunschweig 1967*, p. 32.
Viennot-Bourgin, G. (1954). *Revue Path. Vég. Ent. Agric. Fr.* **33**, 31–45.
Wager, V. A. (1970). "Flower Garden Diseases and Pests". Purnell, Cape Town.
Wallace, J. G. (1948). *Gardeners' Chronicle* **124**, 21.
Weber, Anna (1943). *Gartnertidende 1943*, 1.
Wilcox, H. J. (1961). *Pl. Path.* **10**, 40–41.
Wittman, W. (1972). *Pflanzenarzt* **25**, 70.
Yarwood, C. E. (1947). *Hilgardia* **17**, 241–250.
Yossifovitch, M. (1929). *Revue Path. Vég. Ent. Agric. Fr.* **16**, 235–270.

Chapter 23

# Downy Mildews of Peas and Beans

## G. R. DIXON

*School of Agriculture, Aberdeen, Scotland, UK*

## I. *Pisum* and *Vicia*

### A. Introduction

*Pisum* (pea) and *Vicia* (field and broad bean) form two of the five genera of the tribe *Viciae* within the legume family Papillionaceae (Leguminosae). Both are temperate, Old World genera which have been cultivated since Stone Age times. Their present day importance is as sources of protein-rich seed crops for human and animal consumption. The increasing demands of affluent palates for tender sweet green peas, in particular, and to a lesser extent for young broad beans has caused a virtual revolution in their cultivation over the past 25 years. This has involved decreases in acreages of dried peas and beans and increased production of uniformly unripe crops for quick freezing, dehydration and other forms of processing (Bundy, 1971). A highly sophisticated "vining pea and bean" industry has evolved to satisfy consumer demands using factory-simulated and contract-based techniques for sowing, managing and harvesting the crop, and for controlling its quality (Bland, 1971). As a consequence attention has been paid to very specific pathogen problems and an excellent example of this is pea and *Vicia* bean downy mildew, *Peronospora viciae*.

## B. The organism

The asexual stage of this organism was first described on *Vicia sativa* L. and *Pisum sativum* L. as *Botrytis viciae* by Berkeley (1846). Later Caspary (1855) correctly identified the organism as a member of the order Phycomycetes and transferred the fungus to the genus Peronospora as *P. viciae*. Further de Bary (1863) erected two species for those Peronosporales which infect members of the Papillionaceae viz. *P. viciae* which has reticulate oospores and *P. trifoliorum* which has smooth oospores. Using sporangial dimensions and biological specialization as criteria, Sydow (1921) and Gäumann (1923) separated several species of Peronosporales which infect Papillionaceous hosts giving rise to a number of morphological and biological species which were specialized on single host genera and often on single host species. These authors erected *P. pisi* as the organism infecting *Pisum* spp. *P. viciae* remained as the pathogen of *Vicia* spp. Further complications in nomenclature arose by use of *P. faba* to describe the organism found on broad bean (*V. faba*) (Savulescu, 1948) which was thought to be a sub-species of *P. viciae-sativa* a classification used by Lindquist (1939) and Thind (1942) for downy mildew of vetch (*Vicia* spp.). A further classification, *P. vicicola* has been used for a downy mildew of vetch. This species was originally erected by Campbell (1935) to cover all isolates of *Peronospora* infecting Papillionaceous hosts other than *Pisum* spp. The International Code of Botanical Nomenclature (Recommendation 14A) states that differences in host range are taken to indicate racial rather than specific differences between fungi. Consequently for this present work the term *P. viciae* is used and where authors have used *P. pisi* this is assumed to be a synonym. This agrees with the classification used by Fraymouth (1956) who grouped downy mildew isolates from *Lathyrus pratensis* L., *P. sativum* L., *Vicia sepium* L., *V. hirsuta* Gray and *V. sativa* L. as *Peronospora viciae*. Additionally the most recent definitive description from the Commonwealth Mycological Institute (CMI), Kew, England (Mukerji, 1975) accepts *P. viciae* with a host range of *V. faba* L., *V. sativa* L., *L. sativus* L., *P. arvense* L. with *P. sativum* L. and related forms. This would also include *L. odoratus* L., the sweet or florist's pea, which Beaumont (1951) reported as infected by *P. viciae*.

TABLE 1

Salient references to the distribution of *P. viciae*

| Host | Geographical area | Reference |
|------|-------------------|-----------|
| Broad Bean | England | Glasscock (1963) |
| | Italy | Marras (1963) |
| | Rumania | Savulescu (1948) |
| | Tunisia | Jamoussi (1968) |
| Pea | Argentina | Muntanola (1957) |
| | Australia | Samuel (1931) |
| | | Anon. (1943a) |
| | Canada | Connors and Saville (1952) |
| | Chile | Bertossi (1963) |
| | Denmark | Olofsson (1966) |
| | France | Anon. (1936) |
| | Germany | von Heydendorff (1977) |
| | Italy | Ciccarone (1953) |
| | Kenya | McDonald (1937) |
| | | Anon. (1960) |
| | Netherlands | Hubbeling (1975) |
| | New Zealand | Anon. (1954) |
| | Rhodesia | Anon. (1956) |
| | Spain | Anon. (1943b) |
| | Tasmania | Geard (1961) |
| | | Henrick (1935) |
| | United Kingdom | Anon. (1968) (Summary a) |
| | United States of America | Campbell (1935) |
| *Lathyrus sativus* | India | Mathur (1954) |
| | United Kingdom | Anon. (1968a) |
| | Union of Soviet Socialist Republics | Golubev and Korner (1975) |
| *Pisum arvense* | United States of America | Weimar (1940) |
| Sweet Pea | United Kingdom | Beaumont (1951) |
| Vetch | Argentina | Lindquist (1939) |
| | India | Thind (1942) |
| | United Kingdom | Anon. (1968a) |
| | Union of Soviet Socialist Republics | Leokene and Timofeev (1972) |
| | | Golubev and Korner (1975) |
| | | Turukina and Temnokhud (1977) |

## C. Hosts and geographical distribution

So far no CMI map has been produced for *P. viciae* but its distribution may be judged by the summary given in Table 1 of salient references.

## D. Aetiology of *P. viciae*

This air-borne foliar invading pathogen produces an intercellular mycelium forming filiform, coiled or branched haustoria (Fraymouth, 1956). The latter author commented that it is characteristic of haustoria of *Peronospora* which infect Papillionaceae that when present they are all simple, hyphal, usually unbranched filaments and often covered with a layer of callose. Digitate haustoria were seen by Fraymouth (1956) produced by *P. viciae* similar to those reported by Campbell (1935) for *P. pisi*. Sporangiophores emerge from the intercellular mycelium in clusters of 5–7 from each stoma. These are long, stiff and straight, unbranched for at least two-thirds of their length and 160–750 × 8–13 $\mu$m in size with a main axis 100–450 $\mu$m long which branches 5–8 times dichotomously with the primary branches straight to slightly curved, the upper ones curved and spreading. The ultimate branchlets diverge at obtuse or right angles, are pointed, generally unequal and short, 18–22 × 2–3 $\mu$m each bearing a single sporangium.

The colour of the sporangiophore mat was observed to vary with temperature (Mence, 1971): at 8–16°C it was a normal violet colour while at 4°C it was deep violet and at 20°C almost white. Temperature also affected sporangial size, those formed at 4–8°C were larger than those developing at 12°C or above. Usually the sporangia are oval to elliptical in shape and narrowed towards the point of attachment, 15–30 × 15–20 $\mu$m in size and pale violet to pale grey colour *en masse*. Oospores are readily formed by *P. viciae* particularly in senescent tissue, they are spherical, light brown to deep yellowish pink and 25–37 $\mu$m in size. This may increase to 42 $\mu$m where the oogonial wall persists. The epispore is thick and marked by large reticulations which were initially observed by Thind (1942). There are probably strains of *P. viciae* which are virulent to different host cultivars and species. These are reported to be distinguishable by host specificity tests and by sporangial size (Mence, 1971; Hubbeling, 1975; von Heydendorff, 1977).

## E. Host symptoms

Symptoms caused by *P. viciae* on pea are of three types:
(1) Systemic.

(2) Local foliar, tendril, flower sporulation.

(3) Pod proliferation.

Systemic symptoms are usually most serious on seedlings where they are associated with stunting, rosetting, distortion and overall sporulation on the plant surface. Also extensive foliar lesions may sporulate heavily on the abaxial surface. Systemic invasion leads to a dull mealy growth of this fungus on the entire surface of leaflets, stipules, tendrils and stems. Stipules are greatly reduced in size and curl down at the edges. After the formation of sporangia the host plant withers and dies. Usually the whole plant is killed before flowering and the tissues are crowded with oospores. If sporulation is inhibited by adverse environmental conditions the infected areas turn greyish white and porcellaneous. Seedlings which are systemically invaded virtually cease extension growth at an early stage and such plants mostly die within 2–3 weeks. Less badly crippled plants may produce further branches from the basal nodes. Where systemic colonization results from late infection stunting symptoms may be restricted to the apical zone. But mycelium can spread internally to lower petioles and sub-apical nodes leading to swelling and distortion. Oospores are common in systemically infected tissues either in the presence or absence of sporangial formation. They are found particularly in the pericycle and cortex of stems and petioles.

Localized lesion infections develop from sporangial invasion as yellow to brown blotches on the upper leaf surface with angular areas of fluffy white to bluish cottony mycelium on the under surface (Walker and Hare, 1942; Hagedorn, 1974). Further development of these lesions leads to the formation of chlorotic patches on stipules, petioles and sometimes stems (Kirik and Koshevs'kii, 1976). Young lesions are light green with the upper surfaces, especially at the lesion perimeter blotched with brown. Lesion size varies from 0·2–2 cm diameter, the larger ones often limited by the main leaflet veins. Parts of the leaflet or stipule distal to the infected zone wither and die or where spots are confluent, the whole foliage is killed. Often the pale green lesion is broken by white porcellaneous areas indicative of the presence of oospores. Such symptoms start on leaves 3–4 and then progress up the plant, although on older plants symptoms may be restricted to small necrotic flecks on the laminae which sporulate profusely under humid conditions. Inflorescences and tendrils may also exhibit symptoms of profuse sporulation under conditions of high humidity. On these organs necrosis tends to be limited and there may simply be production of vast numbers of diffuse sporangiophores from the internal mycelium. Oospores may develop in local foliar lesions but are generally absent from the necrotic flecks found on older plants.

Pod and seed infection by *P. viciae* was reported by Smith (1884) and later by Linford (1929). Mycelium was found in the seed coats of infected pods (Melhus, 1931) and Ramsey (1931) identified the oospore stage in pea pod-tissue. Diseased pods exhibit yellow lesions on the outer surfaces with proliferations of the endocarp (Snyder, 1934) accompanied by ovule abortion but sporangia are not usually produced. Pods formed at systemically infected nodes are flattened, yellow and distorted rarely setting seed. Where seeds are infected they have superficial blistering and distortions while oospores may be found in the testas. Inner parchment like linings of pea pods sometimes form intumescences which occasionally contain oospores (Sorauer, 1924). Snyder (1934) described this as the least conspicuous phase of the disease syndrome with yellowish blotches 0·2–2 cm or more in length, slightly sunken and appearing as small islands on the pod. Internally a white felty epithelial proliferation is developed directly beneath the external yellow blotch. This is composed of cells containing chloroplasts with a mat like growth extending into the pod cavity within the lining and there are masses of oospores embedded in the ovary wall. Accompanying the white mycelial mat within the pod cavity is a velvety growth on the pod exterior. Such invasion prevents the ovules from maturing and the fungus is seen to enter the seed coat. Periods of high relative humidity stimulate pod infection. Such invasion was described by Hagedorn (1974) as indiscrete brownish blotches of indeterminate size and shape often, but not exclusively, found near to the dorsal suture. Infected pods are stunted and malformed.

Symptoms caused by *P. viciae* on *V. faba* were described by Glasscock (1963) who studied infections on beans grown in the UK counties of Kent and Sussex on the cultivars Triple White and Seville Longpod. Light grey, greenish spots were found on younger leaves which enlarged into lesions covering more than half of the leaf area. Older lesions developed a light brown mottle, with the whole area becoming brown and tending to desiccate leading to leaflet distortion. Infection was not found on pods and stems. But sporangiophore and sporangia developed abundantly on leaflets as a light grey downy fructification on the undersurface and occasionally on the upper surface. Other outbreaks in the UK are recorded by Moore (1959) in Hampshire and Herefordshire (1922), Lincolnshire (1939) and Yorkshire (1945). The present author identified significant levels of *P. viciae* infection on field bean crops in Cambridgeshire and Lincolnshire in 1977 (Dixon, unpublished data). Observations of infection on broad bean in Tunisia (Jamoussi, 1968) noted the symptoms as characterized by small yellow spots on leaves, stems and pods with violet coloured sporangiophore and sporangial mats. Glasscock (1963) reported that on *V. faba* the sporangia of *P. viciae* measured 15–33 × 13–28 μm (majority 26–31 × 22–24 μm), oospores measured 30–48 μm having a reticulate surface and found particularly in the stem vascular system bordering the medullary cavity. Systemic infection may be found on *V. faba* leading to dwarfing and distortion of either the entire plant or of only a few nodes.

## F. Penetration and colonization

Invasion from sporangia was found by Mence (1971) and subsequently by Singh (1979) to be mainly by direct cuticular penetration. This conflicts with Campbell (1935) who reported penetration as usually taking place through the stomata. A single germ tube developed from each sporangium. Mycelium was confined to host parenchyma tissues and in leaflets passed over and under small veinlets but was limited by larger ones. Young leaves, immediately after emergence were found to be highly susceptible to penetration but became resistant as they matured becoming susceptible again as senescence took place. With older plants only the terminal embryonic leaves remained susceptible. Such apical susceptibility also declined with age such that by the tenth node (flowering node for those cultivars studied by Mence, 1971) sporulation was greatly reduced both in terms of intensity and numbers of plants infected. Sub-apical tissue in these cultivars was highly resistant with sparse sporulation and the pathogen failed to spread downwards into older tissues. Disease incidence increased with inoculum concentration or in younger tissues in which *P. viciae* grew most rapidly. Systemic symptoms developed following colonization of the apical meristem. Mycelium of *P. viciae* grows irregularly in the leaf spongy parenchyma, developing between the upper epidermis and palisade parenchyma. Following systemic infection mycelium develops in the intercellular spaces penetrating the stem, leaf stalks and pods through the veins. Haustoria are most frequently rounded in the leaf mesophyll and filiform in epidermal cells (Koshevs'kii and Kirik, 1979).

## G. Sporulation

Sporangial production by *P. viciae* requires a relative humidity of more than 90% for at least 12 h (Olofsson, 1966). Usually this means the presence of free water and temperatures below 15°C (Allard, 1971). At 20°C sporangium formation ceases and oospores develop; above 25°C the host tissues become necrotic. Light counteracts sporulation in the Peronosporales but advances spore formation in a subsequent dark period (de Weille, 1963). A five-fold increase in sporangium production was found to in plants kept in continuous light as opposed to those kept in continuous dark (Singh, 1979). De Weille (1963) elegantly investigated the causes of high levels of variability of "germinative power" of sporangia from various Peronosporales, including *P. viciae* producing the following guide lines for experimentation:

(1) Samples studied should be greater than 500 propagules.
(2) These should be harvested from a uniform and non-putrefactive host substratum.

(3) Harvests of sporangia should be taken all at once and at the start of each experiment.
(4) The population of sporangia must be statistically homogenous, implying that all propagules are of similar age. Such samples can be obtained by utilizing the periodicity resulting from the effects of light on propagule formation.
(5) The intensity of ultra violet light, atmospheric humidity and temperature in the storage medium must be constant.
(6) For all samples the period of germination should be of equal duration.
(7) Droplet size spectrum and water composition should be identical for all samples.

Sporangium production has been shown to require a light intensity of 4000 lux ($c$. $3.5 \times 10^4$ ergs cm$^{-2}$ s$^{-1}$) for a minimum of 2 h per day at 15°C or 4 h per day at 20°C (von Heydendorff, 1977).

## H. Sporangium release

Maximum sporangium release was shown by Pegg and Mence (1972) under their environmental conditions to occur between 0600–0800 h BST with a peak at 0700 h following 1–2 h insolation. Sporangia produced at below 12°C or above 16°C showed reduced germination (Mence, 1971). Sporangia which were washed from the sporangiophores had a higher percentage germination than those discharged naturally at 60% relative humidity. This suggests that a rain drop dispersal mechanism might be a more effective means of disseminating *P. viciae* in the field compared to passive spore release and air current transmission at low humidities. The latter is the prime factor in determining sporulation. At humidities below 95% the numbers of leaflets with sporulating colonies declined sharply as did sporangial density per unit area (Mence, 1971). If leaves were kept at humidities too low for sporulation the laminae became silver-grey in colour and later chlorotic. But such leaves were able to produce sporangia up to 6 weeks later if transferred to a water-saturated atmosphere. In non-systemically infected plants sporulation is generally the first and frequently the only sign of infection. Sporangia shaken in the dry state from non-systemically infected leaves gave 11–90% germination. The majority lost viability within 3 days of being shed but a few survived for 5 days (Pegg and Mence, 1970).

## I. Sporangial germination

As early as 1935, Campbell showed that 2 h leaf wetness was insufficient for germination of *P. viciae* sporangia, while temperatures of 0–18°C were conducive to disease development but 25°C was unfavourable. Using artificial inoculation techniques, Anon. (1961) found infection to be more severe at 15°C than at 20°C. Sporangial germination took place over the range 1–24°C with an optimum at 4–8°C (Pegg and Mence, 1970). Later studies by von Heydendorff (1977) showed that germination occurred at 1–20°C with an optimum at 5°C. No additives were needed to stimulate germination *per se* but adenine and adenosine increased percentage sporangia which germinated. Sporangia had low viability i.e. 24 h at 20°C, 7 days at 15°C and 31 days at 4°C. Increased physiological age of sporangia was correlated with raised germination rate. Infection was favoured by an average daily temperature of 14–18°C, high precipitation with the maximum development of field epidemics in late May to early June (Kirik *et al.*, 1975). In general infections by *P. viciae* are encouraged by cool moist climatic conditions and disease becomes especially prevalent where coastal mists develop.

## J. Oospore production and seed borne infection

Ramsey (1931) was the first to report *P. viciae* oospores found within the blistered areas in infected pea pods. By sectioning pea seeds containing *P. viciae* mycelium Melhus (1931) found invasion of the intercellular spaces of the seed coat. The oospore stage in infected seed was reported also by Jones and Heald (1932). Seed may become infected in two ways:
1. through the side of the seed from endocarp proliferations originating from infection at the side of the pod, and
2. via the placentae of pods infected along the dorsal suture as indicated by the presence of mycelium and oospores in placental tissues (Campbell, 1935). More recently Koshevs'kii and Kirik (1979) describe oogonia of *P. viciae* as terminal, isolated structures with a pyriform shape when immature becoming spherical at maturity. Spherical oogonia and clavate antheridia are particularly found in systemically infected plants (Campbell, 1935). Oospore germination has not been observed within systemically infected tissue but mycelium can be located in such plants below the cotyledons. The presence of oospores and mycelium in pea testas was confirmed by Mence (1971) but she found no evidence for continuous mycelial invasion of the pedicels and pods. Her results suggested that ovular mycelium resulted from localized pod infections and growth through the endocarp. Seed borne infection has not been conclusively demonstrated as a means of transmission for *P. viciae* but infected seed may fail to germinate or develop. Also systemically infected seedlings which develop from diseased seed may die before emergence. Tests of pea seed from seven different seed batches gave no evidence of seed borne infection (Hagedorn, 1974).

Oospores of *P. viciae* germinated poorly *in vitro* producing a single un-branched germ tube (von Heydendorff, 1977). Variation of substrate, light intensity and temperature did not increase the percentage germination. But oospores were infective up to 2 months after their development, remaining viable for 11 months at low temperature.

Soil borne oospores are a major source of potential systemic invasion of developing pea seedlings (Snyder, 1934; Olofsson, 1966). They are capable of surviving for 10–15 years in soil and can give rise to significant levels of infection.

The infectivity of various sources of inoculum is illustrated in Table 2. This demonstrates that soil borne debris and infected soil can contain viable oospores for considerable periods of time.

TABLE 2

Viability of sources of *P. viciae* inoculum

| Treatment | No. seedlings grown | No. seedlings infected | % seedlings infected |
|---|---|---|---|
| Milled haulm 3 years old | 157 | 36 | 22·9 |
| Milled haulm 6 months old | 180 | 132 | 73·3 |
| Soil taken from plots used to grow infected peas for 4 years | 151 | 27 | 17·8 |
| Uninoculated control compost | 170 | 0 | 0 |

Dixon, unpublished data, from studies at National Institute of Agricultural Botany, Cambridge, UK.

## K. Host–parasite physiology

There have been few studies of this aspect of pea–*P. viciae* interactions. With systemically infected plants a common phenomenon was found to be the occur-rence of an extremely electron-opaque, membrane-bounded hemispherical deposit extending through the host cell wall and into host cytoplasm (Hickey and Coffey, 1977). This material which abutted directly onto the inter-cellular hyphal wall was termed the penetration matrix. Its formation prob-ably resulted from a specific host–parasite interaction. Apparently solid or gel-like material constituted the matrix surrounding digital intracellular haustoria. This membrane bound extra haustorial matrix was found through the penetrated host cell wall and formed a relatively thick layer around haustoria in young shoot tissues. But was much thinner distally around haustoria in mature leaf mesophyll cells. A regularly arranged tubular network of ribosome-free endoplasmic reticula was found in the host cytoplasm of some systemically infected shoot tissues adjacent to the haustoria.

Slight infections by *P. viciae* led to a decrease in host ascorbic acid content, by 13·6%, and sugars, by 13·4% whilst severe infections caused losses of 78·2% and 24·8%, respectively, and could be correlated with reductions of host plant height (Kirik and Koshevs'kii, 1977). Further studies demonstrate the accumulation of vitamin C around the penetration site in resistant cultivars such as Pauli, Orlik and Victoria Heine. In such cultivars oxidation reactions were detected whereas reduction reactions were found in susceptible cultivars such as Chernigovsky 190, Vladovsky 208 and Kubanets 1126 (Peresypkin *et al.*, 1977).

## L. Host resistance and pathogen virulence

### 1. Resistance testing

Two techniques for inoculating *P. sativum* were developed by Ryan (1971) based on the work of Snyder (1934). In the first, finely ground pods taken from diseased plants which contained oospores were incorporated into the growing medium with 0·5 g of inoculum placed around and slightly above the pea seeds. By this method systemic infection developed in 90% of inoculated plants. Pre-germinated pea seeds were used in the second method, these were immersed in a suspension of sporangia prior to sowing. A 48 h germination period followed by immersion for 15 min in a suspension containing $8 \times 10^4$ sporangia ml$^{-1}$ resulted in 95% systemically infected plants. Following inoculation the plants were maintained at 12–15°C in a glasshouse. Inoculation methods to produce non-systemically infected hosts were described by Pegg and Mence (1970). Trays of pea seedlings were inoculated at the first leaf stage, about 10 days after sowing, by atomising a sporangium suspension prepared by shaking sporulating leaves in distilled water which was then filtered through cheesecloth. Following inoculation the seedlings were covered with polyethylene sheet for 20 h in darkness at 12–13°C. After this they were uncovered and kept at a light intensity of 3000 lux (*c.* $2·5 \times 10^4$ ergs cm$^{-2}$ s$^{-1}$) for 12 h photo periods at 12–15°C and 65–90% relative humidity. The trays were re-covered with polyethylene sheet 7 days after inoculation and sporulation appeared 1–3 days later. These simple techniques may be used to provide inoculum for further experiments (Dixon, 1976). Work by Heydendorff and Hoffman (1978) in Germany showed that conidial production by a range of *P. viciae* isolates of different geographical origin was maximal at 15°C with a light intensity of 4000 lux and photoperiodic cycle of 16 h light and 8 h dark. Adult plants grown both under

# PEA DOWNY MILDEW

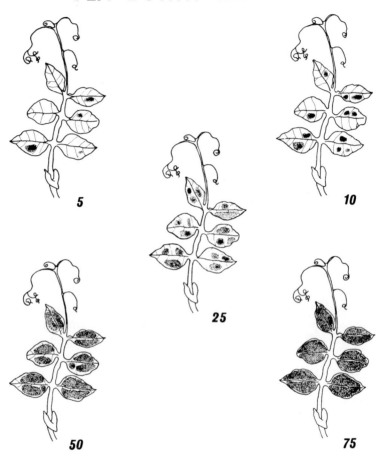

**Percentage    Area    Infection**

Fig. 1.   Key for the assessment of downy mildew (*Peronospora viciae*) on maturing pea plants. This key is designed for the assessment of percentage leaf area infected with downy mildew on maturing pea plants. There are seven categories given in the key: 0 (healthy), 5, 10, 25, 50, 75 and 100% leaf area affected. It is suggested that where necessary users should interpolate values between these figures. For field assessments select 10 sites at random along any one diagonal. Make assessments on 10 plants at each site. Assessments should be made on a growth stage rather than a date basis using the growth stage key:

1 = First true leaf present
2 = Vegetative stage 2·1 = 2nd true leaf
      2·2 = 3rd true leaf
      2·3 = 4th true leaf
      2·4 = 5th true leaf
      2·$n$ = $n$th true leaf

polyethylene sheet structures and in the field may be inoculated using sporangium suspensions. Infection of highly susceptible cultivars such as Cennia or Skagit as pathogen carriers then permits *P. viciae* to invade test cultivars by natural spread (Dixon, 1976).

## 2. Disease assessment

Most workers with *P. viciae* have developed some form of assessment technique. A key (Fig. 1) for assessment of infection on maturing pea plants was developed by Dixon (unpublished data) linked to a growth stage key for pea plants (Webster and Bowring, personal communication). This method measures leaf area covered by sporangiophores. Other workers have looked at intensity of sporulation, for example Matthews and Dow (1970) who recognized categories of resistance and susceptibility in the host on the basis of foliar sporulation: 1 = resistant, 2 = slightly susceptible, 3 = moderately susceptible, 4 = susceptible, 5 = highly susceptible. Simply as a measure of sporulation intensity Singh (1979) identified: 0 = no sporulation, 1 = slight sporulation, 2 = moderate sporulation, 3 = intensive sporulation. This author also devised a useful technique for quantifying sporulation by washing sporangia from leaves and counting them with a Wipple grid. A key which attempted to link assessment of resistance to degree of symptom expression was designed by Hubbeling (1975):

0 = no symptoms visible: very resistant;
1 = local necrosis of leaves: no sporulation: resistant;
2 = limited sporangial production on some leaves: followed by local necrosis: moderately resistant;
3 = abundant sporangial production but confined mainly to inoculated leaves: susceptible;
4 = abundant sporangial production on leaves and stems: often with systemic infection: very susceptible.

In terms of grouping cultivars as resistant or susceptible Hubbeling claimed

---

3 = Bud development
4 = Open flower stage
5 = Pod emergence
6 = Flat pot development
7 = Harvest stage (Webster and Bowring, unpublished data)
Assessments of downy mildew should be made on the uppermost fully expanded and mature leaf on each plant. It is likely that in infected crops there will be foci of systemically infected plants, these should be separately assessed and the recording placed in parentheses after the non systemic foliar recording. For plot assessment it is suggested that all plants in a plot or portion of a plot should be assessed.

that categories 0, 1 and 2 are resistant while 3 and 4 are susceptible (Eucarpia Pea Workshop, John Innes Institute, Norwich, UK, 1976). But the complexity of attempting to differentiate cultivars into arbitrary resistant or susceptible categories was highlighted initially by Olofsson (1966). This worker found that individual cultivars varied in their resistance to systemic, apical and pod infection. Resistance to apical infection was thought to be linked to resistance to infection from oospores. Later it was clearly shown that all three symptom forms: systemic, localized foliar lesions and pod infections must be analysed in any attempt to classify cultivars (Ryan, unpublished data) (Table 3).

TABLE 3

Analysis of symptom forms in relation to the resistance of pea cultivars to *P. viciae*

| | Symptoms | | |
|---|---|---|---|
| Resistance rating | Systemic % | Foliar index | Pod % |
| VR—very resistant | 0–1 | 4–9 | 1–8 |
| MR—moderately resistant | 2–4 | 10–12 | 10–13 |
| SR—slightly resistant | 5–8 | 13–14 | 15–19 |
| SS—slightly susceptible | 9–11 | 15–17 | 20–25 |
| MS—moderately susceptible | 12–15 | 18–22 | 26–33 |
| VS—very susceptible | 16–25 | 23–33 | 35–51 |

Ryan (unpublished date)

As a result of his work Ryan felt that systemic infection was the paramount criterion to be evaluated when testing cultivars. Pod infection was of secondary importance while foliar sporulation had least significance.

## 3. Cultivar resistance

Considerable efforts have been made to evaluate cultivar resistance to *P. viciae*, particularly in UK. Using plants grown under polyethylene and in artificially and naturally infected field trials (Dixon, 1976) the resistance of more than 100 cultivars and breeders lines has been assessed (Table 4). Additionally work at the Processors' and Growers' Research Organization (PGRO), Peterborough, UK; which after 1971 was done in co-operation with the National Institute of Agricultural Botany (NIAB), Cambridge, UK; using naturally infected field sites has given a further putative list of resistant and susceptible pea cultivars (Table 5).

# TABLE 4

Resistance of pea cultivars to downy mildew
(*Peronospora viciae*) as measured in field trials[a]

| Cultivar | Source | 1972 | 1973 | 1974 | 1975 | 1976 | 1977 |
|---|---|---|---|---|---|---|---|
| Acturas | Schafer | — | — | — | — | — | 7 |
| Alfarat | Daeh | — | — | — | — | 3 | — |
| Ajax | Stephenson | 7 | 6 | 7 | — | — | — |
| Aries | Morrison | — | 8 | — | — | — | — |
| Avola | Asgrow | — | 6 | 6 | — | — | — |
| Beacon | Asgrow | 7 | 5 | 5 | 5 | — | — |
| Beagle | Hurst | 8 | 7 | — | — | — | — |
| Betamu | Daeh | — | — | — | — | 4 | — |
| Bodo | Schafer | — | — | — | — | — | 4 |
| Bonus | Asgrow | 5 | 5 | — | — | — | — |
| Canice | Hurst | 7 | 7 | — | — | — | — |
| Celicia | Sharpes | — | — | — | — | 6 | — |
| Cep | S & G | 6 | 7 | 7 | — | — | — |
| Citation | W. Valley | 5 | 5 | 5 | — | — | — |
| Comire | A R Zwaan | 6 | 6 | — | — | — | — |
| Conway | Bro | — | — | — | 8 | — | — |
| Dark Skinned Perfection (Perfected Freezer 70A) | Bro | 8 | 6 | 7 | — | — | — |
| Dart | Asgrow | 6 | 6 | — | — | — | — |
| Dash | Asgrow | — | 7 | — | — | — | — |
| Earl | Asgrow | 6 | 7 | 6 | 8 | 6 | — |
| Elda | Schafer | — | — | — | — | — | 4 |
| Elvira | Nunhem | 5 | 4 | — | — | — | — |
| Famous | Rogers | 8 | 6 | — | — | 7 | — |
| Fek | S & G | 7 | 8 | 8 | 8 | 7 | — |
| Finale | Cebeco | — | — | — | 8 | — | — |
| Findulina | Findus | — | 4 | 5 | 6 | — | — |
| Fraser A | Bro | 6 | — | — | — | 6 | — |
| Freezer 20 | Pureline | 6 | 6 | — | — | — | — |
| Freezer 47 | Pureline | — | — | — | 6 | 6 | 6 |
| Fridol | S & G | 6 | 6 | — | — | — | — |
| Friskey | Sperling | — | — | — | — | 6 | — |
| Frivita | Sperling | — | — | — | — | 8 | — |
| Galaxie | Rogers | 6 | 6 | — | — | 6 | — |
| Greenshaft | Hurst | 8 | 8 | — | — | — | — |
| Hustler | Wrightson | — | — | — | 6 | 7 | 6 |
| Jof | S & G | — | — | — | — | 6 | 4 |
| Johnson's Freezer | Johnson | — | — | — | 7 | 7 | 6 |
| Lacavil | Vilmorin | — | — | 8 | — | — | — |
| Lincoln | Lincoln | 6 | — | — | 5 | 6 | 5 |
| Lorino | RS | — | — | 6 | 6 | — | — |
| Luna | Sharpes | — | — | — | — | 7 | 6 |
| Marlin | Bro | — | — | — | — | — | 6 |
| Martus | Asgrow | — | 5 | — | — | — | — |

TABLE 4 *cont.*

| Cultivar | Source | 1972 | 1973 | 1974 | 1975 | 1976 | 1977 |
|---|---|---|---|---|---|---|---|
| Medalist | Rogers | — | 4 | 6 | 6 | — | — |
| Mid Freezer | Rogers | 8 | — | 7 | 7 | 7 | — |
| Minarvil | Vilmorin | — | — | 6 | — | — | — |
| Miralite | Sharpes | 3 | 3 | — | 3 | 5 | — |
| Mowhawk | Asgrow | 7 | 6 | — | — | — | — |
| Myzar | Clause | 7 | 6 | — | — | — | — |
| Multifreezer | Rogers | 6 | — | — | — | — | — |
| Multistar | Daeh | — | — | — | 2 | 4 | 4 |
| Num | S & G | — | 6 | 7 | 3 | — | — |
| Orbiter | Sharpes | 5 | 5 | — | — | — | — |
| Orfac | Clause | 5 | 5 | — | — | — | — |
| Orion | Morrison | — | 8 | — | — | — | — |
| Ola | Schafer | — | — | — | — | 7 | — |
| Osprey | Unilever | — | — | — | — | 6 | — |
| Parade | Nunhem | 7 | 7 | — | — | — | — |
| Payette | W. Valley | 7 | 7 | — | — | — | — |
| Perfected Freezer 703 | Bro | — | 6 | 6 | 7 | — | — |
| Pirouette | S & G | — | — | 8 | 8 | 8 | — |
| Platinum | Asgrow | 6 | 7 | — | — | — | — |
| Preperfection | Pureline | 6 | 7 | — | — | — | — |
| Puget | Bro | 7 | 7 | — | — | — | — |
| Primagolt | Mansholt | 7 | 7 | — | — | — | — |
| Ralca | R.S. | 6 | 6 | — | — | — | — |
| Rally | Asgrow | — | — | — | — | 5 | 3 |
| Regent | Asgrow | — | — | 6 | — | — | — |
| Relavil | Vilmorin | — | — | 7 | — | — | — |
| Recette | S & G | 5 | 5 | — | — | 5 | — |
| Scout | W. Valley | 7 | 7 | — | — | — | 7 |
| Skagit | Bro | 1 | 1 | 1 | 1 | 2 | 3 |
| Small Sieve Freezer | Rogers | 8 | — | — | — | — | — |
| Sparkle | Rogers | 8 | 7 | — | — | — | — |
| Sprite | Asgrow | 7 | 7 | 7 | 7 | 6 | — |
| Superette | S & G | — | — | 8 | 7 | — | — |
| Superfection | C. Moscow | 5 | 6 | — | — | — | — |
| Suprema | Nunhem | 7 | 6 | 7 | — | — | — |
| Surprise | Hurst | 6 | — | — | — | — | — |
| Swinger | C. Moscow | — | — | — | — | 6 | — |
| Taurus | Morrison | — | 4 | 5 | 5 | — | — |
| Tezieridee | Tezier | 7 | 6 | — | — | — | — |
| Tornade | Schafer | — | — | — | — | — | 5 |
| Trio | Schafer | — | — | — | — | — | 4 |
| Vera | Cebeco | — | — | — | — | — | 6 |
| Vida | Asgrow | 6 | 5 | — | — | — | — |
| Visto | Schafer | — | — | — | — | — | 6 |
| Winfreda | Sperling | — | — | — | — | 7 | — |
| Viking | Gallatin | 8 | — | — | — | — | — |
| Cebeco 601 | Cebeco | — | — | — | 4 | — | 6 |
| Cebeco 608 | Cebeco | — | — | — | 6 | 7 | 6 |

| Cultivar | Source | 1972 | 1973 | 1974 | 1975 | 1976 | 1977 |
|----------|--------|------|------|------|------|------|------|
| D518 | S & G | — | — | 7 | 6 | — | — |
| D5487 | S & G | — | — | — | — | — | 5 |
| D8409 | S & G | — | — | — | — | — | 4 |
| D8416 | S & G | — | — | — | — | — | 4 |
| D9401 | S & G | — | — | — | — | 8 | 7 |
| D9407 | S & G | — | — | — | — | 7 | — |
| D9433 | S & G | — | — | — | — | — | 7 |
| D9457 | S & G | — | — | — | — | — | 7 |
| D9475 | S & G | — | — | — | — | 8 | 7 |
| Pah 78 | Sharpes | — | — | 7 | — | — | — |
| RS 1365 | R.S. | — | — | 4 | 5 | — | — |
| SP 26 | Morrison | — | — | — | 5 | — | — |
| WV 330F | W. Valley | — | — | 4 | — | — | — |
| XPF 70 | Asgrow | — | — | — | — | 3 | 4 |
| XPF 86 | Asgrow | — | — | — | — | — | 5 |
| XPF 87 | Asgrow | — | — | — | — | 6 | 6 |
| 55/69 | Nunhem | — | — | 6 | 6 | 6 | — |
| 56/71 | Nunhem | — | — | — | — | — | 7 |
| 58/69 | Nunhem | — | — | — | — | — | 6 |
| 218/70 | Nunhem | — | — | 7 | — | — | — |
| 274/70 | Nunhem | — | 6 | 6 | 6 | — | — |

[a] Resistance to Downy Mildew measured over the years 1972–1977 on 0–9 scale where 0 = no resistance, 9 = high resistance. — = no data. Total number cultivars tested = 112, from studies at National Institute of Agricultural Botany, Cambridge, UK.

Similarly Ryan (unpublished data) and Matthews and Dow (1970, 1971, 1973, 1974, 1977) have assessed the resistance of a wide range of pea material, the latter authors using principally seedling tests performed under controlled conditions.

Levels of the expression of symptoms caused by *P. viciae* on pea cultivars vary with host age and environmental conditions. These effects are illustrated in Fig. 2 (Dixon, unpublished data) where four cultivars, Earl, Fek, Midfreezer and Skagit, showed maximum symptom development, when inoculated artificially and grown in polyethylene structures, 40–50 days after sowing. Thereafter levels of sporulation as measured by the assessment key shown in Fig. 1 declined. Sporulation on a fifth cultivar Pirouette followed a similar pattern but rose again sharply 70–90 days after sowing. These observations were made in the period March to June when levels of sporulation were probably affected by decreasing relative humidity in the latter part of the assessment period but this does not negate the results obtained with Pirouette in two consecutive years, 1974 and 1975. Also these results demonstrate that relatively resistant cultivars such as Earl, Fek and Midfreezer may be

TABLE 5

Resistance of pea cultivars to *P. viciae* as tested in naturally infected field trials
by PGRO and PGRO/NIAB

| Year | Resistant cultivars | Susceptible cultivars | Reference |
|------|---------------------|------------------------|-----------|
| 1957–1960 | Onward<br>Dark Skinned<br>Perfection (DSP) | Kelvedon Wonder | |
| 1959–1960 | DSP<br>Onward<br>Perfected Freezer<br>Pauli | Kelvedon Wonder<br><br>Lincoln | King and Gane (1965) |
| 1962 | Meteor<br>DSP | Fruhe Kleine Pfalzerin | |
| 1967 | DSP<br>Dik Trom<br>Galaxie<br>Margo<br>Minigolt<br>Pauli<br>Puget | Kelvedon Wonder<br><br>Comire | Anon. (1968b) |
| 1968 | Vedette<br>Preperfection<br>Green Arrow<br>Puget | Orfac<br>Sultan<br>Dart<br>Sparkle | Anon. (1969) |
| 1969 | Green Shaft<br>Preperfection | Orfac<br>Skagit | Anon. (1970) |
| 1970 | Beagle | Bonus<br>Colmo<br>Kelvedon Wonder<br>Martus<br>Parade<br>Platinum<br>Ralca | Anon. (1971) |
| 1972 | Considerable site to site variation in cultivar reaction | | Anon. (1973) |
| 1973 | Cep<br>Cobri<br>Dart<br>Fek<br>Maro | Aldot<br>Mini | Anon. (1975) |
| 1974 | Cep<br>Cobri<br>Earl<br>Fek<br>Num | — | Anon. (1976a) |
| 1975 | Beacon<br>Findulina<br>Lorino<br>Pirouette | — | Anon. (1976b) |

| Year | Resistant cultivars | Susceptible cultivars | Reference |
|------|---------------------|-----------------------|-----------|
| 1976 | Cobri | Banff | |
| | Filby | Danielle | |
| | Florette | Frogel | |
| | Frimas | Heron | Anon. (1978) |
| | Maro | Mantica | |
| | Tessa | Marlin | |
| | | Midiver | |
| | | Vedette | |
| 1977 | Cocquette | Condor | |
| | Primette | Danielle | |
| | Spiket | Heron | Gent and Bingham |
| | | Hiaco | (1978) |
| | | Quincy | |
| | | Rally | |

— = no results.

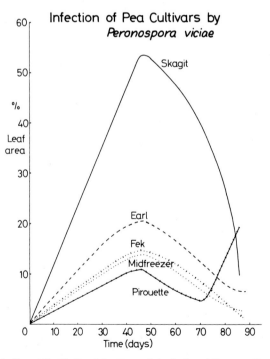

Fig. 2. From studies at the National Institute of Agricultural Botany, Cambridge, UK.

infected to the order of 10–20% leaf area in the early part of the season. The rate at which symptoms develop on individual cultivars varies, and this applies to both resistant and susceptible types. For cultivar evaluation, therefore, assessments need to be made at frequent intervals, every 7–10 days, from the seedling to mature plant stages even where only relative resistance ratings between cultivars are required (Dixon, Eucarpia Pea Workshop, John Innes Institute, Norwich, UK, 1976). Reports from the Soviet Union describe the testing of types of pea cultivars for resistance to *P. viciae* and also testing of vetches, *A. orobis* and *L. sativus* (Leokene and Timofeev, 1972; Golubev and Korner, 1975). Of some 200 lines tested, material from Bulgaria, Portugal, Sweden and the Russian line Zagerskaya exhibited resistance to *P. viciae*.

## 4. Genetics of host resistance

Studies by Matthews and Dow (1973) suggest that *Pisum* spp. possess three genetic mechanisms for resistance to *P. viciae*. These are: a single dominant resistance gene, as found in JI 85 (Lamprecht 1402), two complementary recessive resistance genes, as in JI 411 (Cobri) and a single resistance gene, as in JI 314 (Lamprecht 1363). Earlier it was suggested (Matthews and Dow, 1970) that there was a correlation between the round seeded character of some pea cultivars and resistance to *P. viciae*. So far as the expression of resistance is concerned accessions of cultivars Lincoln, Alaska and their derivatives, are susceptible while accessions of *P. abyssinicum* and *P. humile* are resistant (Matthews and Dow, 1971). Resistance has been identified in *P. elatius* and *P. humile* (von Heydendorff and Hoffman, 1978).

## 5. Pathogen virulence

A report (Anon., 1973) suggested that resistance in pea cultivars varied from site to site possibly indicating differences in pathogen virulence. But an analysis using fitted-constants of the data in Table 4 failed to demonstrate statistically significant differences in the relative resistance expressed by a wide range of cultivars over the years 1972–1977 (Table 6).

A racial classification system has however been developed by Hubbeling (1975) which indicates differences in the virulence complement of *P. viciae* isolates. By this system five physiological races of *P. viciae* are distinguished using five groups of pea cultivar differentials. Similarly work in Germany (von Heydendorff, 1977) based on the pathogenicity of six physiological races to three host differential sets indicates differences in virulence composition between *P. viciae* isolates. An analysis of the origins of the physiological races in Germany indicated that occasionally different races were found together in the same field. Isolates of different origin showed variation in conidial

TABLE 6

Analysis of resistance in pea cultivars to *P. viciae* using fitted-constants

| Year | Mean resistance of cultivars grown | Mean resistance after adjustment for cultivars not grown |
|---|---|---|
| 1972 | 6·12 | 6·51 |
| 1973 | 5·48 | 5·55 |
| 1974 | 6·16 | 6·02 |
| 1975 | 5·62 | 5·81 |
| 1976 | 6·00 | 6·09 |
| 1977 | 5·38 | 5·99 |

| | Analysis of variance | |
|---|---|---|
| | Mean square | Degrees of freedom |
| Cultivars | 8·13 | 29 |
| Years | 0·75 | 5 |
| Residual | 0·56 | 68 |
| Total | 2·72 | 102 |

Analysis based on 30 pea cultivars grown for 3 or more years, 1972–1977, Dixon, unpublished data.

and oospore size and the number of nuclei per conidium (von Heydendorff and Hoffman, 1978).

## 6. *Physiology of resistance*

The resistance of pea plants and of individual leaves to infection increases with age, decreasing as senescence begins (Mence and Pegg, 1971). Resistance is typified as a restriction of fungal growth and sporulation accompanied by a chlorotic reaction in the leaves. Systemic invasion follows infection of meristematic tissue and is induced by inoculation of the apical buds of young plants or the epicotyls and hypocotyls but not roots of germinating seedlings. Plants whose growth was retarded in Mence and Pegg's experiments showed increased resistance to systemic invasion. Pods were found to be externally invaded from sporangia rather than by mycelial growth through the peduncle and pedicel. Where sporangia infected the growing point all subsequent growth became systemically invaded but older parts of the host remained free of infection (Olofsson, 1966).

## M. Spread in field crops

Field infections tend to occur very sporadically and their origins are difficult to demonstrate (Mence, 1971). Spread of *P. viciae* within a crop requires a time equivalent to at least two sporangial generations with temperature and relative humidity suitable for sporulation and infection. In such environments spread is generally rapid but the host crop may develop resistance as it matures with new infections remaining as light foliar spotting. Numbers of sporangia in the air and numbers of new infections are related to both weather conditions and availability of inoculum sources. Within infected crops, development of systemic infections follow a more complex pattern than those developed from sporangial invasion. Mence (1971) found that plants with systemic invasion of the apex ceased growth and died quickly and, since seedlings are most susceptible to *P. viciae*, this caused a loss of stand at an early stage. Hosts with extensive lesions but no dwarfing symptoms were killed but this may have resulted from invasion by secondary foot rotting pathogens such as *Fusarium* spp. Some infected plants were able to recover producing new basal branches and even forming a full seed crop but this has little commercial value because it leads to uneven maturity within the whole crop.

## N. Yield effects

Early workers in USA (for example, Linford, 1928) suggested "*P. viciae* to be one of the most widespread and least important diseases of the pea". On the basis of a presence or absence survey 2·2% of fields were infected but mostly at trace to 2% level of infection with only sporadic systemic infections. Even so, death of the host or forced production of secondary shoots would lead to up to 8% loss of crop. At that time, and not fully investigated since, an important side effect of *P. viciae* invasion was thought to be subsequent invasion by "secondary parasites" such as *Ascochyta pisi*, *A. pinodella*, *Bacterium pisi* and *Cladosporium* spp. The occurrence of *P. viciae* in USA and its economic significance was reviewed by Campbell (1935). In Washington State up to 20% of pods might be culled due to downy mildew infection. It was also reported that "perhaps the greatest importance of the disease is in seed peas, in that oospores of the pathogen which are commonly found on seed from infested fields may serve as a source of infection to plants grown from such seed". Earlier it was also contended that autumn sown crops were those at most risk from *P. viciae* (Jones and Heald, 1932).

   Heald (1932) also reported that pod infections were causing up to 60% loss of crop in the western USA, but a survey by Walker and Hare (1942) in Wisconsin State though identifying large numbers of fields as infected, found the infection to be only at trace levels—the situation discovered by Linford (1929).

   In New Zealand (Anon., 1949, 1950) it was suggested that *P. viciae* was present in large numbers but caused little injury whereas in South Africa, *P. viciae* was noted as a destructive pathogen of crops on the Low Veldt area (Dyer, 1951). In Israel *P. viciae* was found to be present in the early stages of pea growth causing moderate to severe damage (Palti and Stettiner, 1957). In the same year pea crops on Jersey (Phillips, 1957) were totally destroyed by *P. viciae*. In Eire *P. viciae* is recognized as the most important pathogen of pea crops, giving rise to regular devastating epidemics with up to 95% infection of susceptible cultivars such as Kelvedon Wonder (Ryan, 1964). In Sweden losses of up to 30% of pea crops for canning and freezing due to *P. viciae* were reported by Olofsson (1966). In contrast to the results obtained by Walker and Hare (1942) epiphytotic downy mildew occurred in Wisconsin (Hagedorn, 1974) as a single isolated event in which 60–85% of pods were destroyed leading to considerable loss of crop. This was attributed to the contraction of rotations in order to accommodate pea crops more often on the same land and to unusually cool climatic conditions which favoured *P. viciae*. Most recently a survey of German pea crops over the period 1972–1975 established that *P. viciae* infections occurred regularly but did not cause heavy losses and were unable to generate epidemics (von Heydendorff, 1977).

Critical attempts to evaluate the importance of *P. viciae* are few. It appears to have had localized but occasionally severe effects up to the introduction of vining peas in the 1950s. This advent has increased the importance of *P. viciae* due to its tendency to cause irregular maturity and its effect on palatability. Thus Hubbeling (Eucarpia Pea Workshop, John Innes Institute, Norwich, UK, 1976) reported that seeds from downy mildew-infected pods were harder, had lower germinability and bitter flavour. He also thought that overwintered peas were more vulnerable to infection than spring sown crops. But those crops sown in February–March in UK are also more at risk from *P. viciae* than those drilled in April–May (Dixon, unpublished data). Although numbers of infected plants in the field can reach 100% the area of foliage infected per plant was rarely found to be greater than 15% (Pegg and Mence, 1972). Even at the highest levels of infection these authors failed to detect any measurable effects on haulm size or fresh weight of seeds. Where secondary systemic infections were found these would reduce yield of fresh seeds, due to loss in number of podding nodes and in pod size and haulm fresh weight. Where pods were infected this correlated with some reduction in pod yield. The main effects of *P. viciae* infection were to cause variable ripening, whereby tenderometer readings within individual crops became an unreliable indicator of the crop's growth stage, rendering it useless for processing. It is this aspect which over the last 10 years has elevated *P. viciae* to the position of the major pathogen of European vining pea crops. Although only a small proportion of plants within a crop may be infected the effect on uniform ripening means that the whole field may be rejected by the contract

510    G. R. DIXON

processing company leading to a total crop loss for the grower. The decision
to reject a crop is made by the processing company fieldsman and no records
of levels of rejection are made publicly available, therefore, since the rates
of rejection vary according to seasonal variability of peas for vining, it is
difficult to evaluate the full financial impact of *P. viciae*. Availability of
systemic fungicides for control of *P. viciae* may permit quantitative evaluation
of yield losses caused by this pathogen. Using metalaxyl as a seed dressing
treatment Brokenshire (1980) found that with foliar infections of 50% leaf
area infected yield losses due to *P. viciae* were of the order of 25%.


## O. Chemical control

Chemicals such as sulphur dust, copper dusts, lime sulphur and 3-5-50
Bordeaux mixture were ineffective in controlling *P. viciae*. Malachite green
was, however, found to give some control (McWhorter and Pryor, 1937).
Field control was also achieved by applications of calcium nitrate at 500
kg ha$^{-1}$ (Anon., 1961). But tests of a range of 16 chemicals failed to give
either protection against or control of *P. viciae* (King and Gane, 1965).
Herbicidal spraying with amine or ammonium salts of 2,4-di-nitro-ortho-
secondary butyl phenol were found to kill systemically infected pea seedlings
(Gent, 1966). Indeed discontinuation of use of this herbicide may have been
a contributory factor in the upsurge of importance of this pathogen. Sprays
of 2–3 kg maneb or zinc plus maneb per ha at the onset of flowering reduced
by half the numbers of diseased pods (Olofsson, 1966). But in Sweden, at
least, disease intensity fluctuated to such an extent that use of chemical sprays
was only economic in two years out of five. Trials mentioned by Mukerji
(1975) demonstrated the efficacy of maneb, mancozeb, dithane M45 and
thiram for control of *P. viciae*. Trials in Russia applying two sprays of 0·75%
zineb at complete shoot development and flower formation were most
effective for control of *P. viciae* (Kirik and Koshevs'kii, 1978). The importance
of seed treatment as a means of control was emphasized by Mukerji (1975)
who suggested using granosan at 4 kg tonne$^{-1}$ of seed. Seed disinfection
with thiram, at 6 kg tonne$^{-1}$ of seed, or phenthiuram, at 4 kg tonne$^{-1}$ of
seed, was advocated by Turukina and Temnokhud (1977). Hot water treatment for 25 min at 50°C was advised by Chupp and Sherf (1960).

Acylalanine fungicides have recently been developed which possess systemic
activity against Phycomycete pathogens (Urech *et al.*, 1977). Work with
metalaxyl and aluminium ethyl phosphonate used as pea seed dressings has
given significant control of *P. viciae* in some trials (Anon., 1979; Brokenshire,
1980) but only limited effects in others King, 1978; 1979a). Trials to control
*P. viciae* on broad bean (*V. faba*) have shown that sprays of metalaxyl alone
or with mancozeb reduced the number of plants with moderate symptoms

from 50% to 10% and those with severe symptoms from 30% to nil (King, 1979b).

### P. Husbandry control

For efficient control of *P. viciae* it is essential to remove and burn all infected debris which will contain oospores. This is precisely the reverse of the effect achieved by mobile pea viners which macerate the haulm leaving behind an ideal inoculum for further spread of *P. viciae*. Deep tillage and extended crop rotations are also useful techniques for combating this pathogen. Lower infections were found where pea crops succeeded sugar beet or maize rather than winter wheat (Kirik and Koshevs'kii, 1974). Late autumn, and deeper, sowings were also correlated with more severe infections. In order to control *P. viciae* on broad bean (*V. faba*) Jamoussi (1968) advocated at least a 3-year rotation with good soil drainage and use of fungicidal sprays. An essential means of controlling seed borne or seed carried infections of *P. viciae* would be to use only seed raised in arid areas where this pathogen is incapable of developing.

## II. Phaseolus Beans

Several *Phytophthora* spp. are reported as pathogens of *Phaseolus* beans e.g. *P. parasitica* (*P. terrestris*) on snap, dry and kidney beans (*P. vulgaris*) and *P. phaseoli* on lima bean (*P. limensis*). In the literature these pathogens are given the vernacular description of "downy mildews" (Zaumeyer and Thomas, 1957). Since this present work is restricted to *Peronospora* spp. no further comment will be made on these pathogens. There are, however, a few reports of *Peronospora manshurica* as a pathogen of *Phaseolus* spp. This pathogen is dealt with in detail elsewhere (Chapter 24), consequently the following relates solely to the *P. manshurica–Phaseolus* combination. The first report of this combination was made in Anon. (1968a) which cites French (*P. vulgaris*) and runner (*P. coccineus*) beans as hosts to *P. manshurica*. The authority for this reference is given as Sherwin *et al.* (1948) who described seed treatment of soybean (*Glycine max*) to control *P. manshurica* but made no mention of *Phaseolus* spp. A more substantial reference to the ability of *P. manshurica* to infect *Phaseolus* spp. was made by Kenneth *et al.* (1975) following field observations in Israel. The only other worker to discuss this combination was Riggle (1977). In studies of *P. manshurica* on a range of resistant, semi-resistant and susceptible soybean cultivars this worker included what was thought to be a non-host *P. vulgaris* cv. Tendercrop.

Contrary to expectations symptoms developed on the inoculated foliage of Tendercrop. Symptom expression was slight when compared with that found on soybean cv. Acme which has moderate resistance to *P. manshurica* and the pathogen did not become systemic. Some lesions developed small necrotic areas at the centre which is a symptom not found with soybean hosts. Sporulation could be induced on Tendercrop but sporangium production was sparse. Hyphal growth within inoculated leaves was slow and many hyphae ceased to develop and collapsed. Lesion growth was inhibited by the presence of leaf veins but other, undefined, factors were thought to also influence lesion expansion. Numbers of haustoria produced in Tendercrop were similar to those found in susceptible soybeans. It is postulated by Riggle (1977) that there is only a short period in which *Phaseolus* beans are susceptible to *P. manshurica* and this is taken to explain why field infections have not been reported more frequently.

## References

Allard, C. (1971). *Phytiat Phytopharm.* **20**, 23–30.
Anon. (1936). *Ann. Epiphyt. n.s.* **ii**, 381–422.
Anon. (1943a). *Agric. Gaz. N.S.W.* **Liv**, 116–120.
Anon. (1943b). *Public. Estac. Fitopat. Agric. Coruna* **23**, 21–57.
Anon. (1949). *23rd Ann. Rep. Dept. Sci. Indust. Res. N.Z. for 1949.*
Anon. (1950). *24th Ann. Rep. Dept. Sci. Indust. Res. N.Z. for 1950.*
Anon. (1954). *27th and 28th Ann. Reps. Dept. Sci. Indust. Res. N.Z. for 1953 and 1954.*
Anon. (1956). *Ann. Rep. Dept. Agric. Tanganyika for 1955 (Part II).*
Anon. (1960). *Ann. Rep. Dept. Agric. Kenya (Part II).*
Anon. (1961). *Ann. Rep. Instituut voor Plantziektenkundig Onderzoek, Netherlands for 1960.*
Anon. (1968a). *Plant Host–Pathogen Index to volumes 1–40 (1922–1961) Rev. Appl. Mycol.*
Anon. (1968b). *Ann. Rep. Pea Growing Res. Org. for 1967.*
Anon. (1969). *Ann. Rep. Pea Growing Res. Org. for 1968.*
Anon. (1970). *Ann. Rep. Pea Growing Res. Org. for 1969.*
Anon. (1971). *Ann. Rep. Pea Growing Res. Org. for 1970.*
Anon. (1973). *Ann. Rep. Pea Growing Res. Org. for 1972.*
Anon. (1975). *Ann. Rep. Processors and Growers Res. Org. for 1973.*
Anon. (1976a). *Ann. Rep. Processors and Growers Res. Org. for 1974.*
Anon. (1976b). *Ann. Rep. Processors and Growers Res. Org. for 1975.*
Anon. (1978). *Ann. Rep. Processors and Growers Res. Org. for 1977.*
Anon. (1979). Research report for 1978: *Horticulture*, An Foras Taluntais Dublin, Eire 118 pp.
de Bary, A. (1863). *Ann. Sci. Nat. Bot.* **4** Ser. 20, 112–113.
Beaumont, A. (1951). *Gardeners Chronicle* Series 3 **129**, 132.
Berkeley, M. J. (1846). *J. R. Hort. Soc.* **1**, 9–35.
Bertossi, E. O. (1963). *Rev. Univ. Santiago* **48**, 41–56.
Bland, B. F. (1971). "Crop Production: Cereals and Legumes". Academic Press, London and New York.
Brokenshire, T. (1980). *Tests for Agrochemicals and cultivars* **1**, 34–35.

Bundy, J. W. (1971). *In* "Potential Crop Production" (P. F. Wareing and J. C. Cooper, Eds.) 351–361, Heinemann, London.
Campbell, L. (1935). *Bull. Wash. State Agric. Exp. Stn.* No. 318.
Caspary, R. (1855). *Monatsber. K. Preuss. Akad. Wiss. Berl.* p. 333.
Chupp, C. and Sherf, A. F. (1960). "Vegetable Diseases and their Control". Ronald Press, New York USA.
Ciccarone, A. (1953). *Boll. Staz. Pat. Veg. Roma* Ser 3, **10** (1952), 31–37.
Connors, I. L. and Saville, D. B. O. (1952). *30th Ann. Rep. Canadian Plant Disease Survey for 1950.*
Dixon, G. R. (1976). *Ann. Appl. Biol.* **84**, 283–287.
Dyer, R. A. (1951). *Farming in South Africa* **26**, 488–490.
Fraymouth, J. (1956). *Trans. Br. Mycol. Soc.* **39**, 79–107.
Gäumann, E. (1923). *Beitrage Kryptogamenflora Schweiz.* **5**, 1–360.
Geard, I. D. (1961). *Tasm. J. Agric.* **32**, 132–143.
Gent, G. P. (1966). *Misc. Publ. Pea Growers Res. Org. No. 17.*
Gent, G. P. and Bingham, R. J. B. (1978). *Ann. Rep. Processors and Growers Res. Org.* (for 1977), **29**.
Glasscock, H. H. (1963). *Pl. Path.* **12**, 91–92.
Golubev, A. A. and Korner, A. I. (1975). *In* "Immunity of Crops" (V. I. Krivchenko Ed.) *Byulletin Usesoyuznogo Nauchno Issledovatel' skogo Instituta Rastenievodstva imeni N.I. Vavilova No. 50.*
Hagedorn, D. J. (1974). *Pl. Dis. Reptr.* **58**, 226–229.
Heald, F. D. (1932). *Wash. State Agric. Exp. Stn. Ann. Rep.* **42** (Bull. No. 275).
Henrick, J. O. (1935). *Tasm. J. Agric. N.S.* **vi**, 28–29.
Hickey, E. L. and Coffey, M. D. (1977). *Canad. J. Bot.* **55**, 2845–2858.
Hubbeling, N. (1975). *Meded. Fac. Landbouw, Rijks Gent.* **40**, 539–543.
Jamoussi, B. (1968). *Archs. Inst. Pasteur Tunis.* **45**, 117–127.
Jones, L. K. and Heald, F. D. (1932). *Wash. Agric. Exp. Stn. Bull. No. 275.*
Kenneth, R., Palti, J., Frank, Z. R., Anikster, Y. and Cohn, R. (1975). *Volcani Centre (1975) Special Publication No. 36.*
King, J. M. and Gane, A. J. (1965). *Proc. 3rd Br. Insectic. Fungic. Conf.* 162–168.
King, J. M. (1978). *Ann. Rep. Processors and Growers Res. Org. for 1977* p. 18.
King, J. M. (1979a). *Ann. Rep. Processors and Growers Res. Org. for 1978* p. 18.
King, J. M. (1979b). *Ann. Rep. Processors and Growers Res. Org. for 1978* p. 22.
Kirik, N. N. and Koshevs'kii, I. I. (1974). *Zakhist. Roslin No.* 20 92–99.
Kirik, N. N. and Koshevs'kii, I. I. (1976). *Zaschuta Rastenii No. 4.*
Kirik, N. N. and Koshevs'kii, I. I. (1977). *Mikologiya Fitopatologiya* **11**, 332–335.
Kirik, N. N. and Koshevs'kii, I. I. (1978). *Khimiya Sel'skom Khozyaistve* **16**, 71–75.
Koshevs'kii, I. I. and Kirik, N. N. (1979). *Mikologiya Fitopatologiya* **13**, 46–48.
Kirik, N. N., Sidorchuz, V. I. and Koshevs'kii, I. I. (1975). *Selektsiya Semenovodstvo* No. 9, 29–30.
Leokene, L. V. and Timofeev, A. A. (1972). *Selektsiya Semenovodstvo* No. 6, p. 79.
Linford, M. B. (1928). *Pl. Dis. Reptr. Supplement* **59**, p. 91.
Linford, M. B. (1929). *Pl. Dis. Reptr. Supplement* **67**, p. 8.
Lindquist, J. C. (1939). *Physis B. Aires* **xv**, 13–20.
McDonald, J. (1937). *Rep. Dept. Agric. Kenya for 1936* **ii**, 1–12.
McWhorter, F. P. and Pryor, J. (1937). *Pl. Dis. Reptr.* **21**, 306–307.
Marras, F. (1963). *In* "Current state of knowledge of bacterial and fungal diseases of garden plants in Sardinia". Inst. Pat. Veg. Univ. Sassari Italy.
Mathur, R. S. (1954). *Agric. Anim. Husb. Utar Pradesh* **5**, 24–28.
Matthews, P. and Dow, P. (1970). *Ann. Rep. John Innes Institute, Norwich for 1969*, p. 18.

Matthews, P. and Dow, P. (1971). *Ann. Rep. John Innes Institute, Norwich for 1970*, p. 43.
Matthews, P. and Dow, P. (1973). *Ann. Rep. John Innes Institute, Norwich for 1972*, 38–39.
Matthews, P. and Dow, P. (1974). *Ann. Rep. John Innes Institute, Norwich for 1973*, 32–33.
Matthews, P. and Dow, P. (1977). *Ann. Rep. John Innes Institute, Norwich for 1976*, 37–38.
Melhus, I. E. (1931). *Iowa State Coll. J. Sci.* **5**, 185–188.
Mence, M. J. (1971). *PhD Thesis London Univ. UK.*
Mence, M. J. and Pegg, G. F. (1971). *Ann. Appl. Biol.* **67**, 297–308.
Moore, W. C. (1959). "British Parasitic Fungi", Cambridge University Press, London.
Mukerji, K. G. (1975). *Descriptions of pathogenic Fungi and Bacteria No. 455*. Comm. Mycol. Inst., Kew, UK.
Muntanola, M. (1957). *Rev. Agron. Norocste Argent.* **1**, 283–299.
Olofsson, J. (1966). *Pl. Dis. Reptr.* **50**, 257–261.
Palti, J. and Stettiner, M. (1957). *Advisory Leaflet Agric. Consultants Israel No. 20.*
Pegg, G. F. and Mence, M. J. (1970). *Ann. Appl. Biol.* **66**, 417–428.
Pegg, G. F. and Mence, M. J. (1972). *Ann. Appl. Biol.* **71**, 19–31.
Peresypkin, V. F., Kirik, N. N. and Koshevs'kii, I. I. (1977). *Vestnik Sel'skokhozyaistevnoi Nauki* **8**, 25–31.
Phillips, D. H. (1957). *Rep. Mycol. Dept. Rep. States Jersey for 1955*, 31–42.
Ramsey, G. B. (1931). *Pl. Dis. Reptr.* **15**, 52–53.
Riggle, J. H. (1977). *Can. J. Bot.* **55**, 153–157.
Ryan, E. W. (1964). *Res. Rep. Dept. Hort. Agric. Inst. Dublin for 1963*, 105–110.
Ryan, E. W. (1971). *Irish J. Agric. Res.* **10**, 315–322.
Samuel, G. (1931). *J. Dept. Agric. South Australia* **34**, p. 746.
Savulescu, T. (1948). *Sydowia* **2**, 255–307.
Sherwin, H. S., Lefebvre, C. L. and Leukel, R. W. (1948). *Phytopathology* **38**, 197–204.
Singh, H. (1979). PhD Thesis, Newcastle-upon-Tyne University, UK.
Smith, W. G. (1884). "Diseases of Field and Garden Crops". MacMillan, London.
Snyder, W. C. (1934). *Phytopathology* **24**, 1358–1365.
Sorauer, P. (1924). *Handb. Pflkrankh*. Fünfte auflage Band 1. Die nicht parasitaren krankheiten, Paul Parey Berlin.
Sydow, H. (1921). *Mycotheca germanica*. Fasc xxix–xxxiv (No. 1401–1800). *Ann. Mycol.* **19**, 133–144.
Thind, K. S. (1942). *J. Ind. Bot. Soc.* **XXI**, 197–215.
Turukina, N. F. and Temnokhud, M. P. (1977). *Zakhist Roslin No. 24*, 77–80.
Urech, P. A., Schwinn, F. and Staub, T. (1977). *Proc. 1977 Br. Crop Protect. Conf.* 623–631.
von Heydendorff, R. C. (1977). PhD Thesis Fakaltat für Gartenbau und Landeskultur den Technischen Universitat Hannover, Federal Republic of Germany.
von Heydendorff, R. C. and Hoffman, G. M. (1978). *Z. Pflkrank Pflanzenschutz* **85**, 561–569.
Walker, J. C. and Hare, W. W. (1942). *Res. Bull. Wisconsin Agric. Exp. Stn. No. 145.*
de Weille, G. A. (1963). *Neth. J. Pl. Path.* **69**, 115–131.
Weimar, J. L. (1940). *USDA Circular No. 565.*
Zaumeyer, W. J. and Thomas, H. R. (1957). *USDA Tech. Bull. No. 868.*

Chapter 24

# Downy Mildew of Soybean

## J. M. DUNLEAVY

*Department of Botany and Plant Pathology, Iowa State University, Ames, Iowa, USA*

## I. Introduction

The soybean is believed to have originated in Eastern Asia. The wild soybean (*Glycine ussuriensis*) occurring in China, Manchuria, and Korea, is the probable progenitor of *Glycine max* (L.) Merrill, the cultivated soybean. The soybean was an important food crop in China even before the beginning of written history (Piper and Morse, 1923). Commercial production and industrial processing of soybeans began comparatively recently. Japan began importing soybean oil cake in 1894. Soybeans were being crushed for oil in the United States by 1922, but the primary use for the crop was for hay until 1940 (Probst and Judd, 1973). By 1966 less than 2% of the crop was used for forage. Today soybeans are the leading source of the world's supply of vegetable oil. Soybean protein is of excellent quality nutritionally, and the amino acid distribution very closely approximates that required in animal diets (Krober, 1965). In recent years very substantial increases in soybean production have occurred in Brazil, Columbia, Mexico, the United States and the USSR.

Downy mildew of soybeans is caused by *Peronospora manshurica* (Naoum.) Syd. ex Gaum. According to Athow (1973), the disease has been reported from 22 countries and probably is co-existent with the crop. Oospores of the fungus are transmitted on seed, resulting in very effective dissemination. The leaf spot phase of the disease under average weather conditions in Iowa reduced seed production by 8%. Frequent dews can encourage more severe disease development and greatly increase the percentage of seeds infested with oospores. Such seeds are smaller and this phase of the disease can reduce seed yield by an additional 6% (Dunleavy *et al.*, 1966).

## II. Occurrence

Soybean downy mildew was first observed in Kashmir in 1908 by Sydow *et al.* (1912). Wei and Hwang (1939) listed *P. manshurica* among the fungi deposited in the Mycological Herbarium of the University of Nanking between 1924 and 1937. The disease was noted as being of economic importance in China by Sun (1958), and in Manchuria by Miura (1922). After 1950, soybean production in Taiwan increased, and there are several reports of the disease being of significance on soybeans there (Chu and Chuang-Yang, 1961; Han, 1959; and Sawada, 1959).

Downy mildew was first observed on soybeans in the United States by Haskell and Wood (1923). Lehman and Wolf (1924) compared a fungus found on soybean leaves in North Carolina with *P. manshurica* on leaves obtained in Manchuria and observed that the fungi were the same. The disease now occurs throughout the soybean-growing regions of the United States (Athow, 1973; Dunleavy, 1971). The disease is also a problem in the USSR and was first reported as occurring in the Ukraine (Zaianchkovskaia, 1938).

The disease was not observed in Europe until 1942 when it was reported from Denmark (Anon., 1942). It has been reported from Sweden (Gustavsson, 1959), Latvia (Serzane, 1962), England (Kingsley, 1960), Hungary (Voros and Molnar, 1958), Czechoslovakia (Danko, 1962), Romania (Savulescu, 1948), and Yugoslavia (Cuturilo, 1952). The most serious outbreaks reported in Europe have occurred in Yugoslavia (Arsenijevic and Kostic, 1960; Spasic, 1961).

Downy mildew was observed on soybeans in South Africa as early as 1924 (Doidge, 1924), and in India (Kashmir) in 1938 (Mundkur, 1938).

The effectiveness of transportation of oospore encrusted soybean seeds as a means of disease dissemination is well illustrated by reports of *P. manshurica* infecting soybean plants on relatively small islands such as the Ryukyus (Nuttonson, 1952), and Bermuda (Waterson, 1939).

## III. Symptoms

Downy mildew is characterized in its early stages by indefinite, yellowish-green areas on the upper surface of the leaves. A fluffy, grey fungus growth develops on the lower side of these diseased areas (Fig. 1), which is composed primarily of conidiophores and conidia (Fig. 5). Athow (1973) pointed out that this is the best diagnostic character of the disease, and readily differentiates downy mildew from brown spot (*Septoria glycines*) and frogeye

Fig. 1.   Sporulating downy mildew lesions on the lower surface of a soybean leaflet infected by *P. manshurica*.

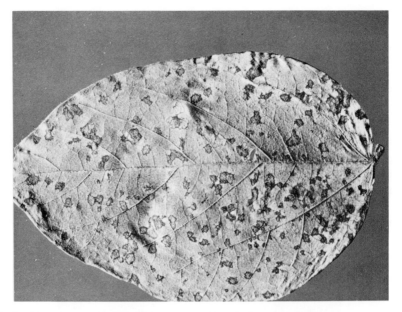

Fig. 2.   Necrotic downy mildew lesions on a soybean leaflet infected by *P. manshurica*.

Fig. 3.   Soybean seed encrusted with oospores of *P. manshurica*.

leafspot (*Cercospora sojina*), which it may resemble at times. Six to eight days after infection, leaf lesions begin to turn yellow, and no longer sporulate on the lower surface, then two to three weeks after infection the lesions turn brown (Fig. 2).

Pod infections may occur and not be evident externally (Sinclair and Shurtleff, 1975). However, when infected pods are examined at maturity the interior pod walls and seed coats are usually covered with whitish masses of mycelium and oospores (Fig. 3). When such seeds are planted, seedlings are infected systemically, and one or both cotyledons may be infected (Fig. 4) as well as one or both unifoliate leaves and most subsequently formed trifoliate leaves. Systemically-infected plants are stunted and leaves are small, thick, leathery, rugose, and tend to be cupped down. These leaves also form the typical downy fungus growth on their lower surface.

## IV. The Pathogen

### A. Morphology

The first recorded observation of downy mildew on soybeans was that of Miura (1922) who identified the causal organism as *P. trifoliorum* var. *manshurica* (Naoumoff, 1914) in Manchuria in 1921. *P. trifoliorum* var. *manshurica* was advanced to a species by Gaumann (1923) and became *P. manshurica*.

Fungus hyphae are coenocytic and range from 7 to 10 $\mu$m wide. During sporulation the fungus emerges from stomata on the lower surface of the leaf and forms conidiophores ranging from 240 to 984 $\mu$m tall, and from 5 to 9 $\mu$m in diameter. Conidiophores may emerge singly or in groups from a stoma. Sterigmata are nearly straight and range from 9 to 13 × 2 to 3 $\mu$m. The hyaline conidia are elliptical (Fig. 5) and have an average size of 24 × 20 $\mu$m. Conidia germinate in water by means of a germ tube which usually develops near the narrow circumference of the conidium (Lehman and Wolf, 1924).

Oospores are globose with the exterior cell walls ranging from smooth to reticulated. The colour of the cell walls ranges from subhyaline through yellow to brown, and the oospores measure from 20 to 36 $\mu$m in diameter (Lehman and Wolf, 1924). Reticulations are difficult to observe, but when the spore wall is stained with cotton blue they are more easily seen. Oospores germinate by means of a germ tube (Dunleavy and Snyder, 1962).

Fig. 4.   *P. manshurica* sporulating on the surface of a soybean cotyledon which developed from an oospore encrusted seed (from Pederson, 1961b).

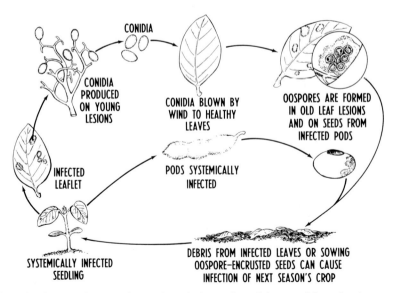

Fig. 5.   Life cycle of *P. manshurica*, the causal organism of downy mildew of soybean.

## B. Life Cycle

The life cycle of *P. manshurica* begins with the infection of germinating seeds by fungus oospore germ tubes. Oospores may contact the seed either from seed coat encrustations resulting from infected pods, or from debris from infected plants (Fig. 5). Lehman (1953a) studied the effect of temperature on systemic infection of seedlings and observed that the rapidity of germination of oospore encrusted seeds affected the percentage of seedlings systemically infected. Encrusted seeds planted in a cold soil (13°C) gave rise to 40% seedling infection, but no infection was obtained in soils at 18°C and above. Either the oospores on the seeds failed to infect the seedlings at 18°C, or the seedlings outgrew the fungus. Dunleavy and Snyder (1962) observed up to 40% systemic infection among volunteer soybean plants in fields that had contained *Peronospora*-infected plants the previous year. Planting seed after the soil has reached a temperature of at least 18°C appears to offer a means of controlling systemic infection and thus reducing the amount of early inoculum in the field.

Until 1942, downy mildew was considered to be strictly a foliage disease, but in that year Johnson and Lefebvre (1942) described seed encrusted with oospores. Atkinson (1944) later suggested the possible systemic nature of the disease which was confirmed by Koch and Hildebrand (1946), and Jones and Torrie (1946), based on histological observations. In a later study of systemic infection, Hildebrand and Koch (1951) noted that growth of the fungus, though not immediately concurrent with that of the plant, finally pervades all plant parts. Some systemically infected plants died prematurely, others exhibited spindly growth and produced few seeds, while still others, although exhibiting foliar symptoms, grew vigorously and produced an amount of infected seed which nearly equalled that of healthy plants. They obtained 37% systemically infected plants in the greenhouse and 13% in the field from oospore encrusted seeds.

Systemically infected seedlings may sporulate in 1 to 2 weeks after emergence, depending upon temperature. The general infection pattern is for all leaves from the cotyledons upwards to become infected but occasionally cotyledons or unifoliate leaves show no infection, and the first symptoms appear on the first trifoliate leaf. Abundant moisture with long periods of high relative humidity and cool nights are the prerequisites for abundant sporulation of *P. manshurica*. Sporulation at 23°C can occur 5–6 days after leaf infection, depending upon other environmental factors. Moisture in the form of rain inhibits sporulation while dew, induces excellent sporulation (Pederson, 1961a).

Conidia produced on systemically infected plants are transported by wind to healthy leaves where they may cause infection provided conditions for conidial germination are favourable. Under ideal conditions for secondary infection, the fungus increases rapidly on susceptible cultivars. Only 8–10 days are required from infection, to conidia production. The disease can thus spread dramatically in areas of intensive soybean production, but is highly dependent upon environmental conditions (Dunleavy, 1969).

P. manshurica overwinters as oospores in leaves and other tissues of infected plants, and as oospores on seed coats of seed produced on systemically infected plants (Jones and Torrie, 1946; Hildebrand and Koch, 1951; McKenzie and Wyllie, 1971). Ploughed-in infected plant debris is a source of infection when healthy seed is sown the following year; and oospore encrusted seeds are an effective means of disease dissemination when seeds are shipped from one production area to another.

## C. Physiological specialization

Physiological specialization in P. manshurica was first demonstrated by Geeseman (1950), who differentiated three collections of the fungus as races 1, 2, and 3 by their reaction on Richland, Illini, and any one of five other cultivars. Lehman (1953b) reported race 4 from North Carolina, and later (Lehman, 1958) proposed an increase in the number of differential cultivars from four to ten for use in describing races 3A, 5, 5A, and 6. Grabe and Dunleavy (1959) established a set of 14 cultivars for differentiating race 7, which was found in Illinois; and race 8 which was found in Indiana, Missouri, and three locations in Iowa. They found race 1 in Arkansas and race 2 in Iowa, and redefined these races on the 14 differential cultivars.

Dunleavy (1971) surveyed the races of P. manshurica in the United States. He obtained 247 samples of seeds from the principal soybean growing regions, and found that 73% of the samples contained oospore encrusted seeds. No encrusted seeds were found in samples from the drier regions (Kansas, Nebraska, and Texas). When 10 or more oospore encrusted seeds were obtained from a sample, they were sown in pots of soil held at 18–20°C. When fewer than 10 oospore encrusted seeds were available, oospores were scraped from the seeds, mixed with 10 parts of finely ground talc, and applied to the epicotyls of germinating soybeans of the same cultivar as that from which the oospores were obtained (Pederson, 1958). Use of this method usually resulted in 5–10% systemic infection. Conidia of the P. manshurica isolates thus propagated, were tested for disease reaction on 72 soybean varieties (Dunleavy, 1970) representing a wide cross-section of germplasm. Fifteen additional races were isolated and studied. These races were assigned the numbers 9 to 23. Five of the 14 cultivars used previously to identify races were eliminated, and two cultivars were added. The cultivars used were: Pridesoy, Norchief, Mukden, Richland, Roanoke, Illini, S-100,

Palmetto, Dorman, Kabott, and Ogden. This set of 11 differential cultivars is currently being used for race identification.

Race 8 was encountered most frequently in the North Central States, and occurred at 40% of the locations from which seed samples were obtained. Races 1, 10, and 12 were each found at seven locations; races 2, 18, and 23 at five locations each, and the remaining 14 races combined, at 19 locations. Race 10 was the race most frequently encountered in the southern portion of the USA. An effort was made to identify as many isolates as possible from Missouri so that more accurate information might be obtained on race distribution within a smaller area. Isolates were obtained from, 30 locations and 11 races were identified. Race 8 was identified from 20% of the locations, and races 12, 18, 23 occurred at 17%, 13%, and 13% of the locations, respectively.

In a later survey covering only the north central portion of the United States, Dunleavy (1977) reported that 60 out of 80 isolates of *P. manshurica* tested were of previously unreported races. These isolates comprised nine new races (24 to 32). Race 25 occurred most frequently and was most widely distributed. Only 25% of the isolates collected were of races previously described, namely 14, 18, 20, and 22.

## V. Pathological Physiology

Light, temperature, humidity, rainfall, dew and wind are involved in sporulation, spore discharge, dispersal, germination, and re-infection. Pederson (1961a) observed that size of leaves, number of conidia produced per unit area of infected leaf surface, and number of days during which sporulation occurred were directly proportional to the duration of daily exposure of plants to light. Conversely, the percentage of leaf area infected was inversely proportional to duration of light. Oospores developed in infected leaf tissue simultaneously with the onset of chlorosis (Fig. 6). They developed earlier and in greater numbers in leaves given short daily exposures to light, than in leaves given long daily exposures.

Wyllie and Williams (1965) reported that daytime temperatures between 20 and 30°C after inoculation had no significant effect on the development of lesions caused by *P. manshurica* on Clark soybeans. They also found that temperatures before inoculation did not affect lesion development. McKenzie and Wyllie (1971) found that there was no sporulation on lesions at temperatures below 10°C, or at 30°C or above.

There have been several studies of the effect of temperature on conidial viability. Pederson (1961b) found that conidia remained viable for at least 19 days stored either aerobically or anaerobically in water at 1°C, and for

524     J. M. DUNLEAVY

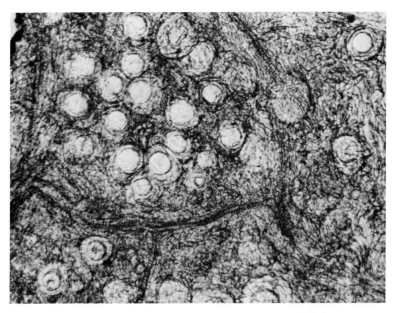

Fig. 6.  *P. manshurica* oospores in soybean leaf cells (from Pederson, 1961b).

Fig. 7.  Entangled conidiophores of *P. manshurica* on the surface of a soybean leaflet after being exposed to a dry atmosphere (from Pederson, 1961b).

at least 14 days stored intact on conidiophores of sporulating leaves at $-14$ and $1°C$. Riggle and Fisher (1973) stored conidia in liquid nitrogen for 108 days and obtained 34% germination *in vitro*.

Concentration of conidia is an important factor in controlling germination. Spores in high conidia concentrations germinate poorly, but germination is improved markedly at low concentrations. Pederson (1960) found that this phenomenon was due to an inhibitor associated with the conidia, which diffused readily in water, was lost after exposure to air for a few days, but accumulated when the conidia were stored anaerobically. An oospore-produced inhibitor active against conidia was reported by Dunleavy and Snyder (1962).

Pederson (1961b) observed that conidial discharge was associated with desiccation of conidiophores. As conidiophores dried, the bases collapsed into flat, ribbon-like forms and rotated in a counter-clockwise direction. Conidiophores often became entangled (Fig. 7), conidiophore tips interlocked, and conidia were rubbed or snapped off with a spring-like action.

Several biochemical changes have been noted in diseased plants. Milliken and Wyllie (1966) observed that 10 days after inoculation, both resistant and susceptible cultivars had a greater than 50% increase in calcium content as compared to the control plants. Milliken *et al.* (1965) found that there was a reduction of nucleic acid synthesis in infected plants, but Tani and Naito (1956) observed that healthy and infected plants did not differ in nitrogen content.

## VI. Pathological Histology

Riggle and Dunleavy (1974) studied the histology of leaf infection of susceptible and resistant soybeans by *P. manshurica*. By 12 h after inoculation, most germinated conidia had developed appressoria on walls of epidermal cells. Penetration pegs forced their way between epidermal cells and into the air spaces between mesophyll cells. Some germ tubes failed to penetrate the epidermis at cell wall junctures. Nearly all appressoria formed on the lower epidermis, and about 10% of the penetrations entered stomata without forming appressoria. The infection process in susceptible and resistant cultivars was similar at 12 h; but by 24 h after inoculation, more hyphal growth and haustoria had formed in the susceptible, than in the resistant cultivar. The fungus made rapid progress in the susceptible cultivar up to 51 h after inoculation. Hyphae frequently grew vertically between palisade cells, forming haustoria in all four adjacent cells. The fungus made slow progress on the resistant cultivar, and at 36 h the fungus in the susceptible tissue penetrated 40% farther than in the resistant tissue. There were 3·1 haustoria

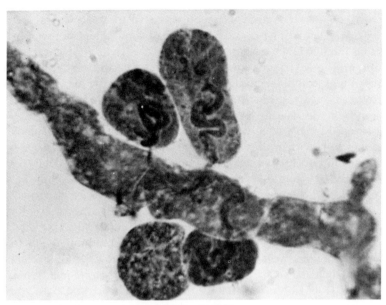

Fig. 8.   Haustoria of *P. manshurica* in soybean leaf cells.

per penetration in susceptible leaves, whereas resistant leaves had only 0·2 haustoria per penetration. After 36 h, there were only 2·7 haustoria mm$^{-1}$ of mycelium in the resistant leaves, but 14 mm$^{-1}$ in the susceptible leaves. Fungus growth of mycelium in the resistant leaves, but 14 mm$^{-1}$ in the susceptible leaves. Fungus growth in resistant leaves stopped at this time. Haustoria vary in size from 1·0 to 3·5 $\mu$m in diameter, and are usually cylindrical, long, curving, tenuous, and branch up to five times within a cell (Ikata and Yamauti, 1941; Peyton and Bowen, 1963; and Riggle and Dunleavy, 1974) (Fig. 8).

In another study infection in young and old leaves was compared (Riggle, 1974). Until the 48th h after inoculation the fungus grew most rapidly on suscepible old leaves, but after 48 h the fungus grew more rapidly on young leaves. Thus, ultimately, the older leaves had the smaller lesions.

Peyton and Bowen (1963) studied the fine structure of hyphae, haustoria, and infected host cells. They reported that extensive lomasomes and perinuclear Golgi apparatus are characteristic of both hyphae and haustoria of *P. manshurica*. They observed that as an intercellular hypha contacts a leaf cell, a haustorium forms as a specialized branch and penetrates the cell, making a hole 1·2 to 1·8 $\mu$m in diameter. The fungal cell is slightly constricted as it passes through the host cell wall. A wall of fungal origin was always present around the entire length of the haustorium, and no plasmodesm-like discontinuties were observed in fungal cell walls. Elaborate local systems of anastomosing tubules occurred in host cytoplasm near haustoria. The tubules were 30–40 mm in diameter, and were not associated with ribosomes. Comparable systems were not found in healthy leaf cells.

## VII. Control

Dunleavy (1970) studied the sources of resistance and susceptibility to 14 races of *P. manshurica*. The disease reaction of these races on 72 soybean cultivars was determined and it was found that Mendota, Kanrich and a proportion of the Kanro plants were resistant to all races studied. Only Bansei was very susceptible to all races of the fungus. Dunleavy and Hartwig (1970) evaluated 300 genotypes for field resistance, but only eight were never found infected. From further greenhouse evaluations with nine races, only Pine Dell Perfection and three plant introductions (PI 171 443; PI 200 527; and PI 201 422) remained resistant.

Bernard and Cremeens (1971) utilized the resistance of Kanrich and Pine Dell Perfection in a study of the inheritance of resistance to downy mildew. They used a backcross technique involving Chippewa and Wayne as susceptible recurrent parents, and Kanrich as the donor of downy mildew resistance. The $F_3$ lines from three backcross populations segregated for monogenic control; with resistant, segregating, and susceptible lines occurring in a 1 : 2 : 1 ratio. The gene controlling resistance was designated *Rpm*. The phenotype of the heterozygote varied from intermediate to resistant depending on the genotype of host and pathogen, and on environmental conditions. The *Rpm* gene was incorporated into Union, a cultivar recently released for use in the central United States. Use of the *Rpm* gene in soybean breeding programmes appears at present to be the most economic route to the control of downy mildew.

There have been few attempts at chemical control of *P. manshurica* during the foliar phase of the disease because of the difficulty in applying sprays or dusts in fields of nearly fully developed plants. Unless the disease is very severe, the expense of aerial application of chemicals for disease control is probably unwarranted. Nagata (1962) recommended Bordeaux mixture and several organo-mercury compounds as useful in control, but the latter products are very toxic and are outlawed in some countries.

A more practical approach to disease control is that of complete ploughing under of soybean refuse and rotation to another crop, or crops, to reduce numbers of oospores in the soil (Sinclair and Shurtleff, 1975). When growers must plant seed contaminated with oospores in soil which contains no oospores, good control can be achieved by use of seed protectants such as Spergon and Fermate (Hildebrand and Koch, 1951).

# References

Anon. (1942). *Tidsskr. Planteavl* **48**, 1–90.

Arsenijevic, G. M., and Kostic, B. (1960). *Zashtita Bilja* **62**, 139–144.

Athow, K. L. (1973). *In* "Soybeans: Improvement, Production and Uses". (B. E. Caldwell, Ed.) 459–489. Am. Soc. Agron., Madison, Wisconsin.

Atkinson, R. E. (1944). *Pl. Dis. Reptr. Suppl.* **148**, 254–257.

Bernard, R. L. and Cremeens, C. R. (1971). *J. Hered.* **62**, 359–362.

Chu, H. T. and Chuang-Yang, C. (1961). *Taiwan Sugar Expt. Stn. Rep.* **25**, 11–25.

Cuturilo, S. (1952). *Zashtita Bilja* **11**, 21–42.

Danko, J. (1962). *Ceska Mykol.* **16**, 119–122.

Doidge, E. M. (1924). *S. Afr. Bot. Survey, Mem.* **6**, 56 pp.

Dunleavy, J. (1969). *In* "Disease Consequences of Intensive and Extensive Culture of Field Crops". (J. A. Browning, Ed.) Iowa Agric. and Home Econ. Exp. Stn. Spec. Rep. 64. 56 pp.

Dunleavy, J. (1970). *Crop Sci.* **10**, 507–509.

Dunleavy, J. M. (1971). *Amer. J. Bot.* **58**, 209–211.

Dunleavy, J. M. (1977). *Pl. Dis. Reptr.* **61**, 661–663.

Dunleavy, J. M. and Hartwig, E. E. (1970). *Pl. Dis. Reptr.* **54**, 901–902.

Dunleavy, J. and Snyder, G. (1962). *Proc. Iowa Acad. Sci.* **69**, 118–121.

Dunleavy, J. M., Chamberlain, D. W., and Ross, J. P. (1966). "Soybean diseases". US. Dept. Agric., Agric. Handb. **302**. 38 pp.

Gaumann, E. (1923). *Beitr. Kryptogramen Schweiz.* **5**, Heft 4.

Geeseman, G. E. (1950). *Agric. J.* **42**, 257–258.

Grabe, D. F., and Dunleavy, J. (1959). *Phytopathology* **49**, 791–793.

Gustavsson, A. (1959). *Op. Bot.* **3**, 1–271.

Han, Y. S. (1959). *Agric. Assoc. China* (Taiwan) *J. New Ser.* **26**, 31–38.

Haskell, R. J., and Wood, J. I. (1923). *Pl. Dis. Reptr. Suppl.* 27, 164–265.

Hildebrand, A. A., and Koch, L. W. (1951). *Scient. Agric.* **31**, 505–518.

Ikata, S., and Yamauti, K. (1941). *Phytopath. Soc. Japan, Ann.* **10**, 326–328.

Johnson, H. W., and Lefebvre, C. L. (1942). *Pl. Dis. Reptr.* **26**, 49–50.

Jones, F. R., and Torrie, J. H. (1946). *Phytopathology* **36**, 1057–1059.

Kingsley, T. (1960). *Pl. Path.* **9**, 38.

Koch, L. W., and Hildebrand, A. A. (1946). *Ann. Rep. Can. Pl. Dis. Surv.* **26**, 27–28.

Krober, O. A. (1965). *J. Agric. Food Chem.* **4**, 254.

Lehman, S. G. (1953a). *Elisha Mitchell Sci. Soc. J.* **69**, 83.

Lehman, S. G. (1953b). *Phytopathology* **43**, 460–461.

Lehman, S. G. (1958). *Phytopathology* **48**, 83–86.

Lehman, S. G. and Wolf, F. A. (1924). *Elisha Mitchell Sci. Soc. J.* **39**, 164–169.

McKenzie, J. R. and Wyllie, T. D. (1971). *Phytopath. Z.* **71**, 321–326.

Millikan, D. F. and Wyllie, T. D. (1966). *Phytopathology* **56**, 890.

Millikan, D. F., Wyllie, T. D. and Pickett, E. E. (1965). *Phytopathology* **55**, 932.

Miura, M. (1922). *Jap. J. Bot.* **1**, 9.

Mundkur, B. B. (1938). Fungi of India. Suppl. I. India Imper. Council Agric. Res. Sci. Monog. **12**, 54 pp.

Nagata, T. (1962). *UN. Food Agric. Organ. Rep.* **1465**, 1–22.

Naoumov, N. (1914). *Bull. Soc. Mycol. France* **30**, 64–83.

Nuttonson, M. Y. (1952). "Ecological Crop Geography and Field Practices of the Ryukyu Islands, Natural Vegetation of the Ryukyus, and Agro-climatic Analogues in the Northern Hemisphere". Am. Inst. Crop Ecology. 106 pp.

Pederson, V. D. (1958). *Proc. Iowa Acad. Sci.* **65**, 146–149.

Pederson, V. D. (1960). *Proc. Iowa Acad. Sci.* **67**, 103–108.

Pederson, V. D. (1961a). *Diss. Abstr.* **22**, 703.

Pederson, V. D. (1961b). Downy mildew of soybeans. Ph.D. thesis. Iowa State Univ.

Peyton, G. A. and Bowen, C. C. (1963). *Am. J. Bot.* **50**, 787–797.

Piper, C. V. and Morse, W. J. (1923). "The Soybean". McGraw-Hill Book Co. Inc., New York.

Probst, A. H. and Judd, R. W. (1973). *In* "Soybeans: Improvement, Production, and Uses".
    (B. E. Caldwell. Ed.) 1–15. Am. Soc. Agric., Madison, Wisconsin.

Riggle, J. H. (1974). *Proc. Am. Phytopath. Soc.* **1**, p. 129.

Riggle, J. H. and Dunleavy, J. M. (1974). *Phytopathology* **64**, 522–526.

Riggle, J. H. and Fisher, J. W. (1973). Proc. Second Int. Cong. Pl. Path., *Am. Phytopath. Soc.*,
    **154**.

Savulescu, T. (1948). *Bucharest Acad. Romana Sect. Sti., Bull.* **30**, 493–498.

Sawada, K. (1959). Descriptive Catalogue of Taiwan Fungi. Part XI. Natl. Taiwan Univ. Col.
    Agr., Spec. Bull. 8.

Serzane, M. (1962). "Plant Diseases. Practical Studies". Riga Latvijas Valsts Izdevniecoba.

Sinclair, J. B. and Shurtleff, M. C. (1975). Compendium of soybean diseases. *Am. Phytopath.*
    *Soc.* St. Paul, Minnesota. 69 pp.

Spasic, M. (1961). *Zashtita Bilja* **63–64**, 57–63.

Sun, S. D. (1958). "Soybean". Moscow. USSR. 248 pp.

Sydow, H., Sydow, P. and Butler, J. (1912). *Ann. Mycol.* **10**, 243–280.

Tani, T. and Naito, N. (1956). *Kagawa Agric. Col.* (Japan) *Tech. Bull.* **7**, 141–143.

Voros, J. and Molnar, B. (1958). *Novenytermeles* **7**, 371–374.

Waterson, J. M. (1939). Annotated list of diseases of cultivated plants in Bermuda. Bermuda
    Dept. Agric., (Bull, 18). 38 pp.

Wei, C. T. and Hwang, S. (1939). *Nanking J.* **9**, 329–372.

Wyllie, T. D. and Williams, L. F. (1965). *Phytopathology* **55**, 166–170.

Zaianchkovskaia, M. S. (1938). *Kul'tur, Trudy* **3**, 5–22.

Chapter 25

# Downy Mildews of Beet and Spinach

## W. J. BYFORD

*Broom's Barn Experimental Station, Higham, Bury St Edmunds, Suffolk, UK*

## I. Introduction

The downy mildews of beet and spinach were initially described as *Peron spora schachtii* Fuckel and *P. effusa* (Grev.) Rabh. (= *spinacea* Laub.) respectively but following the taxonomic study of Yerkes and Shaw (1959) these species, together with others found on members of the Chenopodiaceae, were included in the single species *P. farinosa* (Fr.) Fr. However, many workers had shown that these downy mildews are host specific (Shaffnit and Volk, 1927; Hiura, 1929; Leach, 1931; Cook, 1936; Singalovsky, 1937; Richards, 1939; Darpoux and Durgeat, 1962) and, following further cross-inoculation experiments, Byford (1967a) proposed the *formae speciales P. farinosa betae* and *P. farinosa spinacea*.

Sugar beet downy mildew, which also attacks other cultivated varieties of

*Beta vulgaris*, has been the subject of three major studies (Leach, 1931, 1945; Singalovsky, 1937; Byford, 1966a) and a number of other works. Richards (1939) studied spinach downy mildew in depth but most subsequent work has been on controlling the disease particularly by breeding resistant varieties.

## II. Sugar Beet Downy Mildew

### A. Distribution and economic importance

Downy mildew is widespread on sugar beet and other varieties of *Beta vulgaris* in the temperate regions of the Northern Hemisphere. In France, damaging attacks have occurred since the late nineteenth century (Singalovsky, 1937) and Darpoux *et al.* (1960) reported outbreaks in most years in northern France with almost all plants attacked in some fields. In the 1970s downy mildew became prevalent and damaging in sugar beet seed crops in parts of south-west France causing crop failure in some fields (Lebrun, 1978). In Sweden Björling and Möllerström (1964) reported average infection levels above $1\%$ in 7 years out of 13 with a peak of 30–35% but the disease has been less prevalent in recent years.

Beet downy mildew is important in Eastern Europe, particularly in the USSR. Among many Russian workers who have studied the problem Gorovets and Marshakova (1965) reported that in Byelorussia yields could be reduced by up to 50% and Kazachenko (1967) found 30% of plants were killed by attacks of the disease in June–July. In the USA Leach (1931) reported that the greatest losses caused by downy mildew in Central California were in fields of garden (red) beet grown for seed, where up to 43% of the yield could be lost. Losses also occurred in garden beet root crops, particularly from January to March and in sugar beet root crops that had been sown in late autumn. The decline in importance of the disease after 1954 was attributed to the introduction of resistant varieties McFarlane (1971).

Records of the year to year variations in distribution and incidence of the disease have been made in the UK through monthly counts in randomly selected root crops since 1946 (Hull, 1953), and through counts in stecklings (first year seed plants) and seed crops since 1951 (Hull, 1954). The results of these surveys, summarized by Byford and Hull (1967), showed that the disease was most prevalent where seed crops of sugar and fodder beet were grown together with a high density of sugar-beet root crops. However, in the 21 years from 1946 to 1966 the national average infection exceeded $1\%$ in only 1957 and 1965. It was estimated that the proportion of the total crop lost in England never exceeded $1\%$ and since 1967 downy mildew has been scarce in sugar-beet root crops although occasionally prevalent in some seed crops.

## B. Perennation

The fungus may overwinter as oospores in the soil, as mycelium within seeds, or on living beet plants (seed crops, groundkeepers, wild beets, etc.) either as dormant mycelium in roots or crowns, or by constantly re-infecting fresh plants.

Seed crops are recognized as the main source of infection in many countries (Hochapfel, 1950; McKay, 1957; Krexner, 1959; Pozhar, 1960; Tishchenko, 1966; Byford and Hull, 1967). Other overwintering beet is usually considered less important but Björling and Möllerström (1964) found that ground-keepers were the main source of infection in Sweden after mild winters. In England localized outbreaks have been found arising from groundkeepers in other crops or on old clamp sites and Shukanau (1974) found self sown plants an important infection source in Byelorussia. Where beet is grown as a winter crop, e.g. in Georgia, USSR, downy mildew may overwinter in the root crop itself (Berianidze, 1969).

Byford and Hull (1967) studied downy mildew in beet seed crops in England and showed that the disease builds up most rapidly within seed crops between the end of March and early June. It was also shown that although a few infected plants may be symptomless in mid-winter most of the increase in disease in the spring is due to new infection. The proportion of plants infected with downy mildew in a beet seed crop at any one time is the result of some

TABLE 1

Effect of method of raising sugar-beet stecklings on downy mildew infection in the subsequent seed crop

| Method of raising stecklings | Percentage of plants with downy mildew in June 1959–1966 | | | | | | |
| --- | --- | --- | --- | --- | --- | --- | --- |
| | 1959 | 1960 | 1961 | 1962 | 1964 | 1965 | 1966 |
| Isolated from other beet crops | 1·8 | 0·1 | 0 | 0 | — | — | — |
| Cereal cover, transplanted | 3·0 | 0·1 | 0·6 | 0·4 | — | 1·0 | 4·9 |
| Cereal cover, direct-drilled | 2·5 | 0·5 | 2·2 | 0·9 | 2·1 | 4·8 | 4·5 |
| Mustard cover, transplanted | 0·7 | 0·1 | 6·8 | 2·2 | — | — | — |
| Mustard cover, direct drilled | 15·9 | 0·1 | 3·9 | 8·3 | — | — | — |
| No cover, transplanted | 4·2 | 0·4 | 1·2 | 0·8 | — | 0·1 | 2·8 |

(No cover, direct-drilled crops are grown in different districts from other sugar-beet seed crops).

plants recovering, a few dying and others developing symptoms; a dynamic situation analogous to that found in the root crop.

The method of seed crop cultivation has an important effect on mildew incidence (Table 1). The heavy attacks in crops raised under mustard cover crops are particularly notable. This method, first used widely in 1958, accounted for 42% of all steckling beds in England by 1960 but was abandoned after 1962 because of downy mildew.

The evidence for seed transmission is slender (Leach, 1931; McKay, 1957). Oospores are often found and Leach (1931) reported two infected seedlings when seed was germinated in soil containing oospore-bearing leaf material. Van der Spek (1964) claimed that in Holland oospores were the most important source of primary infection in seed crops and Lebrun (1978) also reported severe mildew attacks, originating from oospores, in transplanted sugar-beet seed crops. However, in most countries crop rotation is normal for the root crop, probably preventing oospores from becoming an important source of infection.

## C. Symptoms

In typical systemic infections of sugar beet, the young leaves of the centre rosette are attacked. They become light green to yellowish in colour, thickened, brittle and distorted, usually with the leaf margins curled downwards. The disease does not spread to leaves already expanded when the first symptoms appear. On some leaves symptoms are confined to the lower part while the distal part remains healthy. Singalovsky (1937) noted varietal differences in the extent of deformity in infected leaves. Spores develop on infected leaves giving them a purplish colour, later becoming buff-grey. Sporulation is most abundant on the underside of the leaf, but under damp conditions spores can also be produced on the upper leaf surface and the petiole. On young plants under damp conditions non-systemic leaf-spot infections may develop and on occasional large plants infection may be confined to a limited area of one expanded leaf.

Eventually the rosette of infected leaves becomes necrotic and dies and in most infected plants healthy leaves are produced, either from the main growing point, or from axillary buds. Recovered plants may be re-infected. The older leaves, which are not invaded by the fungus, often become yellowed, thickened and brittle resembling the symptoms of virus yellows and in the field it may be impossible to determine whether a plant is attacked by one or both diseases.

Second year seed plants may be attacked at all stages of growth: at the rosette stage; during shooting, when the entire seed shoot may be stunted, distorted and covered with spores; or at flowering when flowering shoots may be attacked, or the infection confined to small lateral shoots. Some large roots may produce both infected and healthy seed shoots.

## D. Characteristics of the fungus

### 1. Morphological

The most detailed study is that by Singalovsky (1937), who stated that the germinating conidia penetrate the leaf via the stomata. Coenocytic hyphae 5–6 μm diameter ramify through the intercellular spaces while branched haustoria develop within the host cells. The infected leaf responds hypertrophically with cells expanding to approximately twice their normal size, and the differentiation between palisade and spongy mesophyll tissue, decreases. Conidiophores emerge through the stomata in small groups, of three, or sometimes five. Their length varies from 200 to 500 μm and their diameter from 8 to 12 μm. Conidia average 24·8 × 18·9 μm (Yerkes and Shaw, 1959) as a mean of 15 collections. Oospores are frequently found in infected leaves and measure approximately 35 × 38 μm.

### 2. Physiological

The environmental requirements of *P. farinosa betae* have been studied by Leach (1931), Singalovsky (1937), Darpoux *et al.* (1960), Darpoux and Durgeat (1962) and Byford (1968) and shown to be as follows.

(a) Sporulation

Spores develop between 5 and 22°C, optimum around 12°C. A relative humidity above 70% is needed, and most spores are produced above 85%. Spores are usually produced overnight and there is some evidence of a diurnal rhythm. Continuous light checks spore production which occurs most abundantly when leaves that have been exposed to daylight for 6–8 h are placed in the dark.

(b) Longevity of spores

Cold prolongs the life of spores; their ability to survive falls off above 10°C and few survive more than one day above 20°C. Differences in relative humidity above 60% scarcely affect spore survival. Under favourable conditions some batches of spores can survive for at least 1 week. Spores that survive are able to infect.

(c) Spore germination

Spores can germinate between 0·5 and 30°C, with the optimum between 4 and 10°C. Germination is fastest at 9–10°C.

(d) Infection

Infection occurs between 0·5 and 25°C, most readily between 7 and 15°C. It usually requires at least 6 h, preferably 8–10 h, in the presence of free moisture. The growing point is the most susceptible part of the plant, and after the cotyledon stage it is the only part where infection is likely to occur. Seedlings are most susceptible in the cotyledon stage, and become more difficult to infect artificially after 4 weeks growth at a minimum of 10°C. However, plants are not immune to infection after 8 weeks in the glasshouse and in the field some new infections continue to appear for most of the season.

## E. The host–parasite relationship

Several workers have observed that sugar beet plants infected by downy mildew can recover, and this was often thought to occur when weather is unfavourable to the fungus although Darpoux et al. (1960) reported that in the glasshouse, recovery was independent of temperature or humidity.

The progress of infection and the development of recovery in individual plants was studied in detail in England from 1961 to 1964 (Byford, 1967b). Plants adjoining a source of inoculum (infected seed plants) were marked as they became infected and re-examined regularly to note the development of infection, of recovery, and of re-infection when this occurred. The number of infected plants increased at a near constant rate for at least 2 months from late May but the rate of increase was low, not exceeding an average of 4·6% per week. After mid August few new infections were recorded. However, infection spread rapidly in adjacent stecklings of the same variety showing that the lack of new infections in the root crop was because the plants were resistant not because of unfavourable weather. In 1962 the total percentage of infected plants in the root crop was much less than in the stecklings, showing that not all potentially susceptible plants had been infected.

Table 2 summarizes detailed observations made in 1963 and 1964 on the recovery of infected plants. The results show how the age of the plant affects recovery. Plants infected early in the season recovered most rapidly and completely although many recovered plants were re-infected. Plants infected after the end of June recovered more slowly and more of them failed to recover.

Recovery was influenced less by the weather than by the age of the plant. In 1963 there were only short periods of weather unfavourable to mildew while in July and August 1964 there were prolonged periods of dry warm weather, but the pattern of recovery was similar in both years. Although weather unfavourable to mildew may speed recovery this is of secondary importance. Most recovered plants remained free from active mildew for the rest of the season. Except for plants first infected before the end of May very few were killed by the disease.

Most plants infected early recovered by re-growth from the primary growing point so that by harvest they appeared normal except for their small size, although some developed an elongated neck. The primary growing point was destroyed in many plants infected in late June, July and August and regrowth was from lateral apices, sometimes only a few but often many giving a dense tuft of small leaves. Some plants infected late in the season make little or no recovery. In these, the central leaves die and the disease may then be difficult to distinguish from the symptoms of boron deficiency or the damage caused by lepidopterous caterpillars.

The pattern of systemic infection followed by recovery shown by beet downy mildew suggests that, although infected leaves die prematurely, within

TABLE 2

Recovery of sugar beet plants infected with downy mildew
Experiments at Broom's Barn

| When symptoms appeared | No. of plants in sample | Average time in days for | | % recovered plants re-infected | % not recovered completely | % showing no recovery | % plants dead |
|---|---|---|---|---|---|---|---|
| | | Recovery established | Recovery completed | | | | |
| **1963** | | | | | | | |
| before 15 June | 55 | 14 | 27 | 35 | 0 | 0 | 5 |
| 16 June–2 July | 50 | 19 | 49 | 21 | 2 | 0 | 2 |
| 3–12 July | 59 | 18 | 50 | 34 | 5 | 0 | 2 |
| 13–26 July | 68 | 29 | 70 | 32 | 16 | 0 | 0 |
| 27 July–5 Aug | 53 | 37 | 80 | 32 | 30 | 4 | 0 |
| 6 Aug–10 Oct | 35 | 45 | — | 24 | 47 | 20 | 0 |
| **1964** | | | | | | | |
| before 26 May | 30 | 27 | 45 | 64 | 0 | 0 | 20 |
| 27 May–9 June | 79 | 24 | 39 | 61 | 1 | 0 | 5 |
| 10–17 June | 45 | 25 | 46 | 46 | 7 | 0 | 2 |
| 18–26 June | 33 | 22 | 44 | 41 | 6 | 3 | 3 |
| 27 June–2 July | 32 | 26 | 58 | 19 | 9 | 0 | 0 |
| 3–10 July | 17 | 32 | 56 | 18 | 6 | 0 | 0 |
| 11–20 July | 18 | 38 | 52 | 11 | 6 | 6 | 0 |
| 21–29 July | 21 | 35 | 57 | 10 | 14 | 5 | 0 |

TABLE 3

Effect of downy mildew infection at different times on yield
and sugar content of sugar beet

| Month first infected | Root weight g | Sugar content % fresh weight | Sugar per root g |
|---|---|---|---|
| June | 235 | 15·0 | 34·1 |
| July | 302 | 12·9 | 39·0 |
| August | 511 | 12·8 | 65·4 |
| September and after | 657 | 14·3 | 94·0 |
| Not infected | 674 | 15·9 | 107·2 |

Mean of 4 experiments in England in 1962 and 1963.

the plant as a whole a balance exists between host and pathogen. This suggestion may be compared with the results of Ingram and Joachim (1971) who found that, in callus tissue culture, sugar beet and *P. farinosa betae* formed a stable dual culture that was maintained through successive transfers for almost two years without losing pathogenicity whereas with other downy mildews growing on host callus, the tissues were always killed by the parasite.

## F. Effect on yield

Many workers have attempted to assess the yield loss caused by downy mildew in sugar-beet root crops (Leach, 1945; Cornford, 1954; Möllerström, 1955; Krexner, 1959; Darpoux and Durgeat, 1962; Byford, 1967b; Gorovets and Marshakova, 1965; Čača. 1969), usually considering the effect of infection on individual plants. In most cases healthy plants were compared with infected, either selected at harvest or marked earlier in the season. Substantial falls in root weight and sugar percentage were usually recorded, particularly when infected plants were selected early in the season.

Björling and Möllerström (1964) in Sweden and Byford (1967b) in England compared the yields of plants first infected at different times during the year. Table 3 summarizes the results of four typical trials in England in 1962 and 1963. The greatest loss in root weight was in early-infected plants but sugar concentration was least in plants infected later. Experiments in Sweden gave similar results but the effects were less marked with a maximum sugar yield decrease of 48%.

Although the effect of downy mildew attack on individual sugar-beet plants is well established it is difficult to use these data to estimate crop yield losses. In infected crops, healthy plants can grow at the expense of diseased ones and may compensate for much of the yield lost. When a crop is attacked by downy mildew early in the season most of the yield loss is due to loss of plants, although most are severely stunted rather than killed, and unless many plants are affected or the plant stand is thin or uneven, yield loss will be comparatively small. When plants are infected later in the season, less of the resulting yield loss is likely to be made good by growth of neighbouring healthy plants, and the data for sugar loss in individual infected plants will therefore be a better guide to crop losses. The average loss of sugar yield when beets are first infected in late July and early August is 55%, indicating that 20% infection at this time could cause a crop yield loss of up to 10%.

However, even when using this approximation to assess yield loss, interpreting field counts of downy mildew incidence presents difficulties because of the dynamic situation in an outbreak of the disease. The recovery of almost all plants infected early in the season, accompanied often by their re-infection, means that a typical mildew attacked sugar-beet crop has plants in all stages of infection and recovery, many of which cannot with certainty be said to have been infected. Thus, if the proportion of plants seen to be infected remains constant through June, July and August, it must be inferred that the total number of plants that have been infected is substantially greater. Surveys based on counts of percentage plants infected therefore bear a greater likelihood of error, and must be treated with more reserve when assessing probable losses, than do field counts of diseases where the count of infected plants at any one time gives a cumulative total of infections.

## G. Control

Many workers have reported experiments with protectant fungicides to control beet downy mildew but the control obtained was often poor (Möller-ström, 1955; Cornford, 1959; Byford, 1966b). In particular attempts to control the disease by spraying seed crops during the winter were ineffective (Byford, 1966b; Lebrun, 1978).

In consequence most advice on limiting mildew incidence has concentrated on cultural practices of which the most important is separating sugar-beet root crops from their main source of infection, beet seed crops. In the Soviet Union 1000 m is usually recommended (Sinitsyna, 1964; Tishchenko, 1966). In England most sugar-beet stecklings are raised under cereal cover crops which give some protection from infection (Byford and Hull, 1967). Stecklings not raised under cereal cover must be separated by at least 400 m from other *Beta* seed or root crops. Good crop hygiene is also necessary to limit survival of the fungus on ground keepers in fields and on clamp sites, and proper rotation reduces the risk of infection from oospores in the soil. Both Byford (1967c) and Čača (1969) reported that increasing the spacing between plants increases the proportion infected with downy mildew. Byford (1967c) also reported that infection was increased by high doses of nitrogen fertilizer.

Sugar beet varieties differ in their susceptibility to downy mildew but in areas where the disease is endemic all varieties probably have some non-specific resistance to the disease. Weisner (1964) found that 3–34% of 88 varieties in an experiment were infected and Byford (1967d) reported from 17 to 47% infection in varieties grown under heavy natural infection. Russell (1969) reported an inbred line completely resistant to downy mildew but a race of *P. farinosa betae* able to infect the line was soon found (Russell and Evans, 1972).

Comparisons of the reaction of beet varieties to downy mildew made in the glasshouse may give results which do not agree with counts made on plants in the field (Russell, 1968). To overcome this difficulty in the UK, a "downy mildew nursery" has been established at Trawsgoed in Wales where the damp cool climate favours the disease and there are no commercial beet crops that could be accidentally infected. (Byford, 1966c). Here all established and new varieties are grown under heavy natural infection and plant breeders can evaluate material for resistance.

At the outset it was hoped that this facility would result at least in the resistance of all varieties rising to the level of the best then existing (Byford, 1969). In practice, since downy mildew is not a disease of major importance in the UK there is little commercial incentive to produce resistant varieties. The difference between the most and least resistant varieties in 1979 remains similar to the difference when the Trawsgoed "nursery" was established in 1965. However, the experiment has proved of value first in providing field data on the relative susceptibility of varieties so that the most susceptible can be avoided in areas particularly at risk e.g. near seed crops; secondly it has allowed very susceptible material to be eliminated early in the selection process.

Recently the development of systemic fungicides active against Perono-sporaceae has given promise of direct control of beet downy mildew. Broyakovskaya (1975) reported that editon and antracol showed strong systemic action when applied to the soil. Lebrun and Viard (1979) reported that metalaxyl applied to the seed protected the young seedlings from infection for up to 3 weeks.

## III. Spinach Downy Mildew

### A. The disease

Spinach downy mildew occurs in most countries where the crop is grown, but there is little information on the extent of the losses that it causes. Richards

(1939) estimated that in the USA from 3 to 15% of the crop was unharvested each year because of the disease. Chesters (1943) reported that in addition to field losses, mildew could also attack spinach in transit.

Spots, or large irregular areas of leaves of infected plants become pale yellow and the leaves become curled and distorted. Under humid conditions spores are produced on the under surface of the leaf; sometimes on both leaf surfaces and on the petiole. Infected leaves eventually become water-soaked, then brownish and rot. Beet and spinach downy mildews are morphologically similar but, unlike Singalovsky's (1937) report for beet Richards (1939) found that germinating conidia penetrated spinach leaves through the cuticle at the junction of epidermal cells and never entered through stomata. According to Richards the optimum temperature for spore germination is 9°C and for germ tube elongation 12°C. Spores rapidly lose viability when desiccated or exposed to sunlight. The minimum period between infection and sporulation is 6 days.

In the North-East USA the fungus overwinters mainly on infected plants. (Richards, 1939). Seed lots often contain oospores of *P. farinosa spinacea* and Leach and Borthwick (1934) found downy mildew mycelium in spinach seeds but neither Richards (1939) nor Leach (1931) could demonstrate seed transmission. Oospores are produced and will contaminate infected soil when infected plants are ploughed in. Wright and Yerkes (1950) reported that this was the principal source of primary infection in Washington, USA.

The conditions necessary for the development of an epiphytotic of spinach downy mildew are: plants growing vigorously; abundant condidial inoculum; mean temperature of 5–10°C for a week or more; free water on the plant for 3 h or more; periods of high relative humidity for sporulation and infection (Richards, 1939).

## B. Control

A few workers suggest the use of fungicides. Richards (1939) obtained 70% control with copper oxide but at the cost of some phytotoxicity. Knoppien *et al.* (1954) reported effective use of zineb both under glass and in the open while Kachatryan (1956) recommended soil disinfection and Bordeaux mixture spray.

Since spinach is a market crop with a short growth period protectant fungicides containing copper or other metal based fungicides are of limited use. Rotation and the separation of winter and spring crops is therefore suggested as an alternative (Richards, 1939; Chesters, 1943; Kachatryan, 1956).

In 1950 Smith reported resistant varieties carrying immunity to downy mildew derived from a single dominant gene. Jones *et al.* (1956) reported methods of glasshouse screening for resistance but by 1958 Zinc and Smith reported a second physiological race of the fungus. Smith *et al.* (1962) reported spinach varieties resistant to the new race. The genetics of resistance has recently been reviewed by Eenink *et al.* (1976) who also report finding a third physiological race of the fungus.

The recently developed systemic fungicides active against the Peronosporaceae offer possibilities of improving the control of the downy mildews of both sugar beet and spinach. In particular they are of potential value as a component of integrated control programmes used together with varietal resistance (for spinach), good cultural practices and limiting sources of infection.

### References

Berianidze, M. Sh. (1969). *Proc. Inst. Pl. Prot. Georgian SSR*, 3–6.
Björling, K. and Möllerström, G. (1964). Socker **19**, 17–33.
Broyakovskaya, K. N. (1975). *In* "Effektiv. priemy i sposoby bor'by s boleznyami sakhar. svekly". Kiev, 115–120.
Byford, W. J. (1966a). Ph.D. Thesis, University of London, UK.
Byford, W. J. (1966b). *Proc. 3rd Br. Insectic. Fungic. Conf.*, Brighton UK. 1965, 169–176.
Byford, W. J. (1966c). *Rep. Rothamsted Exp. Stn. for 1965*, 270–271.
Byford, W. J. (1967a). *Trans. Br. Mycol. Soc.* **50**, 603–607.
Byford, W. J. (1967b). *Ann. Appl. Biol.* **60**, 97–107.
Byford, W. J. (1967c). *Pl. Path.* **16**, 160–161.
Byford, W. J. (1967d). *Rep. Rothamsted Exp. Stn. for 1966.* p. 282.
Byford, W. J. (1968). *Ann. Appl. Biol.* **61**, 47–55.
Byford, W. J. (1969). *Br. Sug. Beet Rev.* **38**, 23–26 and 36.
Byford, W. J. and Hull, R. (1967). *Ann. Appl. Biol.* **60**, 281–296.
Čača, Z. (1969). *Ochr. Rost.* **5**, 169–176.
Chester, K. S. (1943). *Pl. Dis. Reptr.* **27**, 708–710.
Cook, H. T. (1936). *Phtopathology*, **26**, 89–90.
Cornford, C. E. (1954). *Pl. Path.* **3**, 82–83.
Cornford, C. E. (1959). *Rep. Rothamsted Exp. Stn. for 1958.* p. 196.
Darpoux, H. and Durgeat, L. A. (1962). Paper submitted to the 25th winter congress of the Institut International de Récherches Betteravières.
Darpoux, H., Durgeat, L. A. and Lebrun, A. (1960). Paper submitted to the 23rd winter congress of the Institut International de Recherches Betteravières.
Eenink, A. H., Groenwold, R. and Raalten, W. J. van. (1976). *Zaadbelangen* **30**, 101–103.
Gorovets, U. K. and Marshakova, M. I. (1965). *Bot. Issled. Minsk.* **7**, 53–59.
Hiura, M. (1929). *Agric. Hort.*, Tokyo **4**, 1394–1406.
Hochapfel, H. (1950). *NachrBl. dt. PflSchutzdienst, (Braunschw.)* **2**, 124–125.
Hull, R. (1953). *Ann. Appl. Biol.* **40**, 603–606.
Hull, R. (1954). *Agric.* **61**, 205–210.
Ingram, D. S. and Joachim, I. (1971). *J. Gen. Microbiol.* **69**, 211–220.
Jones, H. A., McClean, D. M. and Perry, B. A. (1956). *Proc. Am. Soc. Hort. Sci.* **68**, 304–308.
Kachatryan, M. S. (1956). *Bull. Acad. Sci. Arm.* SSR **9**, 111–115.
Kazachenko, R. F. (1967). *Sakh. Svekla* **12**, 35–36.
Knoppien, P., Schmidt, G. J. and Van der Waal, M. A. (1954). *Meded. Dir. Tuinb.* **17**, 228–232.

Krexner, R. (1959). *Pflanzenarzt.* **12**, 114–116.
Leach, L. D. (1931). *Hilgardia* **6**, 203–251.
Leach, L. D. (1945). *Hilgardia* **16**, 317–334.
Leach, L. D. and Borthwick, H. A. (1934). *Phytopathology* **24**, 1021–1025.
Lebrun, A. (1978). Hautes études betteravières et agricoles No. 39. 8–17.
Lebrun, A. and Viard, G. (1979). *Phytiatrie Phytopharm.* **28**, 29–40.
McFarlane, J. S. (1971). *In* "Advances in Sugar Beet Production". p. 422. Iowa State University Press, Ames, USA.
McKay, R. (1957). Field studies on the downy mildew of sugar-beet. Dublin: Irish Sugar Co. Ltd.
Möllerström, G. (1955). *Socker* **11**, 31–36.
Pozhar, Z. A. (1960). *Sug. Ind. Moscow* **34**, 57–59.
Richards, M. C. (1939). *Bull. Cornell Univ. Agric. Exp. Stn.* No. **718**.
Russell, G. E. (1968). *J. Agric. Sci. Cambridge* **71**, 251–256.
Russell, G. E. and Evans, G. M. (1972). *Rep. Pl. Breed. Inst.* 1971, 102–103.
Schaffnit, E. and Volk, A. (1927). *Pflanzenreich* **3**, 1–45.
Shukanau, A. S. (1974). *Vestsi Akad. Navuk BSSR, Biyalag,* Navuk No. 5, 59–62.
Sinitsyna, N. I. (1964). *Sakh. Svekla* **9**, 32–33.
Singalovsky, Z. (1937). *Ann. Epiphyt. n.s.* **3**, 551–618.
Smith, P. G. (1950). *Phytopathology* **40**, 65–68.
Smith, P. G., Webb, R. E. and Luhn, C. H. (1962). *Phytopathology* **52**, 597–599.
Tishchenko, E. I. (1966). *Sakh. Svekla* **11**, 37.
Van der Spek, J. (1964). *In* "Jaarverslag 1963". Inst. voor Plantenziektenkundig Onderzoek, Wageningen.
Weisner, K. (1964). *Z. PflKrankh.* **71**, 271–286.
Wright, C. M. and Yerkes, W. D. (1950). *Pl. Dis. Reptr.* **34**, 28.
Yerkes, W. D. and Shaw, C. G. (1959). *Phytopathology* **49**, 499–507.
Zink, F. W. and Smith, P. G. (1958). *Pl. Dis. Reptr.* **42**, 818.

Chapter 26

# Downy Mildew of Sunflower

## W. E. SACKSTON

*Department of Plant Science, Macdonald Campus, McGill University,*
*Ste. Anne de Bellevue, Quebec, Canada*

## I. Introduction

Sunflower appears to be one of the very few American food plants to have originated in temperate North America; most are from Central or South America (Heiser, 1955). The common sunflower, *Helianthus annuus* L., has three varieties; *H. annuus* var. *lenticularis* (Dougl.) Ckll., the "wild" form, which is widespread in western North America, *H. annuus* var. *annuus*, the "weed" form, which is confined mainly to the middle western United States and *H. annuus* var. *macrocarpus* (D.C.) Ckll., the cultivated form, which is now distributed world-wide and does not persist outside of cultivation (Heiser, 1951).

Developed as a cultivated plant in eastern North America long before the discovery of the New World, sunflower was introduced into Europe as a curiosity and an ornamental by Spanish, French, and English explorers. It was first illustrated and described in Europe by Dodonaeus in 1568 (Heiser, 1955). It reached Russia in the eighteenth century, achieving spectacular success there as a food plant (Heiser, 1955; Putt, 1978). Re-introduced to the Prairie Provinces of Canada and the northern Great Plains of the United States by Eastern European immigrants, it has been grown there on a small scale for over a century (Carter, 1978; Heiser, 1955; Putt, 1978).

Sunflower had long been a major crop in the USSR, Eastern Europe, and in Argentina, but became economically attractive in many other countries after the release in the 1960s of high oil content cultivars developed in the USSR. The discovery of cytoplasmic male sterility (Leclercq, 1969) and of restorer genes (Enns *et al.*, 1970; Kinman, 1970), making possible the efficient production of hybrid seed, resulted in a dramatic increase in sunflower production. It has become the second largest source of edible vegetable oil, exceeded only by soybean (Putt, 1978). It is grown in significant commercial quantities in at least 20 countries in all continents but Antarctica and continues to expand into new areas; production world-wide has reached 10 to 12 M tons (Putt, 1978; Waalwijk van Doorn, 1978).

## II. Host Range, Distribution, and Importance

Any crop grown on so large a scale is bound to suffer from various diseases. Sunflower is no exception, and downy mildew is among its major diseases (Zimmer and Hoes, 1978).

A fungus found by Halsted inducing downy mildew on *Eupatorium purpureum* L. was named *Peronospora halstedii* by Farlow in 1882 (Nishimura,

1922). It was re-named *Plasmopara halstedii* (Farl.) Berl. & de Toni in 1888 (Saccardo, 1888). By 1922 it was regarded as one of the most widely distributed and characteristically American species, which might be expected to occur on almost any member of the Compositae.

Novotelnova (1962) referred to the fungus as a "conspecies", limited to the Compositae, with a host range as follows: Subfamily *Tubuliflorae* AC., tribe *Eupatorieae* Cass., the genus *Eupatorium*. Tribe *Astereae* Cass., genera *Aster, Callistephus, Galatella, Heteropappus, Solidago*. Tribe *Inuleae* Cass., the genus *Gnaphalium*. Tribe *Heliantheae* Cass., genera *Helianthus, Ratibida, Silphium, Xanthium, Bidens, Ambrosia*. Tribe *Cynareae* Cass., the genus *Saussurea*. Subfamily *Liguliflorae* DC, tribe *Cichorieae* Reichb., genera *Scorzonera, Tragopogon*.

Leppik (1966a, b) referred to the "*P. halstedii* complex" on the Compositae. He included the hosts listed by Novotelnova, and added others: Subfamily *Tubuliflorae*, tribe *Vernonieae*, the genus *Vernonia*. Tribe *Eupatorieae*, genera *Ageratum* and *Elephanthopus*. Tribes *Helenieae, Anthemideae, Senecioneae, Calenduleae, Arctoteae, Cardueae, Mutisieae*, genera *Emilia, Centaurea, Cineraria, Clibadium, Coreopsis, Erigeron, Dimorphotheca, Franseria, Iva, Senecio, Verbesina, Venidium, Ximinesia, Zinnia, Chrysanthemum*, and *Petasites*. Bisby *et al.* (1938) reported the fungus also on *Rudbeckia*, not included above.

*P. halstedii* is almost coextensive with the wild *Helianthus* spp. naturally susceptible to it (Orellana, 1970). Most of them occur in the continental USA and some in Canada, but none in Mexico or South America. *P. halstedii* was reported on wild and cultivated *Helianthus* spp. in Canada in early compilations of the fungus flora (Bisby *et al.*, 1938) and was recorded regularly after systematic sunflower disease surveys had been instituted in 1948 (Sackston, 1949; Conners, 1967). Elsewhere in the Americas it has been reported on cultivated sunflowers in Mexico (Fucikovsky, 1976), Santo Domingo (Gomez-Menor, 1936), Argentina (Pontis *et al.*, 1959), Chile (Sackston, 1956) and Uruguay (Sackston, 1957).

Downy mildew was found in eastern Europe first in 1941. Numerous reports summarized by Novotelnova (1966) show that it spread as follows: Yugoslavia 1941, Romania 1946, Bulgaria 1947, Hungary 1949, Czechoslovakia 1954. First reported in Moldavia in 1948, it had, by 1951, begun to spread rapidly through the vast sunflower growing area of the USSR.

Current world distribution of *P. halstedii* on sunflower is shown in Fig. 1; Map 286 of the Commonwealth Mycological Institute which gave the world distribution in 1954. It and subsequent editions (Anon., 1977) list the following confirmed reports, in addition to those given above: Rhodesia, Uganda, Siberia and Central Asian republics in USSR, Turkey, France, Germany, Italy and Poland. They include reports on hosts other than *Helianthus* from Brazil, India, and Kenya, and possibly also for Switzerland, Japan, and

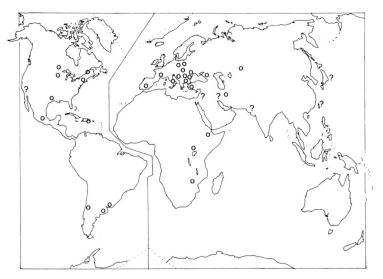

Fig. 1. Current world distribution of *Plasmopara halstedii* on sunflower. Hollow circles
indicate confirmed reports, one per country as given in the text with two or three for the large
land masses of Canada, USA, and USSR. Question marks indicate unconfirmed or doubtful
reports. Equal-area map, scale 1:1000 000, prepared with the assistance of Prof. A. K.
Maconochie, incorporates data from CMI map 286 (Anon., 1977).

Taiwan. Some questionable records based on examination of seed samples
from Ethiopia, Iran, Iraq, Israel, Jordan, Pakistan, Spain, Sweden, and the
United Kingdom, were presented in a map by Leppik (1962) and most were
included in later CMI maps. Occurrence has since been confirmed in Ethiopia
(Hingorani, 1971), in Spain (Jimenez-Diaz, 1973), and in Iran (Acimovic,
in Sackston, 1978). It was not found during intensive surveys by pathologists
in Egypt and Tunisia (Sackston, 1978).

Estimates of the severity of downy mildew attack are usually based on
the observed percentage of plants with typical symptoms of systemic
infection. Severity varies greatly from country to country and with weather
conditions from year to year. Early reports ranged from traces to 95% in
Canada (Sackston, 1954, 1955), traces in Chile (Sackston, 1956), up to 35%
in Uruguay (Sackston, 1957). More recently, up to 90% infection has been
reported in the Red River Valley of USA (Zimmer, 1971). Although first
found in Eastern Europe in 1941, infection up to 90% was observed in
occasional fields in Yugoslavia by 1952, up to 90% in Bulgaria by 1962 and
up to 90% in the USSR by 1965 (Novotelnova, 1966). By 1977 it was rated
a "major" disease in all sunflower producing countries of Europe except
Poland and Italy, where the crop is a minor one (Sackston, 1978).

Severely affected plants which reach maturity yield little or no viable seed.
Losses were estimated to be up to 50% of the potential seed yield in some

Fig. 2. Severely stunted systemically infected sunflowers and apparently healthy plants in a field.

fields in the Red River Valley in 1970; other fields were ploughed down without harvesting (Zimmer, 1971). Destruction of sunflower inflorescences by downy mildew was associated with unusual weather conditions in Hungary in 1975; yield loss equalled percentage infection and was over 50% in some fields (Kurnik *et al.*, 1976).

Quality as well as yield of seed may be affected adversely. Achenes harvested from mildewed plants were significantly smaller, lighter, higher in hull percentage, and lower in oil than those from healthy plants (Zimmer and Zimmerman, 1972). Germinability of seeds from mildew-infected plants was lower than that from healthy plants, and loss in seed vitality was directly correlated with disease intensity (Zazzerini and Raggi, 1974).

## III. Symptoms

Sunflower plants affected by downy mildew may show a wide range of symptoms. Nishimura (1922) noted that five diseased plants at Columbia University in 1918 "were dwarfed and the leaves showed beautiful mosaic areas of light green and dark green". Plants may be stunted to various degrees (Fig. 2). The characteristic green and chlorotic mottling of leaves usually increases in area and intensity from the older lower leaves to the upper

Fig. 3.   Chlorosis spreading from veins on leaf of a systemically-infected plant.

Fig. 4.   Sporulation of *P. halstedii* on lower surface of systemically-infected leaf.

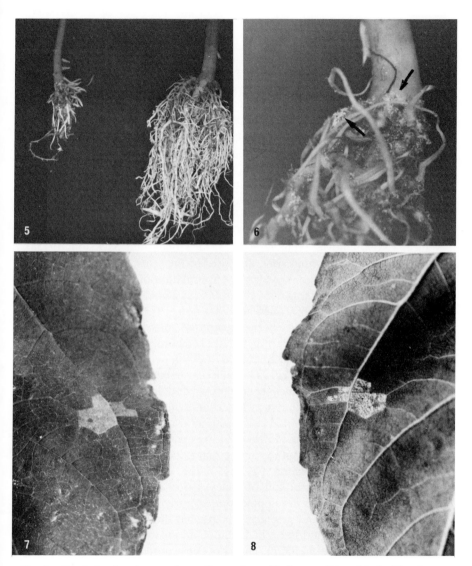

Fig. 5.   Poorly developed roots of a sunflower plant with downy mildew, left; healthy plant, right.

Fig. 6.   Sporulation on roots of symptomless sunflower plant.

Fig. 7.   Localized secondary infection on upper surface of sunflower leaf.

Fig. 8.   Sporulation on lower surface of localized spot of mildew.

younger ones; it appears first along the veins near the petiole and gradually spreads (Fig. 3). Young leaves with chlorosis along the main veins usually become curled. Mottling may occur on the cotyledons as well as leaves of many seedlings soon after emerging in the field. Under moist conditions conidiophores develop on the lower surface of the chlorotic areas (Fig. 4); they may occur rarely on the upper surface as well. The heads of stunted plants are usually stiff and face upwards; they may bear viable achenes (Fig. 2). The roots of affected plants are usually smaller and darker than those of healthy plants (Fig. 5).

Severely diseased plants may live only a few weeks. If they survive, they may be only 10–30 cm tall, with thin stems, poorly developed roots, often completely chlorotic leaves, and a tiny capitulum, 1–5 cm diameter, with no seeds. All the symptoms described above result from systemic infection.

Secondary, localized infection of leaves may occur under certain circumstances (Novotelnova, 1966; Zimmer, 1972; Cohen and Sackston, 1973a). It results in spots of various sizes, delimited by the veins, angular, pale green, and under moist conditions, with white sporulation of the pathogen on the lower surface (Figs 7, 8). It may give rise to systemic infection (Nishimura, 1922; Novotelnova, 1966; Cohen and Sackston, 1973a).

Plants with no apparent external symptoms may harbour the pathogen in or on underground tissues (Fig. 6), and sometimes internally in the stem. Novotelnova (1966) referred to this as the "hidden form" of the disease; Cohen and Sackston (1973a, 1974a) called it "latent infection".

Less familiar symptoms may include brown discolouration at the periphery of the pith in below-ground stem tissues of symptomless plants (Delanoe, 1972), and pith discolouration increasing in area and intensity from a few small spots at the level of the lowest leaves with typical mottling, to discolouration of the entire pith near the plant apex (Sackston, unpublished) (Fig. 9). *P. halstedii* infection may also induce damping-off (Goossen and Sackston, 1968) and basal gall (Zimmer, 1973). An unusual inflorescence infection which stopped development at an early stage, resulting in green, distorted floral parts and green seed coats, but no leaf or stem symptoms, was associated with abnormally high rainfall during the early flowering period in Hungary (Kurnik *et al.*, 1976). The pathogen was present in affected tissues.

## IV. Taxonomy

The fungus collected on *Eupatorium* by Halsted (Nishimura, 1922) was described in 1882 by Farlow as *Peronospora halstedii* (Saccardo, 1888). After revision of *Peronospora*, it was renamed *Plasmopara halstedii* (Farl.) Berl. & de Toni (Saccardo, 1888). In 1907 Wilson (in Nishimura, 1922) transferred

Fig. 9. Discolouration of pith in stem and capitulum of sunflower plant with systemic downy mildew.

the fungus to *Rhysotheca halstedii* (Farl.) but this was not upheld by later writers. Stevens (1913) gave an English description of *P. halstedii* varying only in minor detail from the Latin of Saccardo: "Hypophyllous; conidiophores fasciculate, slender, 300–750 μm, 3–5 times branched, ultimate branchlets 8–15 μm long, verticillate below the apex of the branching axis which is frequently swollen and ganglion-like; conidia oval or elliptic, 18–30 × 14–25 μm; oöspores 30–32 μm, epispore yellowish-brown, somewhat wrinkled." In his discussion, Stevens commented that "This form is quite variable and should perhaps be separated into several distinct species. It is limited to the Compositae, *Helianthus* and *Madia* being the only hosts of economic importance."

*P. halstedii* has been reported on over 80 species in 35 genera and 9 tribes of the Compositae. Fungi in the genus *Plasmopara* usually have a narrow host range. Believing that the pathogen on sunflower was restricted to the

554                                    W. E. SACKSTON

genus *Helianthus*, Novotelnova (1961) renamed it *P. helianthi* Novot. with
three new forms; f. *helianthi*, restricted to annual species of *Helianthus*;
f. *perennis*, on perennial species, and f. *patens*, also on perennial species but
differing slightly in zoosporangial shape. *P. halstedii* was retained for the
fungus on hosts in the tribe *Eupatorieae*, on which it was first described
(Novotelnova, 1962). Many plant pathologists in Europe now use the name
*P. helianthi* Novot. for downy mildew on sunflowers, but pathologists in
North and South America continue to use *P. halstedii* (Zimmer and Hoes,
1978).

Novotelnova reviewed the literature, examined herbarium specimens of
16 genera in six tribes of the Compositae, and inoculated annual and perennial
*Helianthus* species, *Eupatorium*, *Aster*, *Solidago*, and *Erigeron* with the fungus
from cultivated sunflowers. Positive infections resulted only on the annual
species *H. annuus*, *H. debilis*, *H. lenticularis*, and some interspecific hybrids
with annual species as maternal parents. Concluding that *P. halstedii* was
not a single species, but a heterogeneous group which had not been adequately
studied, she differentiated it into ten species, each confined to hosts in only
one tribe of the Compositae. Entities on host plants in the same tribe, but
distinguishable by various characteristics, were named as separate species.
Entities on related host species or genera differing only in spore size or shape,
were grouped in 12 specialized forms (Novotelnova, 1961, 1962). Spore sizes
for the fungus on sunflowers are listed in Table 1.

It is quite possible that *P. halstedii* (Farl.) Berl. & de Toni includes a number
of distinct species, as suggested earlier by Stevens (1913). Failure to infect
perennial species of *Helianthus* with *Plasmopara* from annual sunflowers is
poor evidence for species differentiation in the pathogen, however, as even
highly susceptible cultivars of sunflower may not show symptoms unless
inoculated under very specific conditions (Novotelnova, 1966; Cohen and
Sackston, 1973a; Zimmer, 1975). Differentiating forms on the basis of minor
differences in spore shape or size may be questionable, in view of the great
differences between zoosporangia formed on the leaves and on the roots of
the same plant. Good arguments can be presented for grouping in one species
all the forms similar in morphology, and into special forms those differing
in host range, as was done for *Fusarium oxysporum* by Snyder and Hanson
(1940).

No systematic study has been made of possible cross infection of the various
annual and perennial *Helianthus* spp. in North America with collections of
*P. helianthi* from the respective hosts, although this was suggested by Orellana
(1970). Oogonia and antheridia of the fungus are produced on different
hyphae (Nishimura, 1922), but it is not known if it is heterothallic. Even
so, it might be advisable to include oospore inoculum in a cross-inoculation
study, to exploit the potentially greater variability of sexual spores.

## V. Experimental Methods

### A. Diagnosis

The presence of typical sporangiophores and zoosporangia of *P. halstedii* on entire sunflower plants or detached tissues permits positive diagnosis of mildew in the absence of typical symptoms. Sporulation can be induced by keeping the plants at 100% relative humidity for 12 h or longer. Typical mycelium and haustoria (Fig. 10) may often be demonstrated in the hypocotyls of symptomless infected plants by mounting freehand sections in lactophenol with cotton blue for microscopic observation (Goossen and Sackston, 1968; Viranyi, 1977). Diagnosis of large numbers of specimens can be facilitated by keeping thin discs or longer pieces of hypocotyls or stems in moist petri dish chambers for several days to induce sporulation (Cohen and Sackston, 1973a; Montes and Sackston, 1974; Viranyi, 1977).

### B. Inoculation

The simplest method of inoculating sunflowers with downy mildew is to sow seeds in soil in which diseased sunflowers have been grown (Nishimura, 1922; Young *et al.*, 1929; Novotelnova, 1966; Goossen and Sackston, 1968). Soil can be infested experimentally by adding zoospore suspensions, oospores, leaves containing only mycelium, and various plant parts containing oospores; and on a field scale, by adding infested soil from around diseased plants,

TABLE 1

Spore sizes of the downy mildew pathogen on sunflowers (in $\mu$m)

| Species and form | Oospores | Zoosporangia | | Author |
| --- | --- | --- | --- | --- |
| | | Location | Size | |
| *P. halstedii* | 23–30 | | 19–30 × 15–26 | Saccardo (1888) |
| *P. halstedii* | | leaves | 33–57 × 30–36 | Nishimura (1926) |
| | | roots | 48–60 × 33–56 | |
| *P. helianthi* | | | | |
| f. *helianthi* | 15–32 | leaves | 17–30 × 15–21 | Novotelnova (1961, 1965) |
| | | roots | 36–66 × 29–46 | |
| f. *perennis* | | | 15–27 × 12–21 | |
| f. *patens* | | | 15–24 × 12–21 | |
| *P. helianthi* | 24–37 | leaves | 17–50 × 14–31 | Delanoe (1972) |
| | | roots | 25–78 × 24–36 | |

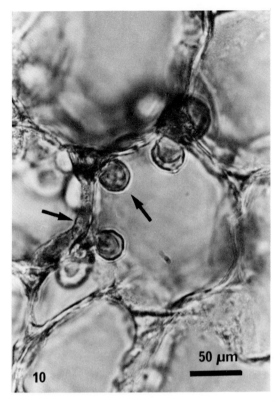

Fig. 10.   Intercellular mycelium and haustoria of *P. halstedii* in cells of sunflower hypocotyl.
Photomicrograph with assistance of Mr D. K. Zura.

and by incorporating stalks of infected plants (Nishimura, 1922; Novotelnova, 1966; Cohen and Sackston, 1974a).

Nishimura (1922) studied penetration by immersing seedling roots in zoospore suspensions. Inoculations on a large scale to screen populations for disease reaction and other studies have been made by drenching emerging seedlings with a zoospore suspension and covering with a polyethylene sheet (Panchenko, 1965; in Novotelnova, 1966). Pregerminating seeds for 1 to 3 days then soaking them in a suspension of zoosporangia and zoospores, before transplanting them into pots on greenhouse benches (Moldovan and Zhivilo, 1965; Delanoe, 1972; Zimmer and Kinman, 1972), the WSI (whole seedling inoculation) method of Cohen and Sackston (1973a), gives the best results under standardized conditions and is the most widely used for routine studies.

Inoculations for specific purposes have been made by placing drops of spore suspension on selected tissues (Novotelnova, 1966), or by using discs or squares of filter paper soaked in a known concentration of zoosporangia for quantitative studies (Cohen and Sackston, 1973a). Inoculum has also been introduced into the pith of sunflower stems at predetermined points and into the inflorescence, using a hypodermic syringe (Montes and Sackston, 1974).

## C. Spore germination

Zoosporangia would germinate in tap water and distilled water, but germination was better in drops of rainwater or dew collected from sunflower leaves (Novotelnova, personal communication), and was improved by adding sucrose to the suspension *in vitro* (Goossen and Sackston, 1968).

Nishimura (1922) stated that "Both the conidia and oospores germinate in the soil by zoospores." Delanoe (1972), however, was the first to report and illustrate germination of oospores, achieved by subjecting them to cold for various periods.

## VI. Factors Affecting Infection

### A. Age of tissues

Age of host tissues is a major factor determining susceptibility of sunflowers to systemic infection by *P. halstedii*. Novotelnova (1966) believed that the critical period for infection through the roots extended from germination of seeds to the formation of 6 to 8 true leaves. Cohen and Sackston (1973a) obtained 24% systemic infection by inoculation at the 2-true-leaf stage, but none at the 4-leaf stage. Delanoe (1972) found that infection dropped from 97% of seedlings inoculated 2 days after germination to 1·5% at 10 days; Zimmer (1975) found it dropped from 100% at 3 days to 0% at 8 days after sowing (still in the cotyledon stage).

Seedlings inoculated by spraying the foliage 5 days after sowing gave 100% infection, but spraying 10 days after sowing gave no infection (Zimmer, 1975). Cohen and Sackston (1973a) obtained 50% infection in plants inoculated at the apical bud in the 8-leaf stage, but none at the 12-leaf stage.

### B. Moisture

Zoosporangia and oospores of *P. halstedii* both germinate to produce zoospores, which require free water to move to infection sites. Zimmer (1971) reported systemic infection up to 90% in fields subjected to heavy late spring rains after sowing, and showed (1975) that if there was enough rain 3 to 14 days after sowing to establish free soil water, the incidence of systemic infection would be high. Conspicuous secondary infections have also been observed in seasons with frequent rains during critical periods of plant growth (Zimmer, 1972; Kurnik *et al.*, 1976; Sackston, unpublished).

## C. Temperature

Novotelnova (1966) found the minimum temperature for zoosporangial germination to be 4°C, optimum 16 to 18°C, and maximum 22°C. Zoosporangia germinated in 15 to 60 min. at 16 to 20°C, remained mobile up to 28°C, and could induce infection between 6 and 26°C. Cohen and Sackston (1973a) found the optimum for infection for both WSI and bud inoculation to be 15°C, with less infection at 20 and 25°C and practically none at 30°C.

## D. Light

Novotelnova (1966) found that light had no significant effect on spore germination or seedling infection. Goossen and Sackston (1968) found that spores produced under continuous light were as infective as those produced in the dark. Inoculated seedlings incubated at low light intensity (about 6000 lux) showed much lower infection percentage than those incubated at 12 000 lux. Apparently lower light intensity permitted the hypocotyls to elongate more rapidly and outgrow the pathogen.

## E. Spore concentration

Cohen and Sackston (1973a) used the WSI method to inoculate with a logarithmic series of suspensions down to 8·5 zoosporangia per seedling. They also used a logarithmic series down to a calculated single zoosporangium per plant for bud inoculations on seedlings with two true leaves, and obtained systemic infection in 67% of the plants. High concentrations of spores in the WSI inoculations induced more damping off than did low concentrations. High concentrations in both inoculation methods resulted in infections at unfavourably high temperatures, and of plants which would have been too old for infection with lower concentrations.

## F. Age of spores

Age of spores was thought to have a marked effect on their viability (Novotelnova, 1966). Zoosporangia survived and remained infective for 14 days in sterilized or pasteurized soil, and 7 days in unpasteurized soil. Spores on leaves remained infective for 3 weeks in a refrigerator at 4°C, and for 14 weeks in a freezer at −20°C; they did not survive 1 week in the laboratory at 20 to 23°C (Goossen and Sackston, 1968).

## G. Hydrogen ion concentration

Sporangia produced on roots germinated well at pH 4·5 to 10. Zoospores tended to burst at pH below 6, but germinated at any pH above 6 which permitted zoosporangial germination. Obviously the pH of any soil in which sunflowers could grow would have little effect on infection by downy mildew (Delanoe, 1972; Delanoe and Hamant, 1972).

## VII. Host–Parasite Relations

### A. Infection sites

Nishimura (1922) stated that *P. halstedii* entered leaves of *H. divaricatus* through the stomata, and that "younger seedlings of the sunflower are easily attacked on their leaves by this fungus" (1926). Novotelnova (1966) observed germination of zoospores on sunflower leaves and penetration of the germ tubes through stomata. Goossen and Sackston (1968), however, reported difficulty in obtaining infection of expanded sunflower leaves, and Delanoe (1972) concluded that it was very rare or nonexistent. The contradictions were explained when Cohen and Sackston (1973a) succeeded in inducing infections by leaf inoculations of plants up to the 6-leaf stage, but not at the 12-leaf stage; they were able to infect older plants by inoculating apical growing points. Zimmer (1971, 1972) reported that secondary infection of sunflower leaves was severe for the first time in North Dakota in fields subjected to heavy late spring rains in 1970, and again in 1971. Allard and Lamarque (1974) and Allard (1978b) concluded not only that secondary infection of young leaves and apical meristems occurred in the field, but that it was responsible for much of the conspicuous systemic disease attributed by others to infection through the roots. Secondary infection of sunflower heads was reported for the first time and reproduced experimentally in the USSR by Tikhonov (1972); it was destructive in the field in Hungary in 1975

(Kurnik *et al.*, 1976). Allard (1978b) obtained infection in only a small percentage of heads by direct inoculation after the outer bracts had unfolded.

Nishimura's (1922) experimental demonstration that zoospores of *P. halstedii* encysted and germinated on root hairs and then penetrated through them and through the uncutinized epidermis of sunflower roots, was the first report of root infection by a fungus in the Peronosporaceae. It was confirmed by various authors (Novotelnova, 1966; Delanoe, 1972; Allard, 1978a; Wehtje and Zimmer, 1978). Wehtje and Zimmer (1978) cut serial sections of entire 3-day-old seedlings after soaking them in a zoospore suspension; they found an average of nine infections in the terminal 1 mm of root tip, 110–600 per mm in the zones of elongation and maturation, but none in the hypocotyl.

Delanoe (1972) had earlier observed marked attraction of zoospores to the elongation and root hair zones of healthy 3-day-old sunflower roots immersed in a spore suspension. Although there was no attraction to 1-day-old roots, inoculation of 1-day-old seedlings gave 100% infection. Similarly, although zoospores in suspension were not attracted to the hypocotyl, Cohen and Sackston (1973a) concluded that invasion of sunflower seedlings was more effective through the hypocotyl than through the roots. Allard (1978b) confirmed invasion through the hypocotyl.

## B. Penetration

Nishimura (1922) observed penetration through the middle lamella between epidermal cells of roots. There was evidence of chemical dissolution of epidermal cell walls and of mechanical pressure in the penetration process (Novotelnova, 1966; Wehtje and Zimmer, 1978; Wehtje *et al.*, 1979). Novotelnova (1966) reported penetration directly by narrow infection hyphae, and also following the formation of appressoria which gave rise to infection hyphae. Wehtje *et al.* (1979) did not mention appressoria. The fungus formed infection vesicles after penetrating epidermal cells; if it grew into another cell, it again formed a vesicle. If it grew into an intercellular space, it continued to produce intercellular hyphae. Branches from these grew into host cells and formed globose haustoria, which seemed to have no effect on host cytoplasm.

The fungus on leaves apparently penetrates through stomata without forming appressoria, and develops intercellularly (Novotelnova, 1966).

## C. Invasion

Melhus (1915) observed abundant mycelium of *P. halstedii* in *H. diversicatus*

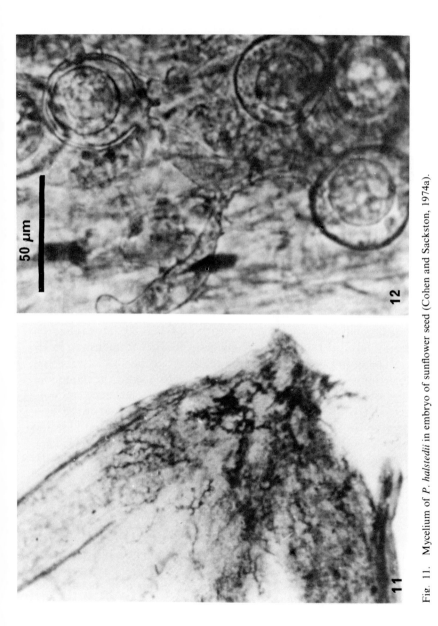

Fig. 11.  Mycelium of *P. halstedii* in embryo of sunflower seed (Cohen and Sackston, 1974a).

Fig. 12.  Oogonia and oospores of *P. halstedii* on inner surface of pericarp of sunflower achene (Cohen and Sackston, 1974a).

50 μm

12

11

(*H. divaricatus*) in all parts of the stem except the fibrovascular bundles. Nishimura (1922) found the pathogen in scalariform vessels of stems and in the vessels of roots in young sunflower plants before mature bundles had been differentiated. It developed well in medullary rays of the stem. It was particularly profuse in the spongy parenchyma of leaves. Novotelnova (1966) stated that the mycelium passes between cells of individual vessels in leaf veins. This would explain the characteristic development of chlorosis following the lines of the leaf veins in systemically infected plants (Nishimura, 1922; Sackston, 1949; Novotelnova, 1966).

Intercellular mycelium of the pathogen is irregular in size and shape, appearing to occupy all the space available to it. It may be present from the roots to the capitulum, and has been reported in petals, pistil, stamens, and seeds (Novotelnova, 1963; Delanoe, 1972). Apparently it cannot develop in the densely packed most active meristem, but can penetrate differentiated embryonic tissues (Novotelnova, 1963).

Novotelnova (1963) found mycelium in the embryo in small seeds from infected sunflower heads (Fig. 11). It was present in the testa of larger seeds, but only rarely in the embryo. Delanoe (1972) found that the fungus was present but that the embryo had disintegrated in seeds infected early in their development. The fungus was present in the testa of 30 to 100%, and in the hulls of 100% of large ripe seeds from infected heads.

Sporophores and zoosporangia of the pathogen form on mycelium which grows out through the stomata of infected leaves (Figs 13 and 14) and on the surface of infected roots in moist soil (Fig. 6). They form also in intercellular spaces, in substomatal cavities in the leaves, in cavities in stems and roots (Nishimura, 1922, 1926; Novotelnova, 1966) and in the capitulum and within seeds (Novotelnova, 1963; Delanoe, 1972; Cohen and Sackston, 1974a). Oospores, described by Saccardo (1888), were found in all host vegetative organs (Nishimura, 1922) but tended to be most plentiful in the roots and in leaves, often just under the epidermis (Nishimura, 1926). They are also formed in seeds, particularly in the hull (Novotelnova, 1963; Delanoe, 1972; Cohen and Sackston, 1974a) (Fig. 12).

## D. Physiology

Symptoms such as stunting and distortion and rugosity of leaves obviously result from interference with growth regulatory systems. Application of gibberellic acid ($GA_3$) to 5-day-old sunflower seedlings immediately after inoculation (Viswanathan and Sackston, 1968) or to plants with 5–8 leaves in field plots (Novotelnova, 1966) reduced severity of downy mildew symptoms and increased height, but to a lesser extent than in healthy plants. The disease followed its normal course after gibberellin treatment ceased.

100 µm

Fig. 13. Sporangiophores and sporangia of *P. halstedii* on surface of sunflower leaf. SEM photograph by Mr Louis Thauvette.

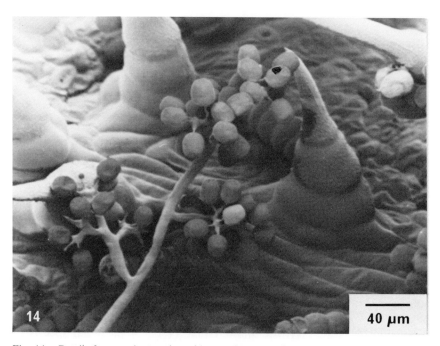

40 µm

Fig. 14. Detail of sporangiophore branching, sterigmata, and zoosporangia. SEM photograph by Mr Louis Thauvette.

A reduction in disease severity was also observed in seedlings treated with indol-3-ylacetic acid (IAA) and with a mixture of IAA and $GA_3$. IAA content was lower in extracts of diseased than healthy seedlings; analysis for $GA_3$ was inconclusive (Viswanathan and Sackston, 1968).

Cytokinin-like activity was higher in root exudates from mildewed than healthy plants. Addition of kinetin to healthy germinating sunflower seeds resulted in dwarfing and its addition to inoculated seedlings had an additive effect to cause even more dwarfing (Viswanathan and Sackston, 1970). Cohen and Sackston (1974b) did not obtain the kinetin effect in seedlings treated later after inoculation.

There was a correlation between area of mildew on leaves and the ability of stem sections to reduce IAA activity in solution; this might explain stunting, and the inhibition of phototropic and negative geotropic responses of mildewed sunflowers (Cohen and Sackston, 1974b). Cohen and Ibrahim (1975) considered that the quantitative changes in phenolic compounds, particularly of scopoletin, found in mildewed sunflowers might be related to reduced IAA content and Cohen (1975) was able to correlate the accumulation of fluorescing material, presumed to be mainly scopoletin, with high levels of peroxidase and $\beta$-glucosidase activity.

## VIII. Resistance

### A. Sources and inheritance

Resistance to downy mildew was observed in sunflower breeding lines produced at the Canada Department of Agriculture Research Station at Morden, Manitoba, and tested in heavily infested soil at Ste. Anne de la Pocatière, Québec (Putt, 1964; Perrault, personal communication). Perennial species of *Helianthus* in infested soil at the All Union Institute for Oil Crop Research (VNIIMK) at Krasnodar in the USSR remained free of downy mildew, although all the cultivated sunflowers tested were diseased (Pustovoit, 1963).

Results of such field tests were valuable but erratic because of variable weather conditions from year to year. Development of laboratory methods for testing seedlings made it possible to screen large populations quickly for sources of mildew resistance (Sackston *et al.*, 1963; Moldovan and Zhivilo, 1965; Pustovoit, 1966; Sackston and Goossen, 1966).

Seedling tests at VNIIMK showed that 11 of 24 perennial species of *Helianthus* were not affected by downy mildew (Pustovoit, 1966). Interspecific crosses between one of them, the hexaploid (2n = 102) *H. tuberosus*, and the diploid (2n = 34) *H. annuus*, yielded hybrids immune to mildew. Two such hybrids, Progress and Novinka, 99% resistant to mildew, were under test in the USSR State Varietal Trials in 1976 (Pustovoit et al., 1976). A mildew-resistant hybrid of H. tuberosus × H. annuus, a diploid line known as HIR-34, was obtained in France (Leclercq et al., 1970).

A sunflower inbred line (S37-388RR) developed in the program of breeding for rust resistance in Canada by Putt, appeared resistant to mildew in seedling tests (Sackston and Goossen, 1966; Goossen and Sackston, 1968). Vranceanu (1970) and Vranceanu and Stoenescu (1970) attributed downy mildew resistance in this and related lines to a single dominant gene designated $Pl$, apparently closely linked to the gene $R_1$ for rust resistance (Putt and Sackston, 1963). From this material they derived the line AD66 for producing mildew-resistant hybrids (Vranceanu and Stoenescu 1970).

Kinman (1968) in Texas released line HA-61 derived from the Canadian line 953-88-3 carrying gene $R_2$ for rust resistance. It proved highly resistant to downy mildew in field and laboratory tests in North Dakota in which S37-388RR and AD66 were susceptible (Zimmer, 1971; 1972). Resistance was conditioned by a single dominant gene designated $Pl_2$, inherited independently of the rust gene $R_2$ (Zimmer and Kinman, 1971; 1972). Vear and Leclercq (1971) found two independent genes for mildew resistance in HA61, both different from gene $Pl$ from Romania; all three were effective in France. Zimmer and Kinman (1972) redesignated the genes $Pl_1$ (in AD66 from Romania), and $Pl_3$ (the second gene in HA-61), both effective in Europe but not in North Dakota, and $Pl_2$, the first gene in HA-61, effective in North Dakota. Vear (1974) found that the gene from H. tuberosus in HIR-34 was different from the first three, and designated it $Pl_4$; it was effective in North Dakota. She also found that some lines with no demonstrable resistance genes were attacked less than susceptible controls in the field, and that others developed symptoms much more slowly than did the controls.

Sunflower breeders are searching for new sources and kinds of mildew resistance. Pogorletsky and Geshele (1975) reported resistance from H. tuberosus in TA 3722 and other sunflower lines at Odessa, USSR. Skoric et al. (1978) found mildew resistance in lines from many sources though the resistance genes had not then been identified. The classic concept that the best place to find resistance to diseases of cultivated plants is in the gene centres of those plants (Vavilov, 1951; Leppik, 1970a), is being applied to sunflower mildew. One hundred and thirty-six accessions of six species of Helianthus, most of them from Texas, were screened for mildew reaction in North Dakota. Twenty-three collections in three species were resistant; the genes have not yet been identified (Thompson et al., 1978).

Some doubts remain about the identity of the four $Pl$ genes. Zimmer (1974) found that resistance in HA-61 and HIR-34 was due to a common gene. Spanish workers (Fernandez-Martinez and Dominguez-Gimenez, 1978) found only one operative gene for resistance in HA-61; they suggested that the second gene (Vear and Leclercq, 1971) might have been lost. Vear (1978) believed that the action of two genes might be demonstrated more readily if resistance were defined as the lack of sporulation on the true leaves rather than cotyledons of inoculated seedlings. Piquemal (1978) found that a "mild" screening test demonstrated two genes for resistance in some progenies, whereas a "severe" test demonstrated only one gene.

## B. Physiological races

Discovery of mildew resistance in cultivated sunflowers was soon followed by the discovery of races in the pathogen population. Although all four *Pl* genes conditioned resistance in most European countries, only $Pl_2$ and $Pl_4$ were effective against the "Red River Valley race" in North Dakota (Zimmer, 1974). Collections of *P. halstedii* attacking host lines with the $Pl_1$ gene, but not $Pl_2$, were reported in Italy (D'Armini *et al.*, 1975), in Odessa, USSR (Pogorletsky and Geshele, 1975), and at Fundulea, Romania (Vranceanu *et al.*, 1978). These reports followed uncomfortably quickly after the expressed concern that the Red River race might be imported into Europe, or that comparable races might arise there (Sackston, 1974; Vear, 1974; Zimmer, 1974).

    Zimmer (1974) found only the Red River race in the main sunflower-producing area of North America, and suggested that the less aggressive European race may have evolved in Europe. This appears unlikely since the first reported resistance, in eastern Canada, was conditioned by the gene now known as $Pl_1$, indicating that the pathogen there was comparable to the European race. At the same time the wide distribution of the Red River race in north central North America shows that it is fit to survive and is not likely to revert to a form with narrower virulence, as postulated for some other pathogens by Van der Plank (1968). It is more likely that the European race was introduced from North America and went unnoticed in small loci until it was brought to areas of extensive sunflower culture with the movement of masses of people throughout the continent in the war and postwar years.

## C. Mechanisms of resistance

Very little is known about the mechanisms of resistance to mildew, but its inheritance, though still somewhat confused, is being worked out. Giant, late-maturing sunflower varieties had less mildew than short, early maturing ones in the USSR (Ventslavovich and Novotelnova, 1963). Plants with *Pl* genes appear to have some form of resistance localized in the hypocotyl and the cotyledonary node. This was shown by Delanoe (1972) who found the pathogen restricted to the lower hypocotyl of inoculated resistant seedlings. Montes and Sackston (1974) found that it sporulated very sparsely on the cotyledons of a few inoculated seedlings of lines with $Pl_1$ or $Pl_4$ genes, on none with $Pl_2$, but on 7 to 14% of seedlings of crosses between $Pl_2$ lines and susceptible plants. When the pathogen was injected with a hypodermic syringe into the first stem internode, it spread to the third or fourth internode of susceptible plants, but though it survived it did not spread beyond the nodes in plants with $Pl_2$ or $Pl_4$ genes.

Viranyi (1978), and Viranyi and Dobrovolszky (1980) also observed mycelium and oospores within hypocotyls of seedlings with the $Pl_2$ gene inoculated by WSI. Wehtje and Zimmer (1978), however, reported that zoospores of the Red River race encysted on seedling roots, but did not penetrate. Wehtje *et al.* (1979) found that a hypersensitive reaction was triggered in the resistant host roots, resulting in death of the zoospore before penetration occurred. Viranyi (1980) could find no mycelium in tissues of seedlings 2 days after inoculation, but after 4 days it was present in both resistant and susceptible seedlings, and a hypersensitive reaction began to develop in hypocotyls of the resistant seedlings.

## IX. Survival and Dissemination

Melhus demonstrated (1915) that the mycelium of *P. halstedii* survived over winter in the rhizomes of perennial *H. diversicatus* (properly *H. divaricatus*) and gave rise to infected shoots in the spring. Nishimura (1922) found that conidia (zoosporangia) and mycelium of the pathogen from annual sunflower could not survive the winter in soil, but that oospores could over-winter and induce infection in seedlings in infested soil next spring.

Oospores appear to be long lived. Air dried soil from a mildew plot retained its infectivity during 14 years storage at greenhouse temperature (C. Perrault, personal communication). Varying recommendations for disease control by crop rotation are based on assumptions that oospores survive for periods from 3 to 10 years, but there are no clear experimental results on their longevity under field conditions.

Conflicting opinions have been expressed on the significance of wind-borne zoosporangia in outbreaks of downy mildew. Delanoe (1972) and Delanoe and Hamant (1972) showed that zoosporangia survived wind transport for at least 1·5 km. They believed that such spores did not induce secondary infection of the foliage, but thought they might wash down into the soil and infect roots. Work in the USSR and elsewhere has indicated that secondary infection could occur through above-ground tissues (Novotelnova, 1966; Tikhonov, 1972; Kurnik *et al.*, 1976) and this was considered an important factor in spread of the disease by Allard and Lamarque (1974) and Zimmer (1972, 1975).

Sunflower and the mildew pathogen are both native to North America, and have spread by various routes to most parts of the world. It was natural, when the disease was reported in country after country where sunflower production became significant, to assume that the pathogen had been introduced with the seed.

In this context Nishimura (1922) proved that early systemic infection arose from oospore inoculum in the soil, and stated that mycelium did not develop in seed of infected plants whereas Young and Morris (1927) were at first convinced that infection on greenhouse plants arose from the seeds but they later changed their opinion (Young et al., 1929).

It was later shown by Novotelnova (1963) that the pathogen was present in a significant percentage of seeds from infected plants. The fungus was present only in the pericarp (hull) and testa of mature seeds of normal size, but in all tissues, including the embyro, of small, nonviable seeds (Novotelnova, 1963; Delanoe, 1972).

Seeds from infected plants gave rise to apparently healthy seedlings and symptoms were observed on fewer than 1 plant per 1000 in three years of study (Novotelnova, 1963), making it difficult to explain the observed rapid spread of mildew to new sunflower areas. It was found, however, that the pathogen sporulated on a high proportion of roots of symptomless plants from infected seed (Novotelnova, 1963; Cohen and Sackston, 1974a). Oospores remaining in the soil served to infect seedlings and induce symptoms of systemic infection in subsequent crops. It is not known why oospores which are present in the pericarp of infected seed give rise to plants with symptomless or "latent" infection. It is possible that they germinate more slowly than oospores that have been in the soil for some time and they then might not release zoospores until the seedling was past the 4-leaf stage. Root invasion at this or later stages gives rise to latent infection (Cohen and Sackston, 1973a; Tikhonov, 1972). Such symptomless plants may produce infected seed (Cohen and Sackston, 1974a; Tikhonov, 1972), explaining the failure of many investigators to detect seed transmission of P. halstedii.

## X. Control

The discovery that P. halstedii invaded sunflower roots and that its oospores could overwinter in the soil (Nishimura, 1922; Young et al., 1929), and observations by farmers and scientists that disease intensity increased with repeated sowings in the same soil, led naturally to the use of crop rotation as a control measure. Recommendations for length of rotation vary from sunflowers no oftener than once in 8 to 10 years, to once in 3 or 4 years. The longer rotations may be practical on the huge collective and state farms of the USSR, but even once in 4 years seems a long rotation to private farmers with small holdings. Most recommendations are based on observations rather than on experimental data (Novotelnova, 1966).

Removal (roguing) of mildewed seedlings during hand thinning to the desired spacing was recommended by various authors (Novotelnova, 1966). Although it might lessen secondary infection by airborne spores, such roguing is becoming increasingly impractical with the widespread use of precision seeders which eliminate the need for hand thinning, and with the increasing cost of hand labour in most parts of the world.

Attempts to escape or lessen infection by varying the date of seeding have not been successful wherever weather conditions are variable. Infection is governed not by the date of sowing but by the temperature and particularly the rainfall which follow it (Novotelnova, 1966; Zimmer, 1975; D'Armini et al., 1975). Where weather conditions are more predictable, sowing when temperatures favour speedy germination and growth so that plants remain susceptible for only a short time, may reduce infection. Agronomic considerations such as heat or drought during seed filling, maturing before early frost, etc., may be critical and determine if varying sowing dates is feasible.

Leppik (1962, 1964, 1970b) used literature references and the presence of P. halstedii on samples of sunflower seed to map the distribution of the pathogen and to propose measures to prevent its further spread. Unfortunately, he accepted mildew development in field plots as proof of seed transmission and unlike later workers (Novotelnova, 1963; Anselme and Planque, 1972; Cohen and Sackston, 1974a), he thought it was easy to demonstrate. He knew that mildew infection was impossible to detect during routine quarantine inspection, however, and recommended that imported seed should be grown in isolation under close observation for at least 2 years to lessen the danger of introducing mildew (Leppik, 1962). His recommendation has been repeated by others (Louvet and Kermoal, 1966; Cohen and Sackston, 1974a) and is a legal requirement in Australia (Morschel, personal communication). It is particularly important in the effort to exclude the Red River race; even cultivars and hybrids resistant to mildew may play a part in its dissemination (Sackston et al., 1976; Viranyi, 1978) and should be included in quarantines or embargoes.

Although treatment of seed with fungicides has been used successfully against many seed-borne diseases, none of the products tested prior to 1966 controlled downy mildew of sunflowers (Novotelnova, 1966). Soil fumigation also gave negative results. Allard (1972) tested about 80 preparations, including systemic fungicides, most of which were ineffective as protectants or eradicants. Streptomycin was the most effective protectant but proved to be phytotoxic. Cohen and Sackston (1973b) found that 10 $\mu$g ml$^{-1}$ of streptomycin sulphate inhibited sporangial germination and infection of sun-flowers, but Wehtje *et al.* (1979) routinely added 500 $\mu$g ml$^{-1}$ each of penicillin and of streptomycin sulphate to their inoculum to control bacterial parasitism of zoospores, and obtained infection.

The systemic fungicides pyroxychlor and prothiocarb, ineffective as seed treatments, apparently eradicated the pathogen if applied to seedlings within 4 days of inoculation (Sackston *et al.*, 1976). Recent work has shown that the systemic fungicide metalaxyl applied as a seed dressing prevents infection of seedlings exposed to massive doses of inoculum (Sackston, 1979; Steiner, 1980; Iliescu, 1980; Garcia-Baudin *et al.*, 1980). It is also effective as a thera-peutant when applied up to a month after infection to plants with symptoms of severe mildew infection (Sackston, unpublished; Jimenez-Diaz and Melero-Vara, unpublished).

The most effective and economically practical control of downy mildew at present is by sowing resistant cultivars. Mildew resistance is dominant. An increasing proportion of the sunflowers now grown are hybrids produced by crossing a cytoplasmically male-sterile line with a fertility-restoring line. It is relatively easy to produce mildew-resistant hybrids by making either of the parents homozygous for resistance.

Practically all the female lines used around the world derive their cyto-plasmic male sterility from the original material discovered by Leclercq (1969, 1971). Fertility restoration has been encountered from various sources, but the pool of genes conferring resistance now in use is very limited. Restorers with the $Pl_1$ and $Pl_4$ genes have been developed and used in Europe (Vranceanu *et al.*, 1974; Vear, 1974; Vulpe, 1976; Burlov and Kostyuk, 1976; Skoric *et al.*, 1978). Restorers derived from HA61 with $Pl_2$ and $Pl_3$ genes are in use by breeders practically everywhere in the world.

The danger of such widespread use of so few genes is obvious. Only two races or race-groups of *P. halstedii* have been described, but the probability is high that others are still to be discovered and that forms with new virulence patterns will arise. A one-step mutation might permit the current Red River race to attack all the cultivars or hybrids with resistance based on $Pl_2$, and possibly $Pl_4$.* If all new hybrids released possessed at least two genes for resistance effective against the mildew in their areas, the probability of attack by new mutant races would be very much reduced. The desirability of multiple genes for resistance has been recognized (Sackston, 1974; Vear, 1974; Zimmer and Kinman, 1972), and efforts are being made to produce hybrids with such resistance (Vear, 1974; Vranceanu *et al.*, 1974).

It is essential to increase the number of available genes for mildew resistance. Current work suggests that there may be extensive untapped pools of genes for disease resistance in populations of wild *H. annuus* in various

---

* What appears to be such a mutation was encountered in two locations in the northern USA in late 1980 (G. N. Fick, personal communication).

parts of its range, and probably also in other annual species (Thompson *et al.*, 1978). Interspecific crosses frequently give sterile hybrids, but techniques such as embryo culture have given encouraging results in overcoming some of the difficulties with sunflowers (Chandler and Beard, 1978).

## XI. Problems and Future Research

The literature on downy mildew of sunflowers is fairly extensive, but only Nishimura (1922, 1926), Novotelnova (1966) and Delanoe (1972) published comprehensive studies, and only Novotelnova dedicated herself to a long-term study of the disease. A great deal remains to be done, some of it experimentation of practical significance, but much of a more theoretical nature.

How long can oospores survive in various types of soil under various climatic conditions in the absence of sunflowers, and still induce disease when sunflowers are sown again? How long can oospores survive on or in infected seed under various conditions of storage, and still serve to infest the soil when that seed is sown? Can readily applicable agricultural practices, or treatments with acceptable chemicals, affect oospore longevity in either situation?

Continued success in breeding for resistance depends on finding new sources of resistance in wild *H. annuus*, and in other annual and perennial *Helianthus* spp. A great deal of taxonomic and of practical interest could be learnt from an extensive, long-term, careful study of the host range of the *P. halstedii* complex based on inoculation of germinating seeds, or other tissues found appropriate by experiment, of many individuals of many species, under controlled and repeatable conditions.

Downy mildew of sunflowers has not been found in California after intensive searching despite massive importations of seed from the Red River Valley where the disease is prevalent (Claasen, Kliesewics, personal communications). The statement by Leppik (1965) that *P. halstedii* was a common disease of sunflowers on the west coast is misleading, based on his system of determining distribution by sowing seed from various sources in field plots in Iowa. If California is indeed free from *P. halstedii*, is this due to climatic conditions, or to some mildew repressive factors in the soil in sunflower areas?

Downy mildew of sunflowers has not yet been found in Australia or South Africa, and stringent embargoes and quarantines have been established to try to keep it out. The Red River race or its equivalent is nominally absent from most of Europe, and efforts are being made to prevent its introduction from the USA. Plant breeders eager to exploit valuable characters of parental materials from the USA, and seed companies eager to exploit the economic potentials of USA hybrids are finding it difficult to import seed from America. If it can be shown that *P. halstedii* is unable to establish itself and infect sunflowers in California, importation of seed which has been increased there for at least two or three years may pose little risk of bringing mildew inoculum with it.

Will climatic conditions over all or part of Australia, North Africa, and South Africa prove more effective than the strictest embargoes or quarantines in restricting the establishment of mildew? Simulations in controlled environment facilities of conditions likely to occur during early development of sunflowers in various regions may be useful in determining the probability of mildew establishment, but they are difficult, time consuming, and expensive.

Sunflower seedling tissues very rapidly become resistant to *P. halstedii*. Heritable resistance to the pathogen is also known. Do these two types of resistance have anything in common? Is the mechanism of resistance to the European race different from resistance to the more virulent Red River race? The answers to these questions, of theoretical interest, also have great significance in the epidemiology of the disease.

What is the significance of secondary infection of the foliage by zoo-sporangia? Some observers believe it has no significant effect on yield. If it occurs sufficiently early, and gives rise to systemic infection as believed by others, it might reduce yield, as well as increase inoculum in the soil. Its possible role in seed infection may be of greatest importance.

What determines if the pathogen will infect the seed? What determines the location and quantity of the pathogen in specific tissues of the seed? How does the location, amount, or form (mycelium or oospores) of the pathogen in the seed determine if the seed is viable, or whether the resulting seedling will present typical mildew symptoms or appear healthy?

Finding the answers to some of these questions will keep many Ph.D. candidates and other researchers occupied for years to come.

### References

Allard, C. (1972). Rapport 1970–1971, CETIOM Inform. Tech. 49–58.
Allard, C. (1978a). *Ann. Phytopathol.* **10**, 197–217.
Allard, C. (1978b). CETIOM Inform. Tech. **62**, 3–10.
Allard, C. and Lamarque, C. (1974). *Proc. 6th Int. Sunflower Conf.* 631–638. Bucharest, Romania.
Anon. (1977). Distribution maps of plant diseases No. 286 (4th edn.). Comm. Mycol. Inst. Slough, UK.
Anselme, C. and Planque, J. P. (1972). CETIOM Inform. Tech. **29**, 61–66.
Bisby, G. R., Buller, A. H. R., Dearness, J., Fraser, W. P. and Russell, R. C. (1938). "The Fungi of Manitoba and Saskatchewan". Nat. Res. Counc. Canada, Ottawa, Canada.
Burlov, V. V. and Kostyuk, S. V. (1976). *Proc. 7th Int. Sunflower Conf.* **1**, 322–326 Krasnodar, USSR.

Carter, J. (1978). *In* "Sunflower Science and Technology". (J. F. Carter, Ed.) Agron. 19. Am. Soc. Agron., Preface. ix–x. Madison, USA.

Chandler, J. M. and Beard, B. H. (1978). *Proc. 8th Int. Sunflower Conf.* 510–515. Minneapolis, USA.

Cohen, Y. (1975). *Physiol. Pl. Pathol.* **7**, 9–15.

Cohen, Y. and Ibrahim, R. K. (1975). *Can. J. Bot.* **53**, 2625–2630.

Cohen, Y. and Sackston, W. E. (1973a). *Can. J. Bot.* **51**, 15–22.

Cohen, Y. and Sackston, W. E. (1973b). *Phytopathology* **63**, 200.

Cohen, Y. and Sackston, W. E. (1974a). *Can. J. Bot.* **52**, 231–238.

Cohen, Y. and Sackston, W. E. (1974b). *Can. J. Bot.* **52**, 861–866.

Conners, I. L. (1967). "An Annotated Index of Plant Diseases in Canada". Can. Dept. Agric. Publ. No. 1251. Ottawa, Canada.

D'Armini, M., Monotti, M. and Zazzerini, A. (1975). *In* "Atti Giornate Fitopatol". 1975, 1–8. Perugia, Italy.

Delanoe, D. (1972). CETIOM Inform. Tech. **29**, 1–49.

Delanoe, D. and Hamant, C. (1972). *Proc. 5th Int. Sunflower Conf.* 152–155. Clermont-Ferrand, France.

Enns, H., Dorrell, D. G., Hoes, J. A. and Chubb, W. O. (1970). *Proc. 4th Int. Sunflower Conf.* 162–167. Memphis, USA.

Fernandez-Martinez, J. and Dominguez-Gimenez, J. (1978). *An. Inst. Nac. Invest. Agrar. Prod. Veg.* **8**, 105–111.

Fucikovsky-Zak, L. (1976). "Enfermedades y Plagas del Girasol en Mexico". Col. Postgr. Esc. Nac. Agric., Chapingo, Mexico.

Garcia-Baudin, C., Melero-Vara, J. M. and Jimenez-Diaz, R. M. (1980). *Proc. 9th Int. Sunflower Conf.* Abstracts p. 21.

Gomez-Menor, J. (1936). In *Rev. Appl. Mycol.* **15**, 657 (1936).

Goossen, P. G. and Sackston, W. E. (1968). *Can. J. Bot.* **46**, 5–10.

Heiser, C. B. (1951). *Proc. Am. Phil. Soc.* **95**, 432–448.

Heiser, C. B. (1955). *Am. Biol. Teach.* **17**, 161–167.

Hingorani, M. K. (1971). *In* "Rep. Inst. Agric. Res., Ethiopia, April 1970 to March 1971", 45–50. Addis Ababa, Ethiopia. In *Rev. Plant Pathol.* **52**, 685 (1973).

Iliescu, H. (1980). *9th Sunflower Conf.* Abstract p. 16.

Jimenez-Diaz, R. M. (1973). *An. Inst. Nac. Invest. Agric. Ser. Prot. Veg.* **3**, 95–105.

Kinman, M. L. (1968). *Proc. 3rd Int. Sunflower Conf.* 131–137. Crookston, USA.

Kinman, M. L. (1970). *Proc. 4th Int. Sunflower Conf.* 181–183. Memphis, USA.

Kurnik, E., Léránth, J., Parragh, J. and Vörös, J. (1976). *Proc. 7th Int. Sunflower Conf.* **2**, 205–209. Krasnodar, USSR.

Leclercq, P. (1969). *Ann. Amélior. Pl.* **19**, 99–106.

Leclercq, P. (1971). *Ann. Amélior. Pl.* **21**, 45–54.

Leclercq, P., Cauderon, Y. and Dauge, M. (1970). *Ann. Amélior. Pl.* **20**, 363–373.

Leppik, E. E. (1962). *Pl. Prot. Bull. FAO* **10(6)**, 126–129.

Leppik, E. E. (1964). *Proc. Int. Seed Test Ass.* **29(3)**, 473–477.

Leppik, E. E. (1965). *Pl. Dis. Reptr.* **49**, 940–942.

Leppik, E. E. (1966a). *Pl. Prot. Bull. FAO* **14(4)**, 72–76.

Leppik, E. E. (1966b). Crops Res. Div., ARS, USDA Pl. Intr. Invest. Paper 8, 1–4.

Leppik, E. E. (1970a). *A. Rev. Phytopathol.* **8**, 323–344.

Leppik, E. E. (1970b). *Proc. Int. Seed Test Ass.* **35(1)**, 3–9.

Louvet, J. and Kermoal, J. P. (1966). C.r. Hebd Séanc *Acad. Agric. Fr.* **52**, 896–902.

Melhus, I. E. (1915). *J. Agric. Res.* **5**, 59–69.

Moldovan, M. Y. and Zhivilo, B. (1965). *All Union Conf. on Immunity Agric. Pl.* **4**, 184–186. Kishinev, Moldavia, USSR.

Montes, F. and Sackston, W. E. (1974). *Proc. 6th Int. Sunflower Conf.* 623–629. Bucharest, Romania.

Nishimura, M. (1922). *J. Coll. Agric. Hokkaido Imp. Univ.* **11**, 185–210.

Nishimura, M. (1926). *J. Coll. Agric. Hokkaido Imp. Univ.* **17**, 1–61.

Novotelnova, N. (1961). *Proc. Sci. Conf. Pl. Prot. 1960*, 129–138. Tallin, Estonia.

Novotelnova, N. S. (1962). *Bot. J. Acad. Sci.* USSR **47(7)**, 970–981.

Novotelnova, N. S. (1963). *Bot. J. Acad. Sci.* USSR **48(6)**, 845–860.

Novotelnova, N. S. (1965). *Bot. J. Acad. Sci.* USSR **50(3)**, 301–312.

Novotelnova, N. S. (1966). "Downy Mildew of Sunflowers". (In Russian). Nauka, Acad. Sci. USSR, Moscow.

Orellana, R. G. (1970). *Bull. Torrey Bot. Club* **97**, 91–97.

Piquemal, G. (1978). *Proc. 8th Int. Sunflower Conf.* 254–257. Minneapolis, USA. (Translation in English) In *Sunflower Newsletter*, Vol. **3(2)**, 9–10 (1979). Int. Sunflower Ass., Zevenaar, Holland).

Pogorletsky, B. K. and Geshele, E. E. (1975). *Genetica* **11**, 21–28.

Pontis, R. E., Feldman, J. M. and Klinger, A. (1959). *Pl. Dis. Reptr.* **43**, 422.

Pustovoit, G. V. (1963). *In* "Oil and Essential Oil Crops". 75–92. (V. S. Pustovoit, Ed.), Agric. Lit. Publishers, Moscow, USSR.

Pustovoit, G. V. (1966). *Proc. 2nd Int. Sunflower Conf.* 82–100. Morden, Canada.

Pustovoit, G. V., Ilatovsky, V. P. and Slusar, E. L. (1976). *Proc. 7th Int. Sunflower Conf.* **1**, 193–204. Krasnodar, USSR.

Putt, E. D. (1964). *Proc. 1st Int. Sunflower Conf.* 6 pp. Coll. Stn., Texas, USA.

Putt, E. D. (1978). *In* "Sunflower Science and Technology" (J. F. Carter, Ed.). Agron. 19, 1–29. Am. Soc. Agron., Madison, USA.

Putt, E. D. and Sackston, W. E. (1963). *Can. J. Pl. Sci.* **43**, 490–496.

Saccardo, P. A. (1888). *Sylloge fungorum.* **7**, 242.

Sackston, W. E. (1949). *Can. Pl. Dis. Surv.* **28**, 31–33.

Sackston, W. E. (1954). *Can. Pl. Dis. Surv.* **33**, 45–48.

Sackston, W. E. (1955). *Can. Pl. Dis. Surv.* **34**, 48–50.

Sackston, W. E. (1956). *Pl. Dis. Reptr.* **40**, 744–747.

Sackston, W. E. (1957). *Pl. Dis. Reptr.* **41**, 885–889.

Sackston, W. E. (1974). *Proc. 6th Int. Sunflower Conf.* 619–622. Bucharest, Romania.

Sackston, W. E. (1978). *Proc. 8th Int. Sunflower Conf.* 7–29. Minneapolis, USA.

Sackston, W. E. (1979). *Sunflower Newsletter* **4**, 7–8.

Sackston, W. E. and Goossen, P. G. (1966). *Proc. 2nd Int. Sunflower Conf.* 40–45. Morden, Canada.

Sackston, W. E., Miah, M. A. Jabbar, Goossen, P. G. and Devaux, A. L. (1963). *Proc. Can. Phytopathol. Soc.* Abstr. **30**, 17.

Sackston, W. E., Romero-Muñoz, F. and Garcia-Baudin, C. (1976). *Proc. 7th Int. Sunflower Conf.* 187–192. Krasnodar, USSR.

Skoric, D., Cuk, L., Mihaljcevic, M. and Marinkovic, R. (1978). *Proc. 8th Int. Sunflower Conf.* 423–426. Minneapolis, USA.

Snyder, W. C. and Hansen, H. N. (1940). *Am. J. Bot.* **27**, 64–67.

Steiner, H. (1980). *9th Int. Sunflower Conf.* Abstr. p. 8.

Stevens, F. L. (1913). "The Fungi which Cause Plant Disease". Macmillan, New York.

Thompson, T. E., Rogers, C. E., Zimmerman, D. C., Huang, H. C., Whelan, E. D. P. and Miller, J. F. (1978). *Proc. 8th Int. Sunflower Conf.* 501–509. Minneapolis, USA.

Tikhonov, O. I. (1972). *Proc. 5th Int. Sunflower Conf.* 156–158. Clermont-Ferrand, France.

Van der Plank, J. E. (1968). "Disease Resistance in Plants". Academic Press, New York and London.

Vavilov, N. I. (1951). "The Origin, Variation, Immunity and Breeding of Cultivated Plants".

(Translated from Russian by K. Starr Chester.) Ronald Press, New York.

Vear, F. (1974). *Proc. 6th Int. Sunflower Conf.* 297–302. Bucharest, Romania.

Vear, F. (1978). *Ann. Amélior. Pl.* **28**, 327–332.

Vear, F. and Leclercq, P. (1971). *Ann. Amélior. Pl.* **21**, 251–255.

Ventslavovich, F. S. and Novotelnova, N. S. (1963). *Sci. Work Kuban Expt. Stn. VIR, USSR* **2**, 123–130.

Viranyi, F. (1977). *Acta Phytopathol. Acad. Sci. Hung.* **12**, 263–267.

Viranyi, F. (1978). *Phytopath. Z.* **91**, 362–364.

Viranyi, F. (1980). *Sunflower Newsletter* **4**, 11–13.

Viranyi, F. and Dobrovolszky, A. (1980). *Phytopath. Z.* **97**, 179–185.

Viswanathan, M. A. and Sackston, W. E. (1968). *Proc. 3rd Int. Sunflower Conf.* 89–91. Crookston, USA.

Viswanathan, M. A. and Sackston, W. E. (1970). *Proc. Can. Phytopath. Soc.* Abstr. **37**, 28.

Vranceanu, V. (1970). *Proc. 4th Int. Sunflower Conf.* 136–141. Memphis, USA.

Vranceanu, V. and Stoenescu, F. (1970). *Probleme Agric.* **2**, 34–40.

Vranceanu, A. V., Stoenescu, F. M., Iliescu, H. and Parvu, N. (1974). *Proc. 6th Int. Sunflower Conf.* 367–371. Bucharest, Romania.

Vranceanu, A. V., Pirvu, N. and Iliescu, H. (1978). *Proc. 8th Int. Sunflower Conf.* 328–333. Minneapolis, USA.

Vulpe, V. (1976). *Proc. 7th Int. Sunflower Conf.* **1**, 296–309. Krasnodar, USSR.

Waalwijk van Doorn, J. J. L. van. (1978). *Proc. 8th Int. Sunflower Conf.* 1–6. Minneapolis, USA.

Wehtje, G. and Zimmer, D. E. (1978). *Phytopathology* **68**, 1568–1571.

Wehtje, G., Littlefield, L. J. and Zimmer, D. E. (1979). *Can. J. Bot.* **57**, 315–323.

Young, P. A. and Morris, H. E. (1927). *Am. J. Bot.* **14**, 551–552.

Young, P. A., Jellison, W. L. and Morris, H. E. (1929). *Science* **69**, 254.

Zazzerini, A. and Raggi, V. (1974). *Sementi Elette* **20**, 21–25.

Zimmer, D. E. (1971). *Pl. Dis. Reptr.* **55**, 11–12.

Zimmer, D. E. (1972). *Pl. Dis. Reptr.* **56**, 428–431.

Zimmer, D. E. (1973). *Pl. Dis. Reptr.* **57**, 647–649.

Zimmer, D. E. (1974). *Phytopathology* **64**, 1465–1467.

Zimmer, D. E. (1975). *Phytopathology* **65**, 751–754.

Zimmer, D. E. and Hoes, J. A. (1978). *In* "Sunflower Science and Technology". (J. F. Carter, Ed.). Agron. 19, 225–262. Am. Soc. Agron., Madison, USA.

Zimmer, D. E. and Kinman, M. L. (1971). *Phytopathology* **61**, 1026.

Zimmer, D. E. and Kinman, M. L. (1972). *Crop Sci.* **12**, 749–751.

Zimmer, D. E. and Zimmerman, D. C. (1972). *Crop Sci.* **12**, 859–861.

Chapter 27

# Downy Mildew of Tobacco

## PIERRE SCHILTZ

*Tobacco Experimental Institute, Bergerac, France*

## I. Introduction

The sudden catastrophic increase in tobacco blue mould in Europe in 1960 and the extensive damage it caused attracted the attention of pathologists all over the world. Circumstances favourable to its development and the inexperience of tobacco growers in coping with the problem resulted in *Peronospora tabacina* causing great damage to crops. In western Europe, the tobacco crop decreased by 27500 tons (CORESTA*, 1960) and the financial loss amounted to US $50 million (Todd, 1961). In 1961 losses again accounted for 75% of the crop in Algeria and 65% in Italy (CORESTA, 1961b).

## II. The Disease

### A. History

Reports of the presence of a blue mould on tobacco were first published at the end of the nineteenth century in Australia. Farlow (1885) gave the name *P. hyoscyami* to a parasite of *N. glauca*, and Harkness (1885) reported the presence of a *Peronospora* sp. on *N. bigelovii*. An outbreak on cultivated tobacco was reported by Cooke in 1891, and Spegazzini (1891, 1895 and 1897 in McGrath and Miller, 1958) described *P. nicotianae* on *N. longiflora*. McAlpine (1900) assumed it to be *P. hyoscyami*, which developed on *N. suaveolens* and *N. glauca*.

Adam (1925) thought it was the same fungus that attacked *N. tabacum* and *N. suaveolens*. This observation was confirmed by Angell and Hill (1932) who also reported that *Hyoscyamus niger* was not susceptible to this fungus. In 1933 Adam gave the name *Peronospora tabacina* to the parasite infecting tobacco.

### B. Taxonomy

In 1935, Clayton and Stevenson stated that the conidia of the blue mould infecting cultivated tobacco did not produce zoospores when they germinated as did the conidia from *P. nicotianae* described by Spegazzini. Clayton and Stevenson (1943) undertook further studies in an effort to differentiate between *P. tabacina* and *P. hyoscyami*; however, they accepted the name *P. tabacina* after Wolf *et al.* (1934), as they observed that *Hyoscyamus niger* was not affected. Kröber and Weinmann (1964b) compared the morphological differences between *P. tabacina* and *P. hyoscyami*. In the same year, Skalicky

---

*CORESTA (Cooperation Centre for Scientific Research Relative to Tobacco) is an international organization instituted in 1956 and which now has 128 members in 54 different countries. The CORESTA Phytopathology Study Group deals with problems relating to blue mould, but the warning service is run by the General Secretariat in Paris.

(1964) deemed these criteria to be insufficient and proposed that *P. tabacina* should be a sub-species of *P. hyoscyami*. More recently, Shepherd (1970) differentiated the three strains of blue mould previously isolated by Hill (1963b) (see Section VI.D) as follows: $APT_2$ which attacks hybrid *N. tabacum* obtained from interspecific crosses (particularly with *N. debneyi* and *N. good-speedii*), would be called *P. hyoscyami* f. sp. *hybrida*; *P. hyoscyami* f. sp. *tabacina* would be retained for $APT_1$ and *P. hyoscyami* f. sp. *velutina* for $APT_3$. Shepherd based his proposal on the morphological dissimilarity observed between $APT_2$ and $APT_3$ and on the differing pathogenicity of the three strains. As yet no other authors have commented on this suggestion.

## C. Geographical distribution

The first appearance of blue mould on tobacco was reported by Cooke in Australia in 1891, where it is considered to be the most important disease affecting tobacco both in seedbeds and in the field. Since 1929 the Common-wealth Scientific Industrial Research Organization (CSIRO) has been responsible for dealing with all problems connected with the disease in Australia.

Blue mould was first reported in the United States in 1921 (Smith and McKenney), but some people believe that Harkness (1885) had already observed the parasite (Stevenson and Archer, 1940) and that *Nicotiana repanda* a wild, spontaneous and susceptible species contributed to the spread of the fungus (Godfrey, 1941; Wolf, 1947).

Blue mould was also reported in Canada and Brazil (1938), in Argentine (1939), Chile (1953), Cuba (1957) and in Mexico (1964).

The parasite was unknown in Europe until 1958, when it was introduced accidentally into England and spread very rapidly. It was found in the Nether-lands and the German Federal Republic in 1959, and in 4 years had colonized the whole of Europe, North Africa and the Near East (Corbaz, 1964). Its progress towards southern Africa was probably stopped by the Sahara, whereas it progressed eastwards and was reported to have reached Cambodia by 1969 (Anon., 1969).

## D. Damage

In Australia, the risk is permanent (Anon., 1961) and *P. tabacina* can occur not only in the field, but also in the seedbed where infection often leads to serious consequences (Hill *et al.*, 1967). In the United States the disease is most often reported on seedbeds but Lucas (1975) states that no serious outbreak occurred there between 1960 and 1974. Losses in the field due to the parasite in one or the other regions of the United States were reported in 1921, 1931, 1932, 1951, 1954 and 1963. In 1979 the United States, Cuba and Canada reported serious losses at a local level owing to abnormally cool weather conditions and heavy rain. The estimated total loss was over US $250 million in the United States and Canada.

In Europe it is one of the most serious diseases affecting tobacco and as is the case in Australia, infection may occur at all stages of tobacco growth. Every year the parasite is reported in at least one of the countries of the Mediterranean basin where it is endemic and where crops are grown from 1 to 3 months earlier than in more northern regions. Every spring, Blue mould appears in European countries with more temperate climates, causing more or less serious damage depending on micro-climatic conditions, the time of appearance of the first outbreaks and the tobacco cultivars being grown.

However, since control of the disease has been organized, notably by CORESTA, the resulting damage has been less serious and has occurred less frequently than in the period 1959–1961. Nevertheless, in 1969 when countries bordering the Atlantic had a cold, rainy spring, losses were estimated at 25% in Morocco, 20% in Spain, 5% in France (15% in south-west France) and 15% in Belgium (Peyrot, 1969).

## III. The Pathogen

### A. Development of the parasite on a susceptible host

The germ tube of the spore only rarely enters through a stoma and usually penetrates directly into the epidermal cells on the upper surface of the leaf (Henderson, 1937). Two distinct phases appear to exist in the intercellular development of the fungus (Schiltz, 1967). During the "expansion phase", which occupies approximately the first 48 h after inoculation, tissues are invaded. During this time the mycelium can grow to a length of 400 $\mu$m, with a cross section of about 2 $\mu$m and the many finger-like haustoria are from 6 to 10 $\mu$m in length with a diameter of 0·5 to 1 $\mu$m (Fig. 1).

Later, between 5 and 7 days after infection, the parasite considerably enlarges its surface in contact with the host, the conidiophores are built up and sexual processes are initiated. This is the "growth phase", characterized by an increase in the cross section of hyphae (to 6–12 $\mu$m in infected N. *glutinosa*) and a proliferation of suckers, which are sometimes branched and may reach a length of 25 $\mu$m. More recently studies have been made by Marte and Caporali (1974) on the nature of infected tissue (presence of the fungus in the apical meristem) and by Kröber and Petzold (1972), Vassiley and Bogo-yavlenskaya (1975) on the ultrastructure of tobacco cells (particularly on the reaction of the plant).

## B. Anomalies of development

Development of hyphae is inhibited when conditions for infection are unfavourable, owing either to action directly affecting the parasite or to certain properties of the host (see also Section IV.A). Most frequently, widespread mycelial expansion is limited while further colonization of already invaded tissues continues and fructification does not occur. The fungus is confined either to an area close to the veins in the case of systemic infection (Mandryk, 1960) or to a necrotic area in the case of a host with genetic resistance (Shepherd and Mandryk, 1967; Schiltz, 1967).

## C. Reproductive organs

### 1. Conidia

In a very susceptible host, sporulation occurs 5–7 days after inoculation. An undifferentiated mycelium is sent out from tissues through a stoma and quickly adopts a tree-like shape (through false dichotomy). A vesicle appears at the end of each branch and fills with protoplasm before being individualized into a spore or conidium. The place of emergence and the form of the fructifications depend on ambient light and moisture, and spores vary greatly in size (from 15 to 25 $\mu$m × 10 to 17). The high density of these hyaline conidia and conidiophores present a blue-grey felt-like appearance that is one of the criteria for identifying the disease (Figs 2 and 3). Conidia contain from 8 to 24 nuclei with a diameter in the range of 1·2 to 1·5 $\mu$m. The nuclear material of these structures is very quickly broken down by the fixing agents used in cytology (Izard and Schiltz, 1963a), which may explain their sensitivity to ultraviolet rays (Angell and Hill, 1932; Shepherd, 1962; Jankowski, 1967b; and Abedi, 1971). This DNA lability is a very important point for the comprehension of variations in pathogenecity (see Section VI.D).

### 2. Oospores

Sexual reproduction takes place within tissues and the phenomenon has been followed *in vivo* (Schiltz, 1967). Gametes are constituted from terminal vesicles which are progressively filed with protoplasm. The oogonia have a cross section of 43–53 $\mu$m and the antheridia 7·5 × 22 $\mu$m. Oospores obtained in the laboratory vary greatly in size (24–40 $\mu$m) and are characterized by their brown colour.

   Zonation occurs before fertilization, during which the oogonial protoplasm is concentrated at the periphery, leaving a fine network in the centre (Fig. 4).

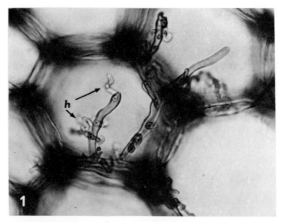

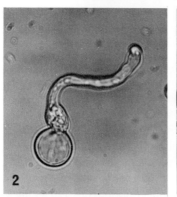

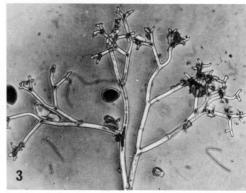

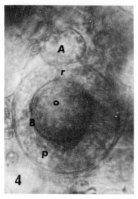

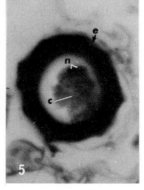

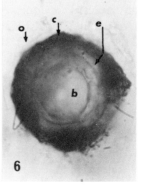

The two fertilization nuclei remain distinct when the structure of the egg appears to be definitely established (Fig. 5) (Schiltz, 1967).

The oospore is characterized by an ooplasm consisting of a single voluminous nucleus containing 4–5 chromophilia granules and a central body previously described in the Peronosporaceae (Dangeard, 1890; Kin Chou Tsang, 1929; Gustavson, 1959). This ooplasm is surrounded by two membranes; the endospore which is thin and slightly tinted and the covering or exospore, a reddish-brown colour and bearing markings (Fig. 6). The oogonial membrane with its wrinkled surface encloses the whole karyogamy, which resembles that already described for *P. parasitica* (Wager, 1900) and *P. effusa* (Kin Chou Tsang, 1929).

# IV. Epidemiology of Tobacco Blue Mould

## A. Symptomatology

Symptoms depend on the genetical constitution and growth stage of the host, and on its physiology which in turn is conditioned by the environment.

### 1. In the seedbed

The fungus can infect very young susceptible plants and does not necessarily cause characteristic symptoms either during its incubation or in the first few days following the appearance of conidiophores (Fig. 7).

Subsequently, tissue disorders initiate the formation of a chlorotic spot that later turns into a necrosis. If blue mould meets any difficulty in the course of its growth either owing to the physiology of the plant or to exterior causes (e.g. high temperature or fungicides products), it may adopt a systemic form. In Australia, the use of insufficiently concentrated benzol vapour helps to mask the presence of the parasite and so *P. tabacina* can cause damage

---

Fig. 1.  Mycelium with haustoria (h).

Fig. 2.  Germinating conidium.

Fig. 3.  Conidiophore.

Fig. 4.  Antheridium (A) and zonating oogonium (B) with networked ooplasm (O) and granulous periplasm (p) which is reduced near antheridium for constituting the "receiver spot" (r).

Fig. 5.  Oospore with two fertilization nuclei (n), an exospore (e) and cytoplasmic corpuscles (c).

Fig. 6.  Ripe oospore with residues of oogonial membrane (o), exospore or covering (c), endospore (e) and central body (b).

in the field without sporulation (Angell, 1957). In France, systemic attacks of the fungus have often been observed on both susceptible and resistant varieties in which young leaves are deformed and growth of the terminal meristem is arrested (Schiltz, unpublished).

## 2. In the field

Various symptoms are found on adult plants. In general, a chlorotic halo appears before conidiophore emergence (Fig. 8). In a very severe case a susceptible plant can be destroyed by blue mould in 3 or 4 weeks, only the stem remains showing blackish patches that reveal a systemic infection (Fig. 9).

In the USA outbreaks are limited to leaf tissues (McGrath and Miller, 1958) whereas in Australia and Europe all forms of blue mould attack are observed (Paddick et al., 1967).

Symptoms are usually fairly characteristic on resistant varieties grown commercially in Europe. In an outbreak of average intensity some chlorotic spots, accompanied by sporulation, are observed on the lower part of the plant and necrotic lesions without fructification occur at the middle leaf stage. Sometimes deformations are occasioned on one or more top leaves, by systemic blue mould (Figs 10 and 11).

## B. Survival of the parasite

P. tabacina can survive as mycelium, and overwinter in the host tissues (Darnell-Smith, 1929; Adam, 1931; Hill and Angell, 1933; Mandryk, 1960; Schiltz, 1967), but overwintering is more usually by means of conidia and oospores.

## 1. Conidia

Conidia are produced in enormous numbers (up to 200 000 to 1 500 000 per

---

Fig. 7. Fructification on the lower side of a young leaf.

Fig. 8. Yellow spots preceding necrosis.

Fig. 9. Very susceptible tobacco destroyed by blue mould.

Fig. 10. and Fig. 11. Puckering and distorting symptoms caused by systemic infection.
(Fig. 10)—on a leaf. Characteristic form observed on resistant varieties, which is followed by a starch accumulation in the upper part of the leaf.
(Fig. 11)—on the top of the plant. These damages are met on resistant or susceptible N. tabacum.

leaf according to Corbaz (1961a), but it is generally agreed that very few survive for very long. In this connection, Hill showed that 1% of conidia remained viable after 131 days in the laboratory at 5°C and at 30–40% relative humidity (Anon., 1963). Kröber (1965) indicated that conidia can survive for 4 to 5 months in dry soil. In 1967, Peresypkin and Zrazhevkaya demonstrated that conidia are inactivated in 12–24 h when they remain on the tobacco plant and in only a few minutes if they are exposed to sunshine, but at 20 to 40% relative humidity, they can remain alive for longer than 2 months. Later Jankowski (1967b) showed that dried conidia were more resistant to heat and Bebiya (1975) found that they remained viable for 113 days on tobacco plant in glass vessels. Bromfield and Schmitt (1967) were able to keep conidia alive for 25 months in a suspension of dimethyl sulphoxide solution at −180°C.

Wolf (1957) estimated that in 1947 conidia were transported over a distance of 1600 km. The capacity of conidia for survival and for dispersing over long distances therefore makes the part they play in epidemiology of the greatest importance.

## 2. Oospores

These are not often implicated in Australian epidemiology, but in the United States, on the other hand, some infections are attributed to oospores as the disease appears more often in seedbeds that have previously contained infected tobacco (Dixon et al., 1935, 1936; Wolf et al., 1936; Wolf, 1939).

Wolf et al. (1934 and 1936) observed the initiation of germination in the laboratory whereas Person and Lucas (1953) originally believed that they had detected the development of sessile zoosporangia, but they later confirmed that it was actually an adherent chytrid, Phlyctochytrium sp. (Person et al., 1955). Despite the difficulty in determining the mode of germination Person (1961) and Pawlick (1961) assumed that oospores play a not inconsiderable part in infections occurring in the spring. Kröber and Weinmann (1964a) succeeded in infecting tobacco by watering it with debris of leaves containing oospores, but similar experiments tried by Uhrin (1966) often failed. Oospores may be found in all parts of the plant (Mihajlova, 1962a, b), their formation being linked with stage of leaf growth (Peresypkin and Markhaseva, 1964) but their cytological structure is often incomplete, and this may be one reason why they do not germinate readily (Schiltz, 1967).

## C. Conditions favourable for the dissemination and germination of conidia

Precise details of the conditions favourable for the germination of oospores are not available and therefore in this section only studies concerning conidia will be considered.

## 1. *Conidia dissemination*

Differentiation of conidiophores and consequently spore morphology is connected with the conditions under which the structures are formed. Conidial size does not depend on the nature of the host (Clayton and Stevenson, 1943), though Smith (1970) found variations in connection with pathogenic strains and temperature. Abnormal structures occurred with high relative humidity (Wolf and McLean, 1940 and Izquierdo Tamayo, 1966) and with pesticide treatments (Racovitza, 1965).

Spore production and liberation depend on diffusion pressure and relative humidity (Cruickshank, 1958) whereas the appearance of the blue felting is subject to temperature and light as well as relative humidity. The ideal temperature for sporulation is from 15 to 23°C though it can occur between 1 to 2 and 30°C (Cruickshank, 1961a).

In order for conidia to appear in the early hours of the morning, the relative humidity of the air must have been 95% for 3 h and the period of darkness must have lasted for a minimum of 1·5 h (Cruickshank, 1963). Spore liberation is dependant on a rise in temperature, a decrease in relative humidity and an increase in insolation (Hill, 1961; Populer, 1962; Jadot, 1966a, b; Uozumi and Kröber, 1967; Jankowski, 1968; Cohen, 1976). Shepherd and Mandryk (1964) claimed that arginine plays a part in the phenomenon of sporulation

## 2. *Spore germination*

Spore germination takes place in a few hours by the formation of a germ tube. It requires a moisture-saturated atmosphere and a suitable temperature. The optimum is 15°C with a minimum of 3·5°C and a maximum of 30°C (Shepherd, 1962). Shepherd and Mandryk (1961, 1962) noted the presence of water soluble substances which are diffused by the conidia and inhibit their germination. Izard (1961), Izard and Chadouteaud (1962) confirmed these observations and suggested that these substances trigger necrotic reactions in the host. These "self-inhibitors" originate solely in the host (Shepherd and Tosic, 1966) but are often also found on the parasite (Schiltz and De Jong, 1967). Shepherd (1962) and Shepherd and Mandryk (1963) reported that riboflavin accelerates germination but this effect has not been observed with French isolates (Izard, personal communication).

One of the inhibitory substances, isolated by Leppik *et al.* in 1972 and named quiesone, has a chemical structure similar to that of $\beta$-ionone, the inhibiting capacity of which varies as a function of blue mould virulence and of the physiology of the host (Schiltz, 1974a).

The most favourable conditions for blue mould can therefore be summarized as an average temperature of 20°C (16°C at night and 24°C during the day), a high relative humidity, overcast weather and moving air that is conducive to dissemination of the parasite (Hill, 1962a).

## V. The Creation of Resistant *N. tabacum*

### A. Hosts to blue mould

#### 1. The Nicotianae

The majority of members of the genus *Nicotiana* are susceptible to infection by *P. tabacina* (Angell and Hill, 1932; Adam, 1933; Clayton and Stevenson, 1943).

Smith-White *et al.* (1936) successfully inoculated 250 varieties of the sub-genus *Tabacum* and 25 of the sub-genus *Rustica*, but found 5 resistant species in the sub-genus *Petunioides* and especially *N. debneyi.* Clayton (1945) showed that the resistance of *N. debneyi* and *N. megalosiphon* appears in 3 or 4 weeks but that *N. exigua* is immune at all stages. Kröber and Massfeller (1961) confirmed the reaction of *N. debneyi* and likewise Hill and Mandryk (1962) did not succeed in infecting *N. exigua* and *N. debneyi.* Bawolska (1963), Ternovsky and Dashkeeva (1964), Kirjukhina (1964), Jankowski (1967a) and Mandryk (1971) indicated that the manifestations of resistance they found were mainly a function of the technique they used. Generally speaking resistant types are found among the *Suaveolentes* originating in Australia. Among *Petunioides*, except *N. exigua* and *N. megalosiphon* (see Section VI.A), there are several sub-species which exist and react differently in the presence of virulent blue mould (Schiltz and Coussirat, 1969).

Because of their susceptibility, varieties of *N. tabacum* are not used as resistant parents in breeding programmes (Clayton and Foster, 1940; Clayton, 1945, 1953). More recently in Europe, similar conclusions were reported during the meeting in Milan of CORESTA (1961a) and by De Baets (1962), Cantillon (1962), Kröber and Massfeller (1962), Sficas (1962), Odic (1963) and Schiltz (1967).

#### 2. Other Solanaceae

Some authors have succeeded in infecting members of the *Solaneae* tribe: *S. lycopersicum esculentum, S. melongena* and *Capsicum annuum* (Wolf *et al.*, 1934), *Physalis alkekangi* and *P. peruviana* (Corbaz, 1960), *Hyoscyamus muticus* and *Petunia hybrida* (Kröber and Massfeller, 1961) and *Hyoscyamus niger* (Schiltz, 1967). These results are not confirmed by Gigante (1962), but it is safe to assume that certain species of the Solaneae harbour tobacco blue mould Kirjukhina (1964), Marcelli and Pannone (1965).

## B. Transfer of resistance

The properties of resistance found among members of the sub-genus *Petunioides* have been incorporated into *N. tabacum* by interspecific hybridization. Clayton (1962) was the first to achieve such a transfer successfully with *N. debneyi*. Lea (1961) reported the use of a larger number of species and recommended the back-cross technique rejected by Clayton. Wark (1962) used a similar process to that used by Lea.

In Europe transfers of resistant genes into commercial varieties were successful from hybrids created between *N. tabacum* and *N. debneyi* using either the amphidiploid form (Özkan *et al.*, 1963; Bailov *et al.*, 1964) or the sesquidiploid form (Izard and Schiltz, 1963b; Bailov *et al.*, 1966). In addition to this, resistant *N. tabacum* was obtained using the amphidiploid *N. tabacum* × *N. suaveolens* (Schiltz, 1967). Furthermore all European breeders have produced intervarietal hybridizations using as the pollen donor a resistant *N. tabacum* obtained in the USA or Australia.

## C. Breeding techniques

Breeding for resistance is facilitated if, in selection trials involving infection with blue mould, a susceptible variety used as control is completely destroyed by the parasite. Screening of resistance is possible in the greenhouse, in the field and in the laboratory under controlled conditions. However, the manifestation of resistance is an evolutive process depending upon the age of the plant and on environmental factors. These phenomena are very important with virulent strains of blue mould and the breeder must take account of all these elements.

### 1. In the field or in the greenhouse

As it is difficult to infect tobacco in the field during the traditional growing season it is advisable; to create a favourable micro-climate by watering and shading (Wittmer, 1962), to work on aftergrowth at the end of the season (Izard *et al.*, 1961b; Schiltz, 1967) and to alter planting-out times for increasing the risk of attack (CORESTA Phytopathology Group). Some authors occasionally infect plants in the greenhouse (Clayton, 1962; Skula, 1962) or combine several systems (Corbaz, 1962).

Assessment of symptoms is difficult as it must be both simple and accurate.

A large number of solutions to the problem have been proposed depending upon the purpose of the experiment e.g. evaluation of the infected area in relation to economic losses (Todd, 1955), determination of resistance manifestations according to either a very simple scale (Ciferri, 1961; Hill and Mandryk, 1962), or the degree of infection (intensity of sporulation) and the type of host reaction (area colonized by the parasite) (Corbaz, 1961b; Izard et al., 1961b) or still more by combining a third assessment, the nature of blue mould development (damage occasioned by its systemic form) (Schiltz, 1974b). Marcelli and Fantechi (1966) gave a disease index for susceptible varieties considering the leaves area infected and the number of conidia per $cm^2$.

## 2. In the laboratory

Following observation of this specific endoparasite under the microscope and characterizing its host–parasite interactions, Izard (1960) devised the "disc test". These discs (14 mm diameter) were cut from tobacco leaves and examined after inoculation. In Australia, Shepherd et al. (1963) removed discs from infected plants (as did Cruickshank and Müller, 1957; and Cruickshank, 1961b) and used a kinetin solution to conserve the lamina. This technique permits the study of resistance mechanisms in relation to environmental factors (Shepherd and Mandryk, 1967) but not the evaluation of genetic properties, because the disc remains influenced by the physiology of the plant on which it originated.

Researching a method for rapid indexing of genetic resistance Schiltz and Izard (1962) presented a technique using plants at the cotyledon stage in defined conditions, without external nutrients, that gave excellent results with the normal lines of spontaneous blue mould found in Europe between 1962 and 1965. The seedlings are inoculated 10 days after germination and the results are given by objective quantification of infection 10 days later. In Europe many workers have used this technique for laboratory testing of new tobacco lines (Pawlick et al., 1963; Racovitza, 1964a, b; Fantechi and Lorito, 1964; Michlewska, 1964; De Baets, 1965; Endemann and Egerer, 1966; Bartolucci, 1966 and CORESTA reports 1963, 1964 and 1965).

As the appearance of virulent lines of the parasitic fungus modified responses to this test, Schiltz (1969) and then Schiltz and Coussirat (1969) showed that genetic resistance to virulent blue mould could be determined in 10 days in seedlings produced under continuous light and nourished with a magnesium salt only (Figs 12 and 13). Applied under these conditions this technique, known as the "cotyledon test" may be used for screening lines for P. tabacina resistance, the final verifications being conducted in the field. This method is of interest to pathologists (Bartolucci et al., 1971) and is not specific to blue mould (Schiltz and Coussirat, 1978). In practice this test is

Fig. 12 (left) and Fig. 13 (right). "Blue mould—cotyledon test" performed under controlled conditions: Fig. 12. Before inoculation: tobacco seedlings at the cotyledonary stage, 10 days after the beginning of the test (about 500 seedlings in a Petri dish of 6 cm diameter). During this time the material is maintained on $Mg(NO_3)_2$, 6 $H_2O$ solution and under continuous light.

Fig. 13. After inoculation: the sporulation occurs and within 10 additional days, the test is complete and the percentage of diseased plants may be calculated.

done on about 1000 seedlings (twice 500), and within 20 days it permits the detection of homozygous populations but not selection from a heterozygous progeny. The "cotyledon test" is suitable for screening many lines chosen from segregating populations. So breeders must adapt their techniques to test the various possibilities (Koelle, 1970).

## VI. Interactions between Resistant Host and Parasite

The notions of resistance and virulence are, by definition, indissociable as the parasite is called virulent if it is able to infect resistant tobacco.

### A. Manifestations of resistance

First demonstrated by Clayton (1945) manifestations of resistance were described from a histological standpoint by Izard *et al.* (1961a), Izard and Schiltz (1962) and by Shepherd and Mandryk (1967). They involve a hypersensitivity reaction, localized in the mesophyll, the epidermis playing no part (Schiltz, 1967). The host's reactions depend on its food supply (Hill, 1962b; Mandryk, 1962; Racovitza, 1966), its stage of development, its genetic resistance and on temperature (see also Section IV) and light. These last three factors are closely connected as they participate in the overall behaviour of the plant in the presence of the parasite. The metabolic activity of the plant represses the expression of resistance genes except in *N. exigua* and *N. megalosiphon* (Schiltz and Coussirat, 1969, 1978), whereas light or kinetin, enhance gene expression (Schiltz *et al.*, 1970b). In some tobaccos photo-synthetic activity is connected with the action of the susceptibility gene (Schiltz *et al.*, 1971).

## B. Heredity of resistant *N. tabacum*

The reactions of the resistant host depend largely on the composition of its genotype. In Europe around 1966, heterozygous hybrids ($F_1$) as well as the tobacco cultivars obtained after repeated back-crosses with the susceptible parent were found to be severely infected with blue mould (Schiltz and Izard, 1966; Schiltz, 1967). For *N. tabacum* with genes from *N. debneyi* a method for assessing the virulence of the parasite and a model of trigenetic heredity *RnRn*, *Rv1Rv1*, *Rv2Rv2*, has been proposed by Schiltz *et al.* (1977a) *RnRn* ensures resistance to normal isolates of the parasite whilst *Rv1Rv1* and *Rv2Rv2* are additional genes contributing to the resistance against virulent lines of blue mould. This proposal satisfies Clayton's data (1962) and complements the information given by Lea (1963), Wuttke (1969), Wark *et al.* (1976) and Palakarčheva (1977).

## C. Determinism of resistance

Some authors have attempted to detect the basis of resistance by biochemical means. Although Schmidt (1971) did not confirm his earlier results imputing an effect of the nucleotide fraction on resistance to blue mould, Edreva *et al.* were able to characterize species by their peroxidase activity and their chlorogenic acid content (Edreva *et al.*, 1970; Palakarčheva *et al.*, 1978). and in studies of host pathogen relations showed a relationship between the parasite and tobacco enzymes (Edreva, 1972; Georgieva and Edreva, 1974).

## D. Blue mould virulence

Symptoms on a resistant host depend in part on the concentration of inoculum (Bartolucci *et al.*, 1970; Schiltz and Izard, 1966), but mainly on the virulence of the parasite.

Three ecotypes of *P. tabacina* were described for the first time in Australia by Wark *et al.* (1960). Subsequently Hill (1963b, 1966a, b), created in the greenhouse, isolates with modified pathogenicity. In Hill's opinion the resistant host acts as a filter, revealing the minor pathogenic lines ($APT_2$ and $APT_3$) that are normally found in the major component ($APT_1$) constituting the initial inoculum. He stressed the importance of this phenomenon in epidemiology and drew European workers' attention to the problem of the possible appearance of virulent strains of *P. tabacina* following the cultivation of resistant tobacco varieties. Elsewhere in Europe Berger (1964), Corbaz (1965) and Govi (1965) observed variations in the pathogenecity of spontaneous blue mould in their countries. Experiments based on Hill's methods make it possible to obtain variants of blue mould with unstable properties (Schiltz, 1967; Abedi, 1971). In 1969 a highly virulent strain was observed in Europe following natural infection of a seedbed of a resistant variety (Schiltz and Coussirat, 1969). Starting from this observation, these authors demonstrated that increased pathogenecity is induced by the inoculation of a genetically resistant host with inhibited gene expression, resulting from the natural influence of young plant metabolism on the manifestations of polygenic heredity (Schiltz *et al.*, 1970a). Subsequently observing that blue mould was more virulent in Europe than in Australia or the USA, these authors attempted to identify the gene governing virulence (*Rn*) in the resistant plant and to explain the phenomena observed in the field (Schiltz *et al.*, 1977a). The growing of resistant varieties on a commercial scale in Europe encourages the development of isolates of the pathogen whose virulence varies as a function of both the weather conditions and the genetical make-up of the host (Schiltz, 1975; Corbaz, 1976, 1979).

## VII. Means of Control

In Australia and Europe the big outbreaks of blue mould most often start with early infection in seedbeds. If this escapes notice the seriousness of the attack depends mainly on weather conditions. Traditional prophylactic measures associated with preventive treatments of seedbeds and in greenhouses are therefore, strongly recommended (Hill, 1963a). Daily fumigations are effected with benzol vapour (Australia) and with paradichlorobenzene (USA), or dithiocarbamate fungicides (maneb, propineb, mancozeb or zineb) are applied twice weekly (McGrath and Miller, 1958). The dithiocarbamates are preferred because of their efficiency and their ease of use (Hitier *et al.*, 1963) whereas antibiotics are not recommended (Hitier *et al.*, 1964). This protection must be continued in the field but the number of treatments depends on weather conditions and on the varieties being grown (Corbaz, 1970).

New endotherapeutic fungicides have been studied for several years, their properties and their efficiacy should soon begin to modify control techniques. At present the family of acylalanines (metalaxyl) applied on seedbeds or in the field as drenches or as sprays has proved superior to the acetamides or phosphites (Avigliano *et al.*, 1976; Avigliano, 1977; Tsakiridis *et al.*, 1977; Schiltz *et al.*, 1977b; O'Brien, 1978 and Johnson *et al.*, 1979). The high efficiency and the specific activity of these new compounds suggests that the parasite could become tolerant of them and it is likely that they will be mixed with orthodox protectant fungicides in many countries.

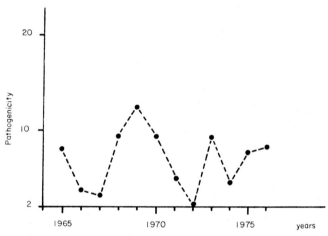

Fig. 14.   Pathogenicity variations of *P. tabacina* isolates spontaneously appeared in south-western France between 1965 and 1976. The pathogenecity is expressed by intensity of symptoms noticed every year on *N. tabacum* hybrid Bel 61–10. These results have been registered on the CORESTA blue mould trap collection transplanted in the field for the studies of the Phytopathology Group. Symptoms are assessed according to the scale (2 to 30) preconized by the group (Schiltz, 1974b).

## VIII.  Conclusion

Tobacco blue mould has been the subject of a vast number of publications the conclusions of which have sometimes been contradictory. North America would seem to be a special case with regard to epidemiology as there, blue mould is principally a seedbed disease, but the experience of 1979 demonstrated that in all the three continents referred to in this chapter the seriousness of outbreaks may be very similar. The most important problem is indubitably the increased pathogenecity of *P. tabacina*. This does not appear to be cumulative in Europe and it is assumed that the variations arising in the parasite in connection with the expression of genes of resistance, whose changes are very important during tobacco growth, may have the effect expected from the theory of successive and simultaneous multilines (Fig. 14). When growing a resistant variety its heredity, agronomic properties and its suitability for local conditions, must be borne in mind. At present disease control must reconcile all known techniques including the use of prophylactic fungicide treatments (particularly on seedbeds) and the growing of resistant varieties (Corbaz, 1978). European experience is of particular importance, and in this respect much of the tobacco growers' success can be attributed to CORESTA. The General Secretariat of this organization runs a warning service and the Phytopathology Study Group has organized collaborative experiments since 1963.

We have made great progress in our knowledge of tobacco blue mould since 1960, and although research workers are not always of the same opinion in some fields, such as in the nomenclature of different isolates, this may well be a good stimulant for future work.

## References

Abedi, H. (1971). PhD Thesis no. 73 Bordeaux Univ. I. p. 79.
Adam, D. B. (1925). *J. Dep. Agric. Victoria* **23**, 436–440.
Adam, D. B. (1931). *J. Dep. Agric. Victoria* **29**, 469–471.
Adam, D. B. (1933). *J. Dep. Agric. Victoria* **31**, 412–416.
Angell, H. R. (1957). *Aust. J. Inst. Agric. Sci.* **23**, 144–147.
Angell, H. R. and Hill, A. V. (1932). *CSIR, Aust. Bull.* **65**, 9–30.
Anon. (1961). *Aust. Tob. Growers' Bull.* **2**, 2–4.
Anon. (1963). *Aust. Tob. Growers' Bull.* **6**, 8–9.
Anon. (1969). *Tobacco US.* **169**(10), 7.
Avigliano, M. (1977). *Rep. Phytopathol. Gr. CORESTA Meet. Bratislava 19–22 Sept.* app. 23.
Avigliano, M., Commino, C. and Sorrentino, C. (1976). *6th Int. Tob. Sci. Congr. Tokyo.*
Bailov, D., Palakarčeva, M. and Daskalov, S. (1964). *Rasten Nauki Sofia 1–7*, 3–16.
Bailov, D., Palakarčeva, M. and Daskalov, S. (1966). *4th Int. Tob. Sci. Congr. Athens.*
Bartolucci, A. (1966). *4th Int. Tob. Sci. Congr. Athens.*
Bartolucci, A., Becherelli, A., Fiaschi, A., Montanari, L. and Testa, F. (1970). *Il Tabacco* **LXXIV–736**, 4–12.
Bartolucci, A., Becherelli, A., Fantechi, F., Fiaschi, A. and Montanari, L. (1971). *Il Tabacco* **LXXV–739**, 29–38.
Bawolska, M. (1963). *Post. Nauk. Roln.* **10**(3), 51–53.
Bebiya, E. A. (1975). *Mikol. Fitopatol.* **9**(1), 51–52.
Berger, P. (1964). *Rep. 8th "Peronospora" Work. Gr. CORESTA Meet. Vienna, 7–8 Oct.*
Bromfield, K. R. and Schmitt, C. G. (1967). *Phytopathology* **57**(10), 1–133.
Cantillon, P. (1962). *Parasitica* **18**(1), 25–38.
Ciferri, R. (1961). *Rep. 9th "Peronospora" Work. Gr. CORESTA Meet. Milan, 16–17 Feb.* app. 8.
Clayton, E. E. (1945). *J. Agric. Res.* **70**, 79–87.

Clayton, E. E. (1953). *Phytopathology* **43**, 239–244.

Clayton, E. E. (1962). *Inf. Bull. CORESTA* **2**, 25–30.

Clayton, E. E. and Foster, H. H. (1940). *Phytopathology* **30**, p. 4.

Clayton, E. E. and Stevenson, J. A. (1935). *Phytopathology* **25**, 516–521.

Clayton, E. E. and Stevenson, J. A. (1943). *Phytopathology* **33**, 101–113.

Cohen, Y. (1976). *Aust. J. Biol. Sci.* **29**, 281–289.

Cooke, M. C. (1891). *Gardeners' Chronicle* **9**, 173.

Corbaz, R. (1960). *Rev. Romande Agric. Vitic. Arboric.* **16**, 101–104.

Corbaz, R. (1961a). *Phytopath. Z.* **42**(1), 39–44.

Corbaz, R. (1961b). *Rep. 2nd "Peronospora" Work. Gr. CORESTA Meet. Milan, 16–17 Feb.* app. 9.

Corbaz, R. (1962). *Inf. Bull. CORESTA* **1**, 9–19.

Corbaz, R. (1964). *Phytopath. Z.* **51**(2), 190–191.

Corbaz, R. (1965). *Rep. 9th "Peronospora" Work. Gr. CORESTA Meet. Interlaken, 7–9 Dec.*

Corbaz, R. (1970). *Rev. Suisse Agric.* **11**(4), 90–92.

Corbaz, R. (1976). *Inf. Bull. CORESTA* **3**(4), 53–56.

Corbaz, R. (1978). *Rev. Suisse Agric.* **10**(5), 155–157.

Corbaz, R. (1979). *Inf. Bull. CORESTA* **1**, 5.

CORESTA report (1960). *1st "Peronospora" Work. Gr. Meet. Lausanne, 4–5 Oct. Inf. Bull. CORESTA* **4**, 2–13.

CORESTA report (1961a). *2nd "Peronospora" Work. Gr. Meet. Milan, 16–17 Feb.*

CORESTA report (1961b). *5th "Peronospora" Work. Gr. Meet. Paris, 22–23 Nov. Inf. Bull. CORESTA* **4**, 5–8.

CORESTA report (1963). *7th "Peronospora" Work. Gr. Meet. Roma, 8–9 Oct.*

CORESTA report (1964). *8th "Peronospora" Work. Gr. Meet. Vienna, 7–8 Oct.*

CORESTA report (1965). *9th "Peronospora" Work. Gr. Meet. Interlaken, 7–9 Dec.*

Cruickshank, I. A. M. (1958). *Aust. J. Biol. Sci.* **11**(2), 162–170.

Cruickshank, I. A. M. (1961a). *Aust. J. Biol. Sci.* **14**, 58–65.

Cruickshank, I. A. M. (1961b). *Aust. J. Biol. Sci.* **14**(2), 198–207.

Cruickshank, I. A. M. (1963). *Aust. J. Biol. Sci.* **16**, 88–98.

Cruickshank, I. A. M. and Müller, K. O. (1957). *Nature Lond.* **180**, 44–45.

Dangeard, P. A. (1890). *Le Botaniste* **2**, 63–149.

Darnell-Smith, G. P. (1929). *Agric. Gaz. N.S.W.* **40**, 407–408.

De Baets, A. (1962). *Parasitica* **18**(1), 8–24.

De Baets, A. (1965). *Rev. Agric.* **10**.

Dixon, L. F., McLean, R. and Wolf, F. A. (1935). *Phytopathology* **25**, 628–639.

Dixon, L. F., McLean, R. and Wolf, F. A. (1936). *Phytopathology* **26**, 735–759.

Edreva, A. (1972). *Acad. Agric. Dimitrov.* **5**(2), 121–125.

Edreva, A., Bailov, D. and Nikolov, S. (1970). *Bulg. Acad. Sci. Agric.* **3**(1), 55–62.

Endemann, W. and Egerer, A. (1966). *Ber. Inst. Tabakforsch.* **13**(1), 5–24.

Fantechi, F. and Lorito, N. (1964). *Il Tabacco* **713**, 241–258.

Farlow, W. G. (1885). *Bot. Gaz.* **10**, 346–348.

Georgieva, I. and Edreva, A. (1974). *Acad. Agric. Dimitrov.* **7**(1), 25–30.

Gigante, R. (1962). *Boll. Stn. Patol. Veg.* **201**, 21–45.

Godfrey, G. H. (1941). *Pl. Dis. Reptr.* **25**, 347–353.

Govi, G. (1965). *Rep. 9th "Peronospora" Work. Gr. CORESTA Meet. Interlaken, 7–9 Dec.*

Gustavsson, A. (1959). *Bot. Not.* **112**, 1–16.

Harkness, H. W. (1885). *Bull. Calif. Acad. Sci.* **1**, 256–271.

Henderson, R. G. (1937). *Phytopathology* **27**, 131.

Hill, A. V. (1961). *Aust. J. Biol. Sci.* **14**(2), 208–222.

Hill, A. V. (1962a). *Inf. Bull. CORESTA* **3**, 15–24.

Hill, A. V. (1962b). *Aust. J. Agric. Res.* **13**(4), 650–661.
Hill, A. V. (1963a). *Inf. Bull. CORESTA* **3**, 6–8.
Hill, A. V. (1963b). *Inf. Bull. CORESTA* **3**, 8–10.
Hill, A. V. (1966a). *Inf. Bull. CORESTA* **1**, 7–15.
Hill, A. V. (1966b). *4th Int. Tob. Sci. Congr. Athens.*
Hill, A. V. and Angell, H. R. (1933). *Aust. J. Coun. Sci. Ind. Res.* **6**, 260–268.
Hill, A. V. and Mandryk, M. (1962). *Aust. J. Exp. Agric. Anim. Husb.* **2**, 12–15.
Hill, A. V., Paddick, R. G. and Green, S. (1967). *Aust. J. Agric. Res.* **18**, 575–587.
Hitier, H., Mounat, A. and Bown, G. (1963). *Ann. Inst. Exp. Tabac. IV* **2**, 377–388.
Hitier, H., Michel, E., Mounat, A. and Bown, G. (1964). *Ann. SEITA, DEE* **1**(2), 115–134.
Izard, C. (1960). *C.r. Acad. Sci. Fr.* **251**, 3063–3064.
Izard, C. (1961). *C.r. Acad. Sci. Fr.* **253**, 2756–2758.
Izard, C. and Chadouteaud, J. (1962). *C.r. Acad. Sci. Fr.* **255**, 1773–1774.
Izard, C. and Schiltz, P. (1962). *C.r. Acad. Agric. Fr.* **31**(I), 110–113.
Izard, C. and Schiltz, P. (1963a). *3rd Int. Tob. Sci. Congr. Salisbury*, 77–79.
Izard, C. and Schiltz, P. (1963b). *Inf. Bull. CORESTA* **2**, 7–10.
Izard, C., Schiltz, P. and Hitier, H. (1961a). *Inf. Bull. CORESTA* **3**, 18–21.
Izard, C., Schiltz, P. and Hitier, H. (1961b). *Inf. Bull. CORESTA* **4**, 19–24.
Izquierdo Tamayo, A. (1966). *Bol. Soc. Esp. Hist. Nat. Sec. Biol.* **64**(3), 321–330.
Jadot, R. (1966a). *Parasitica* **22**(1), 55–63.
Jadot, R. (1966b). *Parasitica* **22**(4), 223–229.
Jankowski, F. (1967a). *Biul. Centr. Lab. Prz. Tyton.* **3**(4), 43–47.
Jankowski, F. (1967b). *Biul. Centr. Lab. Prz. Tyton.* **3**(4), 59–65.
Jankowski, F. (1968). *Biul. Centr. Lab. Prz. Tyton. Krakow* **1**(2), 35–40.
Johnson, G. I., Davis, R. D. and O'Brien, R. G. (1979). *Pl. Dis. Reptr.* **63**(3), 212–216.
Kin Chou Tsang (1929). *Le Botaniste* **21**(1), 93.
Kirjukhina, R. I. (1964). *2nd Coll. Peronospora USSR* Dec. 1962, 35–37.
Koelle, G. (1970). *Der Deut. Tabakbau* **12**(50), 134.
Kröber, H. (1965). *Phytopathol. Z.* **54**(4), 328–334.
Kröber, H. and Massfeller, D. (1961). *NachrBl. Dt. PflSchutzdienst* **13**(6), 81–85.
Kröber, H. and Massfeller, D. (1962). *NachrBl. Dt. PflSchutzdienst* **14**, 82–85.
Kröber, H. and Petzold, H. (1972). *Phytopath. Z.* **74**, 296–313.
Kröber, H. and Weinmann, W. (1964a). *Phytopath. Z.* **51**(1), 79–84.
Kröber, H. and Weinmann, W. (1964b). *Phytopath. Z.* **51**(3), 241–251.
Lea, H. W. (1961). *Inf. Bull. CORESTA* **2**, 21–27.
Lea, H. W. (1963). *Inf. Bull. CORESTA* **3**, 13–15.
Leppik, R. A., Hollomon, D. W. and Bottomley, W. (1972). *Phytochem.* **11**, 2055–2063.
Lucas, G. B. (1975). "Diseases of Tobacco". Biological Consulting Associates, Raleigh, USA.
McAlpine, D. (1900). *Ann. Rep. Dep. Agric. Victoria 1899*, 222–269.
McGrath, H. and Miller, P. R. (1958). *Pl. Dis. Reptr.* Suppl. 250, 1–35.
Mandryk, M. (1960). *Aust. J. Agric. Res.* **11**, 16–26.
Mandryk, M. (1962). *Aust. J. Agric. Res.* **13**(1), 10–16.
Mandryk, M. (1971). *Aust. J. Exp. Agric. and Anim. Husb.* **11**, 94–98.
Marcelli, E. and Fantechi, F. (1966). *4th Int. Tob. Sci. Congr. Athens.*
Marcelli, E. and Pannone, R. (1965). *Il Tabacco* **717**, 33–47.
Marte, M. and Caporali, L. (1974). *Rev. Gen. Bot.* **81**, 277–298.
Michlewska, Ch. (1964). *Buil. Centr. Lab. Przem. Tyton. Krakow*, 57–72.
Mihajlova, P. (1962a). *Bulg. Tjutjun* **1**, 9–12.
Mihajlova, P. (1962b). *Bulg. Tjutjun* **7**, 23–27.
Odic, M. (1963). *3rd Int. Tob. Sci. Congr. Salisbury*, 38–40.
O'Brien, R. G. (1978). *Pl. Dis. Rept.* **62**(3), 277–279.

Özkan, N., Taner, E. and Özyolcular, M. (1963). *Rep. 7th "Peronospora" Work. Gr. CORESTA Meet. Roma, 8–9 Oct.* app. 3.

Paddick, R. G., Hill, A. V. and Green, J. (1967). *Aust. J. Agric. Res.* **18**, 589–600.

Palakarčheva, M. (1977). *Rep. Phytopath. Gr. CORESTA Meet. Bratislava, 19–22 Sept.* app. 18.

Palakarčheva, M., Edreva, A., Cholakova, N. and Noveva, S. (1978). *Z. Pflanzüch.* **80**(1), 49–63.

Pawlik, A. (1961). *Z. PflKrankh. PflPath. PflSchutz.* **68**(4), 193–197.

Pawlik, A., Schmid, K., Sprau, F. and Krauss, E. (1963). *Z. PflKrankh. PflPath. PflSchutz.* **70**(6), 332–339.

Peresypkin, V. F. and Markhaseva, V. A. (1964). *2nd Coll. Peronospora USSR* 1962, 19–23.

Peresypkin, V. F. and Zrazhevkaya, T. G. (1967). *Mikol. Fitopatol.* **1**(3), 235–240.

Person, L. (1961). *Sel'Khozizdad Moscow* **9**, 97–99.

Person, L. H. and Lucas, G. B. (1953). *Phytopathology* **43**, 701–702.

Person, L. H., Lucas, G. B. and Koch, W. J. (1955). *Pl. Dis. Reptr.* **39**, 887–888.

Peyrot, J. (1969). *Rep. Phytopath. Gr. CORESTA Meet. Hammamet Tunisia, 13–17 Oct.*

Populer, C. (1962). *Parasitica* **18**(1), 1–7.

Racovitza, A. (1964a). *Industr. Alimentara Bucarest* **15**(9), 427–430.

Racovitza, A. (1964b). *C.r. Acad. Sci. Fr.* **259**, 648–651.

Racovitza, A. (1965). *Rev. Mycol.* **30**, 52–56.

Racovitza, A. (1966). *4th Int. Tob. Sci. Congr. Athens.*

Schiltz, P. (1967). PhD Thesis no. 203 Bordeaux Univ. p. 145.

Schiltz, P. (1969). *C.r. Acad. Sci. Fr. Paris* **268**(D), 1292–1295.

Schiltz, P. (1974a). *Ann. Tab.* **2**(11), 207–216.

Schiltz, P. (1974b). *Inf. Bull. CORESTA* **1**, 16–22.

Schiltz, P. (1975). *Inf. Bull. CORESTA* **3**(4), 12–19.

Schiltz, P. and Coussirat, J. C. (1969). *Ann. SEITA, DEE* **6**(2), 145–162.

Schiltz, P. and Coussirat, J. C. (1978). *Ann. Biol. XVII* **3**(4), 169–174.

Schiltz, P. and Izard, C. (1962). *C.r. Acad. Agric. Fr.* **6**(VI), 561–564.

Schiltz, P. and Izard, C. (1966). *Ann. SEITA, DEE III Sect. 2*, **2**, 151–155.

Schiltz, P. and De Jong, Ch. (1967). *Coll. Soc. Fr. Physiol. Végét. Bergerac* 1st July.

Schiltz, P., Coussirat, J. C. and Abedi, H. (1970a). *Ann. SEITA DEE* **2**(7), 171.

Schiltz, P., Coussirat, J. C. and Abedi, H. (1970b). *Ann. SEITA DEE* **7**(2), 181–190.

Schiltz, P., Coussirat, J. C. and Abedi, H. (1971). *Ann. SEITA DEE* **8**(2), 185–192.

Schiltz, P., Coussirat, J. C. and Delon, R. (1977a). *Ann. Tab. Bergerac* **2**(14), 111–126.

Schiltz, P., Delon, R., Cazamajour, F., Podeur, G. and Boulogne, R. (1977b). *Ann. Tab. Bergerac* **2**(14), 127–154.

Schmidt, J. A. (1971). *Comm. XIIIth Tabak. Kolloqium.*

Sficas, A. G. (1962). *Rep. 6th "Peronospora" Work. Gr. CORESTA Meet. Budapest, 2–3 Oct.* app. 6.

Shepherd, C. J. (1962). *Aust. J. Biol. Sci.* **15**(3), 483–508.

Shepherd, C. J. (1970). *Trans. Br. Mycol. Soc.* **55**(2), 253–256.

Shepherd, C. J. and Mandryk, M. (1961). *Aust. Tob. Res. Conf. Canberra.*

Shepherd, C. J. and Mandryk, M. (1962). *Trans. Br. Mycol. Soc.* **45**(2), 233–244.

Shepherd, C. J. and Mandryk, M. (1963). *Aust. J. Biol. Sci.* **16**(1), 77–87.

Shepherd, C. J. and Mandryk, M. (1964). *Aust. J. Biol. Sci.* **17**(4), 878–891.

Shepherd, C. J. and Mandryk, M. (1967). *Aust. J. Biol. Sci.* **20**, 87–102.

Shepherd, C. J. and Tosic, L. (1966). *Aust. J. Biol. Sci.* **19**, 335–337.

Shepherd, C. J., Stuart, F. and Mandryk, M. (1963). *Nature London* **197**(4866), 515 pp.

Skalicky, V. (1964). *Acta Univ. Carol.* Suppl. 1964, 25–90.

Skula, K. (1962). *Rep. 6th "Peronospora" Work. Gr. CORESTA Meet. Budapest, 2–3 Oct.* app. 8.

Smith, A. (1970). *Trans. Br. Mycol. Sci.* **55**(1), 59–66.

Smith, E. F. and McKenney, R. E. B. (1921). *US Dept. Agric. Circ.* **174**, 6 pp.
Smith-White, S., MacIndoe, S. L. and Atkinson, W. T. (1936). *J. Aust. Inst. Agric. Sci.* **2**, 26–29.
Stevenson, J. A. and Archer, A. (1940). *Pl. Dis. Reptr.* **24**, 93–103.
Ternovsky, M. F. and Dashkeeva, K. N. (1964). *2nd Coll. Peronospora USSR, Dec. 1962*, 40–50.
Todd, F. A. (1955). *NC. Agric. Exp. Stn. Techn. Bull. III.*
Todd, F. A. (1961). *Tob. US.* **152**(6), 10–15.
Tsakiridis, J., Vasilakakis, Chrisochore, A. (1977). *Rep. Phytopath. Gr. CORESTA Meet. Bratislava, 19–22 Sept.* app. 24.
Uhrin, P. (1966). *Inf. Bull. CORESTA* **4**, 35–38.
Uozumi, T. and Kröber, H. (1967). *Phytopath. Z.* **59**, 372–384.
Vassilyev, A. E. and Bogoyavlenskaya, R. A. (1975). *Mycol. Phytopathol. Acad. CCCP* **9**(6), 468–472.
Wager, H. (1900). *Ann. Bot.* **14**, 263–279.
Wark, D. C. (1962). *Communication of 10.9.1962—C.S.I.R.O. Canberra, A.C.T. Australia.*
Wark, D. C., Hill, A. V., Mandryk, M. and Cruickshank, I. A. M. (1960). *Nature Lond.* **187**, 710–711.
Wark, D. C., Wuttke, H. H. and Brouwer, H. W. (1976). *Tob. Sci.* **XX**(110), 110–113.
Wittmer, G. (1962). *Rep. 6th "Peronospora" Work. Gr. CORESTA Meet. Budapest, 2–3 Oct.* app. 4.
Wolf, F. A. (1939). *Phytopathology* **29**, 194–200.
Wolf, F. A. (1947). *Phytopathology* **37**, 721–729.
Wolf, F. A. (1957). Duke Univ. Press.
Wolf, F. A., McLean, R. (1940). *Phytopathology* **30**, 264–268.
Wolf, F. A., Dixon, L. F., McLean, R. and Darkis, F. R. (1934). *Phytopathology* **24**, 337–363.
Wolf, F. A., McLean, R. and Dixon, L. F. (1936). *Phytopathology* **66**, 760–777.
Wuttke, H. H. (1969). *Aust. J. Exp. Agric. Anim. Husb.* **9**(10), 545–548.

Chapter 28

# Downy Mildew of the Vine

## R. LAFON and J. BULIT

*INRA Bordeaux, Station de Pathologie Végétale, France*

## I. Introduction

Grape vine downy mildew, caused by the fungal parasite *Plasmopara viticola* (Berk. & Curt.) Berlese & de Toni, occurs throughout the world. It attacks mainly the varieties of *Vitis vinifera* L. which constitute the majority of vineyards under cultivation though its development is very much influenced by the climate. The absence of rainfall in spring and summer is a limiting factor with regard to the spread of the disease to vineyards in certain areas (e.g. Afghanistan, California, Chile), as is insufficient warmth during the spring in northern vineyards. Conversely, those regions where it is hot and wet during the vegetative growth of the vine favour downy mildew development (e.g. Europe, South Africa, Argentina, etc.).

## II. Description of the Disease

All the green parts of the vine, namely leaves and tendrils, shoots, inflorescences and bunches, can be affected by the disease. The earlier the attack, the greater the resulting damage, especially to the bunches. In all cases invasion of the tissues by the fungus is accompanied by changes in the colour (yellowing and browning) of the affected organs, and by malformations and necroses. The lesions become covered with white, downy conidiophores only on those parts of the affected organs bearing stomata, and when the weather is wet.

### A. Leaves

The leaves are most vulnerable to attack during active growth, and again at an advanced stage of maturity at the end of summer and in the autumn. A lesser degree of susceptibility to the disease is observed between these two periods.

There are three ways in which downy mildew manifests its presence on the leaves:

(i) When the incubation period has been very brief (4–5 days), the conidiophores of the fungus appear directly on the apparently healthy, green foliar tissues, and there is no obvious underlying lesion.

(ii) After an incubation period of moderate duration (7–10 days), one or more spots with blurred edges are seen to form on the upper leaf surface. These are initially yellowish and oily in appearance ("oil spot" stage), later becoming mottled, and are accompanied by a white, downy felt of conidiophores on the lower surface, especially during wet weather.

(iii) On old leaves the symptoms present a mosaic effect of small, angular blemishes, limited by the veins and yellow or wine-red in colour depending on the vine variety.

Leaf damage frequently occurs, appearing wherever the vine is attacked by the disease. Early attacks lead to destruction of the leaves which accompanies the multiplication of inoculum, whereas later attacks result in reduced ripening of the wood, which limits vine productivity for 2 years (Rives and Lafon, 1972) and is important in the perennation of the fungus.

## B. Branches

The branches are usually attacked during early growth when their maximum length is 10–15 cm, or only at the extremities of the more advanced stages of growth. The infected shoots turn brown, taking on the colour and general appearance of a scalded plant. Very often they curl up in the form of a letter S or become hooked at the tip. The nodes are more susceptible than the internodes which often show only longitudinal brownish streaks at the time of an attack. The tendrils and leaf petioles exhibit similar modifications to those found on the branches.

## C. Inflorescences and bunches

Attacks of downy mildew become apparent on these organs from their initial formation right until ripening.

Brown spots similar to those found on the branches are observed on the inflorescences. If only the stalk is affected the entire organ can dry up and drop off but most often the "grey rot" stage occurs when conidiophores form a whitish down on the inflorescence.

Fig. 1.   Conidiophores on the lower surface of the leaf.

Fig. 2.  Downy mildew symptoms on infected leaf: "oil spots" in summer.

Fig. 3.  Downy mildew symptoms on infected leaf: mosaic effect in autumn.

After flowering the developing bunch remains highly susceptible to grey rot right up until the time when the grapes have attained a diameter of 5–6 mm.

Once this stage has passed, infection of the bunches is rare, probably because the stomata on the grapes and stalk are no longer functional. The symptoms which appear subsequently are due to penetration of the stalk and grapes by mycelium from earlier infections elsewhere. External production of conidiophores no longer occurs, the diseased parts become brown and dry up, and the grapes drop off at the slightest jolt. This is the "brown rot" stage.

Fig. 4.   Damage on bunch: brown rot stage.

The economic significance of the damage caused by downy mildew is apparent during the actual year of the attack, especially if the bunches are affected early on and destroyed. The harvest is reduced but its quality is modified only in the case of very serious attack when the proportion of diseased grapes is high. Downy mildew also causes repercussions in the two years following an attack through successive diminution of the vine shoot's nutritive reserves and reduced flowering.

## III. Taxonomy and Morphological Characteristics

The fungus responsible for vine downy mildew was initially described in 1837 by Schweinitz under the name of *Botrytis cana* Link. Then in concordance with the subsequent evolution of mycological systematics, the fungus was successively called *Botrytis viticola* by Berkeley and Curtis (1848), *Peronospora viticola* by de Bary (1863) and finally *Plasmopara viticola* by Berlese and de Toni in 1888.

*P. viticola* is an obligate parasite and cannot therefore be cultivated *in vitro* on inert nutritive media. It develops intercellularly within the parasitized tissues of the vine in the form of a tubular, coenocytic mycelium, 8–10 μm in diameter, which is furnished with globular haustoria of 4 to 10 μm diameter.

The haustoria gain access to the host cell by invagination of the cellular membrane in which they are ensheathed.

Asexual multiplication is by sporangia, which are ellipsoid, hyaline structures measuring $17–25 \times 10–16$ μm, borne on tree-like sporangiophores (140 to 250 μm) derived directly from the mycelium.

Each sporangium gives rise to zoospores (1 to 6) measuring $6–8 \times 4–5$ μm and equipped with two lateral flagella. The zoospores escape from the side of the sporangium opposite its point of insertion, either by means of an opening in a papilla or by direct perforation of the wall. Liberation always takes place in water and is normally followed by the germination of the zoospores which lose their flagella and produce a germ tube.

Sexual reproduction begins at the end of summer by the fusion of an antheridium and an oogonium derived from the terminal expansion of hyphae. A diploid oospore (winter egg) then forms containing 14 to 16 chromosomes and is enveloped by two membranes, themselves covered by the wrinkled wall of the oogonium. The oospores are 28 to 40 μm in diameter and form in the leaves or possibly throughout the parasitized organ. In spring the oospores germinate in moisture giving rise to a slender germ tube 2 to 3 μm in diameter and of variable length, terminating in a piriform sporangium, $25 \times 35–40$ μm, which in its turn produces about 8–20 zoospores (Viennot-Bourgin, 1949; Galet, 1977).

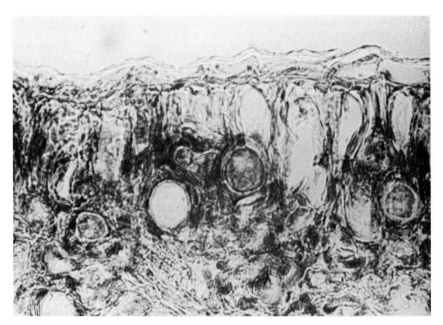

Fig. 5.   Oospores in leaf parenchyma.

## IV. Biology and Epidemiology

### A. Overwintering of the parasite

In those vineyards where the winter period is relatively indistinct, downy mildew may remain in the mycelial state between the bud scales and in diseased leaves which persist on the plant until bud burst (Galet, 1977).

However, the usual form of conservation of the fungus is the oospore, the conditions for the production of which are not completely understood. The time of their appearance may give rise to contradictory observations and their formation between June and October, has been noted.

The survival of oospores in fallen leaves is guaranteed at least until the following spring. It is favoured by the presence of water and cut short by drought or when the leaves are buried deep in the soil. Temperature does not have much effect on the survival of oospores which, according to Gäumann (1950) can resist $-26^{\circ}C$ for 5 days.

According to Tsvetanov (1976) their germinative capacity is maintained until July if they overwinter at a depth of 5 cm or less, and until March at up to 20 cm.

## B. Dissemination

Downy mildew dissemination is achieved in two successive phases: primary infection from oospores followed by spread of the disease by sporangia.

In spring as soon as the temperature attains 11°C the oospores germinate in water or on very moist soil. The zoospores produced are projected on to vine leaves near the soil by rain splash.

The sporangia break off the sporangiophores at a cross-wall of callus that dissolves in water. This means that aerial liberation of sporangia can take place only in moist air. Their transport is ensured by the wind. Corbaz (1972) studied aerial spore production during September and October for two years in a vineyard in Switzerland. He recorded captures of *P. viticola* sporangia ranging from 800 to 2400 m$^{-3}$ with a maximum at about 1800 h. Once separated from the sporangiophore they remain viable in dry air for a maximum of 5 days.

## C. Infection

The infection process, which always takes place in water, generally consists of three phases; germination of the sporangia, followed by that of the zoospores, and finally penetration of the host.

### 1. Germination of the sporangia

This is dependent on their age (the youngest germinating most readily) and also on the presence of water, the temperature and light. Srinivasan and Jeyarajan (1976) showed that the vitality of sporangia under Indian conditions, with day temperatures above 30°C, is conditioned by sunlight. Sporangia harvested during the day between 0800 and 1800 h do not germinate. Those collected at night between 2000 and 0600 h germinate within an hour with the maximum germination among those harvested between 0200 h and 0600 h. Zachos (1959) in Greece reports that exposure of sporangia to the sun for 1 h is sufficient to have a permanent effect on their viability. In France we have noticed in our artificial infection trials that those conidia collected in the evening following a hot sunny day do not germinate.

Sporangia liberate their zoospores exclusively in water and, according to the authors in Galet (1977), this phenomenon occurs between temperatures of 3–9°C and 28–30°C, with an optimum at 22–25°C. Germination requires 10 h at extreme temperatures but only 40 to 60 min at the optimum level.

## 2. Germination of the zoospores

As soon as they have left the sporangium the zoospores swim for a few minutes in the water with a whirling motion. After 20 to 30 min they become motionless, lose their flagella, round off and become enclosed in a membrane i.e. they encyst. Under favourable conditions, notably in persistent moisture, a germination tube is produced approximately 12 min later. It is flexuous, non-septate, and measures 50 to 80 $\mu$m in length.

## 3. Penetration of the host

This takes place almost exclusively via the stomata though other ways of penetration are reported in the literature. These, via the flower stigmas (Boubals, 1958), wounds (Gäumann, 1950), and phylloxera galls (Branas, 1974) are, in our opinion, exceptional. It seems that zoospores are attracted towards the stomata during the mobile phase; and Royle and Thomas (1973) revealed the existence of two stimuli: one chemical in nature, the other physical. The germ tube works its way into the stomatal aperture and occupies the sub-stomatal cavity. Several germ tubes—up to 17 have been counted according to Galet (1977)—may penetrate a single stoma.

### D. Incubation

This is the period between infection and the appearance of the lesion, during which the fungus invades the intercellular spaces of the host. The duration of incubation has been the subject of much research since knowledge of it is essential to the determination of suitable treatment dates. It terminates with the sporulation of the parasite and it is therefore important to apply a protectant treatment to susceptible parts of the vine before this dangerous phase.

Depending on the authors and local conditions of observations, the duration of incubation has been variously reported as from 5 to 18 days with a greater frequency around 7–10 days. It is to some extent affected by air temperature and humidity. Muller (1936) and Baldacci (1973), for example, have published incubation calendars to facilitate the application of preventive treatments.

### E. Sporulation

This is the external manifestation of the parasite at the surface of the host.

It occurs as sporangiophores, which appear at the stomatal ostiole, usually after the other signs (yellowing, browning, etc.) of the disease. In exceptional cases, when the relative humidity is near saturation point and the temperature at its optimum during the days following infection, the conidiophores appear prior to any other sign of the disease.

The formation of sporangiophores is very closely linked to the relative humidity, which must be at 95–100%. Under these conditions and at temperatures between 13 and 27°C the sporangiophores form within a day. At 18–22°C their formation takes only 10 to 12 h, i.e. a single night (Istvanffi, 1914; Ravaz and Verge, 1914). Recently Blaeser and Weltzien (1978) studied the sporulation of *P. viticola* in Germany and they generally confirmed previous observations but specify that a minimum of 4 h darkness is necessary for sporangiophore formation.

## F. Factors affecting the development of epidemics

Vine downy mildew is favoured by all those factors which produce an increase in the moisture content of soil, air and plant. This means that rain is one of the principal contributory factors to epidemics.

Temperature plays a less important role, by retarding or accelerating the development of the disease, but it does not have the determinant characteristic of water. The optimum temperature for development of the fungus is around 25°C, the extremes ranging from 10 to 30°C though we have noted that it can withstand a temperature of 35°C in a saturated atmosphere.

In consequence the most serious epidemics occur when a wet winter is succeeded by a wet spring followed by a hot summer with intermittent rainstorms every 8 to 15 days. The first phase ensures survival of the oospores and their abundant germination in the spring. The second permits the development of the disease and its spread within the vineyard.

The capacity of the plant to harbour the parasite is also important in determining the development of epidemics. As the penetration of the fungus takes place exclusively via the stomata it is essential that they be present and functional if infection is to take place. Active vegetative growth also favours the disease since it is the young actively-growing organs that are the most susceptible. Rainfall thus has a doubly favourable effect disease development. When periods of rain occur in succession they first stimulate the production of young, susceptible shoots (preparatory rains, sensu Capus, 1919) and then ensure their infection (infection rain).

The susceptibility to downy mildew differs relatively little between cultivars in the species *Vitis vinifera* L., which is not only the species most widely grown for the production of quality wine, but is also one of the most susceptible to downy mildew, along with *V. californica* Benth. and *V. arizonica* Engel.

Other species are less susceptible, *V. aestivalis* Michx., *V. labrusca* L., or relatively resistant, *V. berlandieri*, or even highly resistant *V. rotundifolia* Michx., *V. riparia* Michx., *V. cordifolia* Michx., *V. rupestris* Scheele.

## V. The Control of Vine Downy Mildew

### A. Preventive measures

The aim of preventive measures is to inhibit the early development of the disease. Efficient drainage of the soil, for example, will prevent the formation of pools of water which encourage the maturation and germination of oospores. Likewise the destruction of initial infection centres resulting from work at the Wine Technical Institute in the South of France (Anon., 1961) has also proved an efficient measure. Late season pruning of the extremities of diseased branches suppresses young leaves, which are very susceptible to the disease without injury to the rest of the plant.

### B. Chemical treatments

Since the discovery of Bordeaux mixture (Millardet, 1885) vines throughout the world have been treated with chemicals. The treatments are carried out in different ways but the principle is always to apply fungicide to the organs at risk, prior to their infection.

In France a method for determining treatment dates was studied from 1898 onwards by Cazeaux-Cazalet and Capus. It consisted essentially of forecasting the "contamination rains" (Capus, 1919). Later, Darpoux (1943) established the scientific principles for the agricultural warning systems for downy mildew. These are based on: knowledge of the parasite (including conditions for germination of the oospores and duration of incubation), climatic factors (temperatures, rainfall and dew) and the receptivity of the plant (speed of shoot elongation, phenological stages). More recently, other decisive elements have been taken into account by the French Advisory Services (Stations d'Avertissements). These include the persistence of fungicidal action and statistical studies of treatment timing.

The Advisory Service stations distributed throughout the principal vine-growing regions of France inform the viticulturists when treatments should be applied. On average five or six treatments are carried out between the end of April and the beginning of August although there are pronounced regional differences. For example, the Roussillon vineyards usually receive only one or two treatments whereas those in Languedoc require 12 to 15 in certain years in order to be suitably protected. In general special emphasis is placed on treatment before and just after flowering so as to protect the bunches, and at the beginning of August so as to prolong the life of the leaves for as long as possible and thus ensure good ripening of the wood. All the treatment methods presently in use are efficient but air-blast spraying is particularly effective since it deposits droplets of fungicide mixture on those parts of the vine that other methods fail to reach such as the underside of the leaves and inside the bunches (Lafon *et al.*, 1958).

### 1. Preventive surface-acting fungicides

The copper salts are still widely used, notably the sulphate (neutralized with calcium hydroxide = Bordeaux mixture), the oxychloride or the hydroxide. In France cupric compounds are approved at a concentration of 500 g metallic copper per hectolitre although certain commercial Bordeaux mixture formulations are authorized for use at only 300 g copper per hectolitre.

Amongst the organic compounds, mancozeb and folpet (folpel in France) are very widely used on vines and others less widely used include captafol, captan, dichlofluanid mancopper, maneb, metiram-zinc (zineb-etylenebis (thiuram disulphide) mixed precipitation), propineb, zineb. These are authorized for use in various countries in accordance with local regulations.

In another group of protectant materials one or several organic fungicides are mixed with one or more cupric salts. This group, which is very widely used by viticulturists, consists of a great many different combinations the intention of which is to exploit the respective advantages of each of their components while, at the same time, minimizing their disadvantages. Some of these combinations may even demonstrate a synergistic effect (copper oxychloride with zineb, for example).

### 2. Eradicant fungicides with systemic action

In recent years the chemical industry has introduced a number of new fungicides specifically active against certain Phycomycetes and against *P. viticola* in particular. These fungicides have important common advantages in that their penetration of the plant is rapid (less than an hour) and they are not washed off by rain as are the surface-acting fungicides. Furthermore, their action is curative for the first 3 or 4 days following infection (Lafon, 1969).

Certain of these fungicides have shown notable efficacy, particularly: 1-(2-cyano(methoxyimino)acetyl)-3-ethylurea common name cymoxanil aluminium tris(ethyl phosphonate), common name fosetyl-aluminium = phosethyl-Al in France; methyl *N*-(2-methoxyacetyl)-*N*-(2,6-xylyl)-DL-alaninate, common name metalaxyl. These compounds when used for the control of vine downy mildew are always in association with surface-acting fungicides which augment their persistence and broaden their action spectrum.

Cymoxanil is utilized in numerous formulations in association with dithiocarbamates, phthalimides, or copper salts and sometimes with a mixture of several of these compounds. The compound which is always applied at 120 g ha$^{-1}$ produces a very considerable synergistic effect with mancozeb. It is not systemic but penetrates the plant and can inhibit the mycelial development of *P. viticola* in the 3 or 4 days following infection. The persistent action of the combinations is dependent on that of the surface-acting fungicides and is generally from 10 to 12 days.

Fosetyl-aluminium is at present applied to vines only with folpet. The fosetyl-aluminium in this mixture is efficiently translocated to the extremities of the branches and to the bunches right up until fruit-set. This property is manifested in a spectacular manner by the late protection of the branch tips, up to six weeks after the final treatment. Protection of actively growing organs may last for 2–3 weeks and its eradicant activity may be up to 3–4 days after infection. According to Vo-Thi-Hai *et al.* (1979) the activity of fosetyl-aluminium may be due to its ability to stimulate the plant's defence reactions (production of phenolic compounds and phytoalexins).

Metalaxyl, available in a mixture with folpet or copper oxychloride, is systemic and has the additional property of diffusing as a vapour into a moist atmosphere. It is capable of eradicating infection throughout the entire incubation period and even beyond this when applied to sporulating lesions. Its activity persists for 2 to 3 weeks after application.

The systemic and curative properties of these new fungicides should improve the control of downy mildew by permitting greater latitude in the fixing of treatment dates and a reduction in the number of applications. Nevertheless their specificity with regard to *P. viticola* makes them unsuitable for the control of other vine parasites. It is possible to overcome this deficiency to some extent by the use of suitable mixtures but then the reduction in the number of treatments is an obstacle to all-purpose control.

## VI. Conclusions

Vine downy mildew has been invading vineyards throughout the world for a century. During the 2 or 3 decades following its initial appearance its importance led to the initiation of a large number of very careful studies in Europe. Since that time the tendency has been to accept that the disease is now so well-known that the work devoted to it can be relaxed. Downy mildew is still a menace, however, requiring constant and costly preventive control, and is still capable of causing catastrophic damage as it did in French vineyards in 1977.

It is fortunate that chemotherapy is now opening up new horizons especially as this progress coincides with advances in genetic research. This is well on the way to creating resistant varieties of a very high technological standard.

## References

Anon. (1961). "Le Mildiou de la Vigne". *Institut Technique du Vin*, 100 pp.

Baldacci, E. (1973). "Calendario d'Incubazione della Peronospora e Guida alla Lotta Anti-parassitaria nella vita". *Pio Istituto Agricola Vogherese* 23rd edn. C. Gallini Voghera.

Bary, A. de (1863). *Ann. Sci. Nat.* **4** (XX), 125.

Berkeley, M. J. B. and Curtis, M. A. (1848). *In* Ravenel, *Fungi Carol*, Fasc. **V**, 90.

Berlese, A. N. B. and de Toni, J. B. (1888). *In* Saccardo, *Sylloge VII* 1, 239.

Blaeser, M. and Weltzien, H. C. (1978). *Z. Pflkrankh. PflPath. PflSchutz.* **85**, (3/4), 155–161.

Boubals, D. (1958). Thesis. Fac. Sci. Montpellier.

Branas, J. (1974). "Viticulture" Montpellier, 990 pp.

Capus, J. (1919). *Ann. Epiphyties*, **IV**, 162–217.

Cazeaux-Cazalet, G. and Capus, J. (1898). *Rev. Viticulture* **X**, no. 237.

Corbaz, R. (1972). *Phytopathology* **74** (4), 318–328.

Darpoux, H. (1943). *Ann. Epiphyties* **9** (7), 177–205.

Galet, P. (1977). "Les Maladies et les Parasites de la Vigne". Tome 1, Montpellier, Imprimerie du Paysan du Midi.

Gäumann, E. (1950). "Principles of Plant Infection". Crosby Lockwood and Son Ltd, London.

Istvanffi, (Gy.de), (1914). *Bull. Rens. Agric.* (Rome), p. 1454.

Lafon, J., Couillaud, P., Gay-Bellile, F. and Lacouture, J. (1958). *Vignes Vins* **66** 9–12 and 67, 17–22.

Lafon, R. (1979). *Vititechnique* **28**, p. 13.

Millardet, A. (1885). *J. Agric. Pratique* 21st Nov. 1885.

Muller, K. (1936). *Z. Pflkrankh. PflPath. PflSchutz.* **46**, 104–108.

Ravaz, L. and Verge, G. (1914). *Ann. Ec. Nation. Sup. Agron.* Montpellier NS, *XIV* (1914), 169–199.

Rives, M. and Lafon, R. (1972). *Vitis*, bd. **11**, 34–52.

Royle, D. J. and Thomas, G. G. (1973). *Physiol. Pl. Path.* **3** (3), 405–417.

Schweinitz, L. D. (1837). *Synopsis Fungorum*. Am. Boreal. 2663, no. 25.

Srinivasan, N. and Jeyarajan, R. (1976). *Curr. Sci.* **45** (3), 106–107.

Tsvetanov, D. (1976). *Gradin. Loz. Nauka.* **13** (3), 137–143.

Viennot-Bourgin, G. (1949). *In* "Les Champignons Parasites des Plantes Cultivees". (Masson, Ed.) Paris.

Vo-Thi-Hai, Bompeix, G. and Ravise, A. (1979). *C.r. Acad. Sci. Fr.* (Paris) **288**, series D, 1171–1174.

Zachos, D. G. (1959). *Ann. Inst. Phytopath. Benaki* (Athènes) NS, **2** (4), 197–355.

# Subject Index

## A

Acropetal translocation, 313
Acylalanines, 314
  acropetal translocation, 314
  limits of usefulness on peas and beans, 510–511
Adaptation, 250
Additive variation, 267
Adjacent crops,
  significance in disease occurrence, 387–388
Age,
  effect on resistance, 385, 387
Airborne spores,
  threshold concentrations as warnings of infection, 417
Air circulation, 298
Air humidity, 470
*Albugo*, 23
Alfalfa, 356–365
  breeding for downy mildew resistance, 361–363
  screening for disease resistance, 363–365
Alfalfa downy mildew, 356–365, *see also Peronospora trifoliorum*
Alfalfa germplasm, 362–363
*Allium cepa*, 260, 461–471, *see also* onion
Allopolyploid, 267
Allyl-isothiocyanate, 260
Alpha-acid,
  bitterness factor in beer, 400
Alternative hosts, 401
  questionable importance in lettuce, 443–444
Aluminium ethyl phosphites, 415
*Alyssum*, 478–479
Anamorph, 27

Anaphase, 123, 135
*Anemone*, 474–475
Anemophilous fungi, 469
Antagonists, 175
  in elimination of resting spores, 388
Antheridium, 123–124, 219–229, 234–235
  scanning electron micrographs, 219
*Antirrhinum*, 475–477
Apoplastic translocation, 312
Application methods, 317
Appressorium, 438
  adhesion to plant surface, 130
  factors affecting formation, 129
  role of microbes in initiation, 129–130
  thigmotropic mechanism, 129–130, 132
Area dose,
  relation to spore concentration, 92–93
Asexual propagules, 17–28, 241, 444
Asexual reproduction,
  diurnal periodicity, 133–134
Asexual sporophores,
  characteristic in *Bremia*, 423
Asexual sporulation, 165–180, 326–327
  effect of environment on lettuce downy mildew, 441–442
Assessment keys, 454–455, 499–500
Auto-inhibitor, 435
Autoradiography, 150, 214–215
Axenic culture, 22, 160–161, 375

## B

*Basidiophora*, 17–28, 364–370, 376
  wild hosts, 108
Basipetal translocation, 313
Bean, 487–511, *see also* peas and beans, *Vicia* and *Phaseolus*